Quantum Chemistry:
The Challenge of Transition Metals
and Coordination Chemistry

NATO ASI Series

Advanced Science Institutes Series

A series presenting the results of activities sponsored by the NATO Science Committee, which aims at the dissemination of advanced scientific and technological knowledge, with a view to strengthening links between scientific communities.

The series is published by an international board of publishers in conjunction with the NATO Scientific Affairs Division

A Life Sciences B Physics	Plenum Publishing Corporation London and New York
C Mathematical and Physical Sciences	D. Reidel Publishing Company Dordrecht, Boston, Lancaster and Tokyo
D Behavioural and Social Sciences E Engineering and Materials Sciences	Martinus Nijhoff Publishers The Hague, Boston and Lancaster
F Computer and Systems Sciences G Ecological Sciences	Springer-Verlag Berlin, Heidelberg, New York and Tokyo

Series C: Mathematical and Physical Sciences Vol. 176

Quantum Chemistry: The Challenge of Transition Metals and Coordination Chemistry

edited by

A. Veillard

Centre National de la Recherche Scientifique, Strasbourg, France

D. Reidel Publishing Company

Dordrecht / Boston / Lancaster / Tokyo

Published in cooperation with NATO Scientific Affairs Division

Proceedings of the NATO Advanced Research Workshop and the
40th International Meeting of the Société de Chimie Physique:
The Challenge of Transition Metals and Coordination Chemistry
Strasbourg, France
September 16-20, 1985

Library of Congress Cataloging in Publication Data

NATO Advanced Research Workshop on Quantum Chemistry, the Challenge of Transition Metals and Coordination Chemistry (1985 : Strasbourg, France)
Quantum chemistry.

(NATO ASI series. Series C, Mathematical and physical sciences; vol. 176) Published in cooperation with NATO Scientific Affairs Division.

"Proceedings of the NATO Advanced Research Workshop (40th International Meeting of the Société de chimie physique) on Quantum Chemistry, the Challenge of Transition Metals and Coordination Chemistry), Strasbourg, France, September 16–20, 1985" – T.p. verso.

Includes index.

1. Quantum chemistry–Congresses. 2. Transition metal compounds–Congresses. 3. Coordination compounds–Congresses. I. Veillard, A., 1934– . II. North Atlantic Treaty Organization. Scientific Affairs Division. III. Société de chimie physique. International Meeting (40th : 1985 : Strasbourg, France) IV. Title. V. Series: NATO ASI series. Series C, Mathematical and physical sciences; vol. 176.
QD462.A1N16 1985 541.2′8 86–3881
ISBN 90–277–2237–4

Published by D. Reidel Publishing Company
P.O. Box 17, 3300 AA Dordrecht, Holland

Sold and distributed in the U.S.A. and Canada
by Kluwer Academic Publishers,
190 Old Derby Street, Hingham, MA 02043, U.S.A.

In all other countries, sold and distributed
by Kluwer Academic Publishers Group,
P.O. Box 322, 3300 AH Dordrecht, Holland

D. Reidel Publishing Company is a member of the Kluwer Academic Publishers Group

Printed in The Netherlands

CONTENTS

PREFACE

Over the last twenty years, developments of the ab initio methodologies and of the computing capacities have progressively turned quantum chemistry into a predictive tool for molecular systems involving only light elements. The situation appears less advanced for systems containing transition metal elements where specific difficulties arise, like those linked to the quasi-degeneracy of the lowest atomic states. Correlation effects, which are important only for quantitative accuracy in the treatment of molecules made of light elements, need sometimes to be considered even for a qualitative description of transition metals systems (like the multiple metal-metal bond). The treatment of atoms of a high atomic number has necessited the development of model potential methods. These difficulties exacerbate for systems containing several transition atoms : a correct description of the dichromium molecule Cr_2 still represents a challenge to quantum chemists. Yet many advances have been made recently in the theoretical treatment of these systems, despite the fact that our understanding still remains disparate with a variety of models and methodologies used more or less successfully (one-electron models, explicitly correlated ab initio methods, density functional formalisms). For these reasons, a NATO Advanced Research Workshop was organized to review in detail the state-of-the-art techniques and at the same time the most common applications. These encompass many fields including the spectroscopy of diatomics and small aggregates, structure and reactivity problems in organometallic chemistry, the cluster-surface analogy with its implications for heterogeneous catalysis and the description of extended structures.

This conference was patronized jointly by NATO and by the Société de Chimie Physique (40th International Meeting). The organizing committee consisted of Drs. E. Baerends, A. Dedieu, O. Eisenstein, I. Hillier, P. Hofmann, J.-P. Malrieu, C. Minot, M. Pelissier, P. Siegbahn and myself. It was financially sponsored by the NATO Scientific Affairs Division. It also benefited from the financial assistance of the Centre National de la Recherche Scientifique, the Société Française de Chimie, Cray Research France and the Conseil Général du Bas-Rhin. We gratefully acknowledge their generous support and their interest in the meeting.

A. Veillard

ACCURATE STUDIES ON THE STRUCTURE AND REACTIVITY OF TRANSITION METAL COMPLEXES AND CLUSTERS

M. Blomberg, U. Brandemark, I. Panas,
P. Siegbahn and U. Wahlgren

Institute of Theoretical Physics
University of Stockholm
Vanadisvägen 9
S-11346 Stockholm
Sweden

ABSTRACT. Three different applications of CASSCF and CI methods on transition metal systems are described. The first application is concerned with the accuracy of calculated binding energies. For NiH^+ and NiO^+ comparison is made to molecular beam experiments. For $NiCO^+$ and NiCO comparisons are made to measurements of appearence potentials in fragmentation experiments. In the second application comparisons are made to matrix isolation experiments for NiH_2O, CuH_2O and $Ni(C_2H_4)_n$ with n=1,2. The final application is concerned with the modelling of chemical reactions on transition metal surfaces. A strategy for choosing a proper effective core potential for the metal atoms is outlined. Preliminary results for O_2 dissociation on nickel clusters are presented.

1. INTRODUCTION

The present contribution will describe the recent research on transition metal compounds, which has been performed in the quantum chemistry group in Stockholm. This work is part of the more general project with a goal to obtain a fundamental understanding of chemical reactions involving transition metals using methods including electron correlation. This general project started with a detailed study of potential curves and spectrum of NiH [1] and continued with a similar detailed study of NiH_2 [2]. Important chemical implications were found already from this very simple model system. It was found that the ^1D state of nickel binds quite differently to H_2 than the triplet states. The lowest singlet state has a shallow minimum with practically no barrier for formation, whereas the triplet states all have large barriers. The behaviour of the singlet state is characteristic also for what is found experimentally for H_2 addition to nickel complexes and for H_2 dissociation on nickel surfaces. It was therefore concluded that the singlet state is the chemically most relevant state, and in all subsequent calculations on larger systems this is the state that has been mainly considered. It is clear that there is a similarity between the single atom results for Ni + H_2 and for what is observed

A. Veillard (ed.), Quantum Chemistry: The Challenge of Transition Metals and Coordination Chemistry, 1–14.

for H_2 dissociation on surfaces. To study the surface reaction a more realistic model was then chosen with 13-14 nickel atoms representing the surface [3]. The on top dissociation was found to be the most favorable dissociation pathway and indeed showed large similarities to the potential surface for Ni + H_2. The method of calculation and the models used in papers 2-3 can be considered as prototypes for all the other studies on transition metal compounds which have been performed in our group.

The computational methods used are of MCSCF and CI type. The CASSCF method [4] is used for the MCSCF step and should take care of near degeneracy effects. This is by far the most difficult part for the user. For a review of some of the problems which appear see Ref.5. A large part of the computational work is spent in finding a reasonable active space describing all the major chemical effects. When this goal has been achieved the chemistry of the studied problem is in general also understood. It is often found that the addition of only a few configurations to the Hartree Fock configuration can be of a qualitative chemical significance. Notable is for example the change in binding energy for the ground state of NiCO, which is negative by -59 kcal/mol at the Hartree-Fock level but is positive by +21 kcal/mol using a small active space [6]. Findings of this type are probably the most significant contribution from recent accurate calculations to the understanding of transition metal chemistry. After the MCSCF step the addition of dynamical electron correlation is achieved through multi-reference CI methods. In our group we have used the Contracted CI (CCI) method [7], which is much cheaper and achieves nearly the same accuracy as CI without approximations.

One of the major reasons for making CI calculations on small model systems, which in most cases already have been studied using simpler approaches, is of course the potential of CI methods to achieve a very high accuracy. A fundamental question is therefore how high this accuracy is in usual applications. This point has been of major concern in our group the last year and we will below show examples of calculations where the accuracy is taken to its present day limit. It should be noted, however, that the limit for the size of a calculation is changing quite rapidly with time and calculations 10 or a 100 times larger will be possible to perform already within a few years. The full capacity of the faster computers is straightforward to make use of in CI calculations to increase the accuracy. This is one of the main advantages of the open ended CI method as compared to for example X_α methods or semi-empirical methods. Given a 100 times faster computer these latter methods will still give the same error.

To test the accuracy of a CI calculation there are two different approaches. The first approach is to systematically increase the size of the calculation and study the convergence pattern of for example relative energies. The possibility to test accuracy in this way is another principle advantage of open ended ab initio methods. The second approach is the more conventional one of simply comparing to experimental results. The main problem of this approach for simple transition metal compounds is that accurate experiments of relative energies are scarce. An exception is mass spectrometric measurements on ionic compounds. We have compared our results to a few systems of this type. In the case of NiH^+ and NiO^+ a comparison is made to the ion beam studies of Beauchamp and coworkers [8-9]. For $NiCO^+$ comparison is made to the photoionization study of Distefano [10], who determined the appearence potentials for the fragmented positive ions of $Ni(CO)_4$ by electron and photon impact. The binding energy of neutral NiCO is finally compared to the experiments by Stevens et al [11] and Compton and Stockdale [12]. In Ref.12 the appearence potentials of the fragmented negative ions of $Ni(CO)_4$ were measured and these values were combined in Ref.11 with a determination of the electron affinities of the $Ni(CO)_n$ complexes to yield experimental binding energies for the neutral carbonyls. Comparison between the theoretical and experimental results shows that the agreement is very good for NiH^+ and NiCO, but that there are large discrepancies for NiO^+ and $NiCO^+$ - most probably due to errors in the experimental results, however. The results are presented in Section II.

Another area where interesting experiments are being performed on simple transition metal compounds is the area of matrix isolation spectroscopy. Since the experimental information mainly comes from vibrational frequencies, a direct quantitative comparison between theory and experiments is not as straightforward as in the comparison between measured and calculated relative energies, however. A complicating factor is the often non-negligable coupling between different vibrational modes. Calculations of vibrational frequencies in the sytems $Ni(C_2H_4)_n$ and $Ni(CO)_n$ including off-diagonal effects are presently being performed but are not yet completed. Some general qualitative comparisons between matrix isolation results and calculations can, however, sometimes rather easily be made. A few such comparisons will be made in Section III on the systems NiH_2O and $Ni(C_2H_4)_n$. Particularly intriguing was for example the recent finding of Kauffman et al [13] that the Ni atom does not bind at all to H_2O. Since we obtained a binding energy of 24 kcal/mol (for the singlet state) [6] and Bauschlicher [14] obtained 9 kcal/mol (for the triplet state) and 9 kcal/mol (for the singlet state) the experimental result may seem quite surprising. Already the discrepancy between the two theoretical results showed that there is some problem in the calculations, however, and an investigation of the basis sets was already underway when the experimental result appeared. This investigation showed that there are some quite unusual problems in the choice of the basis sets for the metal - H_2O systems. For the complexes $Ni(C_2H_4)_n$ a qualitative comparison between our calculated values and the experiments by Ozin et al [15] shows such large discrepancies that a reinterpretation of the experimental results is suggested. The qualitative picture obtained from our calculations of the metal olefin bond agrees on the other hand quite well with the experiments by Kafafi et al for $Fe(C_2H_4)_n$ [16], which gives some hints of where the experimental problems in Ref.15 lie.

The final area briefly discussed in this presentation is the modelling of chemical reactions on metal surfaces and clusters. The first reaction studied has been the O_2 dissociation. One of the most important findings for the H_2 dissociation on a nickel surface [3], was that the metal d orbitals are absolutely essential for breaking adsorbate bonds in on top positions. In other positions, where multiple sites are active in the dissociation, d orbitals seem less important. For example at four-fold positions it may be possible to model the O_2 dissociation without explicitly treating the d orbitals of the metal atoms. In this crude, but computationally attractive, procedure the metal atoms are therefore treated as single electron atoms using an effective core potential (ECP). A single electron ECP has been developed by Melius et al [17] for the Ni atom. This ECP has been extensively used, yielding remarkably accurate results. Since we are also interested in other metal atoms besides nickel we had to develop our own one electron ECP's. The first part of this project was therefore devoted to finding a strategy for the use of ECP's in cluster modelling and testing out different choices of ECP's with different parameters. It should be noted that some of these parameters, like p- and d- projections, can not be determined from the results for the ground state atom. The values of these parameters are defined from atomic results only if very farfetched excited states are used in the parametrization. A better procedure seems to be to test the parameters on small molecular systems. Our test system has been Ni_5O with O in a four-fold position and with one Ni atom in the second layer. This system is known at the all electron level [18]. We quickly found that describing the full effect of the core orbitals using only model potentials and projection operators was a very risky procedure. Introducing a frozen 3s orbital improves the situation drastically but the results are still rather sensitive to the choice of projection operators. A reasonable convergence can, however, be obtained with for example variations of 0.2-0.3 Å in the oxygen height above the surface, depending on the choice of ECP parameters. If higher accuracy is desired the recommended procedure is to use also a frozen 3p orbital if this can be afforded. The work on the O_2 dissociation is still in progress but some preliminary results are available. These will be presented in Section IV.

2. THE ACCURACY OF CALCULATED BINDING ENERGIES

It is today possible to obtain an accuracy in a computed binding energy of 10 % or better for small molecules consisting of two or three atoms from the first or second rows. For molecules containing transition metals the situation is much less clear, basically due to lacking experimental information. To clarify the situation somewhat, we will in this section make a few comparisons to experiments.

2.1. NiCO

In an earlier study of the nickel carbonyls $Ni(CO)_n$ [6] the comparison to experimental binding energies [11,12] gave a rather confusing trend. For NiCO the agreement seems quite good with a calculated value of 30 kcal/mol and an experimental value of 29±15 kcal/mol. Already for the the second carbonyl in $Ni(CO)_2$ the error in the binding energy is much larger with 27 kcal/mol calculated and 54±15 kcal/mol obtained experimentally. For the total binding energy of $Ni(CO)_4$, which is known thermochemically to be 141 kcal/mol [19], the calculated value of 86 kcal/mol is very poor. It is clear that the basis set used, and possibly also the CI expansion, is not adequate for $Ni(CO)_4$. The large experimental uncertainty for NiCO does not, however, rule out the existence of a large problem already in the calculation on this system. To study this latter possibility we have recalculated NiCO using a much larger basis set. This basis is a (15s,11p,6d,3f) set on nickel contracted to <9s,6p,4d,2f> and for carbon/oxygen a (11s,6p,3d) contracted to <5s,4p,1d>. The previous basis set was a contracted <5s,4p,3d> set for nickel and a <3s,2p> set for carbon/oxygen. The active space in the CASSCF calculations had 6 electrons (Ni s,d_σ and d_π) distributed in 6 orbitals. The CCI calculations correlated the 10 electrons in 3d,4s on nickel and the 2 electrons in the carbon lone pair pointing towards nickel. The adequacy of the active space and the number of correlated electrons was carefully tested using the smaller basis set. The binding energy at the CASSCF level was 20 kcal/mol with the new basis set compared to 21 kcal/mol with the old basis set. The corresponding values at the CCI level was 29 kcal/mol compared to 30 kcal/mol. The small difference between the results and the agreement between theory and experiment therefore lead us to the conclusion that already the small basis set was adequate for treating NiCO. The reason for the failure of the corresponding calculations on $Ni(CO)_2$ and $Ni(CO)_4$ is under investigation.

2.2. NiH^+ and NiO^+

There is a wealth of experimental information on binding energies of positive ions of transition metal compounds through the molecular beam work performed mainly by Beauchamp and coworkers [8-9]. These types of compounds should therefore represent an excellent test ground for the accuracy of ab initio work on transition metal systems. We have for this reason performed calculations on several ionic molecules like NiH^+, NiO^+, $NiCH_2^+$, $Ni(C_2H_4)_n^+$ etc. For the first two of these ions, NiH^+ and NiO^+, we have increased the accuracy as far as we could manage, with somewhat conflicting results.

For NiH^+ it turned out that basis set convergence was rather easily achieved. Already our standard basis sets of double zeta + polarization type gave a binding energy at the CCI level of 36 kcal/mol in reasonably good agreement with the experimental value of 43±2 kcal/mol. Increasing the basis set to (15s,11p,6d,2f) contracted to <11s,8p,3d,1f> for nickel and (5s,1p) contracted to <3s,1p> for hydrogen gave a binding energy of 42 kcal/mol. The situation is actually somewhat better than we had expected.

For NiO^+ the situation is quite different. Our standard basis of double zeta type turned out to be quite inadequate with unacceptable superposition errors. This is in fact a general experience for oxygen in any molecule. A triple zeta description of the 2p orbital is always needed to describe the diffuse nature of this orbital when oxygen becomes negative, which occurs even in a system like H_2O. Going to the large basis set, which for Ni is the same as used above for NiH^+, and for oxygen is a (10s,7p,3d) set contracted to <5s,5p,1d> leads to a binding energy of only 24 kcal/mol compared to the experimental value 45±4 kcal/mol. Diffuse d functions, which were added to improve the polarizability of oxygen (see discussion in subsection c), gave a less than 1 kcal/mol improvement. Diffuse f functions to polarize the diffuse d part of the nickel 3d shell had no effect. After studying the basis set convergence we actually conclude that our value of 24 kcal/mol is most probably an upper bound to the correct binding energy. The reason is that at long distance the Ni^+ ion has a d^9 configuration whereas at shorter distances the d population drops down towards 8 electrons. Extension of the basis set to include more 3d correlation therefore lowers the binding energy. The effect of the two compact f functions for example was a lowering of the binding energy by 7 kcal/mol. We therefore conclude that the interpretation of the experiment is probably wrong. There are two possible explanations for this. The first one is simply that excited d^8s Ni^+ ions are present in the beam. The second explanation is that there is a barrier for the reaction between Ni^+ and O_2 which would move the threshold reaction energy to higher energies. It seems clear that until these problems have been resolved, one should be careful with drawing conclusions based on comparisons to these types of experiments.

2.3. $NiCO^+$

The binding energies of the positive ions of the nickel carbonyls have been measured using another method than the ion beam technique. Instead, electron and photon impact was used on $Ni(CO)_4$ to determine the threshold energies for the formation of the different nickel carbonyl ions. For $NiCO^+$ a binding energy of 48 kcal/mol was reported using this technique [10], which is notably larger than the 30 kcal/mol binding energy found for the neutral NiCO. The first observation made in the calculations on $NiCO^+$ was that the type of binding is entirely different from that of NiCO. The equilibrium bond distance has moved out from 3.28 au in NiCO to 4.03 au in $NiCO^+$, which means that the π bonds are completely broken. This fact combined with the fact that the positive charge exhibits a strong electric field on CO means that for $NiCO^+$ we are in a region where pure electrostatic interactions are dominant. The leading interactions are charge-dipole and charge-induced dipole interactions. Since the dipole moment of CO is known to be sensitive to correlation effects, correlation on CO is likely to be more important than correlation on Ni in sharp contrast to what was the case for NiCO. A few different large calculations were performed. First a set of calculations were performed with the same large basis set as described above for NiCO. At the CASSCF level with an active space including the CO π orbital, which is important for the dipole moment, a binding energy of only 22 kcal/mol is obtained. A CCI calculation which correlated the Ni 3d and 4s and the CO π electrons hardly changed the binding energy, resulting in 26 kcal/mol. This was not too surprising and just confirms the fact that there is no π bonding. In the second set of calculations the nickel ion was replaced by a simple point charge, which lead to a computed binding energy for the same bond distance (Ni-C = 4.00 au) of 28 kcal/mol. This simple model is consequently quite reasonable. In the final set of calculations the same simple model was used but with a very large basis set on CO. This basis is (11s,6p,4d,1f) contracted to <9s,5p,4d,1f> and should mainly improve the polarizability of CO, which could be important for the charge-induced dipole interaction. The CCI calculation gave only a minor change, however, to 31 kcal/mol. The final large discrepancy between theory and experiments therefore remains and we again conclude that that the experiment is most likely incorrectly interpreted. At present we have no explanation for this.

3. COMPARISON TO MATRIX ISOLATION EXPERIMENTS

Matrix isolation spectroscopy do in general not provide direct quantitative information about binding energies, but some qualitative conclusions can sometimes be made anyway. We will in this section discuss two such cases. The first one is the transition metal-water interaction, where for example the experimental evidence is that for Ni-H_2O there is no binding at all. In the second case the information is more indirect. We have made calculations on Ni$(C_2H_4)_n$ with n=1,2 and the results are compared to measurements on these systems and on the corresponding iron complexes.

3.1. The Transition Metal-Water Interaction

As mentioned in the introduction there is a large discrepancy between the two most recent calculations on the Ni-H_2O interaction. For the singlet state we obtained a binding energy of 24 kcal/mol and a bond distance of 3.79 au [6] whereas Bauschlicher obtained 9 kcal/mol and a bond distance of 4.07 au [14]. The main reason for this descrepancy was found to be due to the double zeta description of O 2p in [6], which lead to large superposition errors. The main effect of improving the O 2p description is that the bond distance increases. Once this has occurred the character of the binding has shifted from a σ donating bond to a long range interaction between a dipole on H_2O and an induced dipole on Ni. A further basis set extension shows that Bauschlicher's result is not converged either. Improving the dipole moment and reducing the superposition error on H_2O in a sequence of calculations, constantly decreases the interaction energy down to 2.7 kcal/mol. Once this trend has been established the experimental information that there is no binding at all in Ni-H_2O is not as surprising anymore. Our final binding energy for the singlet state shows that this state is of no relevance for the experiments. The lower atomic energy for the ^3F and the ^3D states of Ni means that the ground state of Ni-H_2O should be a triplet. In our calculations on the triplet states we used a Ni (14s,11p,6d) set contracted to <8s,6p,3d> set. The H_2O basis set gives the Hartree-Fock limit dipole moment, which is 0.78 au compared to 0.73 au experimentally. This basis is for O a (11s,7p,2d) set contracted to <4s,3p,2d> and for hydrogen (6s,1p) contracted to <3s,1p>. The hydrogen p exponent is unusually low, 0.4, and the second oxygen d exponent, 0.2, optimizes the polarizability. Calculations were only performed at the SCF level, which should be adequate to describe the major effects in these types of electrostatic interactions. At this level the water interaction with the d^8s^2 ^3F state is unbound whereas the interaction with the $^3A_1$ component of the d^9s ^3D state is bound by 2.8 kcal/mol. Van der Waals interactions will of course only increase the binding energies. It therefore seems as if at most only ^3F Ni atoms are interacting with H_2O in the matrix. The experiment was performed in an argon matrix which should contain both ^3F and ^3D Ni atoms, however, which means that an additional explanation to the experimental result must be found. We have no suggestion how to resolve this problem presently.

To confirm that the present calculations are reasonably adequate the Cu-H_2O interaction was also calculated with the same type of basis sets and at the SCF level. A binding energy of 0.8 kcal/mol was found at a bond distance of 5.08 au. The calculated bending frequency shift was also calculated and found to be 9 cm^{-1} which can be directly compared to the experimental value 21 cm^{-1}. This means that the magnitude of the interaction is correct but that the difficult van der Waals contribution must be added to obtain quantitative agreement. This is beyond our present interest. The main purpose of the present study was to find the origin of our earlier error and to establish that the transition metal-water interaction is indeed very weak and of an electrostatic nature. To further confirm that our understanding of the interaction is correct we have also established a direct relationship between the polarizability of the metal atom and the size of the binding energy. This follows of course from the leading dipole-induced dipole term.

3.2. $Ni(C_2H_4)_n$

In a matrix experiment on $Ni(C_2H_4)_n$ Ozin et al [15] obtained substantial shifts in the C-C stretching frequencies. A C-C frequency shifted 130 cm^{-1} compared to free C_2H_4 was interpreted as coming from NiC_2H_4. A GVB study indicated that this system should be in a triplet state with no d bonding. Our study on NiC_2H_4 showed on the other hand that the ground state is a singlet with a large d contribution in the bonding [20]. A one dimensional determination of the C-C shift resulted in a value as large as 276 cm^{-1}. The main difference between the two theoretical studies was found to be due to an inadequate d basis set used in Ref.15. Our study also showed that the singlet state is probably unreachable from ground state products in matrix experiments, and that the triplet state is at most weakly van der Waals bound. The line observed by Ozin et al could consequently not be coming from NiC_2H_4 as originally thought. Recently we have also studied the $Ni(C_2H_4)_2$ system in different geometrical configurations. We found the lowest energy for a twisted D_{2d} structure in a singlet state. This system has such a large binding energy that it should be reachable in matrix experiments. From a one dimensional grid of energy values we calculated a harmonic frequency shift of 145 cm^{-1} for the symmetric C-C stretch. This value is rather close to Ozin's shift of 163 cm^{-1} for the bis-olefin complex, but also rather close to his value of 130 cm^{-1} for the mono-olefin complex. To determine which of these values should be assigned to the bis-olefin complex, the shift should be calculated from asymmetric C-C distortions and couplings to for example the CH_2 scissoring mode may have to be considered. Such calculations are presently being performed. In the meantime our calculations can be compared to the recent and more reliable measurements on the corresponding iron complexes [16]. In these measurements only a very weakly bound complex was formed between iron and one ethylene. Since iron is in a high spin state this result can be compared to our calculation on the triplet state of nickel and ethylene, which likewise showed at most only a weak binding. The measurements on iron and two ethylenes on the other hand showed a much stronger bound complex with a C-C shift of 138 cm^{-1}. This agrees with our expectations from the calculations on $Ni(C_2H_4)_n$ that the low spin bis-olefin complex should be observable due to its higher binding energy. In Ref.16 there is an interesting comment stating that in an earlier measurement the bis-olefin complex was erroneously identified as the mono-olefin complex. It was shown that very low concentrations of ethylene are needed for observing the mono-olefin complex. In Ozin's measurements a high concentration of ethylene was used, which makes it very likely that the same mistake has been made in these experiments.

Another confirmation of the reliability of our calculations on NiC_2H_4 can be obtained from a similar study of NiC_2H_2 [21]. In these calculations a bound triplet state with a binding energy of 1 kcal/mol was found. This is in contrast to the triplet NiC_2H_4 result at the same level of approximation, which gave no binding at all. In a recent matrix experiment the existence of the weakly bound π complex of NiC_2H_2 has been confirmed [22].

4. MODELLING OF CHEMICAL REACTIONS ON METAL SURFACES

The study of catalytical processes occuring on metal surfaces is an area of great current interest both experimentally and theoretically. From the ab initio viewpoint this whole area would be beyond reach if it were not for the possibility of modelling at least some of the metal atoms by effective core potentials (ECP's). In the first part of this section we will discuss our strategy and some of our findings concerning the use of ECP's for cluster modelling. In the second subsection we will discuss some preliminary results for O_2 dissociation on transition metal surfaces.

4.1. The Use of ECP's in Cluster Modelling

The type of ECP which is used in our group is the one described for example in Ref.23. Characteristic of this approach is that the outermost core orbitals are included as frozen orbitals which are in turn described with a small basis set. The advantage with this approach is that the part of the ECP which is described by parameterized model potentials and projection operators is well localized. The disadvantage is that basis functions describing the frozen orbitals sometimes have to be added, which then leads tolonger integral times. A typical such ECP for the nickel atom in a metal complex would for example have the 1s,2s and 2p orbitals described by model potentials and projection operators. The 3s and 3p orbitals would be frozen, whereas the remaining 3d, 4s and 4p orbitals would be described by the original untruncated basis set. An original basis set of (12s,8p,5d) would with this ECP be only moderately truncated to (5s,4p,5d). This type of ECP is only useful for metal complexes and for clusters with a few metal atoms at the SCF level (if there are open d orbitals), if one is not willing to spend a very large amount of computer time in the calculations. In realistic studies of reactions beyond the SCF level some additional approximations are needed. We have found the following strategy useful for reducing the computational work.

Level 1. Freeze also the 3d orbitals with an average occupation of each d orbital. This is the highest level of approximation for reasonably sized metal clusters with open d orbitals.

Level 2. Remove the frozen 3d orbitals and include instead the interaction from these orbitals in the model potential and a 3d projection operator. Using this type of ECP it is perfectly feasible to treat also quite large clusters. This is in fact our recommended procedure for treating most systems.

Level 3. Remove also the frozen 3p orbital. This procedure has been extensively tested in our group the last year. The conclusion is that this procedure can be made to work but requires a large amount of testing on realistic systems where higher level results are known. Otherwise this procedure can be dangerous. The computational advantage compared to level 2 is rather limited in most cases and does only concern the integral time and the length of the integral file.

Level 4. Remove also the frozen 3s orbital. The computational advantage obtained at this level compared to level 3 is extremely small, and the procedure becomes very sensitive to the choice of ECP parameters. This procedure is not recommended unless metal atoms far away from the active center are modelled.

An important additional point for obtaining reliable results is that one should be extremely careful in discarding the action of the d orbitals for the metal atoms which are close to the active center. This is particularly true if on top adsorption is modelled. Meaningless results were for example generated for on top dissociation of H_2 on a nickel cluster when the central nickel atom was modelled with a one electron ECP [3]. Also on top binding of CO requires active participation of the 3d orbitals [24]. For the adequacy of treating adsorbates in bridge, 3-fold or 4-fold positions the situation is less clear due to the difficulty to perform adequate comparative calculations including the 3d orbitals. The indications are, however, that the action of the 3d orbitals quickly decreases with increased distances. This means that adsorbates in 3-fold or 4-fold positions are most likely well treated without explicitly incorporating 3d orbital effects.

4.2. O_2 Dissociation on a Nickel Surface

To computationally describe the O_2 dissociation on a nickel surface is a much more difficult problem than to describe the H_2 dissociation on the same surface, which was done in a previous paper. This is simply because when the different bonds in O_2 are broken and formed a variety of important configurations show up and also because intra-atomic correlation effects on oxygen can not be neglected. In a pilot study of on top dissociation of O_2 a very small triangular nickel cluster was therefore used as a first step in a study of this reaction [25]. The on top nickel atom is described with all its electrons and the other two atoms with a one electron ECP of level 3 accuracy. From the experience gained in this study more realistic clusters should then be used. The general picture obtained in Ni_3 + O_2 is that first a molecularly adsorbed state is formed with a negatively charged O_2. The bond distance has increased by 0.4 au compared to free O_2 and the molecule is adsorbed with a binding energy of 24 kcal/mol. Then there is a barrier for finally breaking the O_2 bond, which is at an energy 14 kcal/mol higher than the asymptotic value. The finally dissociated oxygen atoms are strongly bound to the Ni_3 cluster at an energy which is 51 kcal/mol lower than the asymptote with O_2 at long distance.

The 4-fold hollow positions are known to be the equilibrium geometries for the dissociated oxygen atoms on a (100) surface. The height above the surface is around 1.6 au and the binding frequency is 330 cm^{-1} or 420 cm^{-1} depending on the density of O atoms. After some work with the ECP, of the level 3 accuracy described above, this bonding situation is fairly well mimicked with a cluster where all the atoms are treated at the ECP level. For a Ni_5 cluster we obtained a bond height of 1.87 au and a frequency of 260 cm^{-1}. For larger clusters slightly different values are obtained. Overall, the accuracy is similar, but slightly lower, than that obtained by Bauschlicher et al [24,26].

Before studying the on top dissociation again with a larger cluster some other possible dissociation paths were investigated. These were first the dissociation over a bridge position towards 4-fold hollow positions and second the dissociation over a 4-fold hollow position across bridge positions towards 4-fold hollow positions. This latter pathway is indicated to be the preferrable pathway from surface experiments [27]. In the modelling of these reactions all metal atoms are treated with one electron ECP's at the level 3. For the first pathway a Ni_8 cluster was used and for the second pathway a Ni_{11}

cluster. The general picture obtained from these studies is the following. When O_2 approaches the surface there is an initial repulsion due to the contact between the fully occupied π_u shell of O_2 and the tail of the electrons from the nickel surface. At a distance of between 3.5-4.5 au above the surface there is an electron jump and a molecularly adsorbed O_2^- is formed in a local minimum. The O_2 bond distance is here around 2.7 au. The barrier for formation of this state is around 15 kcal/mol and the local minimum is at 10 kcal/mol above the asymptote. When the O_2 bond is stretched further the energy continues to go up to a barrier of 20 kcal/mol before the O_2 bond is definitely broken. For the dissociated oxygen atoms the energy drops quickly down to a positive binding energy compared to the asymptote of between 20-50 kcal/mol depending on cluster size. This binding energy corresponds to oxygen atoms bound by between 60 and 75 kcal/mol to the surface. Un uncertainty of 20% in the calculated binding energy is what realistically can be expected with cluster models of this size. The dissociated oxygen atoms are best described as O^-, with consequently one covalent bond to the surface. The differences and similarities between these dissociation pathways and the pathway found for Ni_3 are quite clear. Particularly noticable is the difference in binding energy for the adorbed O_2^- state. It should further be emphasized that the region outside the molecularly adsorbed state was not investigated in the Ni_3 study. It is quite likely that a barrier in this region exists for this case also.

The experimentally indicated pathway is as mentioned above a dissociation over a 4-fold position [27]. A questionable assumption in the interpretation of the experiments is that the dissociated oxygen atoms can not easily move from one 4-fold to another 4-fold position after dissociation. To test this assumption an oxygen atom was moved between two 4-fold positions on a Ni_8 cluster. A barrier of only 8 kcal/mol was obtained for this motion in the calculation. Such a low barrier means that a dissociative adsorption followed by adatom reordering can not be ruled out, in contrast to what is suggested in Ref.27. When independent adatom reordering is allowed nothing can be concluded about the dissociation pathway directly from the surface experiments.

5. SUMMARY

A few different transition metal systems of current interest in our group have been discussed. It has been shown that for smaller systems it is possible to reach a very high accuracy. The results from the calculations on NiH^+ and the convergence pattern for the binding energy of NiO^+ strongly suggests that the experimentally determined binding energy of the latter system is incorrect. Both NiH^+ and NiO^+ were measured using an ion beam technique. A similar conclusion is drawn for the binding energy of $NiCO^+$, which was determined from appearence potentials in a photoionization study. The experimental dissociation energy of 48 kcal/mol seems much too high. For the interaction between Ni and H_2O very large basis sets for H_2O had to be used to obtain agreement with the matrix isolation experiment. Using standard double zeta basis sets, qualitatively wrong results were generated. In the correct description, there is at most a very weak electrostatic binding, which means that H_2O should not be considered as an electron donor, and H_2O is thus not a suitable model for ligands of this type in transition metal complexes. A neutral model for a pure σ donor is therefore missing presently. The final area discussed in this presentation is the modelling of chemical reactions on metal surfaces. For metal atoms sufficiently far away from the active center, a one electron ECP is recommended which explicitly includes frozen 3s and 3p orbitals. This type of ECP will localize and minimize the parameterized part of the ECP. For on top dissociation, the active metal atom must be treated with all its d electrons. The preliminary results from a study of O_2 dissociation on nickel, indicate barriers of the order of 10-20 kcal/mol for all three investigated pathways. The rather low barrier for motion of oxygen atoms on the surface further indicates the possibility for dissociation followed by adatom reordering as a possible mechanism for reaching the final structure for the oxygen atoms on the surface.

References.

1. M.R.A. Blomberg, P.E.M. Siegbahn and B.O. Roos, Mol. Phys. 47 , 127 (1982).

2. M.R.A. Blomberg and P.E.M. Siegbahn, J. Chem. Phys. 78 , 5682 (1983); and J. Chem. Phys. 78 , 986 (1983).

3. P.E.M. Siegbahn, M.R.A. Blomberg and C.W. Bauschlicher, J. Chem. Phys. 81 , 2103 (1984).

4. B.O. Roos, P.R. Taylor and P.E.M. Siegbahn, Chem. Phys. 48 , 157 (1980).

5. P.E.M. Siegbahn, Faraday Symp. Chem. Soc. 19 , 97 (1984).

6. M.R.A. Blomberg, U. Brandemark, P.E.M. Siegbahn, K. Broch-Mathisen and G. Karlström, J. Phys. Chem. 89 , 2171 (1985).

7. P.E.M. Siegbahn, Intern. J. Quantum Chem. 23 , 1869 (1983).

8. P.B. Armentrout and J.L. Beauchamp, Chem. Phys. 50 , 37 (1980).

9. P.B. Armentrout, L.F. Halle and J.L. Beauchamp, J. Chem. Phys. 76 , 2449 (1982).

10. G. Distefano, J. Res. Natl. Bur. Stand. Sect. A 74 , 233 (1970).

11. A.E. Stevens, C.S. Feigerle and W.C. Lineberger, J. Am. Chem. Soc. 104 , 5026 (1982).

12. R.N. Compton and J.A.D. Stockdale, Int. J. Mass. Spectrom. Ion Phys. 22 , 47 (1976).

13. J.W. Kauffman, R.N. Hauge and J.L. Margrave, J. Phys. Chem. in press.

14. C.W. Bauschlicher, J. Chem. Phys. in press.

15. H. Huber, G.A. Ozin and W.J. Power, J. Am. Chem. Soc. 98 , 6508 (1976); G.A. Ozin, W.J. Power, T.H. Upton and W.A. Goddard III, J. Am. Chem. Soc. 100 , 4750 (1978).

16. Z.H. Kafafi, R.H. Hauge and J.L. Margrave, in press.

17. C.F. Melius, C.L. Bisson and W.D. Wilson, Phys. Rev. B 18 , 61 (1974).

18. C.W. Bauschlicher, J. Chem. Phys. in press.

19. A.K. Fischer, F.A. Cotton and G.J. Wilkinson, J. Am. Chem. Soc. 79 , 2044 (1957).

20. P.-O. Widmark, B.O. Roos and P.E.M. Siegbahn, J. Phys. Chem. 89, 2180 (1985).

21. P.-O. Widmark, G.J. Sexton and B.O. Roos, Theoret. Chim. Acta, in press.

22. Z.H. Kafafi, private communication.

23. L. Pettersson, U. Wahlgren and O. Gropen, Chem. Phys. 80 , 7 (1983).

24. P.S. Bagus, C.W. Bauschlicher, C.J. Nelin, B.C. Laskowski and M. Seel, J. Chem. Phys. 81 , 3594 (1984).

25. M.R.A. Blomberg, P.E.M. Siegbahn and A. Strich, Chem. Phys. 97 , 287 (1985).

26. C.W.B. Bauschlicher and P.S. Bagus, Phys. Rev. Letters 54 , 349 (1985).

27. C.R. Brundle, J. Behm and J.A. Barker, J. Vac. Sci. Technol. A2 , 1038 (1984).

THE IMPORTANCE OF ATOMIC AND MOLECULAR CORRELATION ON THE BONDING IN TRANSITION METAL COMPOUNDS

Charles W. Bauschlicher, Jr.
NASA Ames Research Center
Moffett Field, CA 94035

Stephen P. Walch
ELORET Institute*
Sunnyvale, CA 94087

and

Stephen R. Langhoff
NASA Ames Research Center
Moffett Field, CA 94035

ABSTRACT. An accurate determination of the separation between the low-lying atomic states ($3d^n4s^2$, $3d^{n+1}4s^1$, and $3d^{n+2}$) of the transition metal atoms requires both extensive basis sets and a high level correlation treatment. Since molecular systems containing transition metal atoms often involve admixtures of atomic states, molecular calculations also must be designed to correctly incorporate the dominant atomic correlation effects. In addition, the large atomic exchange energy associated with high spin coupled 3d orbitals requires that a balance between atomic exchange and molecular bonding effects be achieved. These two requirements often necessitate the use of an MCSCF reference wavefunction followed by extensive CI calculations to provide an accurate description of molecules containing transition metal atoms. Several examples will be given including the dipole moment of the $^2\Delta$ state of NiH, which is a sensitive measure of the mixture of $3d^84s^2$ and $3d^94s^1$ in the NiH wavefunction, the dissociation energy of the $X^2\Pi$ state of CuO, which is sensitive to the correct mixture of the $3d^94s^1$ and $3d^{10}$ Cu^+ atomic states, and the bonding in CrO, where an equivalent description of the relative energies associated with the Cr 3d-3d atomic exchange and the Cr-O bond is important. Some examples of these effects in transition metal dimers will also be given.

I. INTRODUCTION

The molecular orbital concept of the bonding has great utility in theoretical chemistry. Such concepts as bond order and dative bonding have meaning when the orbitals can be partitioned into subsets, such as core, bonding, anti-bonding and lone-pair orbitals. Implicit in such a picture is an assumption about the atomic state of the atoms in the system from which the bonding arises. For example, the approximately 130° bond angle in $^3B_1$ methylene involves sp^2 hybridization

* Mailing address: NASA Ames Research Center, Moffett Field, CA 94035

A. Veillard (ed.), Quantum Chemistry: The Challenge of Transition Metals and Coordination Chemistry, 15–35.

on carbon, whereas the approximately 100° bond angle in $^1A_1$ methylene results primarily from unhybridized carbon atom. Hence, the $^1A_1$ state is associated with ^3P carbon, while the $^3B_1$ state involves a strong admixture of ^5S. The fact that the ^5S-^3P separation in carbon atom can be computed quite accurately, is related to the success of theory in accurately determining the $^3B_1$-$^1A_1$ separation in methylene [1,2].

A major difference between transition metal and first-row chemistry arises from the contribution of atomic asymptotes involving variable d occupations [3], namely $3d^n4s^2$, $3d^{n+1}4s^1$, $3d^{n+2}$, $3d^n4s^14p^1$, and $3d^{n+1}4p^1$. The difference in the electron density at the nucleus and the relative size of the 3d and 4s orbitals results in large differential correlation and relativistic contributions to the transition metal atomic states with different 3d and 4s occupations. These problems are well illustrated by comparing the computational requirements for computing an accurate ^5S-^3P separation in carbon with computing the ^3F $(3d^84s^2)$-$^3D(3d^94s^1)$ separation in Ni. For C a double-zeta basis set yields an SCF ^5S-^3P separation that is within 0.02 eV of numerical Hartree-Fock (NHF). At the singles and doubles configuration-interaction (SDCI) level (with the Davidson correction for unlinked clusters in parentheses), the error in the ^5S-^3P separation of carbon is 0.37(0.29) eV using a double-zeta plus polarization basis and only 0.18(0.06) eV using a triple-zeta plus double polarization basis [1]. In contrast, the 3d orbitals of the ^3F and ^3D states of Ni are sufficiently different that three contracted d functions are required to approach the NHF separation [4]. To approach to within 0.1 eV of the SDCI basis set limit, requires at least four contracted d functions, a (4f/3f) basis and a g function [4,5]. Unfortunately, if a correction (0.35 eV) is made for the large differential relativistic contribution [6], this limit is still about 0.4 eV larger than the observed ^3F-^3D separation [3].

Since states with mixed atomic character are quite common in transition-metal chemistry, the orbital optimization step is more crucial. For example, for the ^3P and ^5S states of carbon atom, the error introduced at the SCF level by using orbitals optimized for the other state is 0.1-0.2 eV, whereas it is 4.0 eV for the ^3F and ^3D states of Ni atom. SDCI calculations allow some orbital relaxation, but still lead to an error of about 1 eV, compared to CI calculations with optimal orbitals for each state. This is compared to a 0.002-0.004 eV error for carbon. Hence, for many of the molecular states of transition metals, an orbital optimization step (MCSCF,CASSCF) is essential, because the subsequent CI treatment cannot fully overcome the orbital bias of an SCF level orbital optimization.

Difficulties in describing the atomic states manifest themselves in molecular calculations as illustrated by our calculations [7] of the dipole moment of the $X^2\Delta$ state NiH. We have found the dipole moment to be sensitive to the 3d population and to the quality of the wavefunction. The SCF description is constrained to have $3d^94s^1$ bonding, leading to a large dipole moment of 4 Debye, and a Ni 3d population of 8.89 electrons. The SDCI calculation reduces the population and dipole moment slightly, but cannot overcome the large orbital bias toward $3d^94s^1$. A CASSCF calculation introduces the flexibility to describe bonding from both $3d^84s^2$ and $3d^94s^1$, but mixes in too much $3d^84s^2$ character because of the incorrect atomic separation, and leads to a dipole moment (1.74D) that is smaller than experiment. A multi-reference SDCI that improves the separations of the atomic states leads to a dipole moment of 2.15D, which is in reasonable agreement with the experimental

[8] dipole moment of 2.4±0.1D. Natural orbital (NO) iterations further increase the dipole moment to 2.59D and the 3d population to 8.72, reflecting the mixed atomic character of the wavefunction.

The bonding in molecules containing transition metals involves a variable degree of contribution from both the 3d and 4s orbitals. The ratio of the radial extent of the 4s to 3d orbitals, computed using NHF wavefunctions, varies from 2.03 in Sc to 3.36 in Cu [4]. For the second transition row, the ratio of the 5s to 4d varies from 1.61 in Y to 2.67 in Ag [9]. A smaller ratio, such as for Sc, favors the formation of s-d hybrids, such as observed for ScH [10], whereas for NiH [7,11-12], the 3d is contracted relative to the 4s and is not as strongly involved in the bonding. However, for metal-metal bonds both the 3d and 4s orbitals are involved in bonding, especially for those elements in the left half of the transition row. Because of the larger size of the 4s with respect to the 3d, the 4s bonds must be compressed, weakening the 4s bonding, to allow the formation of the d bonds. At this compromise bond length, the d-d overlaps in the transition metal dimers are substantially smaller than those of C-C bonds (a factor of three smaller for $3d\sigma$ and $3d\pi$ bonds and a factor of 10 smaller for $3d\delta$ bonds). Hence, the smaller overlaps and the large d-d exchange terms lead to weaker bonds. Sophisticated theoretical treatments are needed to balance these weak bonds against the large atomic exchange terms. A prototype example is Cr_2, where a CASSCF procedure correctly describes the atomic exchange, but does not describe the 3d-3d bonding sufficiently well to give a bound inner well [9,13-15].

On a positive note, we find that a SCF or MCSCF description is adequate for dative bonding, where one fragment donates a pair of electrons to the other, since large changes in d-d exchange are not involved. The constrained space orbital variation method (CSOV) method has been developed [16-17] to analyze changes occurring with dative bonding, since Mulliken population analyses often give misleading results.

In the remainder of this manuscript we give a more detailed description of several systems which illustrate how a correct treatment of the atomic correlation leads to an accurate description of molecular systems. Section II contains a description of the basis sets and methods used used to correctly treat the atomic correlation. Sections III and IV contain a description of the calculation of the transition metal hydrides and oxides. Section V contains examples of calculations on the transition metal dimers. Section VI contains a discussion of dative bonding. The conclusions are contained in Section VII.

II. THE $3d^n4s^2$-$3d^{n+1}4s^1$ ATOMIC ENERGY SEPARATIONS

As a first step in understanding transition metal bonding, we have studied [4-5,18] the computational requirements of computing accurate atomic separations. (We use the MOLECULE-SWEDEN [19-20], or BIGGMOLI-SWEDEN [21] program systems on the CRAY-XMP and the Karlsruhe [22] version of the COLUMBUS [23-25] system of programs on the CYBER 205.) For calculations on first-row transition metal atoms, we generally start with the 14s9p5d basis sets optimized by Wachters [26], plus his diffuse 4p-like functions to account for 4s-4p near degeneracy effects. The diffuse p functions are needed in the atom to describe the important s^2 to p^2 excitation in the $3d^n4s^2$ state (about a 0.8 eV correlation effect), and sd to pf correlation. For use in molecular systems, these functions are generally multiplied by about 1.3, and are needed for polarization of the 4s electrons. Since Wachters op-

TABLE I. Energy separations between the $3d^n4s^2$ and $3d^{n+1}4s^1$ states of the first row transition metal atoms.

	Transition metal atoms[a]							
	Sc	Ti	V	Cr	Mn[b]	Fe	Ni[b]	Cu[b]
	^2D-^4F	^3F-^5F	^4F-^6D	^5D-^7S	^6S-^6D	^5D-^5F	^3F-^3D	^2D-^2S
SCF	1.03	0.56	0.15	-1.24	3.34	1.80	1.27	-0.39
NHF	1.01	0.54	0.13	-1.27	3.33	1.80	1.28	-0.37
SDCI	1.68	0.94	0.36	-1.09	2.77	1.19	0.28	-1.37
SDCI+Q[c]	1.77	1.02	0.46	-0.95	2.86	1.25	0.30	-1.33
SDCI(3s3p)	1.33	0.67	0.12	-1.20	2.64	1.09	0.24	-1.35
SDCI+Q(3s3p)	1.51	0.81	0.27	-1.04	2.64	1.13	0.23	-1.33
exp[d]	1.44	0.81	0.25	-1.00	2.14	0.88	-0.03	-1.49
exp-rel[e]	1.32	0.67	0.08	-1.21	1.94	0.62	-0.39	-1.92

[a] The basis set includes two f functions, unless otherwise noted. A positive sign indicates that the state arising from the $3d^n4s^2$ occupation is lower in energy.
[b] Includes only one f function.
[c] Davidson's correction has been included to estimate the importance of quadruples.
[d] Average of m_j's from Ref. 3.
[e] The experimental result minus the differential relativistic contribution to the separation computed from numerical Hartree- Fock wavefunctions, Ref. 6.

timized most of the basis sets for the $3d^n4s^2$ occupation of the atom, the resulting d basis is not sufficiently diffuse to adequately describe the $3d^{n+1}4s^1$ occupation. The most complete study of the additional d functions was carried out by Hay [27], who optimized diffuse d functions for Wachters' primitive set. Once this single diffuse d is added, the computed separations between the $3d^{n+1}4s^1$-$3d^n4s^2$ occupations are essentially the same as the numerical Hartree-Fock values–see Table I. Note that the omission of diffuse d functions in molecular calculations can result in large superposition errors and qualitatively incorrect potential energy functions [17,28]. Hence, it is necessary that basis sets employed for molecular systems provide a balanced description of both the $3d^{n+1}4s^1$ and $3d^n4s^2$ occupations.

For the second transition row, we commonly use the (17s11p8d) basis set of Huzinaga [29] augmented with the additional p and d functions optimized by Walch et al. [30] To reduce the computation effort, effective-core-potential (ECP) parameters and associated valence basis sets can be employed (see e.g. Hay and Wadt [31]). Since differential relativistic effects between states with different d occupations can be large, the ECP approach also offers one alternative for including them. We have also found that the use of first-order perturbation theory to estimate the dominant (Darwin and mass-velocity) terms [32] to be a satisfactory approach, especially when differential relativistic effects are small.

As can be seen from Table I, the Hartree-Fock limit for the $3d^n4s^2$–$3d^{n+1}4s^1$

Table II. Summary of the Ni $3d^8 4s^2$ (^3F)-$3d^9 4s^1$ (^3D) separation, in eV.

basis	SDCI[a]	CPF
(14s11p6d/8s6p4d)	0.47	0.50
(14s11p6d1f/8s6p4d1f)(diffuse)	0.38	0.37
(14s11p6d3f/8s6p4d2f)(tight)	0.17	0.22
(15s12p6d3f/9s7p4d2f)[b](compromise)	0.14	
(14s11p6d4f/8s6p4d3f)(both)	0.10	0.12
(14s11p6d4f2g/8s6p4d3f2g)	0.03	0.06
experiment[c]	-0.03	
experiment-relativistic[d]	-0.39	

[a] Symmetry but not equivalence restrictions were imposed for the ^3F state, while both restrictions were imposed for ^3D. Imposing equivalence restrictions for the ^3F state reduces the theoretical values in Table II by about 0.06 eV.

[b] Also contains a set of diffuse s and p functions for molecular application that are not expected to significantly affect separation.

[c] Average of m_j's from Ref. 3.

[d] Experimental separation minus the relativistic contribution in Ref. 6.

separations are significantly different from experiment, especially for Mn through Cu where there are doubly occupied 3d orbitals. The inclusion of correlation in a basis set without f functions represents an improvement over SCF, but substantial errors remain. 4f functions are very important for both correlating [4,33] and polarizing [34] the d orbitals. The inclusion of correlation in a basis set including f functions substantially improves the results. Including the Davidson correction [35] for higher excitations has up to an 0.1 eV effect. Since the 3s and 3p orbitals lie in the same region of space as the 3d orbitals, one expects some differential 3s3p effect on the separations. For Sc, the $3d^2 4s^1$ state is preferentially lowered by about 0.3 eV. Note that this differential 3s3p correlation effect is also observed in molecular calculations (see later discussion on ScS). The differential 3s3p correlation effect decreases from about 0.3 eV for Sc to almost zero for Cu. Thus, on the right hand side of the row, it is common to neglect 3s3p correlation since it increases the size of the CI expansion dramatically. With the inclusion of 3s3p correlation the importance of Davidson's correction increases to a maximum of about 0.2 eV.

Even though the addition of f functions, the inclusion of 3s3p differential correlation, and a correction for higher excitations all improve the separations, troublesome differences remain once relativistic effects are included. To illustrate the challenge of obtaining better than 0.3 eV accuracy on the separations, we consider in more detail in Table II the particularly troublesome case of the ^3F-^3D separation in Ni atom. The separations computed in basis sets without f functions show sizable errors. The inclusion of the diffuse f (0.4) designed to obtain the sd correlation makes a 0.09 eV improvement. The inclusion of a (21) contraction of a three-term gaussian fit [36] of a STO 4f function picked to maximize the total correlation in the ^3D state, makes a larger, 0.30 eV, improvement in the separation. The effect of both the tight and diffuse functions is almost additive, since the f functions are in different regions of space and obtain different correlation effects. The (3f/2f) compromise set (the most diffuse f being 0.514) is able to obtain the sd correlation as well as a substantial part of the dd correlation, leading to good agreement with the (4f/3f) basis set. The inclusion of a g function has a 0.07 eV effect, similar in magnitude to the 0.09 eV differential sd correlation from the diffuse f. However, if an estimate is made for relativistic effects from NHF wavefunctions, the remaining error in the calculation is quite large, about 0.4 eV. We have also found using first-order perturbation theory that the differential relativistic effect does not substantially change with the level of correlation treatment. Thus one must conclude that most of the remaining error is in the treatment of correlation. However, it must be due to higher excitations that cannot be accounted for in an approximate manner, since neither a Davidson correction nor the coupled pair formalism [37] (CPF) makes much improvement in the separation. Note, however, that 0.4 eV is a small amount of energy compared with the total atomic correlation energy. For example, a 11-electron CPF treatment of Cu ^2S obtains about 12 eV of correlation energy with about 0.9 eV of it being sd correlation and the remaining 11 eV, dd correlation. In our opinion, the most likely possibility for the remaining difference is differential triple excitations. (see also Ref. 38).

III. TRANSITION METAL HYDRIDES

The nature of the bonding in the simplest molecular systems containing the transition metals (TM), the transition metal hydrides [7,10-12,39-44], also reflects the relative energy separations between the $3d^n4s^2$ and $3d^{n+1}4s^1$ atomic states. For example, one finds predominantly $3d^n4s^2$ character for ScH and MnH where $3d^n4s^2$ is well below $3d^{n+1}4s^1$, and predominantly $3d^{n+1}4s^1$ character for CrH and CuH where the $3d^{n+1}4s^1$ asymptote is significantly lower. For other TM hydrides, such as VH and TiH, the ground states involve a considerable mixture of both $3d^n4s^2$ and $3d^{n+1}4s^1$. Later in this section we discuss in more detail the mixed-state character of NiH, and the effect it has on the dipole moment.

Consider ScH where the nature of the bonding is dictated by the fact that the $^4F(3d^24s^1)$ state lies 1.44 eV above the $^2D(3d^14s^2)$ ground state in Sc atom. Hence, the bonding is expected to involve considerable sp hybridization, and the ground state is expected to be one of the six states, $^{1,3}\Sigma^+$, $^{1,3}\Pi$ or $^{1,3}\Delta$, corresponding to the configuration

$$(Sc(sz) + H(1s))^2\bar{sz}^1 \begin{pmatrix} d\sigma \\ d\pi \\ d\delta \end{pmatrix}^1 \tag{1}$$

where sz and $\bar{sz}$ are the bonding and anti-bonding combinations of the 4s and $4p\sigma$ orbitals. Another mechanism is possible for the $^1\Sigma^+$ state, however, namely the formation of a d bond directly to the H(1s) with a concomitant polarization of the 4s electrons away from the ScH bond pair. Previous theoretical work has shown [10] the $^1\Sigma^+$ state to be the ground state with considerable d-bond character and a significantly smaller r_e. The ordering of the triplet states can be predicted solely on atomic considerations. The $^3\Delta$ state is lowest, because it can mix in the $3d\sigma^13d\delta^1$ occupation coming from the pure ^4F component of $Sc(3d^24s^1)$. The $^4\Pi$ state is intermediate since it mixes in the $3d\sigma^13d\pi^1$ occupation corresponding to

TABLE III. Results for NiH Dipole Moment ($X^2\Delta$ state).

Method[a]	Dipole Moment	3d population
SCF	3.98	8.89
SCF/SDCI	3.64	8.83
CASSCF	1.74	8.31
CASSCF/MRSDCI	2.15	8.58
NO/MRSDCI[b]	2.39	8.66
NO/MRSDCI′	2.52	8.70
NO/MRSDCI″	2.59	8.72
exp[c]	2.4±0.1	

[a] At 2.79 Bohr, experimental r_e.
[b] Lowest energy.
[c] Ref. 8.

40% $^4F(3d^24s^1)$ and 60% of the higher $^4P(3d^24s^1)$. The $^3\Sigma^+$ state is highest since it mixes in very little $Sc(3d^24s^1)$ character. In general, we find that the ground states of the TM hydrides are not determined from Hund's rules, but from the mixings of different atomic components of the $3d^{n+1}4s^1$ and $3d^n4s^2$ occupations.

The comparison of ScH with TiH illustrates the competing effects that determine the ground state. In Ti atom the $3d^34s^1$ asymptote is only 0.81 eV higher than $3d^24s^2$, leading to an increased degree of mixing. Also, with two 3d electrons in Ti, the formation of d bonds is less favorable because of the reduced radial extent of the 3d orbital and the loss in d-d exchange energy upon bonding to the 3d orbital. Hence, the ground state is $^4\Phi$ and involves a bond between the Ti 4s4p hybrid orbital and H(1s).

For VH the two asymptotes are quite close together, and the $^5\Delta$ ground state is found to be of mixed character (3d population of 3.42). The $^5\Phi$ state, which is expected to be the ground state based on Hund's rules, lies 0.5 eV above $^5\Delta$, since it cannot mix in any $3d^44s^1$ character. Note that mixed states, such as the $^5\Delta$ state of VH and the $^2\Delta$ state of NiH, are more difficult computationally, since the SCF optimization introduces a strong orbital bias towards one asymptote.

The $^6\Sigma^+$ ground state of CrH is found to be predominantly $3d^54s^1$ like, since the $3d^44s^2$ asymptote is 1.0 eV higher. The resulting 3d population of 4.86 reflects this relatively small mixing of $3d^44s^2$. The large separation (2.14 eV) between the $3d^54s^2$ and $3d^64s^1$ asymptotes in Mn also leads to a nearly pure $^7\Sigma^+$ ground state for MnH (3d population of 5.06). As one proceeds further to the right in the first row, the separation between atomic states again decreases until one reaches Ni where the $^3F(3d^84s^2)$ and $^3D(3d^94s^1)$ asymptotes are nearly degenerate. Hence, the $^2\Delta$ ground state of NiH has again a very mixed-state character. The dipole moment, which has recently been measured for the v=0 level of the $^2\Delta$ state by Gray et al. [8], provides a sensitive measure of the 3d population. In the remaining portion of this section, we focus on the computational requirements of computing an accurate dipole moment for the ground state of NiH.

In Table III we summarize our study [7,45] of the effect of electron correlation on the dipole moment (μ) of the $^2\Delta$ state of NiH using a [9s7p4d2f] contracted gaussian basis for Ni. SCF calculations on the $^2\Delta$ state of NiH force the $3d^94s^1$ occupation of Ni. This description gives a Ni(4s)-H(1s) bond polarized towards H and a μ of about 4 Debye, which is considerably larger than the value of 2.4±0.1 D determined by Gray et al. from the Stark spectrum. The 1-ref SDCI calculation maintains most of the orbital bias of the SCF, reducing μ to only 3.64 D. However, the $3d^84s^2$ asymptote is low lying and contributes to the bonding through formation of sp hybrids. The $3d^84s^2$ contribution has a very small dipole moment, because the unpaired electron resides in the $\bar{sz}$ hybrid on Ni. Hence, μ is a sensitive measure of the mixed-state character (3d population) in the $^2\Delta$ state. To allow the $3d^84s^2$ occupation to influence the orbital optimization step, a CASSCF [46] calculation was performed [7] with the Ni(3dσ, 4s and 4pσ) and H(1s) as active orbitals. The CASSCF calculation is in better agreement with experiment, but overestimates the importance of $3d^84s^2$, giving a dipole moment of 1.74 D and a 3d population of 8.31 electrons. Too much $3d^84s^2$ character is mixed in, because at this level the $3d^84s^2$–$3d^94s^1$ separation is too large. Hence, a high level correlation treatment is required to describe properly the mixed- state character. A multi-reference SDCI calculation from the important references in the CASSCF increased μ to 2.15 D and the 3d population to 8.58. However, even at this level of treatment, there is some bias towards the starting orbitals. To overcome the remaining orbital bias, we carried out a series of natural orbital (NO) iterations from the MRSDCI wavefunction. After three NO iterations, μ is reasonably well converged and in good agreement with experiment. The strong correlation between μ and the 3d population in Table III provides rather strong support for the mixed d character of the $^2\Delta$ state.

Another approach that has been an excellent diagnostic tool in studying transition metals is the coupled pair formalism (CPF). Since this is an iterative approach that includes higher excitations in a size-consistent manner, it is less sensitive to the orbital basis. However, it is a single reference approach, so that the inherently multi-reference character of the $^2\Delta$ state of NiH is a severe test. In Table IV we have summarized our study [45] of the importance of higher angular momentum functions (f and g functions on Ni and d functions on H) on μ at the SCF, SDCI and CPF levels. The dipole moment is rather insensitive to basis set saturation at the SCF and SDCI levels, but at the CPF level μ changes markedly from 0.764 D for a basis set without f functions to 1.749 D for a basis set containing 3f2g on Ni and 2d functions on H. The CPF wavefunction mixes in too much $3d^84s^2$ character, but less so as the $3d^84s^2$–$3d^94s^1$ atomic separation is decreased with basis set saturation, i.e. the d population and μ are directly related to the computed atomic separations at the CPF level, since this treatment is better able to overcome the SCF orbital bias.

Since the CPF is presently limited to a single reference and the SCF orbitals give a poor description, it again seemed worthwhile to perform a series of NO iterations. Using the CPF natural orbitals at r=2.9 Bohr with the 3f basis gave μ values of 1.61, 2.56, 2.73 and 2.90 D on successive iterations. (Such NO iterations have little effect on the single reference SDCI treatment.) When the SDCI NO were used the CPF μ was 2.49 D. All of these results are in reasonably good agreement with the MRSDCI results and experiment, and suggests that the range of validity of the CPF procedure can be increased by a natural orbital procedure. In any case, all of these calculations demonstrate the complexity of computing accurate properties

TABLE IV. Basis set sensitivity to NiH dipole moment.

	SCF	SDCI	CPF
No f functions			
r_e(Bohr)	2.921	2.804	2.896
μ(D)	4.211	3.745	0.764
3d population	8.91	8.83	8.20
D_e(eV)		2.03	2.62
diffuse f function			
r_e	2.928	2.807	2.903
μ	4.153	3.597	0.872
3d population	8.91	8.84	8.23
D_e		2.12	2.74
tight f functions			
r_e	2.927	2.800	2.851
μ	4.188	3.758	1.675
3d population	8.91	8.86	8.42
D_e		2.26	2.71
both f functions			
r_e	2.931	2.809	2.865
μ	4.142	3.679	1.622
3d population	8.90	8.84	8.44
D_e		2.30	2.78
2g's + d's on H			
r_e	2.934	2.798	2.847
μ	4.149	3.697	1.749
3d population	8.90	8.82	8.45
D_e		2.42	2.89

for states with mixed d character, especially when it is exceedingly difficult to get the proper atomic spacings between $3d^n4s^2$ and $3d^{n+1}4s^1$.

IV. TRANSITION METAL OXIDES

The bonding in the transition metal oxides is generally more complex than in the hydrides [47-53], and like the transition metal hydrides changes significantly from Sc to Cu. In this section we consider how the bonding varies with metal atom, focusing on a few representative systems.

For ScO, the ground state is $^2\Sigma^+$, with the lowest excited state more than 15,000 cm^{-1} higher in energy [52,54]. The bonding in the ground state has both an ionic and a covalent component. The ionic component, arising from Sc^+ ($3d^14s^1$)

and O^- (p^5), results in a bond between the Sc $3d\sigma$ and O $2p\sigma$ orbitals, whereas the covalent component, arising from Sc ($3d^24s^1$) and O (p^4), forms a double bond. The $3d^14s^2$ ground state configuration of the atom does not contribute significantly to the bonding because the smaller radial extent of the 3d orbital does not result in effective 3d bonding at the bond length where the optimal 4s overlap occurs. Thus it is better to have only one 4s electron which can polarize away from the O, reducing the repulsion and allowing 3d bonding. In both the ionic and covalent components of the bonding, the donation of the O lone pair electrons into empty 3d orbitals on Sc, gives the molecule some triple bond character. The degree of covalent bonding increases with decreasing separation between the $3d^14s^2$ and $3d^24s^1$ occupations. At the SCF level, the $3d^14s^2$ occupation is favored resulting in a very ionic system. A 1-ref SDCI treatment cannot overcome this orbital bias (a d population of 1.17 and net charge on Sc of 0.74). However, at the CPF level, which is less sensitive to this orbital bias, the d population is significantly increased (1.43) with a corresponding reduction in net charge (0.54). Note that the excited states in ScO are much higher, since they involve promotion of the Sc 4s electrons into empty 4p and 3d orbitals. The nature of the bonding in ScS is also found to be similar to the oxide. For ScS, correlation of the Sc 3s and 3p electrons significantly increases both the 3d population (by 0.3 electrons) and D_e (by 0.3 eV) at the CPF level. This effect is not surprising considering the relatively large effect that 3s3p correlation has on the $3d^24s^1$–$3d^14s^2$ atomic separation (see Table I).

For TiO, the $^3\Delta$ ground state is formed by adding the extra d electron to the empty $3d\delta$ orbital [47,54]. No additional bonding results, since there are no O δ orbitals. In TiO, apart from the low-lying $^1\Delta$ state, the next excited states arise from moving either the open 4s or 3d electron into a 3d or 4p orbital. For VO the extra electron also goes into an empty $3d\delta$ orbital. The small reduction in D_e between ScO and VO, results from a combination of the increased loss of exchange terms in VO, and the greater polarization of the 4s electrons due to the reduced size of the 3d orbital.

Although the bonding in ScO, TiO and VO are quite similar, dramatic changes occur for CrO since the $3d^54s^1$ occupation is lower than $3d^44s^2$. Hence, d electrons are in all symmetries and the O lone pair donation into an empty d orbital is lost unless Cr^+ $3d^44s^1$ mixes into the wavefunction. Our theoretical results [50] for CrO and MoO are summarized in Table V. The bonding in CrO is best understood starting from the ionic limit. There are four candidates for the ground state, namely

$$3d\sigma^1 3d\pi^1 3d\bar{\pi}^1 3d\delta^1 3d\bar{\delta}^1(^6\Sigma^+)O2p\sigma^1 2p\pi^4(^2\Sigma^+) \quad {}^7\Sigma^+, \tag{2}$$

$$3d\sigma^1 3d\pi^1 3d\bar{\pi}^1 3d\delta^1 3d\bar{\delta}^1(^6\Sigma^+)O2p\sigma^2 2p\pi^3(^2\Pi) \quad {}^7\Pi, \tag{3}$$

$$3d\pi^1 3d\bar{\pi}^1 3d\delta^1 3d\bar{\delta}^1(^5\Sigma^+)2p\pi^4(3d\sigma + O2p\sigma)^2 \quad {}^5\Sigma^+, \tag{4}$$

and

$$3d\sigma^1 3d\delta^1 3d\bar{\delta}^1(^4\Sigma^+)O2p\sigma^2(3d\pi + O2p\pi)^4(3d\pi - O2p\pi)^1 \quad {}^5\Pi. \tag{5}$$

The $^7\Pi$ and $^7\Sigma^+$ states have one less bond than the quintet states. For ScO-VO such excited states which eliminate one bond are very high in energy, but in CrO the septet states are helped by the more favorable d-d exchange. The Pauli repulsion

TABLE V. Spectroscopic parameters for the $^5\Pi$, $^5\Sigma^+$, $^7\Pi$ and $^7\Sigma^+$ state of CrO and MoO; the MoO results are in parentheses.

	$^5\Pi$	$^5\Sigma^+$	$^7\Pi$	$^7\Sigma^+$
SCF/CASSCF				
r_e(Å)	1.86(1.80)	1.86(1.82)	1.94(2.07)	1.89(2.09)
ω_e(cm^{-1})	630(560)	630(730)	620(455)	550(400)
D_e(eV)	2.52(2.13)	1.34(1.41)	2.26(1.42)	1.42(0.62)
T_e(cm^{-1})		9500(5800)	2100(5900)	8800(12220)
d population	4.26(4.47)	4.98(4.89)	4.25(4.69)	4.23(4.80)
net charge	0.63(0.59)	0.81(0.82)	0.64(0.64)	0.68(0.74)
SDCI				
r_e(Å)	1.66(1.71)	1.68(1.77)	1.92(2.06)	1.86(2.04)
ω_e(cm^{-1})	820(1035)	750(830)	590(470)	690(540)
D_e(eV)	3.09(3.67)	2.46(2.76)	2.17(1.74)	1.35(0.79)
T_e(cm^{-1})		5100(7300)	7500(15500)	14000(22000)
d population	4.35(4.47)	4.89(4.90)	4.35(4.73)	4.32(4.84)
net charge	0.53(0.53)	0.78(0.73)	0.53(0.54)	0.58(0.64)
EXP[a]				
r_e(Å)	1.615			
ω_e(cm^{-1})	898.4			
D_e(eV)	4.4(5.0)			

[a] Huber and Herzberg, Ref 54.

and electrostatic attraction favor the O^- $2p\sigma^2 2p\pi^3$ occupation over the $2p\sigma^1 2p\pi^4$ occupation, see CuO below. Thus Π states are favored over the equivalent Σ states. Hence, the bonding in CrO can be understood in terms of the competing effects of d-d bonding, exchange and electrostatics.

Since the bonding in the septet states is quite ionic and the d-d exchange is not altered, they are well described by a single reference. In contrast, the $^5\Pi$ and $^5\Sigma^+$ states require a CASSCF treatment to describe the formation of the d bond. However, quantitative results for the quintet states require an extensive SDCI treatment, since lower level treatments do not describe both bond formation and d-d exchange equally well.

At the SCF/CASSCF level, both the $^7\Pi$ and $^7\Sigma^+$ states of CrO are incorrectly predicted to lie below $^5\Sigma^+$. Hence, the change in atomic spin coupling has cost more than has been returned by forming the bond at the CASSCF level. The SDCI calculations, which give a large (0.2Å) bond shortening for the quintet states, gives the correct ordering as $^5\Pi < {}^5\Sigma^+ < {}^7\Pi < {}^7\Sigma^+$ for both CrO and MoO. It is expected that if the level of treatment of correlation were increased further that the bonding would be enhanced still further relative to the d-d exchange, and the more covalent quintet states would be further stabilized.

Although the bonding in MoO resembles that in CrO, the exchange terms are smaller and the d bonds are stronger owing to the larger radial extent of the 4d orbitals. Hence, it is easier to achieve a balance between bonding and exchange. In MoO the correct ordering of states is achieved at the SCF/CASSCF level, and the SDCI bond reduction for the quintet states is 0.09Å compared to 0.2Å in CrO.

Another interesting feature in Table V is the 3d(4d) populations for CrO(MoO). The $^5\Sigma^+$ states have d populations near 5, because a bond is being formed between the metal dσ and O pσ orbitals. However, in the other states of CrO, the dσ electron is promoted to 4s and polarizes away from the O to reduce repulsion, resulting in a relatively small 3d population. For Mo, where the Mo^+ $4d^5$-$4d^4 5s^1$ separation is larger, this effect is smaller and is only pronounced for the $^5\Pi$ state which has a short bond length to allow the dπ bond.

In Mn, $3d^6 4s^1$ is much higher than $3d^5 4s^2$; thus one expects MnO to be ionic. However, it differs from CrO in that Mn^+ $3d^4 4s^2$ cannot mix in to create an empty d orbital. Therefore, since promotion to the 4s and back donation from oxygen into an empty 3d orbital cannot occur, the ground state is $^6\Sigma^+$, corresponding to the CrO $^5\Sigma^+$ state with the additional electron going into the 4s. Since the $3d^n 4s^2$ is also lower for Fe and Co, one expects the ground states of MnO, FeO, and CoO to have similar bonding characteristics as the additional electrons go into the non-bonding 3dδ orbital. This leads to $^5\Delta$ and $^4\Sigma^-$ ground states for FeO and CoO, respectively (also see Ref. 53).

For Ni atom, $3d^9 4s^1$ and $3d^8 4s^2$ are nearly degenerate and for Ni^+, $3d^9$ is below $3d^8 4s^1$. The bonding in the $^3\Sigma^-$ ground state therefore arises predominantly from the ionic configuration Ni^+ $(3d^9)$ π^3 + O^- $(2p^5)$ π^3 with one electron π bonds forming. Some additional covalent interaction with the 3dσ orbital arises from mixing with the $3d^8 4s^1$ asymptote.

Since for Cu, $3d^{10} 4s^1$ is below $3d^9 4s^2$ and for Cu^+ $3d^{10}$ is well below $3d^9 4s^1$, one expects ionic bonding in CuO as well. In fact the separation between the ground $^2\Pi$ and excited $^2\Sigma^+$ states can be understood in terms of the competing effects of Pauli repulsion and electrostatic stabilization [48]. That is O^- $(2p\sigma^2 2p\pi^3)$ versus O^- $(2p\sigma^1 2p\pi^4)$ interacting with Cu^+ $3d^{10}$. This analysis is similar to that used to describe the $^2\Pi$-$^2\Sigma^+$ separation in the alkali oxides [55]. Earlier work [48] at the SDCI level gave for the $X^2\Pi$ state a r_e that was 0.1Å too long and a D_e (1.88 eV) that was about 1 eV smaller than experiment (2.83$\pm$0.15 eV [56]). An estimate of errors due to basis set incompleteness and higher excitations, lead to the conclusion that the experimental D_e was too large, and that 2.4 eV was an upper limit to D_e. Recently, an ECP treatment [57] which includes the d shell into the potential, but includes a d-valence (core valence (CV)) term, yielded a D_e that is only 0.1 eV smaller than experiment. However, extensive work [58] on the alkali fluorides and chlorides suggest that the CV contribution to D_e cannot be this large.

To help understand the shortcomings of the previous SDCI calculations, we have reconsidered [49] the $X^2\Pi$ and $A^2\Sigma^+$ states of CuO at the SDCI and CPF levels using a large gaussian basis set– Cu(15s11p6d4f1g/9s7p4d3f1g) and O (11s7p3d1f/6s4p3d1f). The theoretical results are summarized in Table VI. With our extensive basis set at the SDCI level (plus Davidson correction), the D_e is only 2.12(2.29) eV, and r_e is only marginally improved (0.02Å) compared to previous

TABLE VI. Spectroscopic parameters for the $X^2\Pi$ and $A^2\Sigma^+$ states of CuO.

	Theoretical[a]		EXP[b]
	SDCI	CPF	
A. $X^2\Pi$			
r_e(Å)	1.812(1.792)	1.749(1.721)	1.724
ω_e(cm^{-1})	604(584)	626(606)	640.17
D_e(eV)	2.12[c]	2.73(2.83)[d]	2.83±0.15[e]
μ(a.u.)	2.324	1.683	
$d\mu/dr$(a.u.)	0.728	0.477	
3d population	9.91	9.53	
4s population	0.25	0.71	
4p population	0.13	0.21	
4pσ population	0.07	0.07	
4pπ population	0.06	0.14	
net charge	0.67	0.50	
B. $A^2\Sigma^+$			
r_e(Å)	1.790(1.779)	1.672(1.649)	
ω_e(cm^{-1})	534(520)	622(591)	
T_e(cm^{-1})	8470(9294)	8694(9940)	7865[f]
μ(a.u.)	2.434	2.247	
$d\mu/dr$(a.u.)	0.912	0.621	
3d population	9.95	9.61	
4s population	0.11	0.31	
4p population	0.18	0.41	
4pσ population	0.07	0.07	
4pπ population	0.11	0.34	
net charge	0.72	0.65	

[a] Values in parentheses include a first-order perturbation theory correction for relativistic effects. All populations refer to the Cu atom.
[b] Huber and Herzberg, Ref 54.
[c] The D_e is computed with respect to the ions and then corrected with experimental IP and EA. The D_e computed directly with respect to neutrals is 2.04(2.01).
[d] The D_e is computed directly with respect to the neutral atoms. The D_e computed with respect to the ions, then corrected using the experimental IP and EA, is 2.65(2.96) eV.
[e] Mass spectrometry, Smoes et al., Ref. 56.
[f] Lefebvre, et al., Ref. 59.

work [49]. However the dramatic differences between the CPF and SDCI results indicate that the SDCI calculation retains much of the SCF orbital bias, and does not mix in enough $3d^94s^1$ character. Hence, the CPF mixes in more $3d^94s^1$ character and also allows back donation of the O 2pπ electrons into the empty Cu 4pπ orbital, thereby shortening and strengthening the bond and bringing both r_e and D_e

into good agreement with experiment. The mixing also occurs for the A $^2\Sigma^+$ state as well, thus the computed T_e is not affected by the mixing, and remains in good agreement with experiment [59]. The mixing of atomic states even for Cu, where the atomic separation is quite large, again illustrates the complexity of bonding in transition metal compounds. Note also that there is considerable 3d involvement in the bonding even though the 3d orbital is significantly spatially contracted compared to the 4s orbital.

V. TRANSITION METAL DIMERS

The accurate description of transition metal-metal bonds has proven to be a challenging theoretical problem [9,13-15,60-65]. By analogy with the TM oxides such as CrO, a critical factor is to obtain a balance between exchange and bonding. However, the formation of a d-d bond between two TM atoms requires the elimination of twice as many exchange integrals as for the oxides, and the metal-metal bonds are relatively weak owing to small overlaps [9,13-15,60-61]. Also, just as for the TM hydrides and oxides, mixing of atomic asymptotes can have a pronounced influence on the nature of the bonding. For example, the $^5\Sigma_u^-$ ground state of Sc_2 arises from the mixed asymptote [9,61,65-66], with one atom in the $3d^14s^2$ ground state and one in the $3d^24s^1$ excited state.

Since the d-d exchange terms are large, the formation of one-electron bonds becomes a viable alternative to two-electron bonds. For example, the $^5\Sigma_u^-$ ground state of Sc_2 can be written as

$$3d\sigma_g^1 3d\pi_u^1 3d\bar{\pi}_u^1 4s\sigma_g^2 4s\sigma_u^1 \quad {}^5\Sigma_u^-. \tag{6}$$

Since there is no change in the d-d exchange terms, the CASSCF approach is able to give a reasonable qualitative description of the bonding. The inclusion of more extensive correlation in a SDCI calculation shortens the bond from 2.85 to 2.79Å and increases D_e by 0.5 eV. The computed ω_e (184) is in reasonable agreement with experiment (238.9 cm^{-1} [67]).

For Ti_2, the analogous state is obtained by putting the two additional d electrons into the $3d\delta_g$ orbital, thereby retaining the high spin coupling. Hence, one has a $^7\Sigma_u^+$ state

$$3d\sigma_g^1 3d\pi_u^1 3d\bar{\pi}_u^1 3d\delta_g^1 3d\bar{\delta}_g^1 4s\sigma_g^2 4s\sigma_u^1 \quad {}^7\Sigma_u^+ \tag{7}$$

with five one-electron d bonds. However, the small overlap of $3d\delta$ orbitals leads to little additional bonding. An alternative for the ground state of Ti_2 is the $^1\Sigma_g^+$ state arising from the $3d^34s^1$–$3d^34s^1$ limit and forming three two-electron d bonds

$$3d\sigma_g^2 3d\pi_u^4 4s\sigma_g^2 \quad {}^1\Sigma_g^+. \tag{8}$$

To describe the $^1\Sigma_g^+$ state one must balance the loss in energy from exchange terms against the energy gained forming the two-electron bonds. Hence, this state is harder to describe than the high spin coupled one-electron bond case, and at the CASSCF level the $^7\Sigma_u^+$ state is predicted to be the ground state, with the $^1\Sigma_g^+$ state 0.7 eV higher. At the MRSDCI level the $^1\Sigma_g^+$ state remains 0.4 eV above the $^7\Sigma_u^+$ state. Since the two-electron bonds are more difficult to describe, further stabilization of the $^1\Sigma_g^+$ state is expected as the calculations are improved. If one compares the experimental [67] ω_e (407.9) to the theoretical values for the $^1\Sigma_g^+$ (438cm^{-1}) and the $^7\Sigma_u^+$ (205cm^{-1}) states, it is clear that $^1\Sigma_g^+$ is in fact the ground state. Very large CI expansions will be required to obtain the correct ordering of states. Hence, such methods as the externally contracted CI (CCI), developed by Siegbahn [68] offers promise for such systems.

The ground state of V_2 can be obtained from the $^1\Sigma_g^+$ configuration of Ti_2 by adding the remaining two electrons to the 3dδ orbitals,

$$3d\sigma_g^2 3d\pi_u^4 3d\delta_g^1 3d\bar{\delta}_g^1 4s\sigma_g^2 \quad {}^3\Sigma_g^- . \tag{9}$$

Thus far only CASSCF calculations have been carried out, which underestimate the binding energy, but yield a r_e (1.76Å) and ω_e (564 cm^{-1}) in good agreement with experiment [67,69], 1.76Å and 537.5 cm^{-1}.

The $^1\Sigma_g^+$ ground state of Cr_2 can be formed from the $^1\Sigma_g^+$ state of Ti_2 by adding the four electrons into the empty 3dδ_g orbital. While the CASSCF approach was able to achieve a qualitative balance of the exchange and bonding for Ti_2, in Cr_2 the extra electrons have gone into the weakly bonding 3dδ_g orbitals, with a much higher loss of exchange energy than in Ti_2. Hence, at the CASSCF level, only a long 4s-4s bond is obtained. Note that the SCF goes to the other extreme giving a bond that is 0.42Å too short, because it does not properly describe exchange. To describe the 3d-3d bonding as well as the exchange will take an extensive CASSCF CCI treatment.

Goodgame and Goddard [15] have developed an empirical approach to describe the bonding in Cr_2. Beginning with a CASSCF wavefunction that yields only a weak 4s-4s bond, they correct the $Cr^- Cr^+$ bonding terms based upon the difference between the experimental and computed IP and EA. This approach enhances the bonding and leads to a 3d-3d well. Their method also suggests a shallow outer 4s-4s well, which may have been observed recently in experiment [70]. The nature of this correction is consistent with the contention in this work that a correct description of the atomic correlation must be incorporated into the molecular calculation.

For the $^1\Sigma^+$ state of Cu_2 dimer, theoretical treatments [5,71-76] have been more successful, since now the d shell is full. However, there is still some d involvement in the bonding since a two-electron treatment of the primarily 4s-4s bond, gives a D_e of 1.25 eV, far less than the experimental value of 2.05 eV [77]. The inclusion of s-d correlation improves D_e to about 1.6 eV. Our theoretical studies [78] of the EA affinity of Cu suggests that the remaining difference is differential d-d correlation. The large difference in D_e between the SDCI and CPF treatments shows the importance of higher excitations in obtaining the small differential d d correlation effect. Hence, basis set saturation, higher-order excitations and relativistic effects all increase D_e and decrease r_e. The relativistically corrected CPF D_e value [5] of 1.9 eV is now within 0.15 eV of experiment. This is a relatively small energy compared with the total CPF correlation energy of about 25 eV when 22 electrons are correlated. One finds very little $3d^9 4s^1 4p^1$ character in the wavefunction as suggested by Pauling [79], at either the SDCI or CPF levels.

In analogy with the transition metal oxides, the bonding in the second row transition row dimers is stronger than for the first transition row. This difference results from smaller 4d-4d exchange terms and the more comparable radial extent of the 5s and 4d orbitals. Thus a CASSCF approach for Mo_2 is able to yield a short bond length with 4d-4d bonding [9,80]. The reduced importance of the exchange terms and the stronger bonds for the second transition row results in less computational problems.

The ground state of Nb_2 is found [9] to be $^3\Sigma_g^-$ with the same configuration as V_2. There is also a low-lying $^3\Delta_g$ state arising from the $4d\sigma_g$ to $4d\delta_g$ excitation

$$4d\sigma_g^1 4d\pi_u^4 4d\delta_g^2 4d\bar{\delta}_g^1 5s\sigma_g^2 \quad {}^3\Delta_g \tag{10}$$

only 0.12 eV higher [9]. Since the $d\delta$ orbital has a smaller overlap in V_2 the analogous $^3\Delta_g$ state lies 0.92 eV above the ground state [9]. The CASSCF spectroscopic parameters for the $^3\Sigma_g^-$ ground state of Nb_2 are r_e=2.10Å and ω_e=448 cm^{-1} compared to the experimental value for ω_e [81] of 430 cm^{-1}. These results are in marked contrast to those of Cotton and Shim [62] in which a longer bond length (about 3.0Å) and smaller vibrational frequency (about 100 cm^{-1}) are obtained. In addition, they find many more low-lying excited states. The calculations of Cotton and Shim use valence CI calculations based upon a single set of SCF orbitals. While this approach may work for cases where the nd orbitals are weakly interacting, it is clear that limited CI treatments cannot overcome the SCF orbital bias in cases where metal-metal bonds are formed.

VI. DATIVE BONDING

For the diatomic systems considered in the previous sections, both atoms are contributing one electron to the bond. When both electrons are donated from one atom (fragment) into an empty orbital of another fragment, we refer to this as dative bonding. This type of bonding occurs for many transition metal complexes, see for example Ref. 82. Dative bonding appears to be reasonably well described at the SCF or CASSCF level. For example, SCF calculations show that the axial groups of $Fe(CO)_5$ have strong CO 5σ donation into the empty Fe 3d orbital, whereas the equatorial CO are bonded mostly by π donation [83-84]. While the inclusion of correlation enhances both of these effects, the qualitative features are visible in the SCF wavefunctions. Using the DISCO codes developed by Almlöf and co-workers [85], very large transition metal complexes such as $Fe_2(CO)_9$ can be studied at the SCF level [86].

The most common class of molecules that exhibit dative bonding are organometalic complexes. They are often highly symmetric and have no permanent dipole moments. We have concentrated on understanding the bonding by decomposing it to its atomic origins. When population analyses are used, population artifacts [83] often occur when diffuse 4s and 4p functions are included in the basis. However, diffuse 4s and 4p functions are needed to describe such systems as $Cr(NO)_4$, where they contribute about 0.7 eV to the binding energy [87]. Note that d populations are much more meaningful, since the d orbitals are more spatially compact.

To avoid problems with population artifacts and still be able to decompose the bonding into its atomic components, we have developed a technique for decomposing the wavefunctions based upon energy (and changes in the dipole moment),

Table VII. The CSOV analysis for the $^1\Sigma^+$ state of NiCO with R(Ni-C)=3.477 Bohr.

	Populations							
	4s	dσ	dπ	COσ	COπ^*	E^a_{INT}	ΔE^b	μ^c
FO	1.00	1.00	4.00	10.00	0.00	-2.47	—	—
V(Ni,Ni)[d]	0.79	1.20	4.00	10.00	0.00	-1.03	1.44	-0.28
V(Ni,σ)	0.71	1.24	4.00	10.03	0.00	-0.99	0.04	-0.34
V(Ni,FB)	0.31	1.66	3.68	10.03	0.30	+0.13	1.12	-1.91
V(CO,CO)	0.30	1.66	3.67	10.04	0.30	+0.38	0.25	-1.72
V(CO,σ)	0.36	1.68	3.67	9.90	0.30	+0.46	0.08	-1.69
V(CO,FB)	0.36	1.68	3.66	9.90	0.32	+0.48	0.02	-1.67
CAS[e]	0.55	1.68	3.55	9.75	0.42	+0.88	0.40	-1.79

[a] Energy (eV) relative to Ni and CO at infinite separation.
[b] The difference between successive CSOV steps.
[c] The dipole moment (a.u.) (1 a.u.=2.54 Debye). For comparison μ for CO ($^1\Sigma_g^+$) at the SCF level is -0.25 a.u.
[d] V(X,Y), indicates that the X occupied orbitals are being optimized using the Y variational space. Thus the difference between V(Ni,σ) and V(Ni,FB), allows the Ni to donate into CO π space and is a measure of metal to CO π donation. FB indicates the full virtual basis set is used, see text.
[e] At R(Ni-C)=3.2 bohr, V(CO+3dσ+4s,FB), yields a ΔE of 60% of the difference between the CASSCF and V(CO,FB), Ref. 88.

in addition to the populations. In this constrained-space-orbital-variation (CSOV) approach [16,17] the wavefunction is optimized in a series of steps in which constraints are eliminated. Consider, for example, the bonding in NiCO. The reference energy is taken to be Ni and CO at infinite separation. The CSOV procedure starts by evaluating the energy of the frozen orbital (FO) interaction of the superposition of the Ni and CO fragments. The Ni atom is then relaxed in the space of the Ni basis set, with the CO frozen– denoted V(Ni,Ni) in Table VII. The CO virtual space is then included in the variational space to allow metal to CO donation, which can be subdivided into separate σ and π contributions. The Ni is then fixed in this form and CO is relaxed in the space of the CO virtuals. Finally CO to Ni donation is considered. Since this step-wise relaxation cannot account for any covalent interaction, in some cases it is necessary to allow mixing to occur between the occupied orbitals of the fragments; in this case the mixing of the Ni 4s and 3dσ orbitals with the CO 5σ orbital. If the donations are sufficiently large, a second CSOV cycle is performed. The final energy is compared with an unconstrained SCF or CASSCF energy to determine whether the CSOV treatment has identified all components of the bonding. Since the change in energy and charge distribution are computed in each step, a complete decomposition of the bonding is achieved.

For the $^3\Delta$ state [17] of NiCO, the Ni atom has a $3d^94s^1$ occupation. The state is unbound since the repulsion of the CO 5σ electrons with Ni is not compensated for by the Ni 3dπ donation to the CO 2π^* orbital and the small CO σ donation to the Ni. The CSOV procedure shows that the energy associated with the CO σ donation is far smaller than would be predicted based upon a Mulliken population analysis. Since there is no bond formation, there is no change in the atomic exchange, and the inclusion of correlation makes a small (0.3 eV) increase in the Ni to CO π donation.

The $^1\Sigma^+$ state [17,88-89] of NiCO is bound by about 1.1 eV at the CASSCF level and by 1.3 eV at the SDCI level. In Table VII, a CSOV is given for the $^1\Sigma^+$ state of NiCO at R(Ni-C)=3.477 Bohr, the equilibrium bond length in $Ni(CO)_4$ [90]. The bonding in the $^1\Sigma^+$ state is found to involve some 4s donation to the $3d\sigma$, i.e. mixing in of some $3d^{10}$ character in to the wavefunction. The frozen-orbital repulsion is similar for both the $^3\Delta$ and $^1\Sigma^+$ states. The CSOV analysis shows that in the $^1\Sigma^+$ state, there is movement of the 4s electrons into the $3d\sigma$ orbital to reduce the CO repulsion, accompanied by donation from the Ni $3d\pi$ electrons to the CO $2\pi^*$. The larger Ni relaxation and donation lead to a bound $^1\Sigma^+$ state. There is 0.4 eV of energy not accounted for by the CSOV, which we ascribe to covalent interaction between the CO 5σ orbital and the Ni $3d\sigma$ and 4s orbitals. While this CSOV technique is limited to dative and ionic bonding, it gives insight into the nature of the bonding and helps identify population artifacts and important mixings with other atomic asymptotes.

VII. CONCLUSIONS

The determination of accurate spectroscopic parameters for molecular systems containing transition metal atoms is a challenging theoretical problem. For example, the determination of the $3d^n4s^2$–$3d^{n+1}4s^1$ transition metal atomic separations to chemical accuracy requires extensive basis sets containing f functions and diffuse d functions, as well as an extensive correlation treatment and the inclusion of relativistic effects. Limitations in the treatment of atomic correlation can carry into molecular calculations, since many electronic states involve a mixture of both $3d^n4s^2$ and $3d^{n+1}4s^1$ character. A leading example is the $^2\Delta$ state of NiH, where the dipole moment and 3d population are a sensitive measure of the mixing of the $3d^94s^1$ and $3d^84s^2$ asymptotes.

Another problem, which is most severe for transition metal-metal bonds, is to obtain an equivalent description of the 3d-3d bonding and the exchange interactions. This problem is especially severe for Cr_2 where the CASSCF potential is not bound in the inner well region. This problem is generally less severe for the second row (e. g. for Mo_2) since the 4d-4d bonding is stronger and the exchange interaction is smaller. The large 3d-3d exchange terms in some cases lead to states with only one-electron bonds, for example the $^5\Sigma_u^-$ ground state of Sc_2. For Ti_2, the $^7\Sigma_u^+$ state (one-electron bonding) and the $^1\Sigma_g^+$ state (two-electron bonding) lie close in energy. Even in Cu_2, where there is little mixing of different atomic asymptotes or change in atomic exchange, there are errors of 0.15 eV in D_e. The enormous amount of correlation in the 3d shell means that only a very small percentage change in the total d correlation represents a large absolute error.

Based upon the work on the transition metal atoms, hydrides, oxides and dimers, it is now possible to design a calculation that can correctly describe the mixing of difference atomic asymptotes, and give a correct balance between molecular bonding and exchange interactions. Thus as more powerful computational algorithms and hardware become available, it should be possible to provide quantitative treatments even for system such as the $^1\Sigma_g^+$ state of Cr_2.

REFERENCES

1. C. W. Bauschlicher and I. Shavitt, J. Amer. Chem. Soc., **100**, 739 (1978).
2. I. Shavitt, Tetrahedron, **41**, 1531 (1985) and references therein.
3. C. E. Moore, Atomic Energy Levels, Natl. Bur. Stand. (US) circ. **467** (1949).
4. C. W. Bauschlicher and S. P. Walch, J. Chem. Phys., **74**, 5922 (1981); C. W. Bauschlicher, S. P. Walch, and H. Partridge, J. Chem. Phys., **76**, 1033 (1982).
5. "The importance of diffuse f functions for transition metals", S. R. Langhoff and C. W. Bauschlicher, submitted to J. Chem. Phys.
6. R. L. Martin, and P. J. Hay, J. Chem. Phys., **75**, 4539 (1981).
7. S. P. Walch, C. W. Bauschlicher, and S. R. Langhoff, J. Chem. Phys.,in press.
8. J. A. Gray, S. F. Rice, and R. W. Field, J. Chem. Phys., **82**, 4717 (1985).
9. "Theoretical studies of transition metal dimers", S. P. Walch and C. W. Bauschlicher, "Comparison of Ab Initio Quantum Chemistry with Experiment", ed. by R. Bartlett, D. Reidel Publishing Company, Boston (1985).
10. C. W. Bauschlicher and S. P. Walch, J. Chem. Phys., **76**, 4560 (1982).
11. S. P. Walch and C. W. Bauschlicher, Chem. Phys. Lett., **86**, 66 (1982).
12. S. P. Walch and C. W. Bauschlicher, J. Chem. Phys., **78**, 4597 (1983).
13. S. P. Walch, C. W. Bauschlicher, B. O. Roos, and C. J. Nelin, Chem. Phys. Lett., **103**, 175 (1983).
14. M. M. Goodgame and W. A. Goddard, Phys. Chem., **85**, 215 (1981).
15. M. M. Goodgame and W. A. Goddard, Phys. Rev. Lett., **54**, 661 (1985).
16. P. S. Bagus, K. Herman, and C. W. Bauschlicher, J. Chem. Phys., **81**, 1966 (1984).
17. "The Nature of the Bonding in XCO, for X=Fe, Ni and Cu" C. W. Bauschlicher, P. S. Bagus, C. J. Nelin, and B. O. Roos, submitted to J. Chem. Phys.
18. C. W. Bauschlicher, J. Chem. Phys., **73**, 2510 (1980).
19. J. Almlöf, MOLECULE, a gaussian integral program.
20. P. E. M. Siegbahn, C. W. Bauschlicher, B. Roos, A. Heiberg, P. R. Taylor, and J. Almlöf, SWEDEN, a vectorized MCSCF-direct CI.
21. R. C. Raffenetti, BIGGMOLI, Program No. 328, Quantum Chemistry Program Exchange, Indiana University, Bloomington (1977), with modifications to allow use of the full D_{2h} symmetry.
22. R. Ahlrichs, H.-J. Bohm, C. Ehrhardt, P. Scharf, H. Schiffer, H. Lischka, and M. Schindler, J. Comp. Chem., **6**, 200 (1985).
23. R. M. Pitzer, J. Chem. Phys., **58**, 3111 (1973).
24. H. Lischka, R. Shepard, F. B. Brown, and I. Shavitt, Intern. J. Quantum Chem. Symp. **15**, 91 (1981).
25. R. Shepard, I. Shavitt, and J. Simons, J. Chem. Phys., **76**, 543 (1982).
26. A. J. H. Wachters, J. Chem. Phys., **52**, 1033 (1970).
27. P. J. Hay, J. Chem. Phys., **66**, 4377 (1977).
28. P. S. Bagus and B. O. Roos, J. Chem. Phys. **75**, 5961 (1981).
29. S. Huzinaga, J. Chem. Phys., **66**, 4245 (1977).
30. S. P. Walch, C. W. Bauschlicher, and C. J. Nelin, J. Chem. Phys., **79**, 3600 (1983).
31. P. J. Hay and W. R. Wadt, J. Chem. Phys., **82**, 299 (1985).

32. R. L. Martin, J. Chem. Phys., **78**, 5840 (1983); R. D. Cowan, and D. C. Griffin, J. Opt. Soc. Am., **66**, 1010 (1976).
33. B. H. Botch, T. H. Dunning, and J. F. Harrison, J. Chem. Phys., **75**, 3466 (1981); T. H. Dunning, B. H. Botch, and J. F. Harrison, J. Chem. Phys., **75**, 3466 (1981).
34. A. D. McLean and B. Liu, Chem. Phys. Lett., **101**, 144 (1983).
35. S. R. Langhoff and E. R. Davidson, Intern. J. Quantum Chem., **8**, 61 (1974).
36. R. F. Stewart, J. Chem. Phys., **52**, 431 (1970).
37. R. Ahlrichs, P. Scharf and C. Ehrhardt, J. Chem. Phys., **82**, 890 (1985).
38. C. M. Rohlfing and R. L. Martin, Chem. Phys. Lett., **115**, 104 (1985).
39. P. R. Scott and W. G. Richards, J. Chem. Phys., **63**, 1690 (1975), and references therein.
40. G. Das, J. Chem. Phys., **74**, 5766 (1981).
41. R. L. Jaffe, private communication.
42. M. R. A. Blomberg, P. E. M. Siegbahn and B. O. Roos, Mol. Phys., **47**, 127 (1982).
43. P. S. Bagus and C. Björkman, Phys. Rev. A, **23**, 461 (1981).
44. S. P. Walch, Chem. Phys. Lett., **105**, 54 (1984).
45. C. W. Bauschlicher, S. R. Langhoff, and S. P. Walch, unpublished work on CPF treatment of NiH.
46. P. E. M. Siegbahn, A. Heiberg, B. O. Roos, and B. Levy, Physica Scripta, **21**, 323 (1980); B. O. Roos, P. R. Taylor, and P. E. M. Siegbahn, Chem. Phys., **48**, 157 (1980); P. E. M. Siegbahn, J. Almlöf, A. Heiberg, and B. O. Roos, J. Chem. Phys., **74**, 2381 (1981); B. O. Roos, Intern. J. Quantum Chem. **S14**, 175 (1980).
47. C. W. Bauschlicher, P. S. Bagus and C. J. Nelin., Chem. Phys. Lett., **101**, 229 (1983).
48. P. S. Bagus, C. J. Nelin, and C. W. Bauschlicher, J. Chem. Phys., **79**, 2975 (1983).
49. "Theoretical Study of the $X^2\Pi$ and $A^2\Sigma^+$ States of CuO and CuS", S. R. Langhoff and C. W. Bauschlicher, submitted to Chem. Phys. Lett.
50. C. W. Bauschlicher, C. J. Nelin and P. S. Bagus, J. Chem. Phys., **82**, 3265 (1985).
51. "Theoretical Study of the Low-lying Electronic States of ZnO and ZnS", C. W. Bauschlicher, and S. R. Langhoff, submitted to Chem. Phys. Lett.
52. Unpublished work on ScO, ScS, VO and VS.
53. M. Krauss and W. J. Stevens, J. Chem. Phys., **82**, 5584 (1985).
54. K. P. Huber and G. Herzberg, "Molecular Spectra and Molecular Structure," (Van Nostrand Reinhold, New York ,1979).
55. J. N. Allison and W. A. Goddard, J. Chem. Phys., **77**, 4259 (1982).
56. S. Smoes, F. Mandy, A. Vander Auwera-Mahieu and J. Drowart., Bull. Soc. Chim. Belges, **81** 45 (1972).
57. G. Igel, W. Wedig, M. Dolg, P. Fuentealba, H. Preuss, H. Stoll, and R. Frey, J. Chem. Phys., **81**, 2737 (1985).
58. "Ab initio study of the Alkali and Alkaline-earths mono-hydroxides", C. W. Bauschlicher, S. R. Langhoff, and H. Partridge, submitted to J. Chem. Phys.
59. Y. Lefebvre, B. Pinchemel, J. M. Delaval and J. Schamps, Physica Scripta, **25**, 329 (1982).
60. S. P. Walch and C. W. Bauschlicher, Chem. Phys. Lett., **94**, 290 (1983).
61. S. P. Walch and C. W. Bauschlicher, J. Chem. Phys., **79**, 3590 (1983).
62. F. A. Cotton and I. Shim, J. Phys. Chem., **89**, 952 (1985).

63. B. Delley, A. J. Freeman, and D. E. Ellis, Phys. Rev. Lett., **50** 488 (1983).
64. C. Wood, M. Doran, I. H. Hillier, M. F. Guest, Faraday Symp. Chem. Soc., **14**, 159 (1980), and M. F. Guest, private communication.
65. L. B. Knight, R. J. VanZee and W. Weltner, Chem. Phys. Lett., **94**, 296 (9183).
66. W. Weltner, and R. J. VanZee, Ann. Rev. Phys. Chem., **35**, 291 (1984), and references therein.
67. D. P. DiLella, W. Limm, R. H. Lipson, M. Moskovits, and K. V. Taylor, J. Chem. Phys., **77**, 5263 (1982).
68. P. E. M. Siegbahn, Intern. J. Quantum Chem., 23, 1869 (1983).
69. P. R. R. Langridge-Smith, M. D. Morse, G. P. Hansen, R. E. Smalley, and A. J. Merer, J. Chem. Phys., **80**, 593 (1984).
70. M. Moskovits, W. Limm and T. Mejean, J. Chem. Phys., **82**, 4875 (1985).
71. C. W. Bauschlicher, S. P. Walch, and P. E. M. Siegbahn, J. Chem. Phys., **76**, 6015 (1982).
72. C. W. Bauschlicher, S. P. Walch, and P. E. M. Siegbahn, J. Chem. Phys., **78**, 3347 (1983).
73. P. Scharf, S. Brode, and R. Ahlrichs, Chem. Phys. Lett., **113**, 447 (1985).
74. H.-J. Werner and R. L. Martin, Chem. Phys. Lett., **113**, 451 (1985).
75. B. C. Laskowski, S. R. Langhoff, C. W. Bauschlicher, and S. P. Walch, to be published.
76. K. K. Sunil; K. D. Jordan, and K. Raghavachari, J. Phys. Chem., **89**, 457 (1985).
77. N. Aslund, R. F. Barrow, W. G. Richards, and D. N. Travis, Ark. Fys., **30**, 171 (1965); M. Ackerman, F. E. Stafford and J. Drowart, J. Chem. Phys., **33**, 1784 (1960); also see Ref. 54.
78. C. W. Bauschlicher, S. P. Walch and H. Partridge, Chem. Phys. Lett., **103**, 291 (1983).
79. L. Pauling, J. Chem. Phys., **78**, 3346 (1983).
80. M. M. Goodgame and W. A. Goddard, Phys. Rev. Lett., **48**, 135 (1982).
81. M. Moskovits, private communication.
82. F. A. Cotton and G. Wilkinson, "Advanced Inorganic Chemistry", (Interscience, NY, (1966).
83. C. W. Bauschlicher, and P. S. Bagus, J. Chem. Phys., **81**, 5889 (1984).
84. H. P. Luthi P. E. M. Siegbahn and J. Almlof, J. Phys. Chem., **89**, 2156 (1985).
85. J. Almlöf, K. Faegri and K. Korsell, J. Comp. Chem., **3**, 385 (1982).
86. C. W. Bauschlicher, "On the bonding in $Fe_2(CO)_9$", submitted to J. Chem. Phys.
87. C. W. Bauschlicher, unpublished.
88. C. W. Bauschlicher, Chem. Phys. Lett., **115**, 387 (1985).
89. M. R. A. Blomberg, U. B. Brandemark, P. E. M.Siegbahn, K. Broch-Mathisen and G. Karlstrom, J. Phys. Chem., **89**, 2171 (1985).
90. J. Ladell, B. Post and I. Iankuchen, Acta Crystallogr., **5**, 795 (1952).

THE ELECTRONIC STRUCTURE OF TRANSITION METAL ATOMS AND DIATOMS THROUGH PSEUDOPOTENTIAL APPROACHES.

M. Pélissier, J.P. Daudey and J.P. Malrieu
Laboratoire de Physique Quantique
Unité Associée au C.N.R.S. n°505
118, route de Narbonne
31062 Toulouse Cedex, France

G.H. Jeung
Freie Universität Berlin
FB Chemie, Institut für Physikalische Chemie
Takustrasse 3
1000 Berlin 33, R.F.A.

ABSTRACT. The electronic structure of some first row transition metal atoms, hydrogenated derivatives and homonuclear diatomic is examined from different and interrelated points of view, namely relativistic effects, correlation contributions and possible pseudopotential approaches . The relativity impact is approached by comparing results obtained through pseudopotentials including or not mean relativistic terms. Special emphasis is put on the interplay between relativity and correlation effects on atomic spectra or on bond formation. Comparisons between pseudopotential and all electron approaches are also given.

1. INTRODUCTION

The purpose of this report is to provide some insight in the interplay between correlation and relativity on the electronic structure of transition metal atoms or diatomic molecules containing transition metal atoms. Even for such simple systems, ab-initio calculations carried out by quantum chemists during the past ten years have faced very intricating problems some of them remaining yet unsolved. The origin of these difficulties is now clearly understood and is attributed to the existence of the open shell of d electrons, which is common to all transition metal atoms. For the atoms, the presence of this open d shell is traduced in a special structure of the atomic spectra, namely the existence of two very close electronic configurations (s^2d^n and s^1d^{n+1}). The evolution of the spacing between these configurations is reported in Table 1 which shows two important points. The first one which is experimental is the well-known monotonous decrease of ΔE (s^2d^n-s^1d^{n+1}) starting from Scandium to Copper with a pronounced discontinuity between Chromium and Manganese. The second one, which is very important in view

A. Veillard (ed.), Quantum Chemistry: The Challenge of Transition Metals and Coordination Chemistry, 37–51.

of quantum chemical calculations is the large effect of correlation.

Table 1. Evolution of the energy difference between the s^2d^n and s^1d^{n+1} configurations for the first row of transition metal atoms. First line contains the experimental values of Ref. (1) and second line the numerical non-relativistic Hartree-Fock results of Ref. (2). Third and fourth lines emphasize the relative contributions to the spacing of relativistic corrections and electronic correlation.

	Sc	Ti	V	Cr	Mn	Fe	Co	Ni	Cu
exp.	1.43	0.81	0.26	-1.00	2.14	0.87	0.42	-0.03	-1.49
N.H.F.	1.00	0.54	0.12	-1.27	3.33	1.80	1.53	1.27	-0.37
ΔR[a]	0.12	0.14	0.17	0.21	0.20	0.26	0.30	0.36	0.43
ΔCI[b]	0.31	0.13	-0.04	0.06	-1.39	-1.19	-1.41	-1.60	-1.55

[a] ΔR is the difference between NHF and relativistic Cowan-Griffin (HFR) values of Ref. (2)

[b] ΔCI is the difference between experimental values and HFR results of Ref. (2)

The most spectacular case in this respect is the Nickel atom in which the $s^2d^8(^3F)$ state is more stable by 1.27 eV at the Hartree-Fock level. The correlation contribution almost completely reverses this order and experimentally the $s^1d^9(^3D)$ state is only 0.03 eV above the $s^2d^8(^3F)$ ground state.

Another interesting point is the influence of relativistic effects (ΔR in Table 1) which tends to stabilize the s^2d^n configuration, with an almost monotonous increase along the series from 0.12 eV for Sc to 0.43 eV for Cu. Even if this effect is generaly smaller than the correlation one, it is not negligible and is crucial in many instances.

These simple facts clearly show that quantitative ab-initio calculations on transition metal atoms will need special care since they are compelled to evaluate with great accuracy the energy of two largely distinct configurations. In the s^1d^{n+1} configuration, the d orbitals are more diffuse than in the s^2d^n configuration. This difference in spatial extension requires a large 3d atomic basis set to treat both configurations on an equal footing. A correct description of differential correlations effects will need another extension of the atomic basis set, including f type orbitals for the angular correlation of the d electrons.

In the construction of molecular bonds the above mentioned special features of transition metal atoms induce the same type of complexity. In many cases, there are a lot of different molecular states arising from different atomic configurations in the bottom of the molecular spectra. Very often ab-initio calculations carried with different basis sets and/or different treatment of correlation should give different assignation to molecular states. Another well-known failure of ab-initio calculations is the correct description of multiple bonding in Cr_2, even if partial answer has been given recently to this problem (3).

This quite pessimistic overview of the present state of quantum mechanical calculational approach of transition metal atoms should not dissimulate many progresses made during a recent past which now allow a clear understanding of the electronic structure concurrently with a fair quantitative reproduction of experimental data for a lot of small systems containing one or two metal atoms. For more complicated systems, theoretical studies explicitly including electronic correlation are still very sparse. For transition metal atoms of the second and third rows the situation is even worse due to the larger number of electrons and to the relativity problem.

A very promising way around this increased difficulty is the pseudopotential (or effective core potential) method which was intensively worked out and tested by many groups (4,5) including ours (6). This approach presents two advantages. First, it reduces the electronic problem to valence electrons only and second it provides a simple way to introduce the mean relativistic corrections in an otherwise standard molecular calculation. Pseudopotential calculations have been compared in many occasions with all-electron calculations at the same level for main group elements (7). For transition metal atoms, the situation is a little more complicated and further investigations are still needed. The present contribution will report some recent calculations we made on the specific problem of transition metal atoms using the pseudopotential approach. The first part briefly deals with the methods used to derive the pseudopotential operator and to include mean relativistic corrections. We shall discuss the results obtained for atoms (namely Sc, Cr and Cu) in the second part and for diatomic molecules (CuH, CrH and Cu_2) in the last one.

2. DEFINITION OF (RELATIVISTIC) PSEUDOPOTENTIALS

The prerequisites for all versions of pseudopotential method is first the distinction between core and valence electrons and second the choice of the electronic configuration which will be used to fit the pseudopotential parameters. This is very straightforward for main group elements but is no longer so easy for transition metal atoms. If we choose the experimental ground state as the configuration to be reproduced by the pseudopotential operator, we have no guarantee at all regarding the position of near degenerate low lying configurations until we go further in puting additional constraints in the pseudopotential operator extraction. It is now clear that a quantitatively sucessfull pseudopotential method must consider the electrons in the d shell as valence (active) electrons. Attempts have been made to develop simplified pseudopotentials which include all the 3d electrons in the core, treating the 4s electrons only as valence electrons. The application of such an idea has been mainly restricted (8,9) to the $4s^1[3d^{10}]$Cu atom for which the d shell is completely filled. Nevertheless this possibility of oversimplified pseudopotentials could be usefully proposed for modeling transition metal atoms in large clusters.

The opposite question could also be raised : is it completely justified to include the outermost core orbitals in the pseudopotential ? For instance, in the first row of transition metal atoms, the regions of space occupied by the 3s and 3p shells have a large overlap with the region occupied by the 3d shell. The reproduction of the non local two-electron interaction between those shells by a one-electron operator (reproduction which is a basic assumption for our pseudopotential method) is questionable. Taking into account this uncertainty, we have proposed in some cases two types of pseudopotential operators, one which includes the 3s and 3p electrons in the core, the other one treating them as valence electrons.

2.1. Definition of pseudoorbitals and pseudopotential operators

As soon as these decisive choices are made we follow the method proposed by Durand and Barthelat (6) and extensively tested by Pelissier and Durand (10). The first step is a near Hartree-Fock calculation of the lowest (or of the two lowest) atomic multiplets for which we used the analytic Hartree-Fock program by Clementi et al. (11). The optimized Slater basis sets given by Clementi and Roeti (12) are augmented by more diffuse non-optimized orbitals in order to warrant a very good description of the all-electron valence orbitals $\{\varphi_v\}$ at large distance of the nucleus.

From this set of near Hartree-Fock valence orbitals, pseudoorbitals $\{\phi_v^{ps}\}$ are defined using the following criteria.

a) ϕ_v^{ps} must reproduce as closely as possible the all electron valence orbital φ_v in the valence region, i.e. from r_{max} to infinity (r_{max} is defined as the point where φ_v has its maximum of amplitude).

b) ϕ_v^{ps} must decrease smoothly to zero going from $r = r_{max}$ to $r=0$.

There are some arbitrariness in the above definition of pseudoorbitals ϕ_v^{ps}. A detailed study by Pélissier and Durand (10) has shown that the final results at the molecular level are stable with respect to these variations.

Once the set of pseudoorbitals $\{\phi_v^{ps}\}$ is complete, the Hartree-Fock valence pseudooperator is defined as

$$F^{ps} = -\frac{1}{2}\Delta + W^{ps} + V_{\phi_v} \qquad [1]$$

where V_{ϕ_v} is the (coulomb + exchange) operator associated to the occupied valence pseudoorbitals.

The pseudopotential operator W^{ps} is determined by the conditions :

a) ϕ_v^{ps} is the solution of F^{ps} associated with the all-electron energy level ε_v :

$$F^{ps}\,\phi_v^{ps} = \varepsilon_v\,\phi_v^{ps} \qquad [2]$$

b) ϕ_v^{ps} and ε_v are the lowest solutions of eq. [2].

It is possible at this level to include other conditions for the pseudopotential operator W^{ps}. One may try to reproduce valence levels and pseudoorbitals not only for a single configuration but for the two lowest configurations of the atom. Additionally, a further constraint may be added to get a correct energy spacing between these configurations. In the following parts, such operators will be produced for Sc and Cr atoms.

2.2. Relativistic corrections

For heavy atoms relativistic corrections are essential for core electrons but their relativistic character have important indirect effects on the valence electrons. A very attractive way to include these corrections for the valence electrons, and to simultaneously retain the simplicity of calculations in the L.S. coupling scheme is to build up relativistic pseudopotentials. They are determined exactly in the way described in the preceeding section, the reference calculation being now a relativistic all electron calculation. Recently a similar approach has been used by Hay and Wadt (5) to produce complete sets of relativistic pseudopotentials parameters for the second and third rows of transition metal atoms. The reference calculation they used is a numerical one-component wave function in the formalism proposed by Cowan and Griffin (13). Pitzer et al. (14) have also proposed relativistic effective potentials based on a two components description of atomic Dirac-Hartree-Fock wave functions. From this two components description, it is possible to build an average relativistic effective potential. In the present work we start from an analytical relativistic calculation based on the method developed by Barthelat et al. (15) which approximately includes the mass-velocity and Darwin relativistic terms. Spin orbit corrections can be introduced by a perturbation method and could be also reduced to a pseudopotential operator form (16) but they will not be considered here. This pseudopotential approach has been used for studying the relativistic contributions to molecular bonding in the series XH_m (X = In, Tl, Sn, Pb, m = 1 or 3 for In and Tl, 2 or 4 for Sn and Pb) (17).

3. ATOMIC RESULTS

3.1. Chromium as a test case for different pseudopotential methods

Table 2 resumes atomic tests made with different pseudopotentials for the lowest multiplets of chromium atom arising from $3d^5 4s^1$ and $3d^4 4s^2$ configurations. The pseudopotential corresponding to the [Ne] core is obtained through a relativistic all electron calculation on the $Cr^+[3s^2 3p^6 3d^5 (^5S)]$ atom. The second one is taken from the work by Hay and Wadt and it does not contain relativistic contributions. The third one represents an Ar core and reproduces in a least square sense non-relativistic calculations on both the ^7S and ^5D multiplets of the chromium atom. The last one, also obtained by Hay and Wadt, is extracted on the lowest ^7S multiplet only. The basis set used (6s 6p 6d non contracted GTO functions) is the same for all calculations reported in Table 2. The six d exponents were taken from ref. 18, the six s and the most concentrated four p exponents were optimized in a valence ([Ne]R core) Hartree-Fock calculation of the $3d^5 4s^1 (^7S)$ ground state. Two external p functions were added, assuming that the ratio of their exponents with the corresponding external s functions is equal to the ratio obtained for the last optimized p function.

Comparisons of this set of results with the all-electron non relativistic calculation for the ^7S multiplet indicate that all pseudopotentials readily achieved what they are designed for, namely the reproduction of valence energy levels. The small differences between AE and [Ne]R traduce the well-known contraction effects on the 3s and 3p orbitals, the agreement for the external 3d and 4s being almost perfect for [Ne]NR. In the [Ar]NR column, the agreement is also very good. For [Ar]NR2C, the larger deviation observed is a consequence of the larger constraints imposed to this particular pseudopotential operator.

Table 2. Comparison of one-electron energy levels and energy differences obtained with different pseudopotential methods for Cr atom.

Atomic State	Energy levels	AE	[Ne]R	[Ne]NR	[Ar]NR2C	[Ar]NR
$3d^5 4s\ (^7S)$	ε_{3s}	-3.285	-3.352	-3.275	-	-
	ε_{3p}	-2.051	-2.069	-2.050	-	-
	ε_{3d}	-0.373	-0.372	-0.377	-0.369	-0.372
	ε_{4s}	-0.222	-0.224	-0.222	-0.211	-0.223
$3d^4 4s^2 (^5D)$	ε_{3s}	-3.499	-3.564	-3.491	-	-
	ε_{3p}	-2.248	-2.265	-2.250	-	-
	ε_{3d}	-0.569	-0.565	-0.575	-0.558	-0.597
	ε_{4s}	-0.240	-0.240	-0.239	-0.238	-0.234
$E(^5D)-E(^7S)$		1.24	1.17	1.39	1.34	1.68

AE : Non relativistic all electron results of Clementi and Roeti (ref.

[Ne]R : Relativistic pseudopotential with a Ne-like core, this work

[Ne]NR : Non relativistic pseudopotential, with a Ne-like core of Hay and Wadt (ref. 5)

[Ar]NR2C : Non relativistic pseudopotential with an Argon-like core extracted on the ^7S and ^5D configurations (see text)

[Ar]NR : Non relativistic pseudopotential, with an Argon-like core, of Hay and Wadt (ref. 5)

Energy levels are in a.u., differences are in eV.

As concerning the second $3d^4 4s^2 (^5D)$ multiplet, [Ne]R and [Ne]NR are in good agreement with AE values, the reorganization of the 3s and 3p shells under the modification in the d shell being explicitly taken into account. For [Ar]NR2C, the agreement is also satisfactory which was expected because the reproduction of the valence energy levels for this multiplet was imposed in the determination of the pseudopotential parameters. In the last column, the values obtained with [Ar]NR show a larger deviation from the AE results. Regarding the energy difference between the ^5D and ^7S multiplets, the [Ne]R results (which is to be compared with the Hartree-Fock quasi-relativistic value 1.06 eV of ref. 2) is slightly overestimated by 0.1 eV. Also overestimated are the results obtained with [Ne]NR (0.15 eV) and with [Ar]NR2C (0.10 eV). The reoptimisation of the basis set for the second multiplet (^5D) would decrease the energy difference and improve the result for all cases. The worse result is obtained with the [Ar] core pseudopotential derived from a single all electron configuration. It therefore appears that a satisfactory reproduction of the two lowest multiplets with the same pseudopotential operator cannot be reached by extracting only on the lowest configuration. The solution proposed and used for [Ar]NR2C, even if it were not completely satisfactory (the energy valence levels for each multiplet are not reproduced as exactly as with standard potentials) gives a better compromise.

3.2. Scandium Atomic Spectra

If we choose the [Ar]$3d^1 4s^2$ pseudopotential representation of Scandium, the small number of valence electron allows a near complete treatment of the electronic correlation problem. We will report now results obtained recently by Koutecki (19) and one of us (G.J) with a non-relativistic pseudopotential extracted on the two lowest configurations of Scandium [Ar]$3d^1 4s^2 (^2D)$ and [Ar]$3d^2 4s^1 (^4F)$ for $\ell = 0$ and 2 symmetries and on the [Ar]$3d^1 4s^1 4p^1 (^4F^0)$ for $\ell = 1$ symmetry. The basis set used is composed of 4s 4p 7d primitive GTO optimized in pseudopotential atomic calculations and contracted to 3s 3p 3d GTO (see ref. 19 for a more detailed description of the construction process of the basis set). Electronic correlation is taken into account at a nearly full CI level using the MRD-CI (20) method. In Table 3 are displayed the results obtained for the $d^1s^2(^2D)$, $d^2s^1(^4F)$, $d^1s^1p^1(^4F^0)$ and $d^3(^4F)$ multiplets of Sc, the $d^1s^1(^3D)$ and $d^2(^3F)$ multiplets of Sc^+ and the $d^1s^2p^1(^1D)$ and $d^1s^2p^1(^3D)$ of Sc^-. For comparison, the Hartree-Fock (non relativistic and relativistic) results of Martin and Hay (2) are also given. The overall agreement with the experimental spectra is good, even for multiplet $d^3(^4F)$ for which the occupation in the d shell is different from the occupation in the reference configurations. The comparison between HFNR and HFPP is very stimultating even if there is a small discrepancy for the ionization potential.

Table 3. Multiplets of Sc, Sc^+ and Sc^-

Atomic State		HFNR	HFR	HFPP	CIPP	Exp.
Sc	$3d^24s^1(^4F)$	1.00	1.12	1.08	1.62 (1.64)	1.43
	$3d^24s^14p^1(^4F^0)$	-	-	1.23	1.81	1.96
	$3d^3(^4F)$	4.47	4.65	4.48	4.14 (4.32)	4.19
Sc^+	$3d^14s^1(^3D)$	5.20	5.22	5.52	6.38 (6.40)	6.56
	$3d^2(^3F)$	6.14	6.32	6.35	6.84 (7.02)	7.16
Sc^-	$3d^14s^24p^1(^1D)$	-	-	0.65	- 0.10	- 0.19[b]
	$3d^14s^24p^1(^3D)$	-	-	0.44	0.06	- 0.05

HFNR and HFR : Hartree-Fock non relativistic and relativistic values of ref. (2)
HFPP : Hartree-Fock pseudopotential results of ref. (19)
CIPP : multireference configuration interaction results of ref. (19). Values in parentheses are corrected for relativistic contributions (see text)
Exp. : experimental values of ref. (1) for Sc and Sc^+, of ref. (21) for Sc^-
All values are in eV and represent energy differences with the ground state $3d^14s^2(^2D)$ of Sc.

Supposing as a first approximation, that the coupling between relativistic effects and correlation energy is negligible we can add the relativistic corrections obtained by Martin and Hay at the Hartree-Fock level to the correlated pseudopotential results. The remaining differences between the corrected theoretical spectra and the experimental ones do not exceed 0.2 eV in all cases. They can be attributed first to strong limitations in the basis set (Bauschlicher and Walch (22) reported a decrease by 0.1 eV of the ^2D-^4F excitation energy by adding f orbitals) and second to correlation effects in the 3s-3p shells (which might further decrease this excitation energy by 0.2 eV, see ref. (22)). The results obtained for Sc^- are in good agreement with very recent experimental detection (21), but the energy difference is so small (of the same order of magnitude that the discrepancies discussed above) that no clear cut conclusion can be drawn at present.

3.3. Relation between relativistic corrections and correlation effects on the spectra and ionization potential of Copper atom

In the preceeding section we have made the usual assumption that relativistic effects obtained at the Hartree-Fock level could be directly transferred at the correlated level. Let us now examine the validity of this hypothesis by comparing results obtained for Copper using non-relativistic and relativistic pseudopotentials. They are extracted from non-relativistic and relativistic all-electron calculations according to the prescriptions described in section 2. The 3s and 3p electrons are treated explicitly in both pseudopotentials. The basis set is a

8s 7p 6d 1f GTO basis, where the six d exponents were taken from ref. (18). For s and p functions, the most diffuse parts (four s and four p) are identical to the most diffuse functions used by Werner and Martin (23). The internal exponents (four s and three p) were optimized through an atomic calculation of the $3d^{10}4s^1(^2S)$ to take into account the orbital/pseudoorbital deviation in the core region. The f exponent was taken equal to 1. This primitive basis set was later on contracted to 5s 4p 3d 1f with the scheme 41111/4111/411/1 and the same contraction coefficients were used in all calculations. Results in Table 4 confirm that $\Delta E(s^1d^{10}-s^2d^9)$ splitting and ionization potentials obtained by AEHF calculations are perfectly reproduced by PPHF. The relativistic effect on ΔE (0.38 eV) is also in good agreement with the AE results of Martin and Hay but the relativistic contribution to the IP is half (0.10 eV) of their result.

Table 4. Relativistic contributions to atomic spectra and ionization potential of Copper.

		AEHF	PPHF	$PPMP_2$	WM
ΔE $(s^1d^{10}\rightarrow s^2d^9)$	NR	- 0.37	- 0.39	2.27	1.42
	R	0.06	0.05	1.90	0.94
	ΔR	- 0.43	- 0.44	- 0.37	- 0.48
IP	NR	6.41	6.42	7.07	7.15
$Cu(s^1d^{10})$	R	6.56	6.58	7.28	7.36
$\rightarrow Cu^+(d^{10})$	ΔR	0.15	0.16	0.21	0.21

AEHF : All electron Hartree-Fock calculation, ref. (2).
PPHF : Pseudopotential Hartree-Fock calculation, this work
$PPMP_2$: Pseudopotential Moller-Plesset calculation, this work
WM : All electron CEPA-1 calculation by Werner and Martin, ref. (22)

As previously shown by other authors (2,23), the relativistic contribution stabilizes the $3d^94s^2(^2D)$ configuration which becomes the ground state at the relativistic level.

Correlation effects on ΔE and IP were estimated through a second order Moller-Plesset perturbation calculation (MP_2). This level of perturbation is not sufficient to give a fair quantitative account of correlation energy, even if third order and fourth order contributions partially compensates each other in many cases. Hovewer the results of Table 4 are essentially in agreement with the all-electron calculations of Werner and Martin (23) using the more accurate CEPA (24) method for dealing with electron correlation and a first order perturbation approach for relativistic corrections. The relativistic correction correlated wave functions is slightly larger (25%) than the correction at the Hartree-Fock level for the IP. We also notice a small difference on relativistic effects between correlated and non-correlated wave functions for ΔE, the relativistic correction being decreased by taking into account the electron correlation. This effect is not found by Werner and Martin and may be due to higher order correlation effects namely an increase of the core valence coupling due to the relativistic contraction of the 4s orbital.

4. RELATIVISTIC CONTRIBUTIONS TO BONDING

In order to present a clear analysis of the coupling between relativity and correlation in the formation of bonds with transition metal atoms, we shall report now calculations on transition metal hydrides (CrH and CuH) and on transition metal dimers (Cu_2). In all these calculations, we compare results obtained with non-relativistic and relativistic pseudopotentials. The 3s and 3p electrons are included in both pseudopotentials. Since basis sets and correlation treatment are identical in both calculations, the differences could be only attributed to relativistic effects.

4.1. Chromium and Copper Hydrides

The basis set for Chromium is obtained by contracting the 6s 6p 6d basis set described in section 3.2, using the scheme 3111/3111/411. This contracted 4s 4p 3d basis was completed with one f function (η = 0.75). For Copper, we used the same basis set given in section 3.3. For Hydrogen an optimized 5s GTO basis contracted 311 was completed with two p functions (η = 1.12, 0.2). Results gathered in Table 5 show at first glance that relativistic effects are similar on both hydrides.

Table 5. Relativistic contribution to bonding in CrH and CuH

		SCF(NR)	SCF(R)	CI(NR)	CI(R)	Exp [a]
CrH	Re(Å)	1.759	1.746 (- 0.013)	1.705	1.674 (- 0.031)	1.66
	De(eV)	0.98	0.99 (0.01)	1.84	1.99 (0.15)	-
CuH	Re(Å)	1.569	1.555 (- 0.014)	1.472	1.438 (- 0.034)	1.463
	De(eV)	1.38	1.42 (0.04)	2.26	2.35 (0.09)	2.85

Non relativistic, relativistic and relativistic contribution (in brackets) at the SCF and CI levels for the equilibrium distance and dissociation energy of the CrH and CuH ground states.
[a] Ref. (25).

A more detailed examination indicates that
i) the effect of relativistic corrections on R_e is noticeable producing a 0.03 Å shortening for both molecules, of the same order of magnitude than correlation contributions (at the non-relativistic level, 0.05 Å for CrH and 0.1 Å for CuH)
ii) the direct transferability of relativistic effects for R_e from SCF calculations is not justified at all
iii) for dissociation energies, relativistic effects are indeed very small in all cases. At the SCF level, it is almost negligible. The Residual difference after inclusion of correlation contribution with

the experimental value (only known for CuH) is much more larger than the observed relativistic effects and prevents to conclude definitely even if a larger effect is obtained (0.15 eV for CrH and 0.09 eV for CuH) than at the SCF level.

Our pseudopotential results are close to other reported calculations. For CrH, Walch and Bauschlicher (26) obtained R_e=1.70 Å and D_e=2.10 eV using a CASSCF + CI calculation with a basis set which is similar to ours for the valence orbitals. This also agrees with the older result by Das (27) using a limited MCSCF reference wave function (R_e=1.71, D_e=2.0). Our result confirms that the discrepancies between calculated R_e in ref. (26) and the experimental result can be attributed to relativistic effects. For CuH, the most refined all electron calculation up to now appears in the recent work by Raghavachari et al. (28) who used different correlation approaches in a fairly large basis set (including three contracted d and two diffuse uncontracted f orbitals). Their calculated D_e value for what they claim to be the best approximation (CCD + ST(4)) is 2.70 eV. If we add our relativistic correction (0.09 eV) to this value, the difference with the experimental value (2.85 eV) is less than 0.1 eV. For R_e , the agreement is not so excelent. Our ultimate result (1.438 Å) is too short by 0.025 Å, and if we add the relativistic correction to the MP4 (SDTQ) result of Raghavachari et al. the distance (1.420 Å) is considerably too short (0.04 Å). This fact confirms the underestimation of bond length by MP4 (SDTQ) perturbation theory already noticed in ref. (28).

4.2. Copper dimer

The Copper dimer has been the subject of many theoretical investigations during the past decade, due to its closed shell electronic structure. The bond can be seen as a simple 4s-4s single bond even if it has been shown that 3d electrons play an important role in bonding. This picture has been confirmed by all recent ab-initio calculations (28,29) and we will not be interested by this aspect. Very recently, two groups (23,30) have also studied the effect of relativity on bonding, using a similar approach namely first order perturbation theory on a highly correlated wave function. One of us (M.P.) has also compared (31) some years ago non-relativistic and relativistic pseudopotential calculations on the copper-copper bond in order to get insight in the relativistic effects. In these calculations, 3s and 3p electrons were included in the pseudopotential operator and no f functions were included in the basis set. We will report now some new results obtained with the two pseudopotentials already used for Cu atom and CuH molecule and with the same basis set. In Table 6 are displayed the values obtained at the SCF and MP_2 level. At the non relativistic SCF level the equilibrium distance (2.46 Å) is a little longer than the value corresponding to all electron calculations with similar basis sets in the valence region (2.423 Å in ref. (30) with a set of three contracted to two f functions, 2.44 Å in ref. (23) with four contracted d functions and two non-contracted f functions). The relativistic contraction at the SCF level (0.034 Å) is in agreement with the preceeding pseudopotential results of Pelissier (31) (0.03 Å).

Table 6. Relativistic contributions to bonding in Cu_2

		SCF(NR)	SCF (R)		CI(NR)	CI (R)	
R_e (Å)	This work	2.463	2.429	(-0.034)	2.258	2.217	(-0.041)
	WM	2.44	2.39	(-0.05)	2.27	2.23	(-0.04)
	SBA	2.42	-		2.27	2.24	(-0.03)
D_e (eV)	This work	0.42	0.46	(0.04)	2.07	2.25	(0.18)
	WM	0.51	0.58	(0.07)	1.69	1.80	(0.11)
	SBA	-	-		1.73	1.84	(0.11)

WM : Reference (23)
SBA : Reference (30) (with basis set A)
Experimental results are R_e = 2.22 Å and D_e = 2.05 (reference 25).

After inclusion of correlation, the relativistic contraction is slightly greater (0.041 in very good agreement with the results of ref. (30) and ref. (23)) but the large coupling between correlation and relativistic effects observed for CuH (and CrH) on the bond distance is no longer observed. In that case the lowering of the ionic components in the wave function, mainly Cu^+H^- or Cr^+H^-, due to the CI is likely to be responsible of the apparent coupling (the relativistic effects stabilize the neutral with respect to the positive ion). This means that a first order procedure to include relativistic effects applied to the correlated wave function would give the right answer.

CONCLUSIONS

We have presented here non-relativistic and relativistic pseudopotential results at different levels (SCF or CI). By using the same basis set for non-relativistic and relativistic calculations we have eliminated the ambiguity associated with the choice of basis functions. Even for the first series of transition metal atoms the relativistic effects are noticeable. This fact has been pointed out by many authors for Cu_2 (31,32) and has been confirmed by recent all-electron calculations (23,30). Our approach is in agreement with these findings. For transition metal hydrides, the relativistic effects are also important, at least on the equilibrium distance and we pointed out an apparent large coupling between relativistic and correlation contributions.

This is no longer true for dissociation energies for which relativistic effects are small, both at SCF and CI levels. It appears clearly that residual differences between refined ab-initio calculations and experimental D_e values are not due to relativistic effects but to incomplete basis sets and/or approximations in the correlation treatment.

The comparisons of our results with the experimental fact or with ab-initio calculations show that the direct inclusion of relativistic modifications in a pseudopotential operator is an efficient (and simple) way to deal with relativistic effects. Other simple approaches in which these effects are estimated by first order perturbation theory may face increased difficulties with heavier atoms because they do not take into account possible changes in the correlation energy due to the relativistic modifications of molecular orbitals which is present in our model from the very beginning.

From a general point of view, the preceeding tests fully justify the use of pseudopotentials for dealing with transition metal atoms. The inclusion of the outermost core electrons in the valence shell provides a complete agreement with all-electron results. Furthermore pseudopotentials (for instance the calculations reported here on Sc, the calculations on ScH, ScO, etc, in ref. (19), or the calculations on Cu_2 in ref. (31)) including these electrons in the core give correct results in many cases.

Acknowledgments. The calculations reported before have been performed on the CRAY-1S through a grant attributed by the Conseil Scientifique du Centre de Calcul Vectoriel pour la Recherche.

REFERENCES

(1) C.E. Moore, Atomic Energy Levels (Natl. Bur. Stand., Washington, D.C., 1952) Circular 467, Vols. 2 and 3 (1959)
(2) R.L. Martin and P.J. Hay, J. Chem. Phys. 75, 4539 (1981)
(3) M.M. Goodgame and W.A. Goddard, Phys. Rev. Lett. 54, 661 (1985)
(4) For a recent review about Effective Core Potentials see M. Kraus and W.J. Stevens, Ann. Rev. Phys. Chem. 35, 357 (1981)
(5) Complete sets of pseudopotential parameters are given by P.J. Hay and W.R. Wadt, J. Chem. Phys. 82, 270 (1985)
(6) Ph. Durand and J.C. Barthelat, Gazzetta Chimica Italiana, 108, 225 (1978)
(7) Within our pseudopotential approach, see for instance calculations by G. Trinquier and collaborators on molecular systems containing Si, Ge, P, As, etc : G. Trinquier, J.P. Daudey and N. Komiha, J. Am. Chem. Soc. 107, 7210 (1985) ; G. Trinquier, J.P. Daudey, G. Caruana and Y. Madaule, J. Am. Chem. Soc. 106, 4794 (1984) ; G. Trinquier and J.P. Malrieu, J. Am. Chem. Soc. 103, 6313 (1981); G. Trinquier, J.P. Malrieu and P. Rivière, J. Am. Chem. Soc. 104, 4529 (1982)
(8) G.H. Jeung and J.C. Barthelat, J. Chem. Phys. 78, 2097 (1983) ; F. Illas, J. Rubio and J.C. Barthelat, Chem. Phys. Letters 119, 397 (1985)
(9) H. Stoll, P. Fuentealba, P. Schwerdtpfeger , J. Flad, L.V. Szentpály and H. Preuss, J. Chem. Phys. 81, 2732 (1985) and references there in
(10) M. Pélissier and Ph. Durand, Theoret. Chim. Acta 55, 43 (1980)
(11) B. Roos, C. Salez, A. Veillard and E. Clementi, Tech. Rep. RJ 518, IBM Research (1968)
(12) E. Clementi and C. Roeti, Atomic Data and Nuclear Data Tables 14, 177 (1974)
(13) R.D. Cowan and D.C. Griffin, J. Opt. Soc. Am. 66, 1010 (1976)
(14) Y.S. Lee, W.C. Ermler and K.S. Pitzer, J. Chem. Phys. 67, 5861 (1977)
(15) J.C. Barthelat, M. Pélissier and Ph. Durand, Phys. Rev. A21, 1773 (1980)
(16) C. Teichteil, M. Pélissier and F. Spiegelmann, Chem. Phys. 81, 273 (1983) ; C. Teichteil and F. Spiegelmann, Chem. Phys. 81, 283 (1983)
(17) M. Pélissier and J.C. Barthelat, unpublished results ; M. Pélissier, Thèse d'Etat, Toulouse (1980)
(18) A.K. Rappe, T.A. Smedley, W.A. Goddard, J. Phys. Chem. 85, 2607 (1981)
(19) G.H. Jeung and J. Koutecky, to be published ; contribution by G.H. Jeung in this volume
(20) R.J. Buenker and S.D. Peyerimhoff, Theoret. Chim. Acta (Berl.) 39, 217 (1975)
(21) C.S. Feigerle, R.R. Corderman, S.V. Bobashev and W.C. Lineberger, J. Chem. Phys. 74, 1580 (1981)
(22) C.W. Bauschlicher and S.P. Walch, J. Chem. Phys. 74, 5972 (1981)
(23) H.J. Werner and R.L. Martin, Chem. Phys. Lett. 113, 451 (1985)

(24) W. Meyer, J. Chem. Phys. 66, 4377 (1977)
(25) H.P. Huber and G. Hertzberg, Molecular Spectra and Molecular Structure. IV. Constants of Diatomic Molecules (Van Nostrand Reinhold, New York, 1979)
(26) S.P. Walch and C.W. Bauschlicher, J. Chem. Phys. 78, 4597 (1983)
(27) G. Das, J. Chem. Phys. 74, 5766 (1981)
(28) K. Raghavachari, K.K. Sunil and K.D. Jordan, J. Chem. Phys. 83, 4633 (1985)
(29) C.W. Bauschlicher, S.P. Walch and P.E.M. Siegbahn, J. Chem. Phys. 76, 6015, 1982 ; 78, 3347 (1983)
C.W. Bauschlicher, Chem. Phys. Lett. 97, 204 (1983)
(30) P. Scharf, S. Brode and R. Ahlrichs, Chem. Phys. Lett. 113, 447 (1985)
(31) M. Pélissier, J. Chem. Phys. 75, 775 (1981) ; J. Chem. Phys. 79, 2099 (1983).

SPECTROSCOPIC CHARACTERIZATION OF BONDING IN DIATOMICS AND SMALL TRANSITION METAL AGGREGATES

M.Tranquille
Laboratoire de Spectroscopie Infrarouge
Université de Bordeaux I 351, Cours de la Libération
33405 TALENCE Cédex
France

ABSTRACT.

The transition metal vapors are essentially monoatomics. The first part of the paper is devoted to the methods used to produce small aggregates: gas phase aggregation or matrix isolation in solids at low temperature. The second part describes the spectroscopic methods involved in the characterization of such species with their possibilities illustrated by some results obtained in this field. We tentatively show how these techniques are yielding new electronic vibrational, structural and bonding information. They provide also certain physical and chemical properties of a wide range of metal clusters in different solid supports. In conjonction with gas phase data, it exists now a unique opportunity to determine the strength of metal-support interactions and develop new experiments in which we will able to control the alterations in charge density and structure of a supported metal cluster. In a next future it seems necessary to focus the experiments on ligand-free clusters size ranging from 4 to 6 atoms and also to check the chemical reactivity of the smallest clusters.

1. State of the art

A lot of works are undertaken concerning the synthesis and spectroscopic characterization of ligand-free metal clusters. Molecular beams or cryogenic rare gas matrices are the commonly used techniques and contribute to the understanding of electronic, geometric, dynamic and chemical bonding properties of a wide-range of uni and bimetallic clusters as a function of the nuclearity and metal-type (1). The results of these studies form an important data for checking the quantum chemical methods of calculations for such systems as well as for elucidating some aspects of chemisorption and catalysis.

The theory of naked metal clusters is developped because these aggregates are thought to represent the most chemically and physically

A. Veillard (ed.), Quantum Chemistry: The Challenge of Transition Metals and Coordination Chemistry, 53–65.

realistic bridge between isolated metal atoms and the extended arrays of them found in the bulk. Theoretical studies are oriented on the size dependence of electronic, magnetic, structural, bonding type and reactivity properties. Especially for transition metals a large amount of experimental results is necessary to obtain more accurate models for calculation.

2. Gas phase or condensed media ?

In the ideal collision-free environment of a molecular beam; the metal-cluster can be considered to be truly isolated from substrate-interactions. The spectroscopic methods now capable of selectively extracting direct information from a molecular beam are not numerous (mass spectrometry, UV-visible absorption, laser induced fluorescence), but their extension hold great promise for studying diverse size dependent properties of aggregates in their equilibrium configurations.

Co-condensation of metal atoms or/and clusters with rare gases from either jets or beams at cryogenic temperatures represents the alternative to observe clusters in a more or less interacting medium. A large number of spectroscopic methods can be applied and the long time required for many experiments are easily obtained.

In fact, the particular advantage of the second technique in that the spectra are not reflecting only the clusters properties but also the influence of the environment. So, it is easy to evaluate the role played by metal-support effects which are here simulated by appropriate guest-host interactions.

These interesting techniques make possible a detailed study of particles up to approximately ten atoms size which are of great and pratical importance for the knowledge of spectroscopic, chemical and catalytic properties of supported metal clusters in weakly or strongly interacting environments (2).

3. Metal clusters in the gas phase

In most cases, the production of metallic aggregates is obtained from the gas phase. The equilibrium vapors above heated metals are predominantly atomic, this is particularly true for metals of the first transition series for which 99% of the vapor is formed by atoms.

The mass spectrometry studies have permitted the measurements of bond dissociation energy of almost diatomics under equilibrium conditions (3). This technique has been extended to include measurements of the atomization energies of clusters but recently other methods have been developped in order to produce metal vapors which contain higher proportions of clusters than in the equilibrium state.

- Sputtering techniques (4)
- Vapor phase nucleation by vaporization in presence of inert gas (5)
- Supersonic expansion of either pure metal vapor or in excess of rare gas (6)
- pulsed laser vaporization of metal (7,8)
- photochemical decomposition of organometallic precursors (9).

The clusters obtained from beams and jets have been detected by mass spectrometric measurements but at the present time a few spectroscopic studies have been performed excepted with laser-induced fluorescence technique.

As demonstrated by combination of beams and mass spectrometry on studying alkali clusters, series up to n=15 were produced for $(Na)_n$ and ionization potentials measured. These studies have been extended at seeded beams expansions of Na in rare gas or N_2 carrier gases. In contrast to conventional expansion which produces cluster size distributions with an exponential decay, the seeded beam produces much larger clusters up to several tens with an irregular abundancy distribution (10).

Recent works involve transition metal in molecular beams Fe, Ni (11, 14) and even the refractory metals chromium tungsten and molybdenum (12, 13) to form clusters up to c.a. 25 atoms. The ionization potentials exhibit oscillations as function of the cluster size.

Laser vaporization of graphite substrate within the throat of a pulsed nozzle has been used to produce carbon clusters from 2 to 190 atoms. They were probed by UV laser photoionization of the beam with time-of-flight mass analysis of the resulting photoions (8). It is shown a distinctly bimodal cluster size distribution with all possible sizes between 2 to 30 atoms/cluster but only even clusters above. This effect is tentatively assigned at the presence of carbyne-like clusters structure which is a new high temperature from proposed for the carbon element.

By means of ionization the clusters can be separated with time of flight mass spectrometer and the ionization potential determined. If rotational constants are determined, the structure of molecular species can be achieved.

Such results indicate a promising future for application of supersonic beam and laser spectroscopic techniques. However, it must be recognized that the rather complicated technique is a significant factor to introduce a delay in the expected progress.

An interesting possibility is offered by the combination of the precedent technique and matrix isolation. In this way,Schumacher (15) and Scoles (16) have demonstrated that the gap between the results of these two techniques can be avoided since size-specific absorption spectroscopy in metal cluster beams with mass-selective separation and techniques of collecting in rare-gas solids offers exciting possibilities to study individual particles and also the perturbation of molecules by a host support (SF_6 and CH_3F in argon cluster).

4. Metal clusters in matrices

For the analysis of unstable or reactive species, Porter and Pimentel have created thirty years ago, the so called "rare gas matrix isolation technique" in which a guest molecule is embedded in a host rigid matrix maintained at low temperature.

The advantage of this method is the long time available to study a given sample essentially by spectroscopic methods. Different kinds of information can be obtained changing the matrix material, the temperature and the concentration (stoichiometry, conformations, reaction steps..)

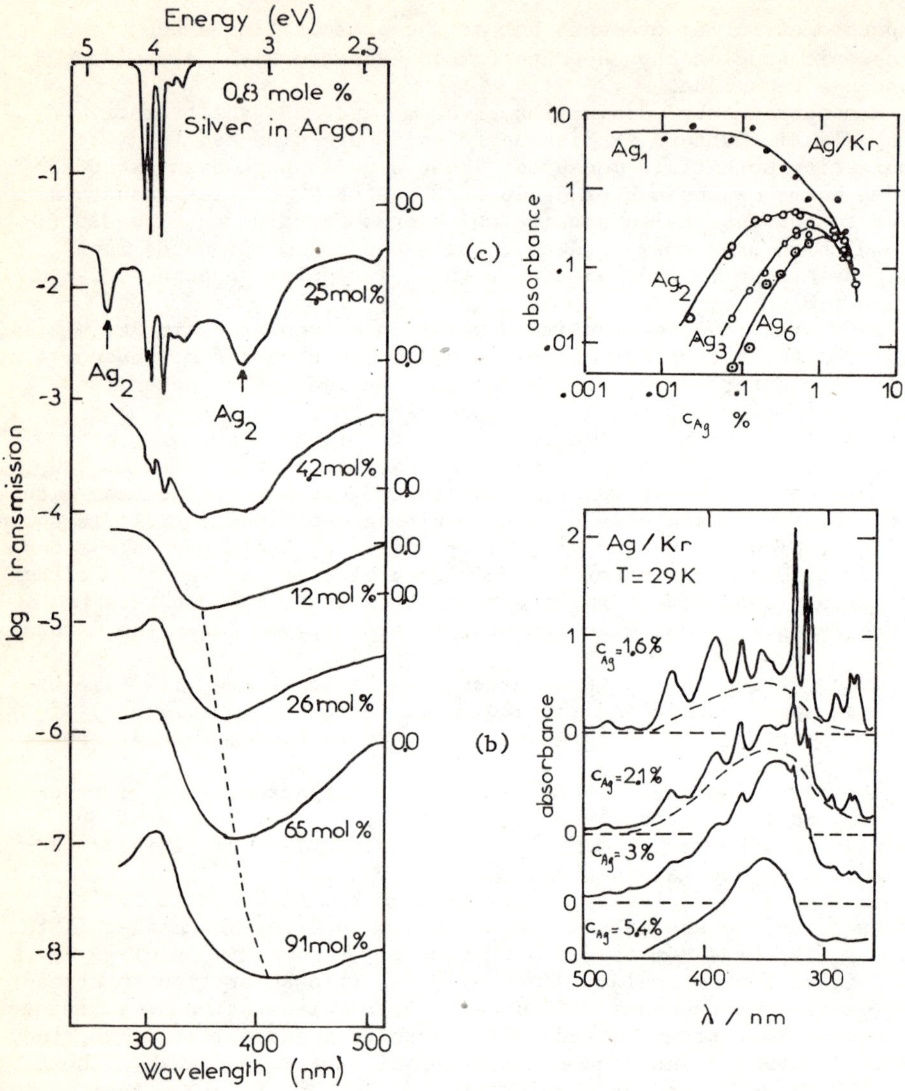

Figure 1 : Optical transmission of matrices containing various metal/gas ratios
a) in argon condensed at 5 K (from ref.20)
b) in krypton condensed at 29 K (from ref.27a)
c) relative yield of Ag_n in Kr as a function of metal concentration. The curves for Ag_2 Ag_3 and Ag_6 were evaluated from the bands at 3.19,2.94 and3.39 eV respectively (from ref.27a).

The original studies on metallic clusters by this method have been reported twenty years ago by Meyer who have observed that under certain conditions a rare gas matrix dopped with metallic atoms such as Mg,Hg , Na contains some aggregates and he observed that the electronic absorption spectra for homonuclear diatomics formed were quite different from those observed for metallic atoms (17).

Prior to deposition all compounds have to be vaporized, implying very high temperatures for metals resistively heated. New methods including electron bombardment and laser evaporation are preferred for refractory elements and often well adapted to produce some aggregates.

The formation of metal aggregates occurs because metals are highly mobile before they dissipate their kinetic energy in the condensing surface layers of the matrix. The aggregation processes occur for a brief period before solidification of the last layers during deposition. It has been demonstrated that this process is most efficient than the annealing of a matrix containing Ni atoms (18). A simple kinetic model for aggregation predicts that for a constant inert gas deposition rate, the ratio of the absorbance of a metal molecule containing n atoms to that of the metal atom should be proportional to the (n-1)th power of the metal deposition rate.

This law has been used in assigning the electronic spectrum arising from aggregates of a given size(19,20,fig.1). The deposition rate of metals is accurately determined using a quartz crystal microbalance (19).

The characterization of a given molecular cluster size can also be done using Resonance Raman Spectroscopy (isotopic substitution) and EPR experiments.

In such experiments, the major problem is due to the distribution of particles with different sizes in the matrix. To select the size of the clusters more sophisticated gas phase evaporation techniques have to be used including molecular jet, ionization chamber and mass spectrometry deflection prior to the formation of the matrix (27).

Ozin and Huber (21) have proposed a method which involves the photoinduced diffusion and aggregation of entrapped metal atoms in a rare gas matrix to produce selected aggregates by narrow-band photoactivation in a strong electronic absorption of metal atoms. Such photoaggregation effects have been observed for Ag,Na,Ni,Cu,Mo, Cr,Rh. The same method applied to the clusters gives a very low yield.

So, the fluorescence and excitation spectra of metal atoms and small clusters reveal a high specificity towards environment effects. Subtle changes in the surroundings, are well detected and depend on guest-host interactions in the ground and excited states. Other aggregates are found non-rigid molecules and their geometry is quite dependant of the matrix medium.

Such effects have to be taken into account especially with selective techniques of analysis which do not permit the simultaneous observation of all species (or states) contained in the matrix. This is true principally for techniques involving resonance conditions.

Thus, in some cases, it is not appropriate to compare directly the results of spectroscopic and structural studies of "immobilized" metal clusters with those from calculations on the corresponding clusters made without external interaction potential.

5. Spectroscopic methods for the study of naked metal clusters in matrices

A variety of spectroscopic methods are currently used to study metal clusters in matrices. In this part, we will describe some studies illustrating the powerful of each technique. Some limitation in the spectrometric measurements is due to the fact that a few laboratories possess more than one technique, and even in this case, it's not always easy to use the same matrix to perform various kind of measurements on it.

5a : U.V.visible absorption spectroscopy.

This technique was the first used to observe matrices containing metallic elements. In fact, the electronic spectrum of an atom trapped in a rare gas matrix is often similar to the corresponding gas phase spectrum. However, some discrepancies exist if we look at the intensity and shape of bands which are broadened and shifted by the interaction potential of the matrix. In some cases, the matrix field induces splittings of the degenerate levels. Displacements and splittings may attain 1000 cm^{-1} and are typical of a given matrix. The spectra in alcanes are broader than in rare gas matrices. For the rare gas series the isolation is better in Xe than in Ne (aggregation is more difficult in xenon) but the perturbation is stronger in Xe which presents a higher polarizability.

When aggregates are present, some other broad bands appear typically different from those of the atoms, in some cases a vibronic structure is observed. The figure 1a represents the evolution of the absorption spectra of silver depending on the metal concentration and shows different steps from the atom absorption to the plasmon transition of the bulk. A comparison between argon and krypton can be made.Using synchrotron radiations such analysis of the spectra can be extended to shorter wavelengths of VUV.

5b : PES : UV - Photoelectron spectroscopy.

This technique is useful to study clusters (and atoms) trapped in rare gas matrices of 50 Å thickness deposited onto a thin film of Ga on a flat copper block as cryosurface. It has been shown by Jacobi for Cu_n clusters (22) and more recently by Kolb for Zn_n clusters (23) that broad emissions quite distinct from those of the metal are observed for different sizes of clusters (fig.2).

In the Cu_n case, although the emission spectra are independent of the matrix gas, emission from the np orbitals of rare gas atoms of the matrix is modified. An additional peak red shifted by 0.8 eV whose intensity depends on the Cu concentration imply that the atomic relaxation for a rare gas atom close to Cu or Cu_n is dominated by relaxation effects in the metal rather than in the surrounding atoms of the matrix.

An enhancement of the signal is obtained by these authors using thicker neon matrices in conjunction with sub-bandgap photoemission (the wavelenght of the exciting light is choosen such that only the isolated species is ionized, but not the matrix).

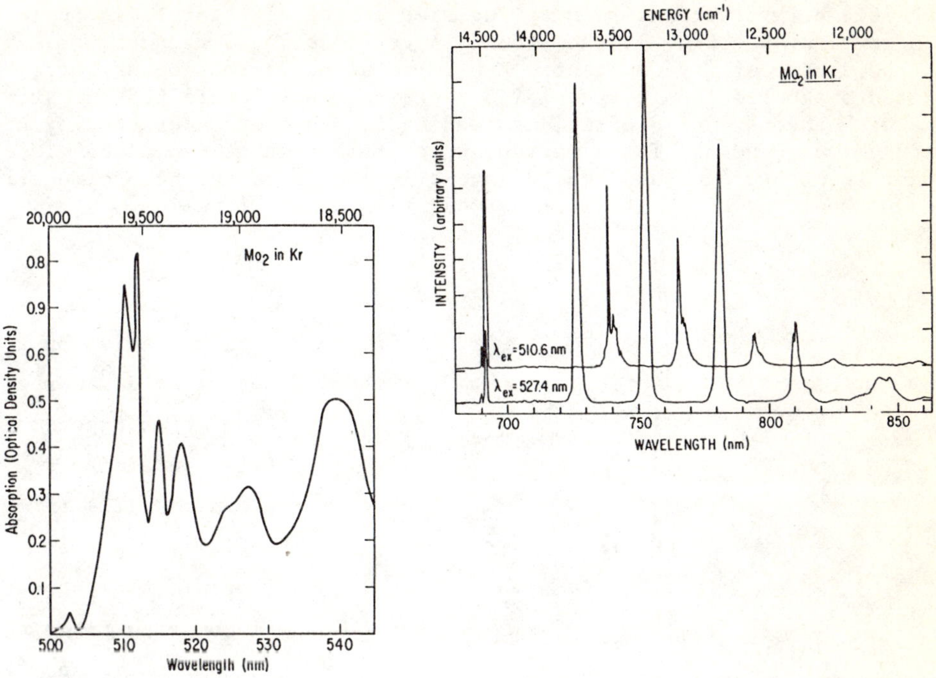

Figure 2 : Absorption and fluorescence spectrum of Mo_2 in krypton matrices at 13 K (from ref.36)

5c : LIF laser induced fluorescence

This technique offers some important advantages over the absorption because narrow bandwith and high intensity of stimulation radiation enable high resolution spectra to be recorded.Where absorption exhibits inhomogeneously broadening, laser excitation may pick out one or a narrow distribution of trapping sites. An illustration of this effect appears on the figure 2 where LIF technique shows unambigously that absorptions at 510.6 and 527.4 nm are due to different sites of Mo_2 moeity in the krypton cage.

The vibrational spacing is due to ground state-vibrational frequencies compared to the informations from absorption spectra which concern the excited states.

Recent investigations of Cu_3 and Mo_2 produced in molecular beams have been made by laser fluorescence and many excited states have been explored.

5d : Raman and Resonance Raman spectroscopy

The high sensitivity of the Raman scattering technique when the laser excitation is tuned into an electronic absorption band make this technique helpful on determining the size of the aggregates (isotopic substitution, see fig.3), the ground state vibrational frequencies and the anharmonicity constants. Experimental determination of force constants and electronic transition energies are valuable for comparisons with theoretical predictions leading to a better understanding of the chemical bonding. The simultaneous contribution of Raman and electronic spectroscopies and bond dissociation energies has been demonstrated in the study of the first row transition metal diatomics.

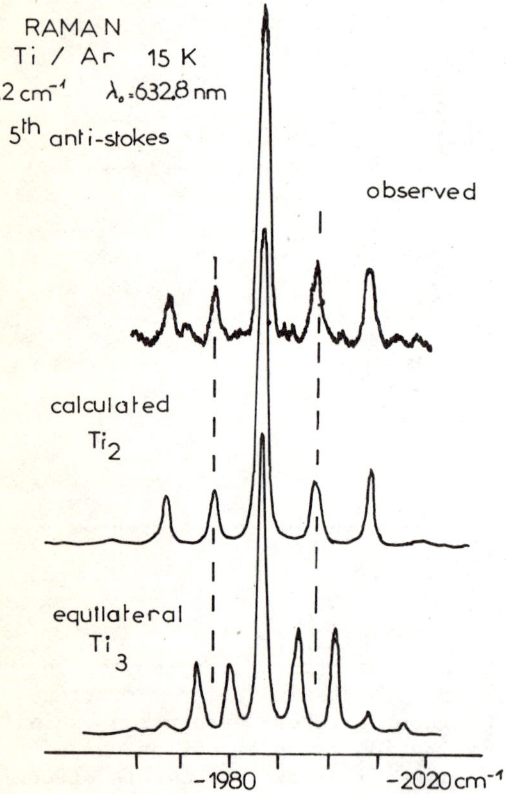

figure 3 : High resolution Resonance Raman spectrum of an argon matrix containing titanium (12K) (from ref.24). Observed isotopic fine structure compared to calculated for Ti_2 and triangular Ti_3 molecules (ref.24).

In the case of V_2 (24) and Pb_2 (25), the Raman spectra in electronic excited states have been recently observed leading to vibrational constants in the corresponding levels.

The electronic Resonance Raman spectrum observed for V_2 molecules in argon matrix make possible the direct measurement of the difference between electronic levels involved. This interesting possibility permits the determination of small shifts not easily obtained by electronic spectroscopy due to the overlapping of the bands.

A lot of Resonance Raman spectra are now available on tri and multi atomic metal clusters such Li_3, Cr_3, Ni_3, Cu_3, Ag_3, Pb_3, Sb_3, Sc_3 ...(26,27). For example, Ni_3 is proposed with a bent structure with an apex angle between 90 and 100° in argon matrix from Resonance Raman experiments. Cr_3 has an apical angle $\sim$75° and Sc_3 is found with equilateral geometry.

However, most of the experiments done on trimers seem to demonstrate that such clusters are in fact non-rigid molecules. Cu_3 in argon and gas phase (molecular beams from Smalley's Group), Li_3 in gas phase, argon and xenon (only one site) are definitely ascribed as dynamical Jahn-Teller molecule. The influence of the matrix seems predominant because from one site to another a molecule can be considered as rigid or not, with one geometry or another. Such conclusions are confirmed by ESR experiments.

5e : FTIR Far-infrared Fourier - transform spectroscopy

Infrared and Raman spectroscopies are both necessary in assigning polyatomic metal clusters vibrations because inactivity or weak activity of certain bands.

The recent improvement in sensitivity of far-IR spectrometry (bolometer detector and interferometry) might incite workers in this field to observe heterodiatomic molecules or tri and tetramers for comparison with Raman results. The early works concern Li-Au and Cr_3 (28) but anything else has appeared in the recent litterature.

5f : ESR Electron spin resonance spectroscopy

After electronic absorption, ESR is probably the most popular of the spectroscopies used in this field. Most of the trimers cited in the 5d section have been studied by this technique. Besides, mixed di and triatomic clusters have been detected (Cr Cu, Cu_2 Ag ...), Some penta and hepta metallic clusters were recently identified (Mn_5, Ag_5, Cu_5, Na_7, K_7, Li_7) (1).

For the K_3 molecule in argon matrix, Lindsay (29) has found two different geometries : one quasi-linear system and in addition, a form exhibiting a dynamic structure which fluctuates between three equivalent bent forms as predicted (and observed subsequently) for the completely fluxional Li_3 molecule. Cu_3 studied in adamantane matrix is found rigid contrary to Jahn Teller distorsion proposed for this aggregate from Raman and fluorescence studies (30).

The Alkali septemers display an ESR spectra presenting large isotropic hyperfine splittings from two equivalent alkali nuclei plus a much smaller splitting from at least five additional equivalent nuclei suggesting a pentagonal bipyramid structure for these M_7 species(29d)with a $^2A''_2$ ground state.

5g : MCD and MLD Magnetic circular (linear) dichroism

As the precedent technique this one gives informations about the symmetry of electronic levels. Besides this assignment of levels the characteristic of MCD is to observe some species undetectable by ESR and to provide information about matrix and atom or molecule interactions (ref.23, page 90/99).

$Cu_{2,3}$, $Ag_{2,3}$, $Mg_{2,3,4}$, Ca_2, Sr_2, Mn_2 and Ni_n isolated in rare gas matrices have been studied by this technique (1). The Mn_2 molecule is responsible of the two temperature dependent transitions observed at 331 and 347 nm in an argon matrix. The MCD spectrum confirms the theoretical prediction that Mn_2 should be a weakly bound antiferromagnetic molecule with a $^1\Sigma_g^+$ ground state according well with the low vibrational frequency of 124 cm^{-1} leading to a 25 Nm^{-1} force constant. The exchange energy for ground state Mn_2 was determined to be -10.3 ± 0.6 cm^{-1} which is identical to the value obtained from ESR experiments (31).

Furthermore, quantitative studies from band moment analysis are capable of providing information on the magnetic moments, spin-orbit coupling, interaction potentials and zero-field splittings.

Recently, it has been shown that magnetic linear dichroism (MLD) of a matrix isolated atom can provide complementary information to that obtained from MCD.

MCD	MLD
transition $\Delta J = 0, -1 \rightarrow$ positive C term	$\Delta J = 0 \rightarrow$ positive C term
transition $\Delta J = +1 \rightarrow$ negative C term	$\Delta J = \pm 1 \rightarrow$ negative bands

The conjunction of the two techniques is a promising tool to assign unambiguously the transitions between spin-orbit, multiplets. The application of the combined MCD/MLD approach has been made for the Fe/Kr and Ni/Ar systems. In the latter, it has been shown conclusively that both $^3D_3$ and $^3F_4$ ground state atoms are present in the matrix (32).

The magnitude of the MLD saturation curve vs, temperature has been found to be a sensitive measurement of the matrix field strength. Large zero field splittings of 20 cm^{-1} overall for Ni/Ar and 45 cm^{-1} overall for Fe/Kr have been deduced (ref.23, p.90).

5h : EXAFS/XANES X-ray absorption in fine structure/near edge structure

It is crucial for the development of theoretical treatments to have clear experimental results on the properties already described in the paper but also about internuclear distances in the diatomics and higher aggregates. From EXAFS measurement, Montano et al. (33) have studied Fe_2 in argon and deduced a bond distance of 1.87 Å. A more recent study including XANES results on Fe_n clusters in neon matrices over a wide range of iron concentration indicates a significant contraction of the Fe-Fe distance from bulk : 2.48 Å (α Fe) to 2.30 Å Fe_n (with n > 5) 2.05 Å around Fe_3/Fe_4 and a limit around 2 Å for Fe_2. The near edge absorptions observed for Fe_n aggregates were tentatively assigned to 1s → 3d, 4s, 4p, np transition but to quantify such data a better understanding of XANES technique would be necessary.

5i : Mössbauer Spectroscopy

Although this technique is restricted to a few number of nuclei (^{57}Fe, ^{119}Sn, ^{125}Te, ^{129}I, ^{151}Eu ...) it is very interesting to have informations about the ground state of the smallest clusters and some magnetic parameters.

For Fe_2 a $^7\Sigma$ electronic ground state has been proposed. Studies have been performed on mixed clusters containing iron Fe_n, Fe_nCu_m, Fe_nCo_m, Fe_nMn_m and Fe_nSn_m. It has been shown that isomer shifts are dependent on the size of the cluster and different from the bulk excepted in copper case where the isomer shift reaches quickly the value of the metal (34). The figure 4 illustrates the dependence from the number

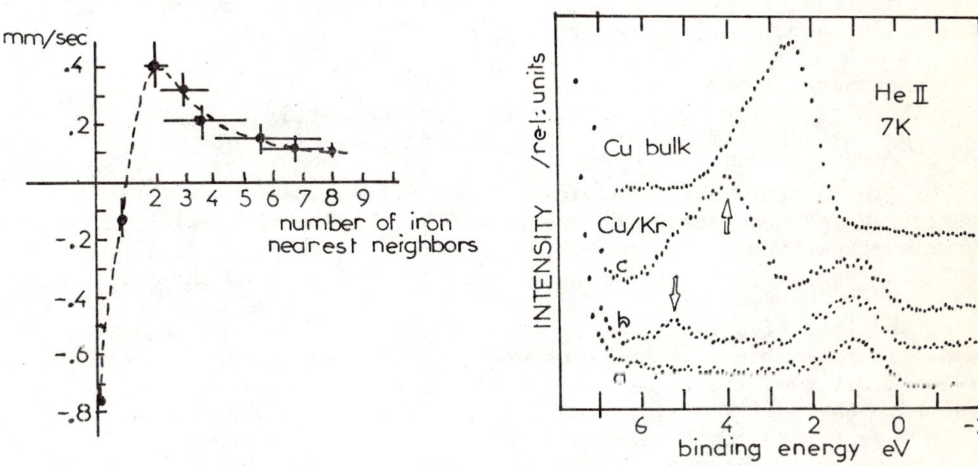

(a) (b)

figure 4 : Cluster size effects on some spectroscopic parameters

a) isomer shift in the Mossbauer spectra (4.2 K) of iron in argon matrices (from ref.34).

b) Emission of Cu in krypton matrices at 7 K under He (II) excitation in photoelectron spectroscopy.a : pure Kr matrix, b : Cu/Kr = 1/1000, c : Cu/Kr = 1/100 (from ref.22).

of nearest iron around a central -Fe atom in argon environment.

In the Fe_nSn_m system two gamma-resonances probes can be used on a single sample of clusters permitting a self-consistent assignment of FeSn, Fe_2Sn, Fe_2Sn_2, Fe Sn_3, Fe_2Sn_3 and $FeSn_4$ molecules, providing geometrical indications about the mixed trimers and giving a relative information on the nucleation processes during matrix condensation (35).

6. Conclusion and prospects

Is the experimentalist ready to offer to theorists the elements they require for calculation ? At the present time, the answer is probably "no". From one hand because a wide variety of experiments are undertaken with a great amount of techniques and from the other hand because the respective preoccupations are not well known from each other.

Such conferences are steps to a better understanding of the mutual possibilities. This brief report was tentatively demonstrating the present potentialities of the spectroscopic techniques involved in this area.

In summary, we have described methods to produce aggregates and to control their size. Two complementary techniques - gas phase and matrix isolation - which have proved their efficiency since they are coupled to several spectroscopic techniques to propose the following informations about small clusters :

- the energy level diagram
- the dissociation energy and ionization potentials
- the influence of an external potential on their energy levels.

For ground state and some excited states, the spectroscopy gives : bond length and bond strength, the geometry, spin-orbit coupling and magnetic properties and also some constants about the dynamical properties of the molecule depending on the interaction with the surroundings.

On the last five years, an impressive array of homo and heterobimetallic molecules of transition elements have been studied especially for the first row. The trimer series seems yet more interesting in respect to their fluxional nature (depending on the environment).

In a next future, the extension of the study to hetero metallic clusters seems preferable although aggregates of greater sizes are also needed to know the evolution of the properties in increasing this parameter.

Finally, over the understanding of the chemical bonding which requires many additional experiments, we have to keep in mind the aim of these studies which is the chemical reactivity of the small clusters. This field is widely open for theory and experiment in order to build the chemistry of the future.

7. References

1 - G.A.OZIN,S.A.MITCHELL, Angew.Chem.Int.Edn., 22,674 (1983).
2 - E.L.MUETTERTIES,M.J.KRAUZE, Angew.Chem.Int.Edn., 22, 135 (1983).
3 - K.A.GINGERICH, Faraday Symp.Chem.Soc., 14, 109 (1980).
4 - N.WINOGRAD,D.E.HARRISON Jr.,B.J.GARRISON, Surface Sci., 78,467 (1978).
5 - H.ABE,W.SCHULZE,B.TESCHE,Chem.Physics, 47, 95 (1980).
6 - J.L.GOLE,G.J.GREEN,S.A.PACE,D.R.PREUSS,J.Chem.Phys., 76,2247 (1982).
7 - V.E.BONDYBOY,J.M.ENGLISH, J.Chem.Phys., 74,6978 (1981).
8 - E.A.ROHLFING,D.M.COX,A.KALDOR,J.Chem.Phys., 81,3322 (1984).
9 - D.G.LEOPOLD,V.VAIDA, JACS, 105,6809 (1983).

10 - M.M.KAPPES,R.W.KUNZ,E.SCHUMACHER,Chem.Phys.Lett., 91,413 (1982).
11 - S.LEUTWYLER,U.EVEN,Chem.Phys.Letters, 84, 188 (1981).
12 - T.G.DIETZ,M.A.DUNCAN,D.E.POWERS,R.E.SMALLEY, J.Chem.Phys., 74,6511 (1981).
13 - D.L.MICHALOPOULOS,M.E.GEUSIC,S.G.HANSEN,D.E.POWERS,R.E.SMALLEY, J.Phys.Chem., 86,3914(1982).
14 - E.A.ROHLFING,D.M.COX,A.KALDOR,K.H.JOHNSON, J.Chem.Phys., 81, 3846 (1984).
15 - E.SCHUMACHER,W.H.GERBER,H.P.HARRI,M.HOFMANN,E.SCHOLL,in Metal bonding and interactions in high temperature systems, ACS Symp., Ser.N° 179, p.83 (1982).
16 - T.E.GOUGH,D.G.KNISHT,G.SCOLES, Chem.Phys.Letters, 97,155 (1983).
17 - L.BREWER,B.MEYER,G.D.BRABSON,J.Chem.Phys., 43,3973 (1965).
18 - J.HULSE,M.MOSKOVITS,J.Chem.Phys., 66(9) (1977).
19 - M.MOSKOVITS,G.A.OZIN, Cryochemistry, Wiley New-York,1976.
20 - T.WELKER,T.P.MARTIN,J.Chem.Phys., 70,5683 (1979).
21 - G.A.OZIN,H.HUBER,Inorg.Chem.,17,155 (1978).
22 - D.SCHMEISSER,K.JACOBI,D.M.KOLB,J.Chem.Phys., 75,5300 (1981).
23 - Proceedings of the Vth Congresson matrix isolated species, p.103, Fontevraud (1985).
24 - C.COSSE,M.FOUASSIER,T.MEJEAN and M.TRANQUILLE,P.DILELLA, M.MOSKOVITS, J.Chem.Phys., 73(12), 6076 (1980).
25 - H.SONTAG,B.EBERLE and R.WEBER,Chemical Physics, 80, 279 (1983).
26 - M.MOSKOVITS,D.P.DILELLA,J.Chem.Phys., 72,2267 (1980) ;77,5263(1982) J.Phys.Chem. 87, 524 (1983).
27 - W.SCHULZE,H.U.BECKER,Chem.Physics Letters, 35,177 (1978) ; H.SONTAG,R.WEBER, J.Mol.Spectrosc., 91,72 (1982) ; Chem.Physics 70,23 (1982).
28 - G.A.OZIN,M.D.BAKER,S.A.MITCHELL,D.F.McINTOSH, Angew.Chem.Int. Edn.Engl., 22,166 (1983).
29 - D.M.LINDSAY et al. , Mol Phys., 32,1199 (1976) ; J.Chem.Phys., 74,959 (1981); Surface Sci., 106,408 (1981); J.Chem.Phys., 80,4761 (1984).
30 - M.MOSKOVITS,Chem.Phys.Letters, 118,111 (1985).
31 - J.C.RIVOAL,J.S.EMAMPOUR,K.J.ZERINGNE,M.VALA; Chem.Phys.Letters, 92,313 (1982); R.J.VAN ZEE,C.A.BAUMANN,W.WELTNER Jr.,J.Chem. Phys., 74,6977 (1981).
32 - M.VALA,M.EYRING,J.PYKA,J.C.RIVOAL,C.GRISOLIA, J.Chem.Phys., 83,969 (1985).
33 - P.A.MONTANO,G.K.SHENOY,Solid State Commun., 35,53 (1980). H.PURDUM,P.A.MONTANO,G.K.SHENOY,T.MORISSON,Phys.Rev.B.25, 4412 (1982).
34 - W.DYSON,P.A.MONTANO,Solid State Communications, 33,191 (1980).
35 - S.SHAMAI,M.PASTERNAK,M.MICKLITZ,Phys.Rev., B26,3031 (1982).
36 - M.J.PELLIN,T.FOOSNAES,D.M.GRUEN,J.Chem.Phys., 74,5547 (1981).

LOW-LYING ELECTRONIC STATES OF ScH, TiH AND VH ACCORDING TO MRD-CI CALCULATIONS

P.J. Bruna[(+)] and J. Anglada
Lehrstuhl für Theoretische Chemie
Universität Bonn
Wegelerstraße 12
D-5300 Bonn 1, Federal Republic of Germany

ABSTRACT. The electronic structure of the low-lying states of ScH, TiH and VH and their corresponding positive ions have been studied using the MRD-CI method. The influence of the correlation of the (3s3p) subshell on the relative position for both the molecular and atomic states as well as the importance of f-type functions have been analysed. The special behavior of some low-multiplicity states will be discussed on the basis of the present calculations and for MnH and FeH using experimental inferences. Finally, the dissociation energies will be presented and compared with experimental data.

1. INTRODUCTION

Little experimental information is yet available for the low-lying electronic states of the simple metal transition hydrides ScH, TiH and VH as well as for their corresponding positive ions. In the case of the neutral species absorption spectra have been reported by Smith (1), but two mains adverse facts, namely the very complex structure and the absence of the knowledge of the symmetry and multiplicity of the ground states, were responsible for a further detailed analysis of the observed spectra. Although the positive ions MeH^+ containing transition metals have recently received more attention with respect to their thermochemical properties from some experimental groups (2), such kind of information is nevertheless missing for the simplest systems ScH and TiH (for both the neutral and positive radicals).

Therefore, the major part of the present knowledge about the electronic states of MeH hydrides stems from ab initio (all electrons or pseudopotentials) studies. The earlier investigations carried out by Scott and Richards (3) were mainly based on SCF calculations and corrected through an estimation of the correlation energies differences using as calibration the corresponding atomic information. Unfortunately

---

(+) Present address: Dept. of Chemistry, Univ. of New Brunswick, Fredericton, N.B., Canada, E3B 6E2

A. Veillard (ed.), Quantum Chemistry: The Challenge of Transition Metals and Coordination Chemistry, 67–78.

due to the high density of electronic states and the larger differential correlation effects (i.e. mixing between diverse Me states), one must take those prior theoretical results with precaution, especially with respect to the relative separations between several electronic states. A more detailed theoretical study on the bonding features of the ground states for the majority of the MeH compounds was carried out by Bauschlicher and Walch (BW) (4), but little attention was thereby paid to the complete manifold of electronic states correlating with the low-lying dissociation channels (5). Only for the VH radical an extensive MCSCF study on the potential curves for numerous electronic states is available (6). To the best of our knowledge, nothing has been reported about the electronic structure of the species ScH^+, TiH^+ and VH^+, with the exception of a short communication reported earlier (7) for the species ScH^+ and TiH^+.

Due to the facts that an H atom acts as a weak perturber of the electronic structure of the heavy atoms Me or Me^+, one can therefore expect that the electronic states of MeH or MeH^+ reflect more or less the electronic configuration of the transition metal atom (8). By these reasons we have analysed first the degree of sophistification of the calculations which are required to predict relative stabilities for the low-lying atomic states with an accuracy of ca. 0.20 eV. We have followed the transition energies between the $\{(sd^{n+1}) - (s^2d^n)\}$ states of the neutral Me and between $\{(d^{n+1}) - (sd^n)\}$ for the positive Me^+ species as a function of the technical conditions of the calculations, that is, as depending on the total number of correlated electrons (i.e. 3d4s vs. 3s3p3d4s) and also on the composition of the AO basis required to achieve the numerical quality mentioned above.

The information gained from this comparative study on the isolated heavier atoms will be very useful for an estimation of the degree of confidence which one can have with respect to the predicted relative positions of the molecular states. In this connection it must be pointed out that in all earlier theoretical works reported for the MeH species the number of electrons correlated was constrained to the outermost (3d4s) subshell only.

The calculations have been carried out using a multi-reference single and double excitation CI (MRD-CI) method. The general methodology of the method, which includes configuration selection and extrapolation techniques, is well documented in the literature (9). The full CI energy is estimated using a generalized expression of the simple Davidson-Langhoff formula (10). The CI calculations have been done for the various states using the corresponding natural orbital (NO) basis, which has been generated from a first series of CI calculations using appropriate SCF-MO's. The use of NO as basis allows not only a better interpretation of the wavefunctions but also a more compact description and consequently a smaller value of the full CI correction. This higher degree of accuracy is necessary in order to obtain relative energy separations which are reliable for spectroscopical purposes, especially for some electronic states differing in the closed-shell $7\sigma^2$ MO. For example, a simple SD-CI treatment leads to an incorrect ordering of the $^4\Delta$ and $^6\Delta$ states of FeH (4). A detailed discussion of the MRD-CI potential curves will be published elsewhere (11).

2. LOW-LYING ATOMIC STATES OF Me AND Me^+ (Me = Sc, Ti, V)

The relative separation between the (sd^{n+1}) and (s^2d^n) states of these atoms was analysed by Botch et al. (12) considering only the correlation of the (3d4s) electrons. BW have also taken into account the correlation of the (3s3p) electrons, but the CI calculations were limited to only one reference configuration (13). As pointed out by those authors, the differential correlation effects differ by about 0.20 eV between the directly calculated energy difference when compared with respect to the corresponding full CI estimates. A multireference treatment is therefore desirable.

The MRD-CI results are summarized in Table 1, whereby the values tabulated correspond to ΔE (error) = E (calc.) - E (expt.), and the transition energies are symbolized by the letter E. The best calcula-

Table 1 Deviations of the calculated transition energies of Sc, Ti and V and their corresponding positive ions from the experimental data (in eV).

	Basis	Sc		Ti		V	
(4s3d)	A	0.30		0.26		0.25	
	B	0.24		0.20		0.14	
	Lit(12)	0.28		0.22		0.16	
$(3s^23p^6)$ + (4s3d)		Sc	Sc^+	Ti	Ti^+	V	V^+
	A	0.60	0.55	0.51	0.54	0.39	0.43
	B	0.32	0.38	0.21	0.24	0.05	0.16
	C	0.24	----	0.24	----	0.16	----
	D	0.18	0.16	0.07	0.11	-0.01	0.04
	Lit(13)	0.04	----	0.00	----	0.02	----
Ionization Potentials[a)]		$Sc \rightarrow Sc^+$		$Ti \rightarrow Ti^+$		$V \rightarrow V^+$	
	A	-0.12		-0.21		-0.23	0.12
	B	-0.08		-0.13		-0.15	-0.11
	D	-0.14		-0.08		0.02	-0.28

a) The ionization process involves the transition $(s^2d^n) \rightarrow (sd^n)$. The second column for $V \rightarrow V^+$ corresponds to $(s^2d^3) \rightarrow (d^4)$, i.e. between both ground states, respectively.

tions are those obtained using a flexible AO basis expanded by two f-type functions and correlating the complete shell n=3. Each one of the optimized exponents for the f components were taken from the literature, namely from the (3d4s) correlation study of Botch et al. (12) and from the work due to BW (13) taking into account the correlation of the (3s3p) subshell. In order to follow the influence of the AO basis quality on Δ E (error), additional calculations were also undertaken for the original basis (14) of Wachters-Hay (Basis A). The intermediate Basis B and C are expansions of the Basis A using the most diffuse f exponents (B) or the most compact ones (C), respectively, while the flexible Basis D contains both f functions.

The SCF method underestimates the separation between the lowest multiplets of the neutral Me species due to the facts that the ground state configuration $4s^2 3d^n$ contains one closed-shell, while the excited state ($4s3d^{n+1}$) possesses a pure open-shell structure. Because the correlation of the $4s^2$ pair is predominantly described by the $4s^2 \rightarrow 4p^2$ excitation, already the simple correlation of the (3d4s) subshell leads to an overestimation of the transition energies, as displayed in the upper part of Table 1. The error of the calculations are lying between 0.14-0.30 eV when one correlates only the outermost electrons, while the inclusion of a semidiffuse f (Basis B) does not change the situation to a larger extent, with the exception of the V atom, which possesses the largest number of d electrons and is therefore more sensitive to the angular $d^2 \rightarrow f^2$ correlation. The present estimations are comparable to earlier MCSCF results (12), while the simple SD-CI treatment (not contained in Table 1) leads to deviations which are in the mean ca. 0.10 eV greater than those of the MRD-CI counterpart.

The next point to be treated is the influence of the ($3s^2 3p^6$) correlation on the relative energies. The values derived with Basis A indicate larger errors, ranging between 0.40 - 0.60 eV for both the neutral and the positive atoms. The larger magnitude of the error considering only the less flexible Basis A is due to the additional fact that the ($3p^2 \rightarrow 3d^2$) excitations are relatively more important for the ground states ($s^2 d^n$) configuration (i.e. there are more pair combinations possible) than for the excited (sd^{n+1}) states, which, complementary, have a small number of d-holes for pair correlations. In order to establish again a balanced description between the lowest electronic states one must add functions at least with f-character, in order to allow excitations of the class $3d^2 \rightarrow 4f^2$ which are more relevant for the excited states rich in d electrons. One notes for both the Basis B and C that the final errors are yet comparable with the results derived with the partial correlation of the (4s3d) subshell, but the best results are those obtained with a double-zeta representation of the f space. It must be pointed out that if one considers the relativistic effects, the error given in Table 1 may be larger by about 0.15 eV, as earlier estimated by Pittel and Schwarz (15), on the basis of an assumed additivity of correlation and relativistic effects.

With respect to the ionization potentials, the influence of the f functions is not as crucial as for the neutral states and this is related to changes only for one of the occupied orbitals of the neutral species, namely one goes from Me (s^2d^n) to Me^+ (sd^n). The largest deviation is obtained with Basis D for the process involving the ground states of V (s^2d^3) and V^+ (d^4), respectively.

3. LOW- AND HIGH-MULTIPLICITY STATES.

The electronic structure of the MeH systems is governed fundamentally by the heavy atom Me. It is well-known that the expectation value for $\langle r\rangle_{3d}$ is larger than for $\langle r\rangle_{3p}$ for the first atoms of the transition metals, being the Sc atom the most noticeable species of them, but already for the V atom the spatial extension of the 3d shell is more or less comparable with the corresponding value of 3p.

The bonding between Me and an H atom takes place through orbitals possessing σ symmetry exclusively. One has formally two extreme possibilities to build up hybrid for bonding, namely by combination of the outermost expanded species ($4s4p\sigma$) atomic orbitals, or through a more compact hybrid involving the ($4s3d\sigma$) atomic species. Depending on the degree of contribution of these extremes, the corresponding potential curves show different equilibrium distances. We will show now that these spectroscopic properties are also connected with some special electronic states of a given symmetry, possessing either high-multiplicity (larger R_e values) or low-multiplicity (and shorter R_e).

The low-lying electronic states of the MeH radicals can be ascribed to the global configuration $6\sigma^2 d^n 7\sigma$ (high multiplicity) or to the $6\sigma^2 d^{n-1} 7\sigma^2$ (low-multiplicity), respectively. The present study as well as findings reported in the literature (4,5,16) agree in the characterization of the open-shell 7σ MO as built up by a mixture ($4s4p\sigma$), while in those states possessing a closed-shell ($7\sigma^2$) structure, the hybridization involved is of the type ($4s3d\sigma$), leading consequently to shorter bond distances when compared with those electronic states possessing the single occupied 7σ MO.

Some examples of high-and low-multiplicity states are displayed in Table 2. The calculations carried out for ScH, TiH and VH indicate that the low-spin states have shorter R_e distances than the corresponding values of the high-spin states. The low-spin $X^1\Sigma^+$ state is precisely the ground state of ScH as well as $X^4\Delta$ state (expt.) for the FeH radical, although the calculations of BW (4) indicated erroneously a most stable $^6\Delta$ state. Finally, the $^5\Delta$ of MnH was placed experimentally at 0.21 eV above the $^7\Sigma^+$ ground state (17), and also possessing shorter bond distances. A $^5\Sigma^+$ could be an alternative assignment (Table 2).

Although the characterization of low-spin states of CrH waits its theoretical or experimental verification, one can note that both CrH and FeH radicals have complementary electronic structures; the high-spin $^6\Pi$, $^6\Delta$ states are associated with the occupation ($d^3 d\sigma 7\sigma$) for CrH or with ($d^5 d\sigma 7\sigma$) for the radical FeH, that is, containing the complementary atomic occupation d^4 and d^6, respectively. The same is true for the corresponding low-spin $^4\Pi$, $^4\Delta$ states given by the configurations ($d^3 7\sigma^2$) and ($d^5 7\sigma^2$), respectively. Earlier calculations on FeH (18) support the existence of a low-lying $^4\Pi$ state with smaller

Table 2 Selected electronic states of MeH possessing a competition between high and low-multiplicity configurations.

MeH	Ground State	High-multiplicity configuration	Low-multiplicity configuration
ScH	$^1\Sigma^+ (..\ 7\sigma^2)$	$^{3,1}\Sigma^+ (d\sigma 7\sigma)$	$..\ 7\sigma^2$
TiH	$^4\Phi\ (\delta\pi\ 7\sigma)$	$^{4,2}\Delta\ (\delta\ d\sigma 7\sigma)$ $^{4,2}\Pi\ (\pi\ d\sigma 7\sigma)$	$..\ \delta\ 7\sigma^2$ $..\ \pi\ 7\sigma^2$
VH	$^5\Delta\ (\pi^2 \delta 7\sigma)$	$^{5,3}\Sigma^-$ {$(\delta^2 d\sigma 7\sigma)$, $(\pi^2 d\sigma 7\sigma)$} $^{5,3}(\Phi,\Pi)\ (\delta\pi d\sigma 7\sigma)$	$..\ \delta^2 7\sigma^2$ $..\ \pi^2 7\sigma^2$ $..\ \delta\pi\ 7\sigma^2$
CrH	$^6\Sigma^+ (\pi^2 \delta^2 7\sigma)$	$^{6,4}\Pi\ (\delta^2 \pi d\sigma 7\sigma)$ $^{6,4}\Delta\ (\pi^2 \delta d\sigma 7\sigma)$	$..\ \delta^2 \pi\ 7\sigma^2$ $..\ \pi^2 \delta\ 7\sigma^2$
MnH	$^7\Sigma^+ (\pi^2 \delta^2 d\sigma 7\sigma)$	$^{7,5}\Sigma^+ (\pi^2 \delta^2 d\sigma 7\sigma)$	$..\ \pi^2 \delta^2\ 7\sigma^2$
FeH	$^4\Delta\ (\delta^3 \pi^2\ 7\sigma^2)$	$^{6,4}\Delta\ (\delta^3 \pi^2\ d\sigma 7\sigma)$ $^{6,4}\Pi\ (\pi^3 \delta^2\ d\sigma 7\sigma)$	$..\ \delta^3 \pi^2\ 7\sigma^2$ $..\ \pi^3 \delta^2\ 7\sigma^2$

R_e value. On the basis of these facts, one expects similar behaviour for the $^4\Pi$, $^4\Delta$ states of CrH but, taking into account the relative stability between the d δ and d π components, one can assume that the $^4\Pi$ state of CrH would probably be more stable than the $^4\Delta$ state, while for the FeH radical the opposite situation is true.

It must be remarked that the interpretation given by Stevens et al. (17) for the low-lying states of MnH and FeH differs from the present model: they have supposed that those states are associated with the structure $(d^{n+1}\ \sigma^2)$ and directly connected with the first excited $(d^{n+1}s)$ atomic state, while the high-spin state has an occupation $(d^n\ \sigma^2\ \sigma^*)$, i.e. coming from the Me $(d^n s^2)$ ground state. In other words, that model predicts a larger d-population for the low- than for the high-spin state, respectively. By contrast, we assign a global $(d^{n-1}$ $6\sigma^2 7\sigma^2)$ structure for the low-spin levels, whereby (d^{n-1}) represents only dδ or dπ components, and the $7\sigma^2$ MO retains d σ character as (dσ 4s) hybrid, that is, with a net-population more or less similar for both states, independently from their multiplicities. According with this interpetation, the results of BW (4) for FeH have indicated net-d populations of 6.08 and 6.26 for the $^6\Delta$ and $^4\Delta$ states, respectively. Although one notes a small change of the d-population between both states, the value reported for the $^4\Delta$ ground state is quite well below that of a formal (d^7) occupation as assumed by Stevens et al. (17).

4. RELATIVE STABILITIES OF SELECTED ELECTRONIC STATES

The main point of this section will be a comparison of the relative stabilities between various molecular electronic states as the technical conditions of the calculations are changed. The electronic states chosen are contained in Table 3, whereby the first column represents the most simple calculation, that is, correlating only the outermost electrons and working with SCF MO's, while in the second column are quoted those results correlating the subshell (3s3p) and using the corresponding NO's as basis for the CI treatment. Finally, the last column is equivalent to the second one but using the more flexible Basis B.

In the case of ScH the SCF procedure predicts a $^3\Delta$ ground state, while the closed shell $X^1\Sigma^+$ state is placed about 0.8 eV above it (compare with Fig. 1 of Ref. 5). Already the correlation of only four valence electrons gives a reversed order of stability, being the $^3\Delta$ placed at 0.25 eV above the ground state, in agreement with a value of 0.29 eV predicted by BW (16) under similar conditions of calculations. The correlation of the (3s3p) electrons increases this energy separation to a final value of 0.56 eV, a result which remains unchanged after expansion of the AO basis. The second $^1\Sigma^+$ state is found at 2.17 eV and this excitation energy agrees quite well with the observed absorption energy of 2.20 eV (1). The $2^3\Delta$ ($d\delta d\sigma$) state, with a global (d^2) occupation, is relatively little affected in its relative position along changes in the technical treatments.

The results for TiH can be summarized as follows: the low-spin states $^2\Delta$ and $^2\Pi$ are poorly described by the SCF method (see Fig. 2 of Ref. 5), but again the simple correlation of only five valence electrons remedies this situation. The calculated potential curves for these states confirm the expectations discussed in Section 3, namely both have shorter R_e values and anomalous high-low multiplicity splittings. The $2^4\Pi$ and $2^4\Phi$ states, both associated with an atomic (sd^3) occupation, are more or less placed at the same relative energies for the three treatments, denoting again a less influence of the f function on the relative position between low-lying molecular states.

Finally, the VH radical shows similar features as discussed before for ScH and TiH, namely that the expansion of the AO basis through f functions produces energy lowerings in the total energy which are comparable to the corresponding atomic states; nevertheless the relative position between the (s^2d^3) and (sd^4) ($2^5\Delta$) states remains practically unchanged after an AO basis expansion, at least for the low-lying electronic states near the corresponding equilibrium distances. The low-spin states $^3\Sigma^-$, $^3\Pi$, $^3\Phi$ are found to follow the general expectations given in Section III, namely with another electronic configuration

Table 3 Calculated relative energies for selected electronic states (in eV).

		Basis A		Basis B
		(4s3d)	(3s3p4s3d)	(3s3p4s3d)
ScH	$X^1\Sigma^+$ $(7\sigma^2)$	-0.25	-0.56	-0.57
	$^3\Delta$ $(d\delta\, 7\sigma)$	0.00	0.00	0.00
	$^1\Sigma^+$ $(7\sigma\, 8\sigma)$	1.55	1.61	1.60
	$2^3\Delta$ $(d\delta\, d\sigma)$	1.68	1.71	1.58
TiH	$X^4\Phi$ $(\delta\, \pi\, 7\sigma)$	0.00	0.00	0.00
	$^2\Delta$ $(\delta\, 7\sigma^2)$	0.27	0.07	0.22
	$^4\Delta$ $(\delta\, 7\sigma 8\sigma)$	0.42	0.25	0.45
	$^4\Pi$ $(\delta\, \pi\, 7\sigma)$	0.21	0.24	0.26
	$^2\Pi$ $(\pi\, 7\sigma^2)$	0.45	0.25	0.42
	$2^4\Phi$ $(\pi\, \delta\, 8\sigma)$	1.31	1.36	1.27
	$2^4\Pi$ $(\pi\, \delta\, 8\sigma)$	1.27	1.43	1.34
VH	$X^5\Delta$ $(\delta\, \pi^2 7\sigma)$	0.00	0.00	0.00
	$^5\Pi$ $(\pi\, \delta^2 7\sigma)$	0.20	0.21	0.39
	$^5\Sigma^-$ $(\delta^2 7\sigma 8\sigma)$	0.56	0.49	0.68
	$^3\Sigma^-$ $\{(\pi^2 7\sigma\, 8\sigma)\}$	0.69	0.84	0.92
	$^5\Sigma^+$ $(\delta^2\, \pi^2)$	0.87	0.95	0.81
	$2^5\Delta$ $(\pi^2\, \delta\, 8\sigma)$	1.15	1.27	1.22

than the corresponding one assigned to the neighboring high multiplicity states and with potential curves showing minima at smaller R_e.

In summary, an accurate description of the relative positions between high and low-multiplicity states by CI methods requires the use of a multireference scheme. Due to the extra closed-shell $(7\sigma^2)$ the low-spin states possess larger correlation energy. A simple SD-CI treatment gives an unbalanced description of both states, since the high-spin states are well described already by one reference configuration. In the case of the low-spin states, however, the inclusion of other reference configurations, for example, $(7\sigma^2) \rightarrow (4p\pi^2)$ leads finally to an equivalent description, as seen by an increase in the Σc_i^2 and consequently, to smaller full CI corrections.

5. ELECTRONIC STRUCTURE OF THE IONS ScH^+, TiH^+ and VH^+

It was pointed earlier (7) for the pair ScH-TiH^+ and TiH-VH^+ that the corresponding lowest dissociation channels differ in their electronic structures, and therefore a similar observation must be valid for the respective molecular states. For example, while the electronic states $^2D\ (s^2d)$, $^4F(sd^2)$ and $^4F(d^3)$ are placed at 0.00, 1.43 and 4.18 eV in the case of the neutral Sc atom, one finds for the Ti^+ ion another sequence, namely 3.09, 0.00 and 0.11 eV, respectively. A similar comparison between $^3F(s^2d^2)$, $^5F(sd^3)$ and $^5D(d^4)$ of the neutral Ti atom indicates relative energies of 0.00, 0.81 and 3.55 eV against an actual manifold of 4.00, 0.34 and 0.00 eV for the monopositive V^+ ion, respectively. Summarizing, a given MeH^+ species shows a quite different pattern of low-lying states as assigned to a MeH radical possessing the same number of electrons, a finding which reflects the small resemblence which exists between the relative stability and ordering of the electronic states of Me(Z) in comparison with $Me^+(Z+1)$.

The MRD-CI calculations predict a smaller value for the relative energies between the low-lying states of MeH^+ than for the corresponding neutral MeH compound, and thus find a weaker splitting of the d-shell of Me^+ due to interaction with an H atom relative to the situation observed for the neutral systems (7). For example, one finds that the $^2\Sigma^+(6\sigma^2 7\sigma)$ and $^2\Delta\ (6\sigma^2 d\delta)$ states of ScH^+ are quasi degenerate in the best treatment (i.e. correlating eleven electrons and including one f function), while the situation is more complicated for the TiH^+ radical, in which one finds that already three states, namely $^3\Sigma^-$ $(\sigma^2 d\pi^2,\ \sigma^2 d\delta^2)$ and $^3\Phi$, $^3\Pi(\sigma^2 d\pi d\delta)$ are all placed within a narrow energy region of only 0.00 - 0.10 eV, also correlating the complete 3s3p subshell and working with NO expansions in the flexible Basis B.

The ground state of VH^+ is predicted here to be $X^4\Delta(6\sigma^2 d\delta\, d\pi^2)$ and followed very closely by the multiplets $^4\Pi\,(6\sigma^2 d\pi\, d\delta^2)$, $^4\Sigma^-$ $(6\sigma^2 7\sigma d\delta^2,\ 6\sigma^2 7\sigma\, d\pi^2)$ and $^4\Phi\,(6\sigma^2 7\sigma d\,\pi d\delta)$, the latter possessing a T_e value of only 0.22 eV. For the lowest quintet states of the neutral VH radical one finds a similar ordering as for VH^+, but the highest $^5\Phi$ state is about 0.70 eV less stable than the $^5\Delta$ ground state.

The present work also shows that the low-lying electronic states of VH^+ are described by a global (d^3) occupation, that means, the lowest states prefer a structure related with the second dissociation channel, and furthermore, the existence of a similarity between the second pattern of quartet states of TiH (also connected with its second $^5F(sd^3)$ limit).

Comparing the equilibrium bond distances between similar states of MeH and MeH^+, one finds a general contraction after ionization of the outermost 7σ MO of the neutral systems. The actual bond lengths of the MeH^+ are in the mean 0.07 Å shorter when compared with the low-lying states of MeH. Finally, it is of interest to point out again that in the first member of this family ScH^+ the opening of the semicore $(3s^2 3p^6)$ is primordial to stabilize the $^2\Sigma^+(6\sigma^2 7\sigma)$ state relative to the $^2\Delta\,(6\sigma^2 d\delta)$ state when compared with the simplest treatment correlating only the outermost three valence electrons.

6. DISSOCIATION ENERGIES

In Table 4 are quoted the calculated D_e values corresponding to the full CI energies obtained correlating the $(3s^2 3p^6)$ electrons within the Basis B. The present results are equivalent to the findings in earlier ab initio studies of BW (4), but one notes that both theoretical works overestimate the magnitude of D_e when these are compared with the recommended experimental data (19).

The D_e for the VH system was determinated through reaction between V^- and a series of organic compounds, the acidity of which are well established and used as reference standards for H^+ abstraction. The question arises if this hydride was experimentally detected in the ground or alternatively in one of its various low-lying excited states.

Table 4 Calculated D_e for ScH, TiH and VH in eV. Values in parentheses correspond to MeH^+.

	ScH	TiH	VH
This work	2.43 {2.39}	2.00 {2.31 ± 0.03}	2.33 {2.15}
BW (4)	---- ----	2.12 ----	2.30 ----
D_o (19) (recommended)	2.06 ± 0.09	1.65 ± 0.13	1.64 ± 0.13 {2.18 ± 0.3}

Using our theoretical value D_e(VH) = 2.33 eV and after inclusion of zero point effects, one can estimate that the magnitude of the acidity which is necessary to produce VH in its $^5\Delta$ ground state must be 353 kcal/mol, while for the first excited $^5\Pi$ state the reaction occurs for a somewhat smaller threshold of 344 kcal/mol. According to general expectations, it is well-known that the ab initio methods invariably underestimate the magnitude of the dissociation energies, and therefoe it follows that the above estimations for Δ H (acid) both are lower bound values. Unfortunately, the experimental determinations have been carried out using several reference compounds but all of them possessing an acidity lying within the 338 ± 4 kcal/mol energy region. On the basis of these facts one can suppose that the experimental conditions were probably unable to detect of the VH radical in its $^5\Delta$ ground state.

Taking into consideration that the negative ion V^- in the X^5D state correlates with the molecular $^5(\Pi, \Delta, \Sigma^+)$ states of VH, it could be possible that the experimentalists have observed both the $^5\Pi$ state (with a $D_o \approx 1.85$ eV) or most probably the remaining $^5\Sigma^+$ state, possessing a D_o of ca. 1.43 eV, and which is detectable if one works with ΔH (acid) $\simeq$ 335 kcal/mol, that is, just within the energy region used in the laboratory (19).

The earlier estimations derived for D_e on the basis of pseudopotentials (20) or SCF calculations (3) are all smaller than both the ab initio results contained in Table 4. A recent study on TiH carried out by Wedig et al. (21), using a more elaborated pseudopotential treatment, leads to a D_e comparable to that of the present work.

The dissociation energies for the positive ions are quite similar to the corresponding values of the neutral species, which, on the other side, suggest that the IP of the neutral hydrides are expected to differ by less than 0.20 eV with respect to the corresponding atomic ionization potentials. Finally, taking a quoted value of 6.2 $\pm$ 0.3 eV for the IP of VH (19), one can deduce a D_o (VH^+) = 2.18 $\pm$ 0.3 eV, which agree quite well with the present theoretical estimates and indirectly confirms the theoretical values assigned to the dissociation energy of the neutral VH radical in its ground state.

7. SUMMARY AND CONCLUSIONS

If the correlation energy of only the outermost electrons in the transition metal is taken into account the lowest transition energies are overestimated by 0.15-0.30 eV. The inclusion of the (3s3p) subshell into the correlation treatment requires the use of at least two f functions in order to obtain errors less than 0.20 eV.

The low-spin states ($6\sigma^2 d^{n-1} 7\sigma^2$) in the MeH radical are characterized by equilibrium distances which are shorter than those of the high-multiplicity ($6\sigma^2 d^n 7\sigma$) states. These geometrical features are simply related to the hybridization (4s3dσ) characterizing the closed-shell ($7\sigma^2$) in comparison with the spatially expanded (4s4pσ) combination associated with the open shell 7σ MO, respectively. The low-spin state is the ground state of ScH ($X^1\Sigma^+$) and appears again as the ground state in this family of compounds in the FeH radical with a $X^4\Delta$ ($6\sigma^2\pi^2\delta^3 7\sigma^2$).

For all other systems between TiH and MnH it is known that the ground state is in each case a high-multiplicity state, although the general pecularities of the low-spin states (i.e. shorter R_e values and anomalous high-low spin separations) are also observed (or expected) for several of the low-lying electronic states. For MnH there are experimental evidences for the existence of a low-spin quintet (assumed to be a $^5\Delta$ state and placed at 0.21 eV above the $X^7\Sigma^+$), although a $^5\Sigma^+$ ($6\sigma^2\pi^2\delta^2 7\sigma^2$) state is suggested here as an alternative assignment.

Inclusion of the (3s3p) subshell does not change the relative position between the low-lying electronic states of TiH and VH, while in ScH the $X^1\Sigma^+ - {}^3\Delta$ separation is increased from 0.25 eV to 0.56 eV after opening of the (3s3p) core. An additional expansion of the AO basis by one f function gives practically the similar relative stabilities (maximum deviation of 0.20 eV) as obtained working with a smaller number of correlated electrons.

The present study finds that the low-lying electronic states of the MeH^+ ions are contained in a smaller energy region when compared with the similar electronic states of the corresponding neutral systems. For this reason it is somewhat difficult to make a definite assignment for the symmetry of the ground state of ScH^+ and TiH^+, whereby the relative separation between the states under competition is less than 0.10 eV, that is, smaller than the degree of accuracy which one can expect from non-relativistic calculations using a limited AO basis.

The dissociation energies for both the neutral and positive ions are all found in the 2.00-2.40 eV energy region and thus the IP's for an MeH radical is very similar to the value of the constituent heavy atom. The present results for D_e (MeH) agree with earlier ab initio predictions, while the larger deviation obtained for the VH radical when compared with recent experimental data can be explained if one takes into consideration that the experimental conditions were unable to detect the VH radical in its $^5\Delta$ ground state.

ACKNOWLEDGMENT We thank Profs. S.D. Peyerimhoff and R.J. Buenker for helpful discussions. One of us (J.A.) thanks the DFG and CIRIT, Spain, for financial support. The computer time was provided by the Computing Center of the University of Bonn, which is gratefully acknowledged.

REFERENCES

1. R.E. Smith, Proc. Roy. Soc. London, A 332 (1973) 113
2. a) R.G. Pearson, Chem. Rev. **85** (1985) 41 b) P.B. Armentrout, L.F. Halle and J.L. Beauchamp, J. Am. Chem. Soc. **103** (1981) 6501
3. P.R. Scott and W.G. Richards, J. Phys. B **7** (1974) 500, L 347 and 1679
4. S.P. Walch and Ch.W. Bauschlicher, J. Chem. Phys. **78** (1983) 4597
5. J. Anglada, P.J. Bruna, S.D. Peyerimhoff and R.J. Buenker, J. Mol. Structure (Theochem) **93** (1983) 299
6. G.A. Henderson, G. Das and A.C. Wahl, J. Chem. Phys. **63** (1975) 2805
7. J. Anglada, P.J. Bruna, S.D. Peyerimhoff and R.J. Buenker, J. Mol. Structure (Theochem) **107** (1984) 163
8. C.J. Cheetham and R.F. Barrow, 'Adv. in High Temperature Chemistry', ed. L. Eyring, Vol. 1, pp. 7 (Academic Press, N.Y., London, 1967)
9. R.J. Buenker, in 'Studies in Physical and Theoretical Chemistry', ed. by R. Carbo (Elsevier Sc. Publ. Comp., Amsterdam, 1982), pp. 17
10. R.J. Buenker, S.D. Peyerimhoff and P.J. Bruna, in 'Computational Theoretical Chemistry', ed. by I.G. Csizmadia and R. Daudel (NATO adv. Sc. Inst. Ser., D. Reidel, Dordrecht, 1981) pp. 55
11. J. Anglada and P.J. Bruna, to be published
12. B.H. Botch, T.H. Dunning and J.F. Harrison, J. Chem. Phys. **75** (1981) 3466
13. Ch.W. Bauschlicher, S.P. Walch and H. Partridge, J. Chem. Phys. **76** (1982) 1033
14. a) A.J.H. Wachters , J. Chem. Phys. **52** (1970) 1033
    b) P.J. Hay, J. Chem. Phys. **66** (1977) 4377
15. B. Pittel and W.H.E. Schwarz, J. Phys. B **11** (1978) 769
16. Ch.W. Bauschlicher and S.P. Walch, J. Chem. Phys. **76** (1982) 4560
17. A.E. Stevens, C.S. Feigerle and W.C. Lineberger, J. Chem. Phys. **78** (1983) 5420
18. P.R. Scott and W.G. Richards, J. Chem. Phys. **63** (1975) 1690
19. a) L. Sallans, K.R. Lane, R.R. Squires and B.S. Freiser, J. Am. Chem. Soc. **107** (1985) 4379 b) R.R. Squires, idem, 4385
20. G. Das, J. Chem. Phys. **74** (1981) 5766
21. U. Wedig, M. Dolg, H. Stoll and H.W. Preuss, this volume

ENERGY-ADJUSTED PSEUDOPOTENTIALS FOR TRANSITION-METAL ELEMENTS

Ulrich Wedig, Michael Dolg, Hermann Stoll,
and Heinzwerner Preuss
Institut für Theoretische Chemie, Universität Stuttgart
Pfaffenwaldring 55, D-7000 Stuttgart 80, West Germany

ABSTRACT. Various ways of generating pseudopotentials from atomic excitation and ionization energies are discussed in the case of Ti and Cu. Molecular results are given for TiH , TiO and CuH , CuO , Cu_2 .

1. INTRODUCTION

Accurate ab-initio calculations for transition-metal compounds are possible today, but the computational effort is very high (cf. |1| for Cu_2, e.g.). The difficulties may be alleviated, without significant loss of accuracy, by using pseudopotentials (cf. |2| for Cu_2 , e.g.). Pseudopotentials can be generated from atomic data in various ways, e.g. by adjustment to properties of suitable SCF reference states like orbital energies and orbital densities |3 to 8| , or by adjustment to excitation and ionization energies which are usually taken from experiment |9 to 11|. The latter variant has some advantages: total energy differences are physical observables while orbital energies are not; the reference to experimental data allows an easy incorporation of relativistic and core-valence correlation effects into the pseudopotentials. In the following, we discuss merits and shortcomings of energy-adjusted pseudopotentials for two representative transition-metal atoms, Cu and Ti. The following issues are investigated, in particular: the choice of the core and the selection of the reference states for fitting, in the case of Ti; the inclusion of the $3d^{10}$ shell into the pseudopotential and the separation of SCF , relativistic and core-valence correlation effects, in the case of Cu . The pseudopotentials are applied i) to the evaluation of atomic properties not used for fitting (e.g. densities), ii) to a number of diatomic molecules (TiH , TiO , CuH , CuO , Cu_2), both at the SCF and valence CI level.

2. COPPER

Cu has a closed $3d^{10}$ shell, and the simplest approximation is to treat it as a one-valence-electron atom. The reference states for fitting the

A. Veillard (ed.), Quantum Chemistry: The Challenge of Transition Metals and Coordination Chemistry, 79–89.

pseudopotential are then the nl←4s states. One has two options now: frozen-core SCF or experimental term energies can be used for the nl←4s states. With the first choice, a pseudopotential V_{SCF} results which should reproduce frozen-core all-electron results. The following functional form is suitable for fitting V_{SCF} :

$$V_{SCF} = -\frac{Q}{r} + \sum_{l,i} C_{li} \exp(-\gamma_{li} r^2) P_l \quad . \tag{1}$$

Here P_l is the projector on angular momentum l; the radial functions multiplying P_l are intended to describe deviations of the Fock potential for the valence electron from the point-charge approximation, and to represent the Pauli repulsion of the valence electron by occupied inner-shell orbitals.

When adjusting to experimental energies, one implicitly incorporates non-frozen-core effects. Such effects are non-negligible for Cu already at the SCF level (~0.1 eV for the 4s ionization energy), the contribution of core-valence correlation is large (~1 eV for the 4s IE). Non-frozen-core effects have been found to be adequately described by means of semi-empirical polarization potentials V_{pol}, for alkali atoms |12,13|

$$\begin{aligned} V_{pol} &= -\frac{1}{2}\alpha_D (\vec{f}(g^{-1}\vec{r}))^2 , \\ g(r) &= (1-\exp(-\delta r^2))^{1/2} . \end{aligned} \tag{2}$$

Here α_D is the dipole polarizability of the core, $\vec{f}(\vec{r})$ is the field at the site of the core generated by the valence electron at position $\vec{r}$, and g is a cut-off function. V_{pol} may be applied to one-valence-electron atoms, in general, and

$$V = V_{SCF} + V_{pol} \tag{3}$$

should then be a pseudopotential suitable for reproducing experimental energies.

For the Cu atom, accordingly, the parameters C_{li}, γ_{li} and δ in (1) and (2) may be fitted in such a way that V yields experimental ionization energies of the nl←4s states, while $V_{SCF} = V - V_{pol}$ yields frozen-core SCF (Koopmans) ones |14|. There are two options again: if Hartree-Fock Koopmans energies are used, V_{SCF} is a non-relativistic pseudopotential (V^{nr}_{SCF}); with Dirac-Fock (spin-orbit averaged) energies as input, it implicitly contains relativistic effects (V^{r}_{SCF}). Details of the fitting procedure (which leads to radially nodeless pseudo-orbitals for the lowest valence states of each angular quantum number l), and numerical values of the parameters for V, V^{r}_{SCF} and V^{nr}_{SCF} are given in |14|.

When applying these pseudopotentials to systems with more than one electron or core (Cu^-, Cu compounds), two points have to be taken into account: i) The Cu^+ core is rather large, and there is a non-negligible overlap of such cores in Cu_2, e.g. For the interaction of Cu^+ with other cores or nuclei, the point-charge approximation is insufficient, therefore (the concomitant error for Cu_2 is 0.3 eV). Improved interaction potentials have been determined from frozen-core SCF calculations for the ion pairs Cu^+ -Cu^+, Cu^+-$H^+$and Cu^+-O^{6+} |14,15|, and the differences to the point-charge potentials have been fitted to the form

$$V_{core} = B\ \exp(-\beta r) \quad . \tag{4}$$

ii) The field $\vec{f}$ to be inserted in V_{pol} (eq.2) is now that which is generated at the site of a given core by all valence electrons and the surrounding cores (or nuclei). Two-electron contributions appear in V_{pol}, therefore, if more than one valence electron is present, and the molecular V_{pol} is not simply a superposition of atomic polarization potentials, even in the case of a single valence electron.

Let us now consider some representative results. We begin with results obtained using V_{SCF}. Table I shows radial expectation values for the 4s orbital density of the Cu atom |16|. The deviations from accurate all-electron calculations are ≲1 %. The relativistic contraction of the Cu 4s orbital is well reproduced (to ≲15 %). Note that no information on densities was used for fitting the pseudopotentials.

	$<r>$	$<r^2>$	r_{max}
V^{nr}_{SCF}	3.319 (3.331)	13.00 (13.08)	2.595 (2.597)
V^{r}_{SCF}	3.241 (3.262)	12.39 (12.55)	2.533 (2.540)

Table I: Expectation values $<r>$, $<r^2>$, and position r_{max} of the outer maximum of the radial density of the Cu 4s orbital (in au) |16|. Values in parentheses are from all-electron Hartree-Fock and Dirac-Fock calculations |17|.

The simplest system with more than one valence electron is Cu^-. V^r_{SCF} leads to the following values for the electron affinity of Cu |18| (all-electron results |19|, with relativistic correction |14|, in parentheses): -0.001 (0.04) eV at the SCF, 0.74 (0.76) eV at the valence CI level.

	Cu_2		CuH		CuO	
	R_e	D_e	R_e	D_e	R_e	D_e
SCF	4.54 (4.52[a])	0.52 (0.58[a])	3.00	1.32	3.55 (3.54[b])	2.65 (2.71[c])
valence CI	4.56 (4.54[d])	1.29 (1.36[d])	3.02	2.32	3.58	2.67

Table II: Bond lengths R_e (in au) and dissociation energies D_e (in eV) from calculations with V^r_{SCF}, at the SCF and valence CI level, for Cu compounds |15,18|. All-electron values in parentheses: [a]Ref.20; [b]Ref.21; [c]Ref.21, with correction for IE(Cu) as described in Ref. 15; [d]Ref.22, with relativistic correction from Ref. 20.

Results for the Cu_2, CuH, and CuO molecules, obtained with V^r_{SCF}, are shown in Table II |15,18|. Bond lengths are accurate to < 0.05 au, dissociation energies to < 0.1 eV. Relativistic corrections for Cu_2, as determined from (separate) calculations with V^{nr}_{SCF} and V^r_{SCF}, are (at the SCF level, all-electron values |20| in parentheses) -0.10 (-0.09) au for R_e, and 0.09 (0.07) eV for D_e.

We now turn to results with $V = V^r_{SCF}+V_{pol}$ (eq.3). Calculated values for the electron affinity of the Cu atom, the bond lengths and dissociation energies of the Cu_2 and CuH molecules, and the ionization energy of Cu_2 are compared to experimental data in Table III. Various methods of treating core polarization and valence correlation are considered.

	Cu	Cu_2			CuH	
	EA	R_e	D_e	IE	R_e	D_e
expt.	1.23	4.19	1.99	7.89	2.76	2.85
a)	1.07	4.24	2.05	7.90	2.81	3.10
b)	1.13	4.28	1.95	8.09	2.88	2.62
c)	0.85	4.37	1.47	7.68	2.95	2.38
d)	0.84	4.38	1.45	7.69	2.99	2.28
e)	0.93	4.30	1.58	7.71	2.87	2.41

Table III: Electron affinity EA of Cu (in eV), bond lengths R_e (in au) and dissociation energies D_e (in eV) of Cu_2 and CuH, and adiabatic ionization energy IE (in eV) of Cu_2, determined by various methods (a...e, cf. text) |14,18,23|. Comparison is made to experimental values (first line, for references cf. |14|).

A simple superposition of atomic V_{pol} (eq.2), combined with a two-electron valence CI, gives quite acceptable results (line a in Table III): the deviations from experiment are $\lesssim 0.2$ eV for energies, ~0.05 au for bond lengths |23|. The quality of these results is probably due to a cancellation of errors, however: for CuO, a D_e value is obtained with this approach which is more than 2 eV too large |24|. A proper inclusion, at the SCF level, of polarization terms depending on the internuclear distance and of the two-electron part of V_{pol} (cf. above), combined with a density-functional approximation for valence correlation, yields results of homogeneous quality (line b in Table III), with errors of $\lesssim 0.2$ eV for energies and ~ 0.1 au for bond lengths |14|. Unfortunately, the inclusion of the coupling between core polarization and valence correlation, i.e. use of the molecular V_{pol} at the CI level, deteriorates the agreement with experiment (line c in Table III): bond lengths become too large by ~0.2 au, binding energies too small by 0.4...0.5 eV |18,23|. Changing the functional form of the (more or less arbitrary) cut-off function in (2), to

$$g(r) = 1 - \exp(-\delta r^2) \quad , \tag{5}$$

has only a marginal influence on the results (line d in Table III)

|18,23| *). More important is the effect of supplementing the dipole polarization potential (2) by a quadrupole term:

$$V = -\frac{1}{2}\alpha_D(\vec{f}(g^{-1}\vec{r}))^2 - \frac{1}{12}\alpha_Q \sum_{j=1}^{3} (\frac{\partial \vec{f}}{\partial r_j}(g^{-1}\vec{r}))^2 \qquad (6)$$

where α_Q is the quadrupole polarizability of the Cu^+ core and the same cut-off function (5) is used for dipole and quadrupole terms (line e in Table III): the errors are reduced to ~0.1 au for distances and 0.3... 0.4 eV for energies |18,23| *). The improvement is mainly due to the different angular behaviour of the electron-core and electron-electron cross terms in the quadrupole term (compared to that in the dipole term). Further improvement may be possible by removing the restriction of equal cut-off parameters in the two contributions to (6). We feel, however, that in this case, where a determination of the partitioning between the dipole and the quadrupole terms of V_{pol} is aimed at, the Cu(nl←4s) one-electron reference data may not be sufficient for fitting the pseudopotential; they should be supplemented by data from a system where non-zero cross terms are present (Cu^-, e.g.).

3. TITANIUM

For Ti, the 3d electrons have to be attributed to the valence shell. The largest possible core is then Ar-like, and Ti is treated as a 4-valence-electron atom. When adjusting the parameters of the pseudopotentials V_{SCF} (eq.1) or V (eq.3) to SCF or experimental data, this is most easily done using reference data of Ti^{3+}, since a one-electron problem is involved here only. (A one-electron fit is the usual way of generating energy adjusted pseudopotentials.) We have performed such a fit to ionization energies of Ti^{3+}(nl←4s) states, at the SCF level, using non-relativistic all-electron values |25| as input data. (For a fit to experimental data, see |26|.) The resulting V_{SCF} pseudopotential (V^1_{SCF}) |27| was then applied in SCF calculations for a variety of states of Ti^0, Ti^+, Ti^{2+}. The term differences to the Ti^0 ground state are compared to SCF all-electron values |25| in Fig. 1a |27|. The deviations are seen to be large (up to ~3 eV). (A similar effect was observed in |26|.) The errors are essentially controlled by the d occupancy: with each additional d electron, the pseudopotential values are shifted upwards by ~0.05 au with respect to the all-electron ones. One of the reasons for these errors is apparent: the 3d orbitals are localized in the same spatial region as the 3s and 3p core orbitals. These core orbitals are expected, therefore, to undergo non-negligible changes with increasing 3d orbital occupancy, due not only to changes in the screening of the nucleus by the 3d orbitals but also to changes in the s-d and p-d exchange interactions. Thus, the core of Ti^0 is different from that of Ti^{3+}.

In order to correct for part of the deficiencies of the one-valence-electron pseudopotential fit described above, we repeated this fit with

*) The results in Table III are self-consistent, in contrast to those of |18| where V_{pol} was included to first-order perturbation theory only.

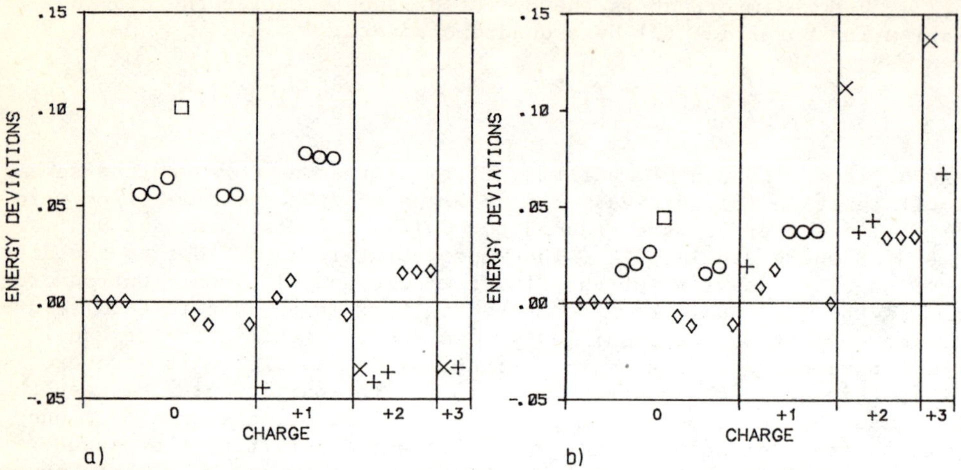

Fig.1: Deviations of pseudopotential |27| and all-electron |25| SCF results for term differences of various states of Ti^{n+} (n=0..3) to the Ti^0 ^3F ground state (in au). The states are classified according to their d occupancy: d^0 (x), d^1 (+), d^2 (◇), d^3 (○), d^4 (□). The pseudopotentials are of the V_{SCF} type fitted to Ti^{3+} (nl←4s) energies, a) without frozen-core restriction (V^1_{SCF}), b) with fixed Ar-like core taken from Ti^0 ^3F (V^2_{SCF}).

modified input data: the ionization energies of the Ti^{3+}(nl←4s) states were taken from frozen-core calculations |25|, the frozen core being that of the Ti^0 ^3F ground state. The resulting V_{SCF} pseudopotential (V^2_{SCF}) |27| should be more appropriate than the previous one for treating the neutral Ti atom. In fact, as seen from Fig.1b, the errors for Ti^0 and Ti^+ are substantially reduced (from ~3 to ~1 eV); for Ti^0, the influence of a unit change of d occupancy becomes ~0.02 (instead of 0.05) au. Thus, the situation is improved, but the errors are still unacceptably large, much larger than those originating from the frozen-core approximation (cf. below). The reason may be attributed, for the main part, to the description of the 3d-4s valence interaction. This interaction depends critically on the functional form of the (nodeless) 4s pseudo-orbital for small r (core region). The inner part of the 4s pseudo-orbital, on the other hand, is not really fixed by the fit of the pseudopotential to the energies of ns valence states ($n \geq 4$) whose $\langle r \rangle$ values are much larger than the core radius.

This defect can be removed by abandoning the one-valence-electron fit of the pseudopotential. Instead to ionization energies of one-valence-electron states, the pseudopotential may be adjusted to (total) valence energies of many-electron states, preferably those of the neutral atom, where s and d orbitals are occupied simultaneously. We have done such a fit for Ti, at the SCF level, using V_{SCF} (eq.1), with SCF energies of 19 states of Ti^0 and Ti^+ as reference data |27|. The fit was done in a least-squares sense, the number of reference data being larger than the number of parameters present in V_{SCF}. The resulting numerical values of the parameters of this pseudopotential (V^3_{SCF}) are given in Table IV.

	l	C_{l1}	γ_{l1}	C_{l2}	γ_{l2}
a)	0	9.19169	0.86	0.0083009	0.172
	1	17.99772	0.86	-0.032600	0.172
	2	-9.50431	1.6	-0.15137	0.32
b)	0	159.41799	13.01	18.07082	5.862
	1	95.99668	12.46	9.14874	5.217
	2	-17.45966	15.35	-0.88885	4.980

Table IV: Parameters of pseudopotentials V_{SCF}; a) fit to valence energies of Ti^0 and Ti^+ (Ar core), b) fit to ionization energies of Ti^{11+}(nl←3s) (Ne core).

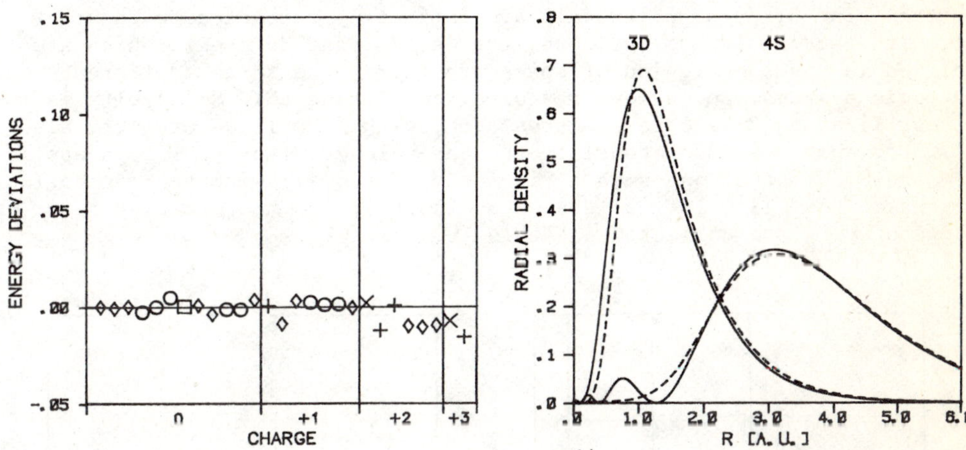

Fig.2: a) Same as Fig.1, but V_{SCF} fitted to (total) valence energies of Ti^0 and Ti^+ states (V^3_{SCF}) |27|. b) Radial densities (in au) of the 3d and 4s orbitals of the Ti^0 ^3F ground state. Solid lines denote all-electron results, broken lines pseudopotential ones with V^3_{SCF}.

The deviations from the reference data, together with deviations from Ti^{2+} and Ti^{3+} terms not used for fitting, are depicted in Fig.2a. The maximum error is 0.1 eV for the Ti^0 states, 0.4 eV for all the Ti^{n+} states considered. For comparison, the corresponding errors in all-electron calculations with a fixed $KL3s^2 3p^6$ core (taken from Ti^0, ^3F) are 0.3 eV for Ti^0, and 2.6 eV for Ti^{n+} ($n = 0,..,3$) |27|; errors in all-electron calculations with a limited basis set of double-zeta quality are $\lesssim 0.5$ eV for Ti^0, $\lesssim 0.9$ eV for Ti^{n+} ($n \leq 3$) |27|. Thus, the pseudopotential errors are of roughly the same magnitude as frozen-core and basis-set errors in all-electron calculations (cf. also |28,29|). A further comparison is possible between our pseudopotential and that of Topiol et al |6|, which was adjusted to orbital energies and densities of Ti^0 states. The accuracy is $\lesssim 0.3$ eV for Ti^0 and Ti^+ states, with the Topiol pseudopotential, but rather large deviations of up to 2.5 eV, as in the frozen-core approximation, appear for Ti^{2+} and Ti^{3+}. Our pseudopotential is adjusted to energies alone and does not contain density information. In Fig.2b, the 3d and 4s orbital densities computed with this pseudopotential for the Ti^0 ^3F ground state are compared to all-electron results. The deviations are non-negligible, in particular for 3d; the differences in the <r> expectation values are 0.09 au (6.1%) for 3d, and 0.05 au (1.3%) for 4s. The 3d errors illustrate the difficulty of describing the core-3d interaction, in a transition-metal atom, for a variety of states, by a single local potential. (Note, however, that the 3d and 4s errors are of different sign, in the valence region; thus, the total valence density is in better agreement with the all-electron one than the individual orbital densities.)

The pseudopotentials discussed for Ti up to now simulate the effect, on the valence electrons, of the Ar-like Ti^{4+} core. The difficulties met with this kind of pseudopotentials are intimately connected to the fact that, with this choice of the core, the 3s, 3p, and 3d shells which are localized in the same region of space are attributed to two different electronic systems. An obvious way of circumventing this difficulty is to treat Ti as a 12-valence-electron atom, with a Ne-like core (cf. also |30|). For such a pseudopotential, the one-valence-electron fit is applicable again. We have done such a fit for V_{SCF} (eq.1), with the ionization energies of the $Ti^{11+}(nl \leftarrow 3s)$ states as input |27|. The resulting parameters of V^4_{SCF} are collected in Table IV.

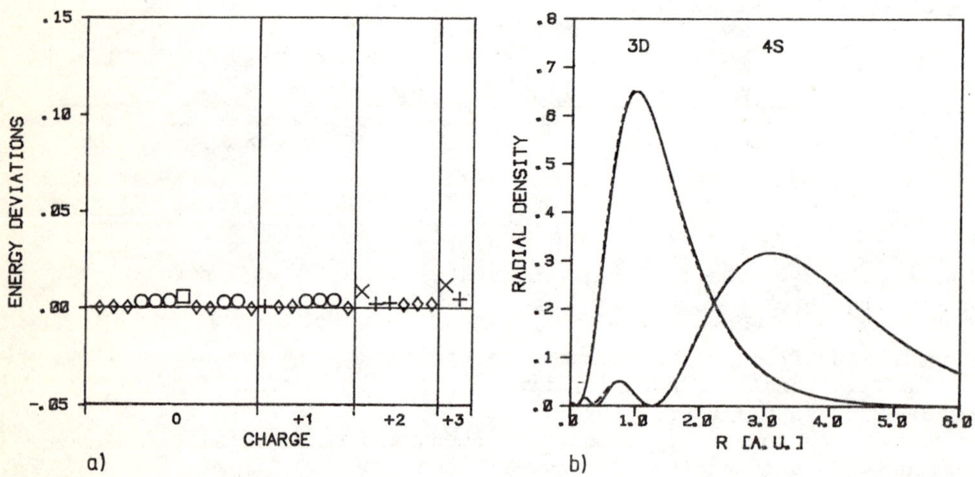

Fig.3: same as Fig.2, but V_{SCF} fitted to $Ti^{11+}(nl \leftarrow 3s)$ states (V^4_{SCF}). The deviations, at the SCF level, from all-electron reference energies

|25| of a variety of Ti^{n+} ($n \leq 3$) states, are given in Fig. 3a |27|. The pseudopotential errors are $\lesssim 0.3$ eV ($\lesssim 0.15$ eV for Ti^0), and comparable thus to those with the pseudopotential V^3_{SCF} described above. In contrast to the V^3_{SCF} potential, however, orbital densities are now in excellent agreement with the all-electron ones, as seen from Fig.3b.

We now turn to molecular results, for TiH and TiO, with the two pseudopotentials which performed best in the atomic case, V^3_{SCF} and V^4_{SCF}. V^3_{SCF} describes the large-core case (3s, 3p orbitals attributed to the core), V^4_{SCF} the small-core case (3s, 3p in the valence shell). Accordingly, more reliable results are expected with V^4_{SCF}, but V^3_{SCF} might be preferable for economic reasons. We note that both V^3_{SCF} and V^4_{SCF} are adjusted to non-relativistic data, but relativistic effects are probably small, anyway, for Ti compounds. Adjustment to experimental data (and inclusion of V_{pol} (eq.2)) should not lead to too much changes, either, in particular for the highly charged Ti^{12+} core. For the Ti^{4+} core, moreover, the polarization picture should not be too appropriate for describing core-valence correlation (3p-3d, e.g.).

Our calculations for TiH and TiO have been done both at the SCF and valence CI level, using the program MELD |31|; the pseudopotential calculations with V^3_{SCF} and V^4_{SCF} are supplemented, for comparison, by corresponding all-electron (AE) ones. The following Gaussian basis sets have been used for the AE case: (13s7p5d) |32|, contracted to double-zeta [8s4p2d], and augmented by diffuse (1s2p1d) and polarization functions (1f), for Ti; (9s5p)/[4s2p] |33|, augmented by 1p and 1d, for O; (5s1p)/[3s1p] |31| + 1d, for H. The same basis sets are employed in the V^4_{SCF} case, while a smaller (5s4p5d)/[3s3p2d] + (1s1d) + 1f valence basis |27| is used for Ti in the V^3_{SCF} calculations. For the CI, ~5000 (TiH) and ~10000 (TiO) spin-adapted configurations are included, single and double substitutions with respect to the SCF ground-state configurations; excitations from the Ti $KL3s^2 3p^6$ core are not accounted for. The coefficient of the SCF determinant is ~0.95 for both TiH and TiO. The influence of valence double excitations not included in the CI can be estimated to 0.0005 eV (TiH), and 0.2 eV (TiO), by means of second-order perturbation theory. With the Davidson correction |34|, the effect of higher excitations is assessed to 0.1 eV (TiH), and 0.7 eV (TiO).

Our results |27| for ground-state properties of TiH ($^4\Phi$) and TiO ($^3\Delta$) are summarized in Table V. It is seen that the V^4_{SCF} pseudopotential values are nearly identical to the all-electron ones ($\Delta R_e \lesssim 0.01$ au, $\Delta D_e \leq 0.04$ eV), both at the SCF and valence CI level. For V^3_{SCF}, the deviations are somewhat larger ($\Delta R_e \leq 0.07$ au, $\Delta D_e \leq 0.09$ eV), but the results are still acceptable. In contrast to that, the pseudopotential of Topiol et al |6|, which may be compared to our V^3_{SCF} one, proves to be much less reliable; for TiO, it yields $R_e = 2.02$ au and $D_e = 12.4$ eV, at the SCF level, using a DZ basis |27|. (A similar collapse, with Ar-core pseudopotentials, has been reported for ScO |35|.) A comparison of our calculated values to experimental data is possible for TiO |36|. The agreement is very good for R_e, but our CI value for D_e, determined by separate calculations for the molecule and the atoms, is too small (by 25%). Estimating the effect of neglected doubles and higher substitutions, D_e is improved but still too small by 16%. Some further improvement (by ~0.2 eV) is possible by including excitations from the 3s and 3p orbitals in the CI |27|. In our

		TiH ($^4\Phi$)			TiO($^3\Delta$)		
		R_e	D_e	μ	R_e	D_e	μ
SCF	a)	3.58	1.58		3.02	2.59	
	b)	3.50	1.60		3.02	2.60	
	c)	3.51	1.58		3.02	2.56	
CI	a)	3.47	1.90 (1.85)	2.2	3.04	5.23 (5.87)	2.9
	b)	3.42	1.94 (1.89)	1.9	3.05	5.15 (5.78)	2.8
	c)	3.42	1.95 (1.91)	2.0	3.05	5.14 (5.78)	2.8
expt.					3.06	6.87	

Table V: Bond lengths R_e (in au), dissociation energies D_e (in eV), and dipole moments μ (in D) of Ti compounds, calculated at the SCF and valence CI level |27|. The pseudopotentials a) V^3_{SCF}, and b) V^4_{SCF} are used; line c) denotes all-electron results. The CI values are variational, numbers in parentheses include corrections (cf. text). Experimental data are from Ref.36.

TiO calculations, an excess charge of 0.54 is found on the O atom; a very similar value (0.55) has been reported in |37|. For TiH, experimental data are not available. The agreement of our results with those of previous calculations |38 to 40| is fair (R_e: 3.51 au (|38|, SCF), 3.61 au (|39|, pseudopotential MCSCF); D_e: 1.57 eV (|38|, SCF), 1.75 eV (|40|, SCF), 1.7 eV (|39|, MCSCF), 1.95 eV (|40|, CI).

4. CONCLUSIONS

Energy-adjusted pseudopotentials provide a simple and reliable means of reducing the computational effort in ab-initio calculations for transition-metal compounds. With the pseudopotentials simulating Ne-like cores (3s, 3p in the valence shell), virtually no loss of accuracy occurs with respect to all-electron calculations, for 3d elements. With larger cores (Ar-like or Cu^+-like for elements with open or closed 3d shell, respectively), the results are still in good agreement with all-electron SCF or valence CI calculations. (The errors are $\lesssim 0.05$ au for bond lengths, and $\lesssim 0.1$ eV for dissociation energies.) It is essential, however, that the pseudopotentials be fitted to energies of neutral or near-neutral species, in this case. A difficulty remains, though, for the large-core pseudopotentials, and this is the description of core-valence correlation. A polarization ansatz has proven to be useful, for the Cu^+ core, but the accuracy is inferior to that at the SCF/valence CI level.

REFERENCES

1 P.Scharf, S.Brode, and R.Ahlrichs, Chem.Phys.Lett. 113 (1985) 447
2 M.Pelissier, J.Chem.Phys. 75 (1981) 775
3 C.S.Ewig and J.R.Van Wazer, J.Chem.Phys. 63 (1975) 4035

4 L.R. Kahn, P.Baybutt, and D.G.Truhlar, J.Chem.Phys. 65 (1976) 3826
5 J.C.Barthelat, Ph.Durand, and A.Serafini, Mol.Phys. 33 (1977) 159
6 S.Topiol, J.W.Moskowitz, and C.F.Melius, J.Chem.Phys. 68 (1978) 2364
7 V.Bonifacic and S.Huzinaga, J.Chem.Phys. 60 (1974) 2779
8 U.Wahlgren, Chem.Phys. 29 (1978) 231
9 M.E.Schwartz and J.D.Switalski, J.Chem.Phys. 57 (1972) 4125
10 T.C.Chang, P.Habitz, and W.H.E.Schwarz, Theoret.Chim.Acta 44 (1977) 61
11 H.Preuss, H.Stoll, U.Wedig, and Th.Krüger, Int.J.Quant.Chem. 19 (1981) 113
12 W.Müller, J.Flesch, and W.Meyer, J.Chem.Phys. 80 (1984) 3297
13 P.Fuentealba, H.Preuss, H.Stoll, and L.v.Szentpály, Chem.Phys.Lett. 89 (1982) 418
14 H.Stoll, P.Fuentealba, M.Dolg, J.Flad, L.v.Szentpály, and H.Preuss, J.Chem.Phys. 79 (1983) 5532
15 G.Igel, U.Wedig, M.Dolg, P.Fuentealba, H.Preuss, H.Stoll, and R.Frey, J.Chem.Phys. 81 (1984) 2737
16 P.Fuentealba, H.Stoll, L.v.Szentpály, P.Schwerdtfeger, and H.Preuss, J.Phys. B 16 (1983) L 323
17 J.P.Desclaux, At.Data Nucl.Data Tables 12 (1973) 311
18 H.Stoll, P.Fuentealba, P.Schwerdtfeger, J.Flad, L.v.Szentpály, and H.Preuss, J.Chem.Phys. 81 (1984) 2732
19 C.W.Bauschlicher, S.P.Walch, and H.Partridge, Chem.Phys.Lett. 103 (1984) 291
20 H.J.Werner and R.L.Martin, Chem.Phys.Lett. 113 (1985) 451
21 P.S.Bagus, C.J.Nelin, and C.W.Bauschlicher, J.Chem.Phys. 79 (1983) 2975
22 C.W.Bauschlicher, S.P.Walch, and P.E.M.Siegbahn, J.Chem.Phys. 78 (1983) 3347
23 M.Dolg and H.Stoll, unpublished results
24 C.Igel-Mann and H.Stoll, unpublished results
25 C.Froese-Fischer, Comp.Phys.Comm. 14 (1978) 145
26 T.C.Chang, P.Habitz, B.Pittel, and W.H.E.Schwarz, Theoret.Chim.Acta 34 (1974) 263
27 U.Wedig and H.Stoll, in preparation
28 L.Pettersson, and U.Wahlgren, Chem.Phys. 69 (1982) 185
29 M.Pelissier and E.R.Davidson, Int.J.Quantum Chem. 25 (1984) 483
30 L.G.M.Pettersson, U.Wahlgren, and O.Gropen, Chem.Phys. 80 (1983) 7
31 program written by L.McMurchie, S.Elbert, S.Langhoff, and E.R. Davidson, Washington University, Seattle
32 I.Hyla-Krispin, J.Demuynck, A.Strich, and M.Benard, J.Chem.Phys. 75 (1981) 3954
33 T.H.Dunning, J.Chem.Phys. 53 (1970) 2823
34 S.R.Langhoff and E.R.Davidson, Int.J.Quantum Chem. 8 (1974) 61
35 O.Gropen, U.Wahlgren, and L.Pettersson, Chem.Phys. 66 (1982) 459
36 K.P.Huber and G.Herzberg in: Molecular Spectra and Molecular Structure (Van Nostrand, New York, 1979) vol. IV
37 C.W.Bauschlicher, P.S.Bagus, and C.J.Nelin, Chem.Phys.Lett. 101 (1983) 229
38 P.R.Scott and W.G.Richards, J.Phys. B 7 (1974) 1679
39 G.Das, J.Chem.Phys. 74 (1981) 5766
40 J.Anglada, P.J.Bruna, S.D.Peyerimhoff, and R.J.Buenker, J.Mol.Struct. 93 (1983) 299

SPECTROSCOPIC STUDIES OF COPPER COMPLEXES

J.C. Barthelat, M. Hliwa, G. Nicolas, M. Pélissier and F. Spiegelmann
Laboratoire de Physique Quantique (U.A. au CNRS n°505)
Université Paul Sabatier
118, route de Narbonne
31062 Toulouse Cedex, France

ABSTRACT. Non-empirical pseudopotentials derived from relativistic atomic calculations are used in theoretical studies of the CuH and $Cu(C_2H_4)$ spectra. Five states of Σ^+ symmetry dissociating into the $d^{10}s^1$ and d^9s^2 configurations were investigated for the copper-hydride molecule. Spectroscopic constants present an overall good agreement with experimental data. The complex behavior of the potential curves in the intermediate region is analyzed through a diabatic description. The $X^1\Sigma^+$ ground state exhibits a strongly ionic character at short internuclear distances. The ground state of the copper-ethylene complex arises from a weak interaction (0.36 eV) between ethylene and the copper atom in its $d^{10}s^1$ configuration. Three bound excited states involving the $d^{10}p^1$ configuration of the copper atom are considered. The calculated vertical transition energies are in good agreement with the observed UV-visible spectrum.

1. INTRODUCTION

The group IB metal atoms represent border-line cases of transition-metal atoms. Their ground state atomic configuration exhibits a d^{10} closed shell, and this is only for problems involving atomic excited states (with a d^9 configuration) that they actually act as transition metals.

Therefore spectroscopic studies of copper containing complexes will be faced with typical difficulties inherent in accurate ab initio calculations of potential energy curves for transition metal containing molecules, i.e. the change in the d shell occupancy from one state to another, and the vicinity of the first low-lying excited states. The copper atom is a typical example with four states $d^{10}s^1(^2S)$, $d^9s^2(^2D)$, $d^{10}p^1(^2P)$, $d^9s^1p^1(^4P)$ lying experimentally within a 5 eV range (1).

It is well known that the effect of electron correlation can be dramatic on such splittings, and so it is of importance to perform very precise CI calculations giving the correct position of the possible avoided crossings and the order of magnitude of the avoidments.

However these large CI description of excited states require a

A. Veillard (ed.), Quantum Chemistry: The Challenge of Transition Metals and Coordination Chemistry, 91–100.

very large number of configurations and the adiabatic treatment does not easily allow a qualitative understanding of the physical phenomena all along the potential curves. In order to analyze the physical content of the wave functions, one may try to construct nearly diabatic potential curves (2). This diabatic description gives a direct insight on the nature of the bonding.

Here, we report on an attempt to solve the problem mentioned above for the low-lying states of the CuH molecule and the Cu-C_2H_4 complex.

2. COMPUTATIONAL DETAILS

The argon-like core electrons of the copper atom are taken into account through a non-empirical relativistic pseudopotential operator (3) whose parameters were taken from Pelissier's study of copper dimers (4). In all cases valence atomic basis sets of Gaussian orbitals (GTO) have been optimized by minimization in an atomic SCF calculation using the same pseudopotential operator. The copper basis set is constructed from three 1s and five 3d primitive Gaussians and two supplementary 2p orbitals are added as polarization functions. The three s GTO are kept uncontracted. For the 3d copper orbital, the contraction leads to a triple-zeta basis set in the case of CuH, and a double-zeta one in the case of Cu(C_2H_4).

The preliminary SCF molecular calculations are made with PSHONDO (5), a modified version of the HONDO program package (6) including pseudopotentials.

3. LOW-LYING Σ^+ STATES OF THE CuH MOLECULE

In order to use the same set of molecular orbitals (MO) for both d^{10} and d^9-type states, we built an appropriate set of MO's from the SCF MO's. Starting from a localized d set obtained by an atomic calculation of the $d^{10}s^1(^2S)$ state the d set originating from the $d^9s^2(^2D)$ excited state is orthogonalized to the former one leading to a "d* set"

$$|d^*\rangle = \left[|d^9\rangle - S|d^{10}\rangle\right]/(1-S^2)^{1/2}$$

$$\text{where} \quad S = \langle d^9|d^{10}\rangle$$

The other occupied and virtual MO's of CuH are then orthogonalized by a projection technique to the "d+d* set". Formally a complete CI in the "d+d* set" would be compulsory to describe the d^{10} and d^9 configurations at the same level of accuracy. However due to the large overlap between d^9 and d^{10} orbitals, CI including only single and selected double and triple excitations in that space lead to satisfactory results (7).

The CI step is carried out using a three-class version (8) of the CIPSI algorithm (9). This is a multireference variation + perturbation theory method similar to the improved MRD-CI scheme of Buenker and Peyerimhoff (10). The first class S of determinants includes the leading configurations for the desired excited states, i.e. the two singlet and two triplet Σ^+ states generated by the Cu $d^{10}s^1(^2S)$+H(^2S) and Cu $d^9s^2(^2D)$+H(^2S). As the ionic forms Cu+ $d^{10}(^1S)$+H$^-(^1S)$ and

Cu^+ $d^9s^1(^{1,3}D)+H^-(^1S)$ are expected to play a role in the nature of the bonding, the corresponding configurations are added to the subspace S. The determination of S is achieved according to an iterative process. All determinants whose coefficients in the first order perturbed wave function are greater than a threshold of 0.07 are included in this reference space, which corresponds to a dimension of 187 determinants. An intermediate class M of medium contributions is selected (threshold 0.01), which corresponds to a dimension of 1500 determinants. The diagonalization is achieved within S+M

$$P_{S+M} = \sum_{K \in S \cup M} |K\rangle\langle K|$$

$$\text{and} \quad P_{S+M}\, H\, P_{S+M}\, |\psi_m^{S+M}\rangle = E_m^{S+M}\, |\psi_m^{S+M}\rangle$$

The remainder determinants belong to a third class Q. Only the first class S is active to generate the interacting configurations of Q. This latter is treated through a Rayleigh-Schrödinger perturbation theory up to the second order

$$\varepsilon_m^{(2)} = \sum_{I \notin S \cup M} \frac{|\langle m|H|I\rangle|^2}{E_m^{(0)} - E_I^{(0)}}$$

$$\text{where} \quad E_m^{(0)} = \langle\psi_m^{(0)}|H|\psi_m^{(0)}\rangle \quad \text{and} \quad E_I^{(0)} = \langle I|H|I\rangle$$

Finally, the last perturbative energy contribution (which is always overestimated) is corrected in order to obtain improved estimates closer to the exact solutions. This correcting factor is taken as

$$\lambda_m = \frac{E_m^{S+M} - E_m^{S}}{\varepsilon_m^{(2)M}}$$

We present in Fig.1 the resulting Σ^+ adiabatic potential energy curves. In addition of the four states dissociating into the $d^{10}s^1$ and d^9s^2 copper configurations, we have studied one state more because it appears to be strongly interacting with the others. Note that these five potential curves are determined for a copper-hydrogen distance ranging from 2.25 to 10.0 a.u. by using only one set of MO's obtained as mentioned above. We consider that asymptotic limits are reached for a copper-hydrogen distance of 10 a.u. The calculated energy gap between the $d^{10}s^1$ and d^9s^2 copper configurations (1.65 eV) is in good agreement with the experimental value (1.49 eV).

The spectroscopic constants for the lowest five Σ^+ states of CuH are summarized in Table I. They present an overall good agreement with the experimental data when available. However our calculations give slightly too long equilibrium interatomic distances and too shallow wells. Note that the (1) $^3\Sigma^+$ state is not observed experimentally, probably because of its very flat shape.

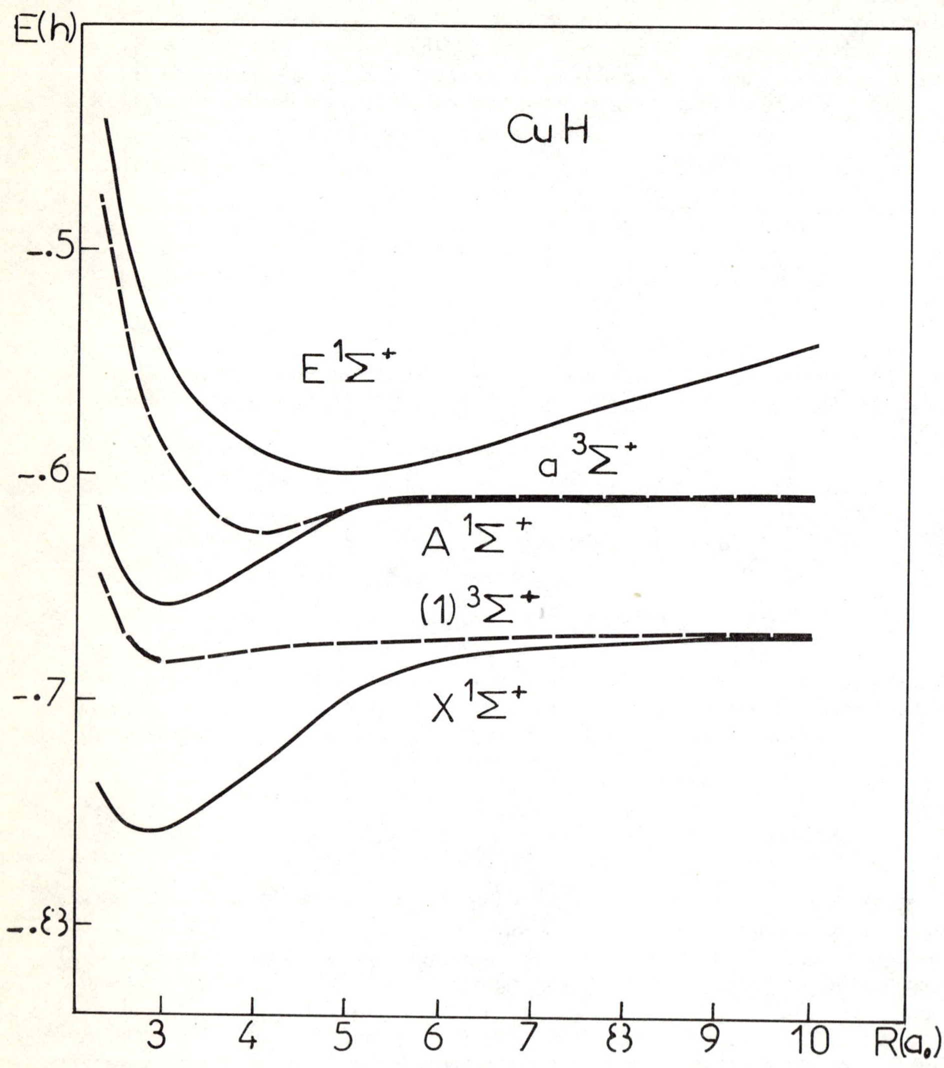

Figure 1. CI potential energy curves for the low-lying Σ^+ states of CuH. The energies are relative to -50.0 a.u.

TABLE I. Spectroscopic constants for the lowest Σ^+ states of CuH. The experimental data [a] are in parenthesis.

state	r_e (A)	D_e (eV)	ω_e (cm^{-1})	B_e (cm^{-1})	Te (cm^{-1})
X $^1\Sigma^+$	1.49 (1.46)	2.36 (2.86)	1232 (1941)	7.6 (7.9)	0 (0)
A $^1\Sigma^+$	1.61 (1.57)	1.33 (1.44)	1506 (1698)	6.5 (6.9)	21588 (23434)
E $^1\Sigma^+$	2.62 (2.34)	- -	738 (574)	2.5 (3.1)	34396 (39299)
(1) $^3\Sigma^+$	1.66	0.37	1341	6.2	16064
a $^3\Sigma^+$	2.18	0.51	1626	3.6 (5.9) [b]	28256 (26365) [b]

[a] Ref. 11.
[b] These experimental values are very uncertain.

In order to more precisely visualize the physical content of the wave functions all through the avoided crossing region, we construct diabatic states by means of a criterion proposed by Cimiraglia et al. (12). An a priori guessed set of wave functions ϕ_I^{Ref} is defined as reference configurations, and the diabatization criterion will be to maximize the sum of the overlaps between the transformed diabatic states ψ_I^D and the corresponding reference configurations

$$\sum_I |\langle\psi_I^D|\phi_I^{Ref}\rangle| \quad \text{maximum}$$

Our ϕ_I^{Ref}'s consist in non-orthogonal VB functions describing the following ionic configurations

$$Cu^+\ d^{10}(^1S) + H^-(^1S)$$

$$Cu^+\ d^9s^1(^{1,3}D) + H^-(^1S)$$

Fig. 2 displays the resulting diabatic curves. They show up the complex behaviour of the potential curves in the intermediate region, in particular the mixing between the neutral and ionic contents of the three $^1\Sigma^+$ wave functions. Moreover the X $^1\Sigma^+$ ground state exhibits a strongly ionic character at short distances.

Unfortunately, this diabatic description is not unique ; it depends on the choice of the atomic reference states. Therefore this diabatic interpretation has to be handled cautiously.

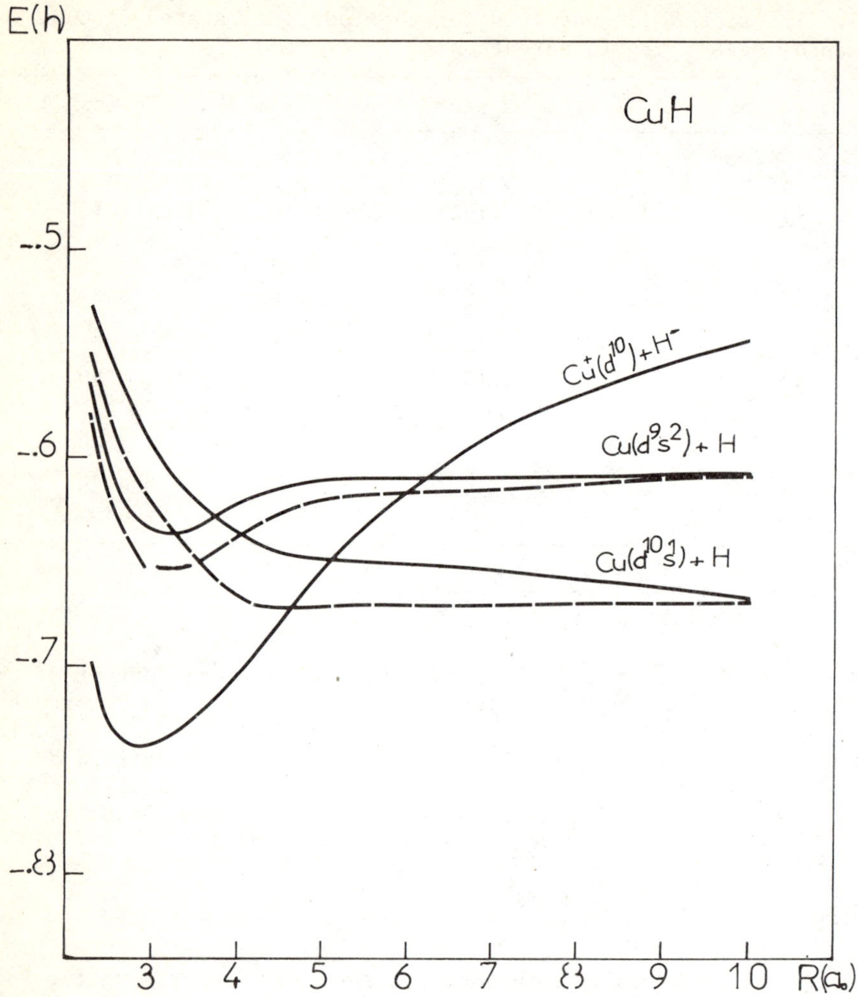

Figure 2. Diabatic potential energy curves for Σ^+ states of CuH. The energies are relative to -50.0 a.u.

4. OPTICAL SPECTRUM OF THE Cu-C_2H_4 COMPLEX

The problem of the theoretical interpretation of the UV-visible spectrum of the copper-ethylene system appears to be very different in nature. In the perpendicular conformation (C_{2v} symmetry) of the Cu-C_2H_4 complex nine states correlate to the $C_2H_4(^1Ag)$ ground state and the Cu(^{2}S), (^{2}D), and (^{2}P) atomic states. The $d^{10}s^1$ configuration generates a $^2A_1$ state while the $d^{10}p^1$ configuration lying at 3.80 eV above the ^{2}S gives rise to three states of A_1, B_1 and B_2 symmetry. Two states of symmetry A_1 and one state of each symmetry A_2, B_1 and B_2 originates from the $^2D(d^9s^2)$ atomic state.

The d^9s^2 copper configuration involves a rather diffuse s^2 closed shell leading to a repulsive interaction with the ethylene electronic distribution. The five states coming from this configuration should be

i) strongly repulsive

ii) nearly degenerate since the overlap between the d holes and the π MO of ethylene is weak and introduces only a slight splitting.

On the contrary, no repulsion occurs in the $^2B_2(p_y^1)$ or $^2B_1(p_x^1)$ states which may therefore induce stronger bonding, involving copper $\leftrightarrow$ ethylene electronic delocalization and/or charge transfer. The $^2A_1(p_z^1)$ state should be higher than the two last ones due to a repulsion between the p_z and π electrons. The three states coming from the $d^{10}p^1$ copper configuration are spectroscopically active while those coming from d^9s^2 are rather unimportant since the $d^{10}s^1 \rightarrow d^9s^2$ transition is dipolarly forbidden. However avoided crossings are expected between these repulsive states and the states dissociating into the $d^{10}p^1$ configuration. Indeed we need information about the position and the magnitude of these avoided crossings, at least for a symmetry.

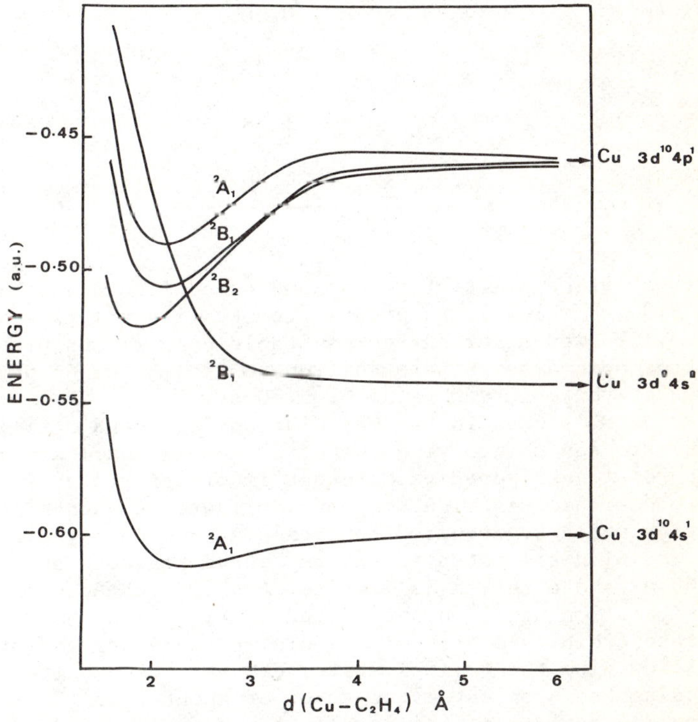

Figure 3. CI potential energy curves for the low-lying states of the $Cu\text{-}C_2H_4$ complex. The energies are relative to -63.0 a.u. The weakly avoided crossing between the $^2B_1(d^9s^2)$ and $^2B_1(d^{10}p_x^1)$ is not pictured.

Fig. 3 gives the potential energy curves obtained by CI calculations. Due to the repulsion between the singly occupied 4s copper orbital and the π MO of ethylene, the $^2A_1(s^1)$ state appears as a weakly bonded Van der Waals complex. The three states $^2B_2(p_y^1)$, $^2B_1(p_x^1)$ and $^2A_1(p_z^1)$ are bound and lie at about 2 eV above the $^2A_1(s^1)$ state which appears to be the ground state of the copper-ethylene complex. Binding energies and equilibrium copper-ethylene distances for these four states are listed in Table II. The states related to the $d^{10}p^1$ copper configuration present deeper wells than the one of the ground state and allow shorter copper-ethylene distances.

TABLE II. Copper-ethylene equilibrium distances and binding energies for the low-lying states of the $Cu-C_2H_4$ complex.

symmetry	$d_{Cu-C_2H_4}$ (Å)	Binding energy (eV)
$^2A_1(p_z^1)$	2.17	0.90
$^2B_1(p_x^1)$	2.20	1.25
$^2B_2(p_y^1)$	1.93	1.58
$^2A_1(s^1)$	2.37	0.36

The dissociation limit of the 4s and 4p configurations is calculated to be 3.95 eV which is in good agreement with the experimental value (3.80 eV). However for further suitable comparisons with experimental spectroscopic data we have shifted in Fig. 3 the 4p potential energy curves at the observed value of 3.80 eV.

The set of MO's used in our calculations has been optimized with a Nesbet (13) procedure involving the $d^{10}s^1$ copper configuration. Going from $d^{10}s^1$ to d^9s^2 configurations induces important reorganization of the d and s electronic distributions and delicate balance of the correlation energies would be required to obtain a good energy difference. The calculated $^2B_1(d^9s^2)$ potential energy curve presents an asymptotic limit at 0.66 eV above the $^2A_1(s^1)$ state. A shift to the correct asymptotic separation has been introduced in Fig. 3. As expected the $^2B_1(d^9s^2)$ state exhibits a strongly repulsive character and presents an avoided crossing with the $^2B_1(p_x^1)$ state. However the magnitude of this avoided crossing has been estimated to be of about 0.05 eV. It is so weak that one may neglect it in a preliminary analysis and consider only the "diabatic" curves as a quite reasonable approximation.

According to the Frank-Condon principle, three vertical transitions can be calculated from the ground state to the three attractive potential energy curves reported in Fig. 3. Even if the crossings with the copper (d^9s^2) - ethylene repulsive states occur at the copper - ethylene equilibrium distance of the $^2A_1$ ground state, the absorption spectrum should not be significantly affected due to the weakness of the avoidment.

Our calculated vertical transition energies are compared with the experimental spectrum observed by Ozin et al. (14) in Table III. These authors have reported the UV-visible spectra of the products of the low-temperature cocondensation reaction of Cu atoms with C_2H_4 in an argon matrix. They have identified two intense absorption bands at 2.95 eV and 3.25 eV attributed to the $Cu(C_2H_4)$ species. Our results lead to reasonably assigned these bands to excitations of the unpaired electron mainly localized on copper from the 4s toward the 4p orbitals.

TABLE III. UV-visible spectra of the $Cu-C_2H_4$ complex (in eV).

X_α Calculations a)		CI Calculations (present work)		Experiment b)
$5a_1(d_{z^2}) \rightarrow 6a_1(s)$	3.21	$^2A_1(s) \rightarrow {}^2A_1(p_z)$	3.39	3.25
$6a_1(s) \rightarrow 3b_2(\pi^*)$	1.80	$^2A_1(s) \rightarrow {}^2B_1(p_x)$	2.97	2.95
	-	$^2A_1(s) \rightarrow {}^2B_2(p_y)$	2.84	-

a Ref. 15.
b Ref. 14.

Note that the X_α calculations of McIntosh et al. (15) lead to a too weak one-electron transition energy for the visible band, and do not allow to correctly assign the optical transitions.

5. CONCLUSION

The overall satisfactory results obtained for the two examples presented in this work are very incentive. They show up that a good accuracy can be achieved on potential energy curves involving excited states of transition metal complexes.

6. REFERENCES

1. C.E. Moore, *Atomic Energy Levels*, NSRDS-NBS Circular 35, Vol. II, Washington, D.C. (1971)
2. F. Spiegelmann and J.P. Malrieu, *J. Phys. B : At. Mol. Phys.* 17, 1259 (1984)
3. Ph. Durand and J.C. Barthelat, *Theoret. Chim. Acta* 38, 283 (1975)

J.C. Barthelat and Ph. Durand, *Gazz. Chim. Ital.* **108**, 225 (1978)
4. M. Pélissier, *J. Chem. Phys.* **75**, 775 (1981)
5. J.P. Daudey, private communication
6. M. Dupuis, J. Rys and H.F. King, *J. Chem. Phys.* **65**, 111 (1976)
7. M. Pélissier and E.R. Davidson, *Int. J. Quant. Chem.* **25**, 483 (1984)
8. S. Evangelisti, J.P. Daudey and J.P. Malrieu, *Chem. Phys.* **75**, 91 (1983)
9. B. Huron, P. Rancurel and J.P. Malrieu, *J. Chem. Phys.* **58**, 5745 (1973)
10. R.J. Buenker and S.D. Peyerimhoff, *Theoret. Chim. Acta* **35**, 33 (1974)
R.J. Buenker, S.D. Peyerimhoff and W. Butscher, *Mol. Phys.* **35**, 771 (1978)
11. K.P. Huber and G. Herzberg, *Molecular Spectra and Molecular Structure*, Vol. **4**, : *Constants of Diatomic Molecules*, New-York : Van Nostrand (1979)
12. R. Cimiraglia, J.P. Malrieu, M. Persico and F. Spiegelmann, *J. Phys. B : At. Mol. Phys.* **18**, 3073 (1985)
13. R.K. Nesbet, *Rev. Mod. Phys.* **35**, 552 (1963)
14. G.A. Ozin, H. Huber and D. McIntosh, *Inorg. Chem.* **16**, 3070 (1977)
15. D.F. McIntosh, G.A. Ozin and R.P. Messmer, *Inorg. Chem.* **19**, 3321 (1980)

SCANDIUM ATOM INTERACTING WITH DIATOMIC GROUPS

G. H. Jeung
Physics Department
Dan-Kook University
Cheon-An, Choong-Nam-Do
330-00 South Korea

ABSTRACT. With a small number (three) of valence electrons, the Scandium atom shows some typical aspects of the transition-metal chemistry. In this work the chemical bonding of Sc with three representative diatomic groups is studied theoretically. The interaction with CO leads to repulsive, moderately attractive and strongly attractive bondings according to the d^1s^2, d^2s^1 and d^3 atomic configurations of Sc respectively. The single radical electron of CN interacts with Sc very similarly as in ScH. The ground state of Sc-CH coming from the two quasi-degenerate states of CH is analyzed. A possible dissociation of H_2 is also shown by calculating the energy balance in a Sc+H_2 approach. A non-empirical HF pseudopotential and a multi-reference single and double CI is used for this purpose. The concepts of the atomic configuration and of the Mulliken population are used upon the natural molecular orbitals to understand the chemical bonding between the Sc atom and the diatomics.

1. INTRODUCTION

The problem of molecular adsorption on the metal surfaces has been a subject of great interest to understand the mechanism of the catalytic reactions. Both the experimental and the theoretical efforts in this field are ever increasing. The latter method generally comprised some simple model systems like (2-dimensional metal layers)+admolecules or (metal clusters)+admolecules. Although it may seem to be farther from the reality, even the (single metal atoms)+admolecules have also been considered. These seemingly very simple systems involved either the alkali atoms or the transition-metal atoms empirically known as good catalysts. In particular the adsorption of the CO group has attracted most attention recently in relation to the Dewar-Chatt-Duncanson theory[1]. It involved the Fe, Ni, Cu, Pd, Rh and Pt atoms or clusters[2~6].

Due to the large difference of atomic characteristics (the size of valence electron distribution, the valence electron orbital energies, the ionization energies, etc.) existing among the transition-metal elements, a given diatomic group interacts differently with different metal species. In this work, for the first time the (Sc atom)+(CO, CN, CH and H_2) systems

A. Veillard (ed.), Quantum Chemistry: The Challenge of Transition Metals and Coordination Chemistry, 101–117.

are treated in a non-empirical manner, to see what electron reorganizations occur within the Sc atom and within the diatomic groups, and to see what binding may be possible between the metal atom and the diatomic groups. Each of these diatomic groups has a particularity of its own. The CO has a closed-shell structure with a relatively high binding energy, the CN has an unpaired radical electron of high reactivity, and the CH has two nearly degenerate states as the lowest states. The H_2 molecule has been approached to see whether a scission of the H-H bond is energetically favorable. The smallest possible number of valence electrons of Sc atom among the transition-metal atoms and its different atomic configurations (occupancies of 3d, 4s and 4p atomic orbitals) which lie quite closely over the ground state make this atom particularly interesting in theoretical point of view in spite of the fact that the chemistry of Sc is not well known in this respect.

To calculate the energetic relative stabilities, a valence electron SCF using a non-empirical HF pseudopotential is followed by a multi-reference single and double CI. To understand the chemical bonding between the Sc atom and the diatomic groups the atomic configurations in molecule and the nature of molecular orbitals(MO) are analyzed using the concept of the Mulliken population on the natural molecular orbitals.

2. CALCULATIONAL METHOD

For the valence electron calculation, a non-empirical HF pseudopotential of Barthelat and Durand[7] is used. The pseudopotential parameters for the atoms concerned in this work are reported before together with the valence basis sets[8~9]. For Sc, 3s2p6d Gaussian type functions contracted to 2s1p2d atomic basis functions are used. The valence pseudopotential calculation using this basis gave good relative atomic energies for the states which may play important roles in triatomics[9]. The valence basis for N, C and O is 4s4p Gaussian type functions contracted to 2s2p atomic basis functions. For H, 6s2p Gaussian type functions are contracted to 3s2p atomic basis functions.

The multi-reference single and double CI is perfomed using the MRD-CI program[10]. In this program a set of most important Slater determinants are chosen iteratively after one or a few diagonalizations in the (sub)-space of the zeroth-order, singly- and doubly-substituted determinants usually beginning from the SCF determinant. The final diagonalization is done either directly when the matrix dimension is relatively small, or more frequently by an extrapolation technique when the dimension is too big to find the full MRD-CI energy.

This type of calculation cannot be too trusted quantitatively with high precision in particular where small energy differences are involved and for long internuclear distances ($\gtrsim$5.5 bohr) where the intra-atomic and intra-molecular correlation effects become more important than the intermolecular correlation. But the essential chemical feature (repulsive or attractive, strong or weak, neutral or ionic, etc.) can surely be deduced from this calculation. The analysis of the electron distribution has been done using the natural orbitals(NO) obtained by diagonalizing the bond order matrix based on the CI wavefunction.

3. Sc + CO

3.1. Calculational detail

The single SCF determinant is employed to obtain the single and double substituted determinants and the energy lowering due to the higher-than-double (mainly quadruple) substituted determinants is estimated through an approximated formula used by Langhoff and Davidson[11]. At first the isolated CO molecule has been calculated to test the pseudopotential and the basis employed for the intermolecular problem. The resulting equilibrium characteristics (R_e=2.20 bohr, ω_e=2050 cm^{-1}, μ_e=0.02 Debye) are in good agreement with the experimental data (respectively 2.132 bohr[12], 2169.81 cm^{-1} {12}, 0.112 Debye[13]).

In Sc-CO, the six lowest doublet and quadruplet states are chosen in SCF calculation: $^2\Sigma^+$, $^2\Pi$, $^2\Delta$, $^4\Sigma^-$, $^4\Pi$ and $^4\Delta$ states. The single SCF determinant has the weight superior to 88% in every states in the internuclear distance interval of 3.5-5 bohr, and no other single determinant has the weight greater than 1.5%. Analogously the Sc-OC (O facing Sc) approach has been calculated for the lowest $^2\Sigma^+$, $^2\Pi$, $^2\Delta$, $^4\Sigma^-$, $^4\Pi$ and $^4\Delta$ states.

For further reference the NO's of CO at 2.2 bohr are expanded in atomic orbitals(AO) in terms of the occupation number and the sign of orbital coefficient in TABLE I.

TABLE I Natural orbitals of CO($X^1\Sigma^+$) developed in atomic orbitals and their occupation numbers at 2.2 bohr. The relative signs are taken from the coefficients. In atomic units. Only the occupation numbers superior to 3.5% are included.

Orbital	C			O			Total
	s	p_σ	p_π	s	p_σ	p_π	
1σ	0.22			1.74			1.99
2σ	-0.24	-0.36			1.38		1.98
3σ	-1.45	0.51		0.05			1.97
1π			1.03			2.89	3.92

3.2. Spectroscopic property

The adiabatic potential energies of Sc-CO keeping the CO distance to 2.2 bohr are reported in FIGURE 1. The three doublet states $^2\Sigma^+$, $^2\Pi$ and $^2\Delta$ correlating to the ground state approach Sc(^2D, $3d^14s^2$)+CO($X^2\Sigma^+$) are all repulsive, the $^2\Pi$ being the least repulsive and the $^2\Sigma^+$ the most repulsive. Two states $^4\Sigma^-$ and $^4\Pi$ correlating to the ground state CO and the excited Sc(^4F, $3d^24s^1$) are rather strongly attractive the $^4\Sigma^-$ being lower of the two. The $^4\Delta$ state has a well situated much higher than the other quadruplet states. The wavefunction of $^4\Delta$ indicates that this state can be adiabatically correlated to the Sc $3d^3$(^4F).

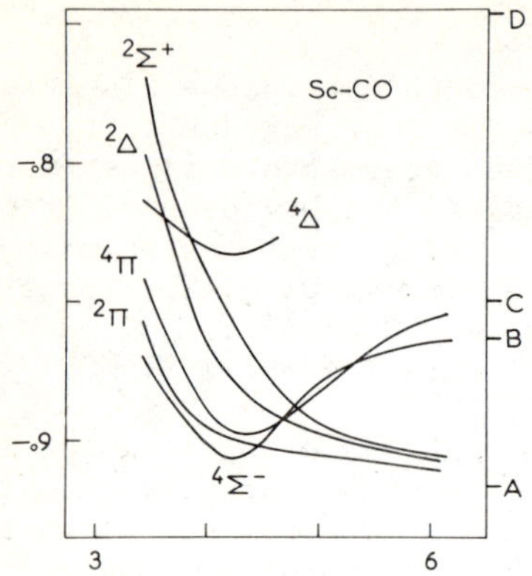

FIGURE 1 A section of the potential energy surfaces (collinear) of Sc-CO for C-O distance 2.2 bohr. A(Sc $4s^23d^1$, ^2D), B(Sc $4s^1 3d^2$, ^4F), C(Sc $4s^14p^13d^1$, $^4F^0$) and D(Sc $3d^3$, ^4F) are dissociation limits with CO($X^1\Sigma^+$). In atomic units.

The estimation of dissociation energies for the bound states is not easy due to the size inconsistency of the SD-CI. The comparison of the equilibrium point energies with that of the longest calculated internuclear distance (6bohr) gives the lower bound to the dissociation energies of 1.14 eV for $^4\Sigma^-$ state and 1.13 eV for $^4\Pi$ state. The geometry optimization relaxing the CO distance and a larger CI may situate the $^4\Sigma^-$ and $^4\Pi$ state wells near the ground state dissociation assymptote.

3.3. Electron distribution in Sc-CO

To understand the electron distribution of Sc-CO, the NO's are developed in terms of the occupation number and the sign of the expansion coefficients in TABLE II. In all attractive states $^4\Sigma^-$, $^4\Pi$ and $^4\Delta$, the 2π(Sc-CO) orbital is responsible for bonding between Sc and C atoms. The d_π AO of Sc is used for this purpose where a slight polarization by using p_π AO occurs. This orbital is anti-bonding for C and O atoms. For repulsive doublet states the 1σ, 2σ, 3σ and 1π orbitals are only slightly changed from the corresponding orbitals of the isolated CO molecule. For attractive quadruplet states the 2σ(CO) and 3σ(CO) are significantly rearranged by mixing with each other to lower the repulsion between the 3σ(CO) lone pair electrons and the approaching Sc 4s electron. The Sc 4s electron is polarized away to the backside to form the 4σ(ScCO) in $^4\Sigma^-$ and $^4\Pi$ states or is put in a d_δ orbital in $^4\Delta$. The valence electron densities for NO's of ScCO are drawn in FIGURE 2.

TABLE II Natural orbitals of ScCO developed in atomic orbitals in terms of the occupation numbers. The relative signs are taken from the orbital expansion coefficients. The C-C distance is 2.2 bohr and the Sc-C distance is 4 bohr. The one δ electron in $^4\Pi$ and $^4\Delta$ states are not included. For the dipole moment the Sc→C→O direction is taken positive. In a.u.

State	Orbital	Sc					C			O			Total	Dipole moment
		s	p_σ	p_π	d_σ	d_π	s	p_σ	p_π	s	p_σ	p_π		
$^4\Sigma^-$	1σ						0.22	0.16		1.57	-0.04		1.99	
	2σ				-0.05		-1.18	0.12		0.16	0.45		1.99	
	3σ	0.04			0.06		0.24	-0.69		0.08	0.84		1.98	
	4σ	0.82	-0.15										1.00	
	1π								1.06			2.86	3.93	
	2π			0.10		1.35			0.36			-0.16	1.97	
	total	0.90	0.19	0.10	0.13	1.38	1.65	0.96	1.46	1.84	1.33	3.06		-0.80
$^4\Pi$	1σ						0.23	0.18		1.52	-0.05		1.99	
	2σ	-0.05			-0.07		-1.23	0.18		0.15	0.29		1.99	
	3σ				0.04		0.14	-0.60		0.15	0.99		1.99	
	4σ	-0.81	0.15										1.00	
	1π								1.08			2.84	3.93	
	2π			0.10		0.29			0.44			-0.16	1.00	
	total	0.90	0.19	0.11	0.15	0.30	1.62	0.97	1.56	1.82	1.34	3.04		-1.02
$^4\Delta$	1σ						0.20	0.10		1.68			1.99	
	2σ	-0.05			-0.05		-1.24	0.08		0.10	0.45		1.99	
	3σ	-0.06			-0.05		-0.20	0.74			-0.89		1.98	
	1π								1.05			2.91	3.95	
	2π			0.18		1.08			0.50			-0.21	1.98	
	total	0.11	0.04	0.18	0.16	1.09	1.65	0.93	1.57	1.81	1.35	3.12		-2.95

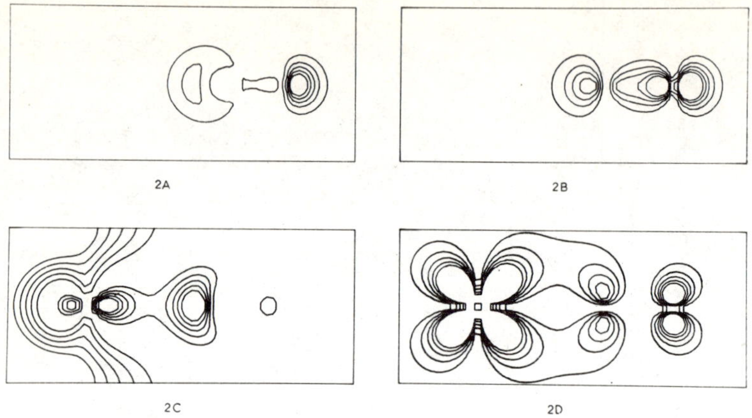

FIGURE 2 Electron density contour maps for the natural orbitals of Sc-CO in $^4\Sigma^-$ state. The C-O distance is 2.2 bohr and the Sc-C distance is 4 bohr. A(2σ), B(3σ), C(4σ), D(2π).

The population analysis is consistent with the dipole moment indicating an electron migration from Sc to CO for the $^4\Sigma^-$ and $^4\Pi$ states. The $^4\Delta$ state can be characterized by a larger charge transfer resulting in a large dipole moment. This facility recalling the $^2\Delta$ state of ScC or ScO may be explained as in such diatomic cases in terms of the lower ionization energy of Sc $3d^3$ [9,14]. Using the experimental values the ionization energy of Sc $3d^3 \rightarrow Sc^+\ 3d^2$ ($^4F \rightarrow {}^3F$) is 3.02 eV which is smaller than other possible ways: $3d^1 4s^1 4p^1 \rightarrow 3d^1 4s^1$ ($^4F^o \rightarrow {}^3D$) of 4.60 eV, $3d^2 4s^1 \rightarrow 3d^1 4s^1$ ($^4F \rightarrow {}^3D$) of 5.13 eV, or $3d^1 4s^2 \rightarrow 3d^1 4s^1$ ($^2D \rightarrow {}^3D$) of 6.56 eV.

Another argument to explain the difference of the potential energy curves coming from different s- and d-AO occupation may be found in recent studies[2~6]. There the repulsive character of the s(metal atom)-AO with 3σ(CO)-MO has been discussed to explain the repulsiveness of the states where s-AO is occupied. Inspite of the fact that the 4s-AO of Sc is very diffuse leading to a weak resistance to penetration, the outward polarization of 4s(Sc) to 4σ(ScCO) and the outward rearrangement of the 3σ(CO) to 3σ(ScCO) may be translated as a manifestation of this repulsive character.

The changement of the 2σ and 3σ in isolated CO to the 2σ and 3σ in ScCO involves the use of a small portion of Sc AO. This can be interpreted as a σ donation in donation and back-donation hypothesis[1]: 0.22, 0.24, and 0.31 electron for respectively $^4\Sigma^-$, $^4\Pi$ and $^4\Delta$ states. The radical unpaired electron (4σ in $^4\Sigma^-$ and $^4\Pi$, 1δ in $^4\Delta$) uses only the Sc AO's thus meaning no σ-backdonation. The 2π orbital has a CO component, i.e. a backdonation of 0.52, 0.59 and 0.73 electron for respectively $^4\Sigma^-$, $^4\Pi$ and $^4\Delta$ states.

The SCF picture of the $^4\Sigma^-$ state is schematized in FIGURE 3. Here the weakly polarized 1σ and 1π orbitals appear to be slightly destabilized. The Sc π electrons are stabilized through a bonding with CO and the 4s electron is stabilized by the polarization. A net electron transfer from Sc to C and O occurs in ScCO. In the $^4\Sigma^-$ state the C and O gains respectively 0.07 and 0.23 electron, and in the $^4\Pi$ state they gain 0.15 and 0.20 electron respectively. In the $^4\Delta$ state the C atom gains 0.15 electron and the O atom gains 0.28 electron.

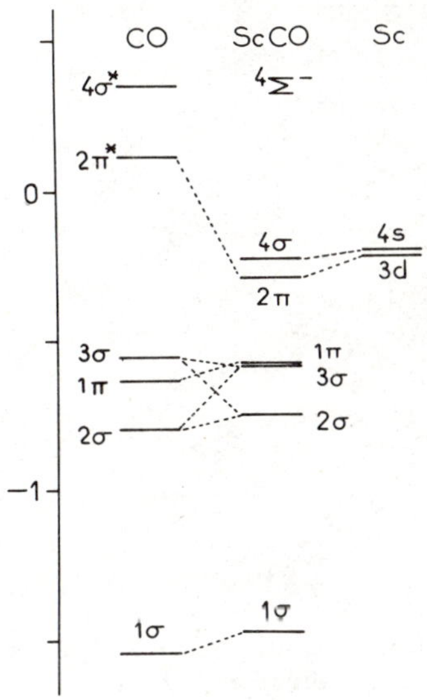

FIGURE 3 Correlation diagram for the SCF one-electron energies. In atomic units.

?.4. Sc-OC

The collinear approach of CO group to Sc atom on the side of O has been calculated for comparison to the approach C facing to Sc. The resulting potential energy curves of the lowest $^2\Sigma^+$, $^2\Pi$, $^2\Delta$, $^4\Sigma^-$, $^4\Pi$ and $^4\Delta$ states are drawn in FIGURE 4. The general feature of this approach is qualitatively the same as in Sc-CO case. The attractive states $^4\Sigma^-$ and $^4\Pi$ are quantitatively less stable than the corresponding states in Sc-CO. The $^4\Delta$ state has the dissociation energy of 1 eV when the quadruple correction energies are compared.

The bonding 2π orbital shows a backdonation of about 0.21 electron for $^4\Sigma^-$ and 0.85 electron in $^4\Delta$. The σ donation is very small: 0.06 in $^4\Sigma^-$ and 0.11 in $^4\Delta$. This result shows the important role of 3σ(CO) lone pair electrons for σ donation in the Sc-CO approach. The orbitals essentially belonging to the CO group (1σ, 2σ, 3σ and 1π) keeps nearly the same composition as the corresponding ones in isolated CO and they are slightly stabilized by the approaching Sc atom to the O side. This may be explained by the topologically stable electron distribution of CO 3σ orbital which doesn't interfere with the approaching 4s(Sc) electron. The energies of the 2π and 4σ orbitals are nearly the same as the 3d energy respectively.

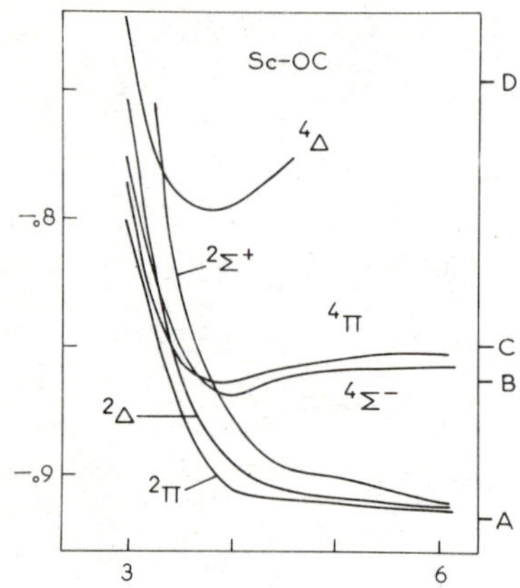

FIGURE 4 A section of the potential energy surfaces of Sc-OC (collinear) for C-O distance 2.2 bohr. For A, B, C and D see the caption for Fig. 2. In atomic units.

4. Sc-CN

4.1. Calculational detail

A test calculation has been done for the isolated CN group at first to see whether the pseudopotential and the basis are appropriate for its description and to use as a reference. The lowest two states $X^2\Sigma^+$ and $A^2\Pi_i$ are studied in full MRD-CI using 4 and 3 reference configurations respectively. The resulting equilibrium constants are R_e=2.29 bohr (the experimental value[12] is 2.214), ω_e=1955 cm^{-1}(2069.59) and μ_e=0.591(C^+N^-)au for the ground state. The corresponding values for the $A^2\Pi_i$ state is 2.42 bohr(2.331), 1666 cm^{-1}(1812.56) and 0.025 au(C^+N^-). The calculated well-to-well transition energy $X^2\Sigma^+ \rightarrow A^2\Pi_i$ is 10.4×10^3 cm^{-1} versus the experimental 9245.28 cm^{-1}. This set of values seems to be satisfactory for further study in Sc+CN.

To find the interaction of the CN group with a Sc atom, only a collinear approach C facing Sc has been studied here. The internuclear distance of C-N is fixed to 2.29 bohr which is the calculated ground state bond distance. The lowest singlet and triplet states of Σ^+, Π and Δ symmetry have been found in SCF calculation. The double CI has been performed using the SCF solution MO's for each spin-spatial symmetry states except in the $^3\Delta$ state where the MO's of the $^1\Delta$ state is used. The doubly excited configurations which give the second-order perturbation correction of more than 10 μ hartree have been included for diagonalization of CI matrix. The largest dimension of the matrices are in the range of from 2500 to 3000.

4.2. Spectroscopic properties

The adiabatic potential energies of the $^1\Sigma^+$, $^3\Sigma^+$, $^1\Pi$, $^3\Pi$, $^1\Delta$ and $^3\Delta$ states are recorded as functions of the Sc-C distances in FIGURE 5. All these states appear to be close-lying to each other, the $^3\Delta$ state being the lowest inspite of the use of the SCF MO's of $^1\Delta$ symmetry. The comparison of the ScCN molecular energies and the sum of the separate energies of Sc and CN shows that the lowest state $^3\Delta$ is stable by at least 3.4 eV. The real dissociation energy may be larger than this number considering various factore unfavoring the molecular energies calculated in the present way. No experimental evidence is available for this molecule according to the author's knowledge. Looking at the Fig. 5, one can find a close resemblance of this set of potential curves with that of ScH except for the $^1\Sigma^+$ state. In fact, as will be discussed in the next section, the bonding between Sc-CN and that of Sc-H are very similar.

4.3. Electron distribution of ScCN

The NO's of the ScCN for the Sc-C distance of 4.5 bohr and the C-N distance of 2.29 bohr are summarized in TABLE III in terms of the AO occupation numbers and the signs of coefficient. For the comparison the NO's of free CN($X^2\Sigma^+$) at 2.29 bohr are also tabulated in TABLE IV. In ScCN the 1σ and 1π orbitals which have negligible components in Sc AO's represent the strong C-N bonding. These orbitals are slightly polarized counterparts of the bonding orbitals (1σ and 1π) in the isolated CN molecule.

The unpaired radical electron of CN(2σ) is used for the bonding with Sc to form the bonding 3σ(ScCN) orbital. In this orbital about 0.5-0.6 electron remains in the Sc site and about 1.5-1.4 electron remains in the C site. Remembering that the unpaired orbital of CN(2σ) uses the C AO's with 0.82 occupation and the N AO's with 0.21 occupation in free CN, it means that about 0.7 electron is transfered from Sc to C.

The antibonding electron of CN(3σ) where about 0.70 electron resides in the C site and about 1.25 electron resides in the N site is redistributed in ScCN(3σ). This deformation effect is an electron transfer of about 0.2 electron from C atom to N atom. So the 3σ orbital of ScCN has more of N lone pair electron character than the antibonding character in free CN. This orbital is still slightly antibonding for C-N and weakly bonding for Sc-C.

The gross result is that the Sc atom gives 0.7 electron to the C atom via a strongly ionic bonding orbital and simultaneously the C atom gives 0.2 electron to the N atom via a conversion of the antibonding electron to

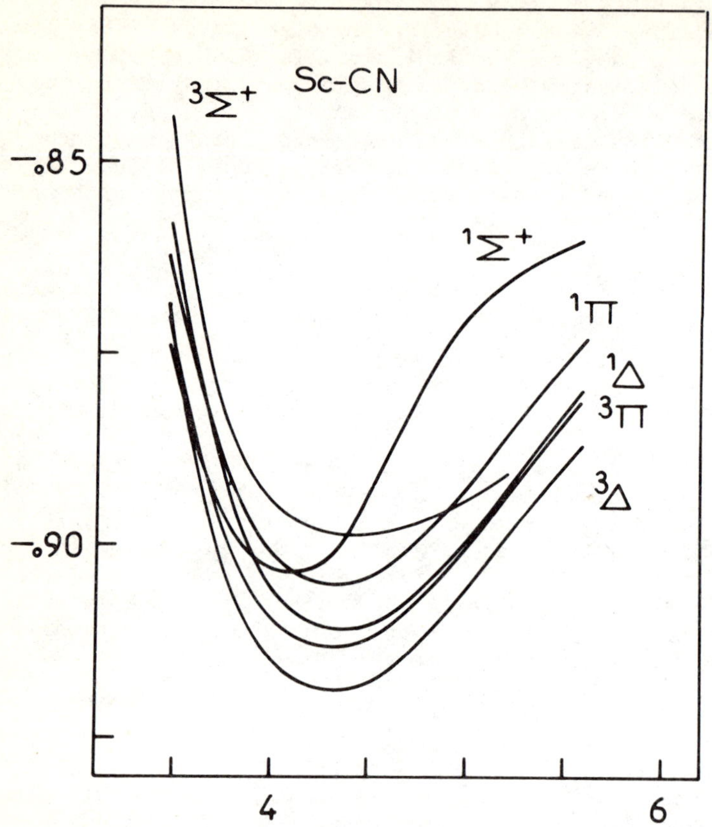

FIGURE 5 A section of the potential energy surfaces (R_{C-N} = 2.29 bohr) of ScCN. In atomic units.

TABLE III Natural molecular orbitals of ScCN developed in atomic orbitals and their occupation numbers. The relative signs are taken from the orbital expansion coefficients. The C-N distance is 2.29 bohr and the Sc-C distance is 4.5 bohr. For the dipole moment the Sc→C→N direction is taken positive. In atomic units.

State	Orbital	Sc					C			N			Total	Dipole moment
		s	p_σ	p_π	d_σ	d_π	s	p_σ	p_π	s	p_σ	p_π		
$^1\Sigma^+$	1σ						0.50	0.06		1.39			1.99	
	2σ						-0.31	-0.15		0.30	1.22		1.98	
	3σ	-0.22			-0.05		-0.61	0.94		0.11	-0.04		1.97	
	4σ	-0.90	0.09		0.88								1.90	
	1π								1.75			2.10	3.90	
	total	1.16	0.15		1.03		1.44	2.99		1.80	3.43			-3.14
$^3\Sigma^+$	1σ						0.65			1.30			1.99	
	2σ	-0.09	-0.05		-0.09		-0.49	1.03		0.19			1.98	
	3σ						-0.32	-0.15		0.29	1.23		1.98	
	4σ	-0.78	0.20										1.00	
	1π								1.76			2.08	3.90	
	total	0.90	0.30		1.05		1.48	3.00		1.79	3.43			-1.78
$^3\Pi$	1σ						0.51			1.44			1.99	
	2σ						0.43	0.14		-0.20	-1.21		1.98	
	3σ	0.10	0.05		0.09		0.54	-0.99		-0.14	0.07		1.98	
	4σ	-0.78	0.19										1.00	
	1π								1.75			2.10	3.89	
	2π					0.99							1.00	
	total	0.90	0.28		1.15		1.49	2.95		1.78	3.44			-2.42
$^3\Delta$	1σ						0.54			1.44			1.99	
	2σ						-0.35	-0.20		0.18	1.25		1.98	
	3σ	-0.09			-0.09		-0.58	0.98		0.18			1.98	
	4σ	0.74	-0.25										1.00	
	1π								1.76			2.09	3.89	
	1δ												1.00	
	total	0.84	0.32		1.13		1.49	2.99		1.80	3.42			-1.99

the lone pair. The total populationanalysis using the Table III is consistent with this picture. The strong bonding of ScCN may thus be explained by the formation of a strongly ionic σ bonding between Sc and C, and the disappearance of the antibonding electrons between C and N.

TABLE IV Natural molecular orbitals of $CN(X^2\Sigma^+)$ developed in atomic orbitals and their occupation numbers at 2.29 bohr. The relative minus and implicit plus signs are taken from the coefficients.

Orbital	C			N			Total
	s	p_σ	p_π	s	p_σ	p_π	
1σ	0.38	0.26		1.27	-0.08		1.99
2σ	0.35	-0.47		0.05	0.16		1.02
3σ	-0.70			0.44	0.81		1.95
1π			1.68			2.14	3.81
Total	1.43	2.52		1.76	3.28		

As the structure of the bonding orbital and the N lone pair orbital is similar for all the states, the slight differences of the potential energies can be explained by the different spatial distributions and the spin exchange possibilities of the nonbonding electrons or the unpaired electrons of Sc. In this respect the interaction between Sc and CN is very similar to the Sc-H interaction (see Ref. 9). Conclusively, the classical formular of the sort:

$$:\mathrm{Sc} \rightarrow \mathrm{C{\equiv}N}: \quad (^1\Sigma^+) \qquad \text{and} \qquad \dot{\cdot}\mathrm{Sc} \rightarrow \mathrm{C{\equiv}N}: \quad (\text{other states})$$

can be regarded as faithful pictures of the valence electron structure. The electron densities of the bonding orbital (3σ) and the non-bonding radical σ electron (4σ) of the $^3\Delta$ state are drawn in FIGURE 6.

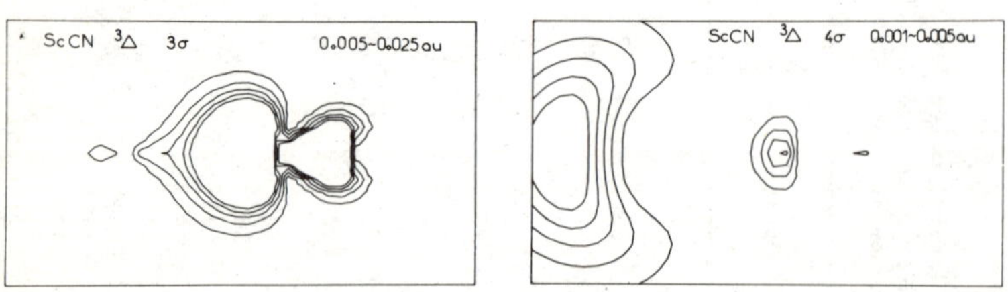

FIGURE 6 Electron density maps for the 3σ and 4σ orbitals of ScCN ($^3\Delta$). R(Sc-C)=4.5 bohr, R(C-N)=2.29 bohr.

5. Sc-CH

The $X^2\Pi_r$ and $^4\Sigma^-$ states experimentally known as the lowest states in isolated CH molecule is calculated at first to find their equilibrium characteristics. The SD-CI energies augmented with the quadruple correction energies give the spectroscopic constants resumed in TABLE V. As the $^4\Sigma^-$ state has not been well characterized until now, we can't judge the caculational precision for this state, but the good comparison with the experimental data of the calculated values for the $X^2\Pi_r$ state gives a reason to rely upon the calculational result for $^4\Sigma^-$.

TABLE V Calculated spectroscopic constants of CH($X^2\Pi_r$ and $^4\Sigma^-$) compared to the experimental values.

State	Method	R_e(bohr)	ω_e(cm^{-1})	D_e(eV)	T_e(cm^{-1})	μ_e(au)
$X^2\Pi_r$	SCF	2.11	2971	2.28	3350	0.64
	CI	2.15	2781	3.39	0	0.53
	Exper	2.117^α	2858.5^α	3.631^α	0	0.57^β 0.35^γ
$^4\Sigma^-$	SCF	2.04	3346	2.69	0	0.19
	CI	2.08	3157	3.12	2180	0.16

α(Ref. 15), β(Ref. 16), γ(Ref. 17).

For the Sc-CH approach, the C-H distance has been fixed to 2.15 bohr, the calculated bond distance of the $X^2\Pi_r$ state. Here only the lowest two states $^3\Pi$ and $^5\Pi$ chosen from a preliminary SCF and SDCI calculation are studied. For the $^5\Pi$ state the single SCF determinant is used as the zeroth-order wavefunction and the single- and double-substituted determinants whose energy lowerings are greater than 5 μ hartree are included in the final diagonalization involving some 2600 determinants. For the $^3\Pi$ state apart from the the SCF determinant a doubly-substituted determinant is included as the reference zeroth-order wavefunction. The final diagonalizations for this state involved all determinants interacting with this zeroth-order wavefunction through the matrix elements greater than 10 μ hartree including some 1500 determinants.

The two lowest states of isolated CH molecule can be represented as

$$:\dot{C} - H \quad (X^2\Pi_r) \qquad \text{and} \qquad \cdot\dot{\underset{\cdot}{C}} - H \quad (^4\Sigma^-)$$

according to the NO's (see TABLE VI). The σ bonding of the former shows a slightly ionic character whilst in the latter it is not the case as both the AO population and the dipole moment indicate. The $^4\Sigma^-$ state is more stable than the $^2\Pi_r$ state in SCF calculation in agreement with the Hund's rule. But the inclusion of the correlation effect through CI lowers the latter state. In ScCH approach keeping the C-H distance at 2.15 bohr (in fact changement of this distance doesn't leads to any significant energetic and bonding character), a $^3\Pi$ state appears to be the lowest whose minimum energy point occurs at 3.51 bohr. A $^5\Pi$ state has a minimum at 4.24 bohr situated higher than the former state by 0.70 eV after CI.

The electron distribution in ScCH may be schematically represented as follows:

$$\cdot\widehat{Sc \rightarrow} \underset{\bullet}{C} - H \quad (^3\Pi) \qquad \text{and} \qquad \cdot Sc \rightarrow \underset{\bullet}{\overset{\bullet}{C}} - H \quad (^5\Pi)$$

The NO's of the $^3\Pi$ can be expressed using the AO population and the NO coefficients as :

(1σ) $0.93\ s(C) + 0.14\ p_\sigma(C) + 0.82\ s(H) - 0.05\ p_\sigma(H)$: 1.97

(2σ) $-0.11\ s(Sc) - 0.13\ d_\sigma(Sc) -0.65\ s(C) + 0.83\ p_\sigma(C) + + 0.26\ s(H)$: 1.96

(3σ) $0.76\ s(Sc) - 0.21\ p_\sigma(Sc)$: 1.96

(1π) $0.12\ p_\pi(Sc) + 1.48\ d_\pi(Sc) + 1.42\ p_\pi(C) + 0.03\ p_\pi(H)$: 2.89

This state has a CI dipole moment of 0.14 au (Sc^+). The estimated interaction energy between Sc and CH is greater than 2.60 eV.

TABLE VI Natural orbitals of CH ($X^2\Pi_r$ and $^4\Sigma^-$) at 2.15 bohr.

State	Orbital	C			H			Total
		s	p_σ	p_π	s	p_σ	p_π	
$X^2\Pi_r$	1σ	0.27	0.47		1.17	-0.05		1.97
	2σ	1.55	-0.48					1.94
	1π			0.96			0.03	0.99
$^4\Sigma^-$	1σ	0.86	0.21		0.86	-0.04		1.96
	2σ	-0.43	0.47		0.09			1.00
	1π			1.96			0.03	1.99

6. Sc-H_2

A collinear and a perpendicular approaches of Sc atom to H_2 has been studied using the SCF and a MRD-CI. Here the H-H distance is kept to 1.6 bohr near to the experimental value of the isolated H_2. The lowest doublets correlating to the ground state collision are very repulsive, while the lowest quadruplet states correlating to collision between the ground state H_2 and the first excited state of Sc $3d^24s^1(^4F)$ are less repulsive. In fact the quadruplet states are so flat for Sc-H distances greater than 4 bohr, that they may have bound attractive wells when a full geometry optimization is considered. These intermediate excited states may be interesting in reaction dynamics.

A linear symmetric arrangement H-Sc-H with the H-Sc distance of 4 bohr is also calculated. The lowest doublet state appeared to be $^2\Delta_u$ whose total energy is slightly lower than the Sc+H_2 energy by at least 0.2 eV. The lowest quadruplet state has a significantly higher energy than the Sc+H_2 dissociation. It is noteworthy that in making the bivalent linear triatomics the ScH moietie is prepared in $^3\Delta$ state preferentially to the $^1\Sigma^+$ state due to the evidently favorable spatial electron distribution for addition. The NO's of the $^2\Delta_u$ state can be represented as:

$$(1\sigma_g) \quad 0.64\ s(H) + 0.42\ s(Sc) - 0.23\ d_\sigma(Sc) + 0.64\ s(H) \quad : 1.95$$

$$(1\sigma_u) \quad 0.81\ s(H) + 0.31\ p_\sigma(Sc) + 0.81\ s(H) \quad : 1.94$$

$$(1\delta_g) \quad : 0.99$$

Considering that the Sc-H energy in isolated ScH molecule is about 1.95 eV when the same basis as here is used, the absolute stability of the H-Sc-H to arbitrary dissociation means that there exists a cooperative stabilization effect as in H-Sc-Li molecule[17]. This fact is interesting and it may help an understanding for the H_2 dissociation on the metal surfaces. The closed-shell ScH_3 molecule with D_{3h} symmetry is calculated to be stable relative to H_2+Sc+H in SCF level although it is not stable relative to H_2+ScH supporting an all-electron SCF calculation of Stevens and Lipscomb[18].

7. CONCLUDING REMARK

Due to a simplicity of the atomic structure and a small number of valence electrons, the Sc atom shows a clear and pedagorgical example of the transition-metal atomic characteristics. The interaction with CO shows the typical variety of the low-lying electron states depending upon different atomic configurations used in the molecular building. The repulsive, mederately attractive and strongly attractive molecular states according to the s^2d^1, s^1d^2 and d^3 atomic occupancies may be explained by the penetration repulsion of 4s atomic orbital(metal) with 3σ(CO) molecular orbital, and the easier electron transferability to mainly $2\pi^*$(CO) molecular orbital of the d^3 and d^2s^1 than the d^1s^2 due to lower ionization energies of the formers.

The remarkable similarity between the one hydrogen electron and the radical σ electron of CN could be explained simply with the SCF approximation except for the $^1\Sigma^+$ state which is stabilized by a correlation effect and which has a relatively shorter bond distance than the other states.

The elctron distribution of the lowest states of ScCH molecule shows a closer matching with the $^4\Sigma^-$ state than with the ground state $^2\Pi_r$ of the isolated CH molecule.

The scission of the H_2 molecule when it is approached to the Sc atom gives an explanation of the possible catalytic dissociation. The interaction of Sc atom with the CN, CH and H_2 molecule could be easily described in a Lewis type valence electron pairing scheme.

ACKNOWLEDGEMENT

A part of this work has been supported by Deutsche Forschunsgemeinschaft through the project SFB-6. Many interesting discussions with Prof. J. Koutecky, Drs. G. Pacchioni and D. Papierowska-Kaminsky are gratefully acknowledged. Many useful suggestions by Drs. J.P. Malrieu and M. Pélissier are also gratefully acknowledged.

REFERENCES

1) M.J.S Dewar, Bull. Soc. Chim. Fr. 18 (1951) c71; J. Chatt and L.A. Duncanson, J. Chem. Soc. (1953) 2939
2) S.P. Walch and W.A. Goddard III, J. Am. Chem. Soc. 98 (1976) 7908
3) A.B. Rives and R.F. Fenske, J. Chem. Phys. 75 (1981) 1293
4) P.S. Bagus and B.O. Roos, J. Chem. Phys. 75 (1981) 5961; P.S. Bagus, C.J. Nelin and C.W. Bauschlicher Jr., Phys. Rev. B 28 (1983) 5423; P.S. Bagus, K. Hermann and C.W. Bauschlicher Jr., Ber. Bunsenges. Phys. Chem. 88 (1984) 302
5) G. Pacchioni, J. Koutecky and P. Fantucci, Chem. Phys. Lett. 92 (1982) 486
6) H. Basch and D. Cohen, J. Am. Chem. Soc. 105 (1983) 3856
7) J.C. Barthelat and Ph. Durand, Gazz. Chim. Ital. 108 (1978) 225
8) Manual of "Workshop on Pseudopotential", Laboratoire de Physique Quantique, Université Paul Sabatier, Toulouse, France, Oct. 5-9, 1981
9) G.H. Jeung and J. Koutecky, to be published ; see also the Appendix
10) R.J. Buenker, S.D. Peyerimhoff and W. Butscher, Molec. Phys. 35 (1978) 771; R.J. Buenker and S.D. Peyerimhoff, Theoret. Chim. Acta (Berl.) 39 (1975) 217; 35 (1974) 33
11) S.R. Langhoff and E.R. Davidson, Int. J. Quantum Chem. 8 (1974) 61; see also E.R. Davidson and D.W. Silver, Chem. Phys. Lett. 52 (1977) 403; P.E.M. Siegbahn, Chem. Phys. Lett. 55 (1978) 386
12) K.P. Huber and G. Herzberg, Molecular Spectra and Molecular Structure Vol. 4, Van Nostrand (1979)
13) R.C. Weast, Handbook of Chemistry and Physics, Chemical Rubber Co., Cleveland (1971)
14) J.N. Allison and W.A. Goddard III, Chem. Phys. 81 (1983) 263
15) G. Herzberg and J.W.C. Johns, Astrophys. J. 158 (1969) 399
16) D.H. Phelps and F.W. Dalby, Phys. Rev. Lett. 16 (1966) 3
17) E.A. Scarl and F.W. Dalby, Can. J. Phys. 52 (1974) 1429
18) H.O. Beckmann, G. Pacchioni and G.H. Jeung, Chem. Phys. Lett. 116 (1985) 423
19) P.E. Stevenson and W.N. Lipscomb, J. Chem. Phys. 50 (1969) 3306

APPENDIX

TABLE A1 Basis set effect on the atomic full CI energy differences relative to Sc $3d^1 4s^2 (^2D)$. In parentheses are SCF energy differences. In eV.

Basis	$d^2s^1(^4F)$	$d^1s^1p^1(^4F^o)$	$d^3(^4F)$	$d^1s^1(^3D)$
I	1.52 (1.08)	1.81 (1.23)	4.14 (4.48)	6.38 (5.52)
II	1.61 (1.13)	1.94 (1.29)	4.46 (4.66)	6.19 (5.27)
III	1.48 (1.11)	1.81 (1.23)		6.47 (5.61)
IV	1.52 (1.07)	1.83 (1.25)		6.44 (5.59)
V	1.38 (0.93)	1.75 (1.23)		6.37 (5.61)
→ VI	1.44 (1.07)	1.80 (1.25)	4.62 (4.78)	6.36 (5.59)
Exper *	1.43	1.96	4.19	6.56

* C.E. Moore, Atomic Energy Levels, U.S. Nat. Bur. Standard (1971)

I: 4s2p7d→3s(211)2p3d(421), II: 3s2p6d→3s2p$\overset{\downarrow 411}{\text{3d}}$, III: 3s2p6d→2s(21)2p2d(51), IV: 3s2p6d→2s2p2d(42), V: 3s2p6d→2s1p2d(51), VI: 3s2p6d→2s1p2d(42): basis used in this paper.

TABLE A2 Basis set effect upon the energy, the dipole moment and the charge shift for $ScH(^3\Delta)$ at 3.7 bohr (in atomic units)

Basis	ΔE(SCF)	ΔE(CI)	Dip.mom.	Charg.sh.
A	0	0	0.90	0.45
B	0.009	0.011	0.90	0.52
C	0.013	0.015	0.89	0.51
→ D	0.020	0.069	0.88	0.50
E	0.038	0.193	0.87	0.50
F	0.043	0.199	0.85	0.41
G	0.095	0.538	0.87	0.16
H	0.104	0.571	0.94	0.26

A: Basis I of Table A1(for Sc)/6s2p→3s(411)2p(for H),
B: 4s2p6d→3s(211)2p2d(42)/6s2p→3s2p, C: Basis IV/6s2p→3s2p,
D: Basis VI/6s2p→3s2p: basis used in this work,
E: Basis VI/6s1p→3s1p, F: Basis VI/5s1p→2s1p,
G: Basis VI/6s→3s, H: Basis VI/5s→2s

THE NATURE OF THE BONDING IN THE TRANSITION METAL TRIMERS

Stephen P. Walch[a]
Eloret Institute
Sunnyvale, CA 94087
and
Charles W. Bauschlicher, Jr.
NASA Ames Research Center
Moffett Field, CA 94035

ABSTRACT. To gain a qualitative understanding of the electronic structure of small metal clusters, we have carried out ab initio calculations on selected transition metal trimers and their positive ions. Here we present SCF/CI calculations for the $^1A_1'$ state of Ca_3 arising from three ground state ($4s^2$) Ca atoms, CASSCF/CCI calculations for the low-lying states of Sc_3 and Sc_3^+, and SCF/CI calculations for the ground state surface and excitation energies of Cu_3, all for near equilateral triangle geometries. The bonding in Ca_3 arises by 4s→4p promotion and leads to an atomization energy of about 0.5 eV for R(Ca-Ca) = 7.5 a_0. For Sc_3 the 4s bonding is similiar to that in Ca_3, but the 3d electrons are also strongly bonding leading to a $^2A_2''$ ground state with an atomization energy of about 1.0 eV and R(Sc-Sc) = 5.75 a_0. The bonding in Cu_3 arises from the $4s^1 3d^{10}$ state of the Cu atom and involves mainly the 4s electrons, leading to a $^2E'$ state which is strongly Jahn-Teller distorted. Based on the Ca_3, Sc_3, and Cu_3 calculations and preliminary calculations on Ti_3^+, the bonding in the other transition metal trimers, including V_3^+ and Cr_3^+ is also discussed.

I. INTRODUCTION

In this paper we discuss the electronic structure of the transition metal (TM) trimers. These calculations include all the valence electrons and also include extensive electron correlation. The most useful result of these studies is the qualitative picture which emerges of the nature of the metal-metal bond in these bare metal clusters. This simple qualitative picture of the electronic structure provides a start toward understanding reactivity of these metal clusters and provides insights into catalysis, while comparison of TM dimers, trimers and possibly higher clusters provides insights into the electronic structure of bulk transition metals and their interaction with other atoms and molecules.

Because of the numerous practical applications of transition metals, small metal clusters are currently of considerable experimental interest. Experimental data for the TM trimers are becoming available from matrix ESR [1,2,3] and Raman studies [4,5] and from gas phase spectroscopy [6], while larger clusters are being

[a] Mailing address: NASA Ames Research Center, Moffett Field, Ca. 94035.

A. Veillard (ed.), Quantum Chemistry: The Challenge of Transition Metals and Coordination Chemistry, 119–134.

studied on inert supports [7,8,9,10] and in the gas phase [12,13]. For a review of the current state of both theory and experiment for small TM molecules the reader is referred to the review article by Weltner and VanZee [11].

Matrix ESR experiments directly provide information on the electronic spin state and through the nuclear hyperfine structure (HFS) provide information on the orbital character of the open-shell orbital. Limited geometric information can also be inferred e.g. for Sc_3 all the Sc atoms are found to be equivalent thus implying an equilateral triangle geometry [1]. The Raman experiments provide vibrational frequencies, but often questions arise as to the identity of the carrier. In addition both of these matrix techniques share the difficulty that matrix effects, while usually small, can invert the ordering of closely spaced states. This has been documented e.g. for Ag_3 [2]. The gas phase spectroscopic work, in principle, can provide more detailed information, but for trimers and larger clusters, only Cu_3 [6] has been successfully studied. The experimental difficulties arise because of the high density of electronic states for these molecules and because for the neutral molecules a mixture of clusters (i. e. monomer, dimer, and trimer etc.) are present in the molecular beam. Experimental studies in the near future therefore will be made on the positive ions of the metal trimers, since these molecules may be mass selected. Thus, the ESR experiments are being used to study the neutrals, and gas phase spectroscopy is expected to provide information on the positive ions; while theory can provide a link between these experiments, since it can study either neutrals or positive ions.

Studies of metal clusters on inert supports [7,8,9,10] are currently able to treat clusters as small as six atoms. Thus, this method is not currently applicable to the TM trimers, but as experimental techniques are refined the method shows promise of being applicable to clusters which are small enough to study by computational methods. Currently these methods yield information about the average metal-metal distances and the change in the metal-metal distances with adsorbate coverage. Generally only saturation coverages are used and no adsorbate site information is obtained. Thus, theory can give considerable physical insight if it can identify the nature of the adsorbate sites.

The reactivity of bare metal clusters has been studied by several groups. Among these are studies by Geusic, Morse, and Smalley of the reactivity of Nb clusters with H_2 [12] and by Parks, Liu, Richsmeier, Pobo, and Riley of the reactivity and stochiometry of the end products in the reaction of H_2 with Fe clusters [13]. Both of these studies show a dramatic variation of reactivity with cluster size. However, these experiments provide no geometric information and it remains a challenge to theory to understand the geometries of these small clusters and the variations in reactivity. As an extension of the work described here studies have been started of the reaction of H atom with Sc_3.

There have been several previous all electron studies of the Cu_3 molecule [14,15,16]. SCF studies [14,15] vary as to whether the linear geometry [14] or equilateral triangle geometry [15] of Cu_3 is lowest, while CI studies [16] using a [2s1d] contracted basis set, which in effect eliminates most of the 3d shell correlation, place the distorted equilateral triangle structures lowest, but are not accurate enough to clearly establish the ground state geometry of this molecule. No previous calculations exist for Sc_3, or for the remaining TM trimers of the first transition row except for Ni_3 which has been studied at the SCF level using a small basis set in conjunction with an ECP [17]. We should also mention here the work of Richtsmeier, Dixon, and Gole on the alkali trimers and the trimers of the group IB metals using the diatomics in molecules (DIM) method [18]. While this method has been successful in studying alkali metal trimers and group IB TM trimers, it is not clear that this approach would be applicable for those TM trimers where 3d bonding is important, or that it could be used to study reactivity.

In this paper we discuss complete active space SCF [19]/ externally contracted [20] configuration interaction (CASSCF/CCI) calculations for the low-lying states of Sc_3 and Sc_3^+ [21] and SCF/CI calculations for Ca_3 [21] and Cu_3 [22]. For Cu_3 we also discuss calculated vertical excitation energies which lead to a new assignment of the upper state in the resonant two photon ionization spectrum [6]. From comparison of the bonding in the Ca_3, Sc_3, and Cu_3 molecules, we are able to predict general trends for the TM trimers, much as we observed for the TM dimers [23].

Section II discusses qualitative features of the bonding in the TM trimers including predicted trends for other TM trimers. Section III discusses the basis sets and other technical details of the calculations. Section IV discusses the calculated results for Sc_3, Sc_3^+, and Cu_3 and compares the results to experiment. Finally, Section V presents the conclusions from this work.

II. Qualitative Discussion.

As with the TM dimers [23], the nature of the bonding in the TM trimers changes with position in the transition row. The two principle factors controlling the electronic structure are i)the relative (n+1)s and nd orbital sizes and ii) the $(n+1)s^2nd^m \rightarrow (n+1)s^1nd^{m+1}$ atomic energy separations. As discussed in ref. 23, the ratio $\langle r_{(n+1)s}\rangle/\langle r_{nd}\rangle$ increases monotonically from 2.03 for Sc to 3.36 for Cu. Thus, we expect that nd bonding will be more favorable for the Sc end of the row than for the Cu end of the row. For Sc the $4s^23d^1$ state of the atom is lowest with the $4s^13d^2$ state about 1.4 eV higher [24]. Given the large excitation energy, the ground state of Sc_3 arises from three atoms in the $4s^23d^1$ state. The large 4s-4p population here (approximately six electrons) leads to relatively long metal-metal bond lengths, but because the ratio $\langle r_{4s}\rangle/\langle r_{3d}\rangle$ is still small for Sc, the 3d electrons still form bonds leading to a doublet ground state. The atomic excitation energy decreases monotonically from Sc to Cr which has a $4s^13d^5$ ground state. This change in the atomic excitation energies is associated with increased 4s $\rightarrow$ 3d promotion for Ti_3, V_3, and Cr_3. The lower 4s-4p populations lead to shorter metal-metal bond lengths, which compensates for the contraction of the 3d shell with respect to the 4s shell, and allows 3d bonding to remain important through Cr_3. For Mn, $4s^23d^5$ is again lowest due to the large exchange interaction associated with a half filled 3d shell, but as for the left half of the row, the excitation energy decreases monotonically from Mn to Cu, which has a $4s^13d^{10}$ ground state. Thus, Mn_3 once again has a large 4s-4p population and large metal-metal distances. Due to the contraction of the 3d shell with respect to the 4s shell for the right half of the first transition row, 3d bonding is no longer important and Mn_3 is predicted to have a high spin ground state. For Cu_3 the 3d shell is closed and contracted with respect to the 4s leading to a ground state arising from three atoms in the $4s^13d^{10}$ state.

In order to understand the bonding arising from three $4s^2$ atomic configurations, we first consider the bonding in the Ca_3 molecule. For Ca_3 the lowest atomic limit is for three ground state atoms (^1S or $4s^2$). SCF/CI calculations have been carried out for this limit for equilateral triangle geometries. Here the SCF curve is repulsive, but the CI curve shows a weak bond (about 0.5 eV) at R Ca-Ca = 7.5 a_0. At this distance, the Mulliken populations from the CI calculation are

4s=5.24, 4p=0.61, and 3d=0.15. At R Ca-Ca = 6.5 a_0, the Mulliken populations are 4s=4.58, 4p=1.08, and 3d=0.33. These results indicate a significant amount of 4s $\rightarrow$ 4p promotion as well as some 4s$\rightarrow$ 3d promotion for shorter R Ca-Ca. For Ca_3 this formation of sp hybrids apparently does not lead to strong bonding; however, the large 4p populations at smaller R indicate that the predominate bonding mechanism here is 4s$\rightarrow$ 4p promotion, as compared to the Ca_2 molecule which does not show the large 4p population and where a large part of the bonding is due to Van der Waals terms arising out of the 4s$\rightarrow$ 4p near degeneracy effect [25]. The larger importance of 4s $\rightarrow$ 4p promotion for the trimer as compared to the dimer probably results from the formation of three bonds for the trimer which compensates for the 4s $\rightarrow$ 4p promotion energy, while the dimer with only one bond can not compensate for the promotion energy. Similar results have been observed by Bauschlicher, Bagus, and Cox [26] for Be_4, Mg_4, and Ca_4.

For the Sc_3 through Cr_3 molecules the 3d electrons contribute strongly to the bonding. In order to understand the 3d electron configurations of these molecules, we discuss next the symmetry and bonding properties of the 3d like CASSCF orbitals. Plots of CASSCF orbitals for the Ti_3^+ molecule are given in Fig. 1. Here we denote the molecular orbitals as $3d\sigma$, $3d\pi'$, $3d\pi''$, and $3d\delta'$ to indicate the corresponding atomic symmetry with the symmetry classification given for 3d functions on each atom where the sigma direction is along the axis connecting the given atom with the center of the molecule. In D_{3h} symmetry $3d\sigma$ and $3d\delta'$ transform as a_1' and e', $3d\pi'$ transforms as a_2' and e', $3d\pi''$ transforms as a_2'' and e'', and $3d\delta''$ transforms as a_1'' and e''. As for the TM dimers, we find that the $3d\sigma$, $3d\pi'$, and $3d\pi''$ atomic orbitals have moderate overlaps in the molecule and form the bonding orbitals, while the $3d\delta'$ and $3d\delta''$ orbitals have small overlaps and are essentially non-bonding. Here we see that the best bond orbitals are $3d\sigma a_1'$ and $3d\pi'' a_2''$, because these combinations of the atomic orbitals have no molecular nodal planes between pairs of atoms. These same atomic orbitals also lead to $3d\sigma e'$ and $3d\pi'' e''$ orbitals which contain one molecular nodal plane, and are thus not favorable as bonding orbitals. However, these orbitals are important correlating orbitals, since they introduce molecular nodal planes between pairs of atoms. The $3d\pi' a_2'$ orbital is anti-bonding with nodal planes between all pairs of atoms, while the $3d\pi' e'$ orbitals contain one molecular nodal plane and are thus less favorable than $3d\sigma a_1'$ or $3d\pi'' a_2''$, but are still bonding between pairs of atoms. Finally, the $3d\delta$ orbitals remain very atomic like due to the small overlaps of these atomic orbitals. Thus, these orbitals are essentially non-bonding.

For Sc_3 we find the low-lying states arise from two atomic limits i) three ground state $4s^2 3d^1$ atoms and ii) two ground state $4s^2 3d^1$ atoms and one excited state $4s^1 3d^2$ atom. This is to be expected based upon the results for Sc_2 [27],[28], where it was shown that the ground state arises from one atom in the $4s^2 3d^1$ state and one atom in the $4s^1 3d^2$ state. For both of these atomic limits the bonding involves 4s to 4p promotion leading to sp hybrid bonding from the 4s shell, in addition ii) also involves 4s$\rightarrow$3d promotion leading to a 3d population of about 4.0. For Sc_3 the 3d orbitals also contribute to the bonding. Because of the smaller size of the 3d orbitals compared to the 4s orbitals the formation of 3d bonds results in a much shorter Sc-Sc distance in Sc_3 as compared to the Ca-Ca distance in Ca_3. This is shown in Fig. 2 where one sees that the $^2E'$ state of Sc_3 has a minimum, for D_{3h} geometries, for R Sc-Sc= 5.5 a_0, while the Ca_3 potential curve is strongly repulsive at this distance. This result implies a strong bonding effect from the 3d shell; although, the size mismatch of the 4s and 3d orbitals results in repulsive 4s-

4s interactions at the short R Sc-Sc distances needed for formation of good 3d-3d bonds, and results in a weak net bonding.

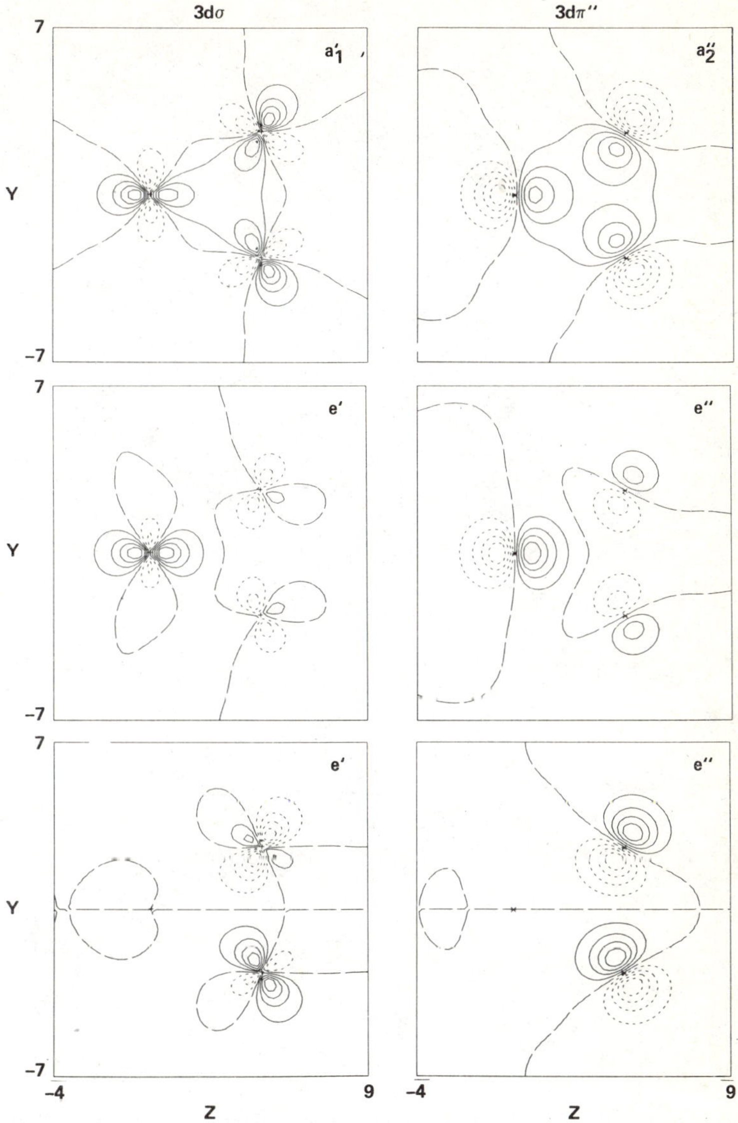

Figure 1a. Contour plots of selected orbitals for the $1^6A_1'$ state of Ti_3^+. The orbitals are plotted in the molecular plane except for the $3d\pi''$ orbitals which are plotted 1.0 a_0 above the molecular plane. Positive contours are denoted by solid lines, negative contours are denoted by dotted lines, and nodal surfaces are denoted by dashed lines.

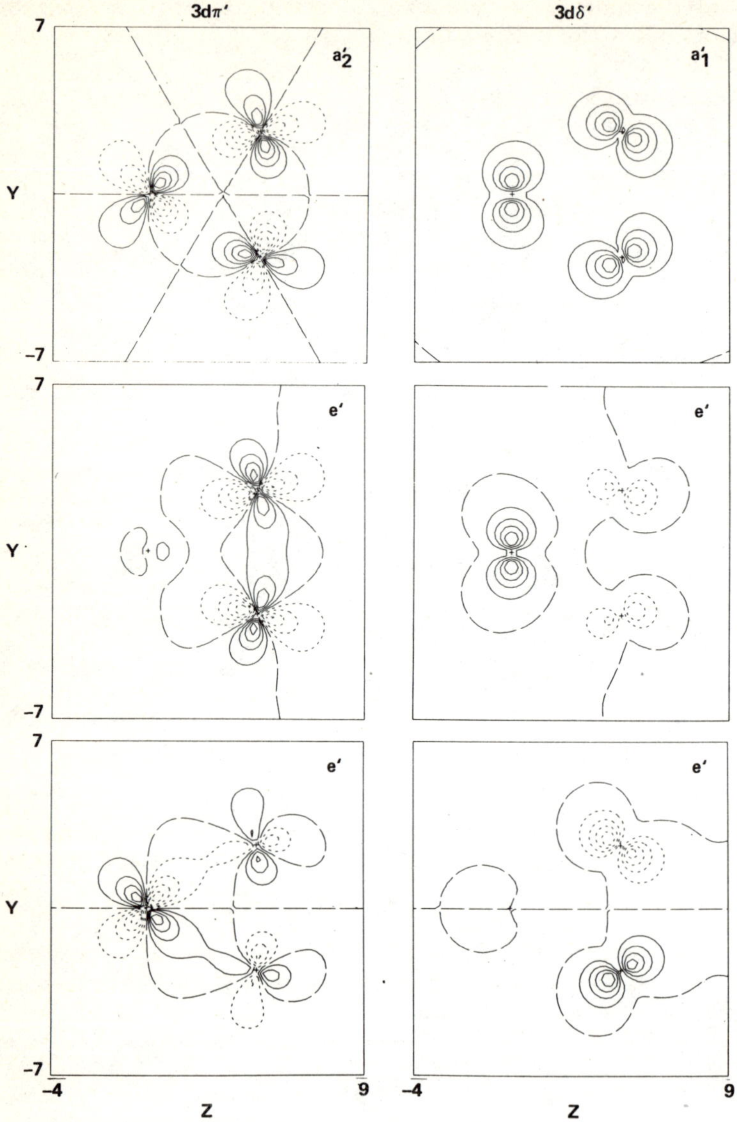

Figure 1b. Contour plots of selected orbitals for the $1^6A_1'$ state of Ti_3^+. The orbitals are plotted in the molecular plane. The conventions are the same as in Fig. 1a.

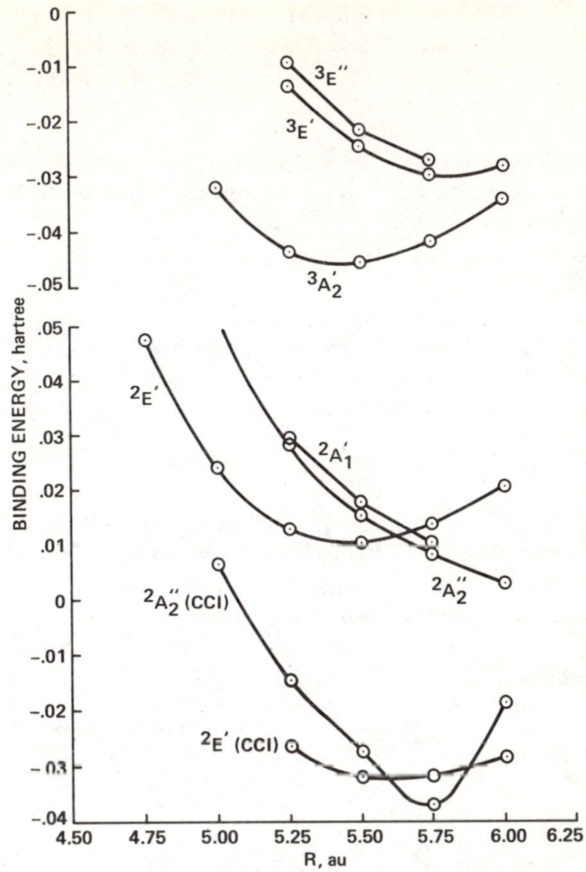

Figure 2. Calculated potential curves for selected states of Sc_3 and Sc_3^+ for equilateral triangle geometries. The binding energy is plotted as a function of R(Sc-Sc). The curves are from CASSCF calculations except for the $^2A_2''$ and $^2E'$ states where contracted CI curves are also shown. The zero of energy for Sc_3 is taken as the sum of three ground state Sc atoms, while for Sc_3^+ the zero of energy is two ground state atoms and one Sc^+ atom in the $4s^1 3d^1$ state.

From the discussion above one expects the principle 3d bonding orbitals for Sc_3 to be $3d\sigma a_1'$ and $3d\pi'' a_2''$. The configurations of the low-lying states here are:

$$^2A_1' = 4sa_1'^2\ 4se'^4\ 3da_1'^1\ 3da_2''^2$$
$$^2A_2'' = 4sa_1'^2\ 4se'^4\ 3da_1'^2\ 3da_2''^1$$
$$^2E' = 4sa_1'^2\ 4se'^3\ 3da_1'^2\ 3da_2''^2$$

The populations, for CASSCF wavefunctions at equilateral triangle geometries with R Sc-Sc= 5.5 a_0, are: 4s=4.55, 4p=1.17, 3d=3.24, and 4f=0.04 for the $^2A_2''$ state and 4s=3.89, 4p=1.24, 3d=3.83, and 4f=0.04 for the $^2E'$ state. Thus, the wavefunctions contain a considerable admixture of 4p character in agreement with the discussion above. The ground state is found to be $^2A_2''$, but the $^2E'$ state is only about 0.1 eV higher. These results indicate that the removal of one 4s electron and the formation of one additional 3d bond leads to additional bonding equivalent to the excitation energy $4s^23d^1 \rightarrow 4s^13d^2$ of 1.44 eV.

For Sc^+, the promotion energy $4s^13d^1 \rightarrow 3d^2$ is much less than $4s^23d^1 \rightarrow 4s^13d^2$ for Sc atom [24], thus, for Sc_3^+ one expects a higher 3d population to be favored, and the ground state of the ion is expected to be derived from the $^2E'$ state of the neutral. By analogy to Cu_3, where the 4se′ orbital is only weakly bound, the lowest state of Sc_3^+ arises by removal of one electron from the 4se′ orbital. This leads to a $^3A_2'$ state with the electronic configuration:

$$^3A_2' = 4sa_1'^2\ 4se'^2\ 3da_1'^2\ 3da_2''^2$$

For the Ti atom, the $4s^13d^3$ state is at 0.81 eV compared to 1.44 eV for the corresponding excitation in Sc [24]. Thus, for Ti_3 one expects 4s→3d promotion to be more important than for Sc_3. Preliminary calculations for Ti_3^+ indicate two important groups of states. The first group of states arises from two Ti atoms in the ground state and one Ti^+ in the $4s^13d^2$ state which gives a 3d population of about six. These states are high spin (spin=8) and have an R(Ti-Ti) of about 7.5 a_0 for equilateral triangle geometries. The second group of states has a 3d population of about seven (4s-4p population of about four). The wavefunctions for these states are characteristic of 3d bonding, and these states are stable at much shorter bond lengths R(Ti-Ti) = 5.0 a_0. One representative configuration from this group of states is:

$$1^6A_1' = 4sa_1'^2\ 4se'^2\ 3d\sigma a_1'^2\ 3d\delta' a_1'^1\ 3d\pi'' a_2''^2\ 3d\delta' e'^2$$

This configuration is similar to the $^3A_2'$ state of Sc_3^+ with the extra 3d electrons in 3dδ′ derived levels. Since the 3dδ′ levels have very small overlaps, they are non-bonding. An alternative configuration is:

$$2^6A_1' = 4sa_1'^2\ 4se'^2\ 3d\sigma a_1'^2\ 3d\delta' a_1'^1\ 3d\pi'' a_2''^2\ 3d\pi' e'^2$$

One might expect this configuration to be lower, since the 3dπ′e′ levels are bonding between pairs of atoms. However, this configuration involves atomic configurations with two electrons in the 3dπ orbitals which is not favorable for the Ti atom and this intraatomic coupling effect leads to comparable energies for the two configurations.

For V_3 and Cr_3 one expects the lowest state to arise from the three atoms in the $4s^13d^{n+1}$ state. The low-lying configurations here are expected to arise from the $1^6A_1'$ configuration of Ti_3^+ by first adding electrons to the bonding 3dπ′e′ levels, and then adding electrons to the non-bonding 3dδ″ levels. A likely ground state configuration for V_3^+ is:

$$^9A_2'' = 4sa_1'^2\ 3d\sigma a_1'^2\ 3d\pi''a_2''^2\ 3d\pi'e'^2\ 3d\delta'a_1'^1\ 3d\delta'e'^2\ 3d\delta''a_1''^1\ 3d\delta''e''^2$$

This configuration has a 4s-4p population of two. This allows for a shorter metal metal bond due to removal of repulsive 4s-4s interactions and compensates for the contraction of the 3d shell with respect to the 4s shell. For Cr_3^+ the extra electrons would go into $3d\pi'$ and $3d\delta$ derived orbitals. This model would result in increased bonding for V_3^+ as compared to Ti_3^+, but it is not clear whether the binding energy for Cr_3^+ or V_3^+ will be larger due to the competition between formation of 3d bonds and loss of atomic exchange energy.

Mn_3 with a $4s^2 3d^5$ atomic ground state is expected to be like Ca_3 with the bonding arising from 4s→4p promotion. Here we do not expect 3d bonding to be favorable because of the contraction of the 3d orbitals with respect to the 4s orbitals. For the remaining elements of the first transition row, we expect mainly 4s bonding arising from mixtures of the $4s^2 3d^n$ and $4s^1 3d^{n+1}$ atomic states with the 3d electrons high spin coupled. There is some experimental evidence for this viewpoint from ESR studies which show that while Mn_2 has a singlet ground state but low-lying high spin states [30], the ground state of Mn_5 is high spin (spin=25) corresponding to all the 3d electrons high spin coupled [30].

The bulk cohesive energies(in eV) [29] for the first few elements of the first transition row are Ca(1.825), Sc(3.93), Ti(4.855), V(5.30), Cr(4.18), and Mn(2.98). These values are consistent with the 3d orbital configurations for the transition metal trimers discussed above in that: i) Sc_3 shows 3d bonding but Ca_3 does not, ii) the 3d bonding for the trimers is predicted to reach a maximum for V_3, and iii) Mn_3 is predicted to show no 3d bonding. A similiar comparison has been made between the binding energies of the TM dimers and the bulk cohesive energies by Moskovits et al. [4]. Here Ca and Mn dimers show anomalously low binding energies compared to bulk metals. This is expected, based on the studies of the TM trimers, since the binding mechanism for the trimer involves 4s→4p promotion and this is expected to persist for the bulk metal, but the dimer does not show 4s→4p promotion and is more weakly bound.

For Cu_3 the ground state of the atom is $4s^1 3d^{10}$ with the closed 3d shell contracted with respect to the 4s. Thus, the bonding in Cu_3 is found to involve mainly the 4s electrons [22], much like an alkali trimer. However, the 3d electrons are not inert since correlating them results in a decrease in R Cu-Cu of about 0.30 a_0. The Mulliken populations for Cu_3 from a CI calculation at the optimal geometry for D_{3h} symmetry (the ground state geometry here is Jahn-Teller distorted but near equilateral triangle) are 4s=2.72, 4p=0.47, and 3d=29.80 in agreement with the qualitative picture above. This result is similiar to results obtained by Basch, Newton, and Moskowitz [17] who found the lowest state of Ni_3 to arise from three atoms in the $4s^1 3d^9$ state.

III. Calculational Details

While these calculations could have been carried out at the all-electron level, accurate effective core potentials (ECP's) have been developed by Hay and Wadt [31] which make it possible to eliminate the Ne core with very little loss of accuracy. Here the 3s and 3p core levels as well as the 4s, 4p, and 3d valence levels are included in the calculation. Gropen, Wahlgren, and Pettersson [32] have pointed out that inclusion of the entire n=3 shell in the valence space is necessary since the 3p and

3d orbitals have similar spatial extents, e.g. removing the entire Ar core has lead to collapse problems for ScO.

The basis set used in the Sc_3 calculations is the basis set reported by Hay and Wadt except that the 4p like functions were replaced by Wachters' [34] 4p functions multiplied by 1.5 to make them suitable for describing 4s→4p correlation. The s and p basis sets were contracted (2111) based on the 3s and 3p atomic orbitals, respectively. Normally, one augments the 3d basis by the Hay diffuse 3d function [35]. However, here the most diffuse 3d function is 0.0812 compared to the Hay diffuse 3d of 0.0588. A diffuse function selected based on an even tempered criterion was tried but it had very little effect on the energy of either the $4s^2 3d^1$ or $4s^1 3d^2$ state of the Sc atom and it was concluded that the valence d basis set was sufficiently diffuse. In order to describe polarization of the 3d bonds a single set of 4f functions was added as a two term fit of an STO(ς=1.6). The Slater exponent here was optimized from an all electron CISD calculation for the $4s^1 3d^2$ state of the atom. The final basis set is (5s5p5d2f)/[4s4p3d1f].

The Ca atom basis set was constructed in the same way as the Sc basis set and the Hay and Wadt ECP with the 3s and 3p included in the valence was used as for Sc. The 3d functions here were taken from the work of Pettersson, Siegbahn, and Ismail [36]. The primitive basis set here is 5d which is contracted (311). No 4f basis functions were added to the basis set, since there are no occupied 3d orbitals. The final basis set is (5s5p5d)/[4s4p3d].

Several tests were carried out for the Sc basis set/ECP used here. Atomic SCF calculations using the ECP gave a $4s^2 3d^1 \rightarrow 4s^1 3d^2$ excitation energy of 1.01 eV compared to 1.03 eV from NHF. Also all-electron calculations on the $4s^1 4p^1 3d^1$ state of the Sc atom showed that multiplying the 4p exponents by 1.5 raised the energy of this state by only 0.03 eV; this result is important since we expect strong 4s→4p promotion in Sc_3. Molecular calculations were also carried out for ScO and Sc_2. The ScO results were in good agreement with the all-electron results reported by Pettersson, Wahlgren, and Gropen [33]. The ECP calculations for the $^5\Sigma_u^-$ state of Sc_2 compared well with all-electron results [37] (errors of -.04 a_0, -18 cm^{-1}, and 0.00 eV for R_e, ω_e, and D_e, respectively).

The basis set used for Cu_3 is essentially the Wachters' basis [34] for the ^2S state $4s^1 3d^{10}$ of the Cu atom. This basis set is augmented by the Wachters' 4p functions multiplied by 1.5 and the Hay diffuse 3d function [35]. The inner s and p functions are contracted in a segmented fashion using contraction III as recommended by Wachters, while the 4s and 4p functions are uncontracted, and the 3d functions are contracted (411) leading to an [8s6p3d] basis set. For Cu_2, addition of 4f functions to the basis set decreased R_e by about 0.04 a_0 [38]. This is viewed as a small effect compared to the relativistic effects which are being neglected and these functions were not included here. The small importance of 4f functions for Cu_2 is a reasonable result given that the bonding involves mainly the 4s electrons of Cu, but is in contrast to results for those transition metal dimers which involve nd bonding where the 4f functions have a large effect [39,40,23].

CASSCF/CCI calculations were carried out for Sc_3. Here the active space in the CASSCF included those molecular orbitals derived from 4s, 3dσ, and 3dπ'' atomic orbitals. This choice is based on the discussion in section II. While the 4p should in principle also be included in the active space, these functions mainly contribute to the 4s→4p near degeneracy effect, which may be adequately treated in a subsequent CI calculation. The above atomic orbitals lead to molecular orbitals of a_1' and e' from 4s, of a_1' and e' from $3d\sigma$, and of a_2'' and e'' from $3d\pi''$. This gives a CASSCF calculation of nine electrons in nine orbitals which is small enough that it was possible to do the full CASSCF calculation, i.e. no orbital occupation constraints are necessary. Calculations were carried out at this level in order to determine the low-lying states. Subsequent calculations were then carried out with the number of electrons in a' and a'' orbitals constrained to be the same as in the dominant configuration. While this constraint results in a very substantial reduction in the number of configurations, it has very little effect on the total energy.

The reference configurations for the CCI calculations included the dominant configurations involving correlation of the 3d derived levels. For the $^2A_2''$ state five reference configurations were used consisting of the SCF plus $3da_1' \rightarrow e'$ double excitations(two configurations) and interpair terms between $3da_1'$ and $3da_2''$(two configurations). For the $^2E'$ state seven reference configurations were used. These were the analogous configurations to the five for the $^2A_2''$ state plus the double excitations from $3da_2'' \rightarrow 3de''$(two configurations).

The calculations for Cu_3 were SCF/CI calculations including Davidson's correction as an estimate for higher excitations.

The calculations were carried out using the MOLECULE [41]- SWEDEN [42] system of programs.

IV. Discussion.

From Fig. 2 one sees that the $^2E'$ state of Sc_3 has a minimum in the CASSCF curve near 5.5 a_0. This state is derived from two ground state Sc atoms($4s^23d^1$) and one excited state atom($4s^13d^2$). The $^2E'$ state is bound with respect to this excited atomic limit but is about 0.3 eV above the lowest atomic limit. The $^2A_2''$ and $^2A_1'$ states are slightly higher in energy in this region and also the CASSCF curves are repulsive as was the case for the Ca_3 SCF potential curve. The CCI potential curve for the $^2A_2''$ state is bound by about 1.0 eV for a D_{3h} geometry with R Sc-Sc= 5.75 a_0. An interesting feature of the $^2A_2''$ CCI curve is a pronounced shoulder in the short R region. This feature is due to a curve crossing with another state of the same symmetry, possibly derived from the same atomic limit as the $^2E'$ state. The $^2E'$ CCI curve has a more normal shape and is calculated to be less than 0.1 eV above the $^2A_2''$ state.

The ESR spectrum assigned to Sc_3 [1] is clearly due to Sc_3, based on the presence of the HFS lines and approximate intensity ratios expected for interaction with three equivalent Sc nuclei. From the magnitude of the nuclear HFS interaction it is concluded that the singly occupied orbital contains less than 3 % 4s character. These results are consistent with the calculated $^2A_2''$ ground state, since this state has an equilateral triangle geometry and the singly occupied orbital is predominately 3d like. The $^2E'$ state is found to Jahn-Teller distort, and the singly occupied orbital is mainly of 4s-4p character. both of which are inconsistent with the ESR results. For these reasons we conclude that the the $^2A_2''$ state is the ground state of Sc_3.

Moskovits, DiLeila, and Limm [4] have carried out resonance Raman studies of Ar matrices containing Sc. They report vibrational frequencies at 246, 145, and 151 cm^{-1} which they assign to Sc_3. Upon warming, the two lower frequencies coalesce into one frequency at 150 cm^{-1}. This behavior is interpreted as the removal of a matrix induced asymmetry in the Sc_3 molecule in the originally deposited matrix. As these authors point out, the assignment of these lines to Sc_3 is based on comparison to other systems where the carrier of the spectrum could be established by isotopic substitution, but that this is not possible for Sc_3, due to only one stable isotope of the Sc atom. We calculate symmetric stretch frequencies of 513 cm^{-1} for $^2A_2''$ and 204 cm^{-1} for $^2E'$. The large difference in force constants between these two molecular states probably results from the steeper inner wall for the $^2A_2''$ state as a result of the larger 4s population. Assuming that $^2A_2''$ is the ground state of Sc_3, as suggested by a combination of the ESR results and the computations, the large discrepancy between our calculated symmetric stretch frequency of 513 cm^{-1} and the experimental value of 246 cm^{-1} suggests that the experimentally observed frequencies do not arise from a Raman transition which terminates in the $^2A_2''$ state. However, it is possible that the final state is $^2E'$. This would imply that the separation between $^2A_2''$ and $^2E'$ is very small. The $^2E'$ state need not be thermally populated. Alternative explanations are that the Raman experiment is not observing Sc_3 or that the computed vibrational frequency for the $^2A_2''$ state is too large. However, computed vibrational frequencies tend to be smaller than experiment and to increase as the calculation is improved. Also electronic structure calculations give vibrational frequencies in good agreement with experiment for Sc_2, Ti_2, V_2, and Cu_2 [23].

For Cu_3 we calculate near equilateral triangle geometries to be lowest. The ground state configuration for equilateral triangle geometries is $4sa_1'^2 4se'^1$ which arises from three ground state $4s^1 3d^{10}$ Cu atoms. The calculated vibrational frequencies for the symmetric stretch mode are 208 cm^{-1} for Cu_3 and 214. cm^{-1} for Cu_3^+. It is appropriate to compare the computed Cu_3^+ frequency to the upper state symmetric stretch frequency of 243 cm^{-1}, because we identify the upper state in the transition observed by Morse et al. [6] as a Rydberg state. DiLella et al. report an experimental symmetric stretch frequency of 354 cm^{-1} for Cu_3 [5], but this experiment has been questioned by Morse et al. [6] and by Weltner and VanZee [11].

Fig. 3 shows the Jahn-Teller effect for the $^2E'$ state of Cu_3. Here the Cu-Cu-Cu angle is varied for R (Cu-Cu) fixed at the optimum equilateral triangle value (4.60 a_0). From Fig. 3 one sees that the $^2E'$ state is a maximum on the potential surface with distortion to C_{2v} symmetry leading to a lowering in the energy for both $^2A_1$ and $^2B_2$ symmetries. The $^2A_1$ symmetry favors acute angle geometries, while the $^2B_2$ symmetry favors obtuse angle geometries. Preliminary calculations indicate that the $^2A_1$ minimum is a saddle point for the global potential surface, while the $^2B_2$ minimum is expected to be a true minimum. The $^2B_2$ surface shows a marked asymmetry leading at larger angles to a linear $^2\Sigma_u^+$ state, which is found to be about 0.26 eV higher than for the $^2B_2$ minimum.

From analysis of the hot bands observed in the resonant two photon ionization spectrum of Morse, Hopkins, Langridge-Smith, and Smalley [6], it is concluded that the ground state of Cu_3 is strongly Jahn-Teller distorted, with Jahn-Teller vibrational levels at 15 cm^{-1} and 100 cm^{-1} above the vibrationless level, respectively. The $^2B_2$ ground state is also consistent with the matrix ESR work of Howard, Preston, Sutcliff, and Mile [3], which has been interpreted as an obtuse angle structure with most of the spin density on the end Cu atoms. Note that the observed spin density is consistent with a $^2B_2$ state, but not with a $^2A_1$ state which would have substantial spin density on the central Cu atom. The small separation between the

$^2A_1$ and $^2B_2$ C_{2v} minima is also consistent with the work of Kernisant, Thompson, and Lindsay [2], which shows that for Ag_3 the $^2A_1$ and $^2B_2$ states are inverted by matrix effects.

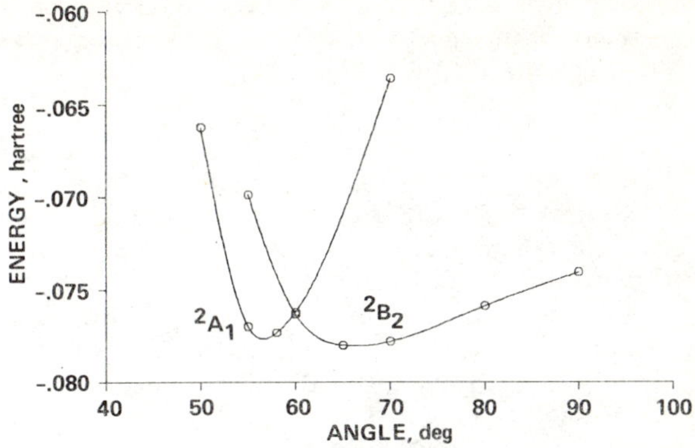

Figure 3. Calculated potential curves for two C_{2v} cuts through the ground state potential surface of Cu_3. Two sides are fixed at R Cu-Cu = 4.60 a_0 and the Cu-Cu-Cu angle is varied.

As discussed in detail elsewhere [22] the computed ground state surface for Cu_3 does not include relativistic effects and has non trivial size consistency and basis set superposition errors. Preliminary calculations by Laskowski et al. [43] using a relativistic ECP and treating the size consistency problem with CPF [44] indicate that the bond lengths will be slightly contracted and the separation between the $^2A_1$ and $^2B_2$ C_{2v} minima may be reduced, but none of the major features of the computed surface will be changed. These effects has also been noted for Cu_2 by Martin [45], by Scharf, Brode, and Ahlrichs [46], and by Werner and Martin [47].

The positive ion of Cu_3 is $^1A_1'$ arising from $4sa_1'^2$. The best estimate of the ionization potential is 5.10 eV based on the experimental excitation energy to the $^2A_1'$ state and the computed I.P. of this state. Experimentally this quantity is not well known but has been bracketed to be greater than 4.98 eV and less than 5.98 eV [6].

Several low-lying states of Cu_3 have been calculated. All but the first few of these are Rydberg states. The upper state in the transition observed by Morse et al. [6] at 2.3 eV has been assigned as $^2E''$ based on i) significant anharmonicity in the upper state vibrational levels which can be fit by assuming a weakly Jahn-Teller distorted upper state and ii) observation of selection rules prohibiting the radiative coupling of the $^2A_1'$ and $^2A_2'$ vibronic levels of the ground state with the vibrationless level of the upper state. However, Morse et al. point out that the upper state levels can also be fit to vibrational frequencies of 149 cm^{-1} and 243 cm^{-1} for a non Jahn-Teller distorted molecule, but with significant anharmonicities. Our calculations support the latter assignment. We calculate a $^2A_1'$ state of 3s Rydberg character at 2.14 eV, which we predict to be about 0.1 eV higher if relativistic effects are included [43]. The lowest state of $^2E''$ symmetry is a 3d Rydberg state calculated to be at 3.32 eV. 3d$\rightarrow$4s transitions were also considered. Although these states are only about 2.0 eV up for linear geometries, the bending potential is quite steep and we predict that they would be much higher than 2.3 eV for equilateral triangle geometries. Given the computed energetics and the fact that the transition observed by Morse et al. is a vertical transition, we assign the upper state as $^2A_1'$. This assignment is also consistent with the observed vibronic selection rules. The presence of anharmonicity for at least the bending modes of this state is expected due to a low-lying linear configuration which leads to observed large anharmonicity in the $^2B_2$ component of the Jahn-Teller distorted ground state.

V. CONCLUSIONS

The bonding in the first row transition metal trimers is found to involve both the 4s and 3d electrons. The detailed nature of the bonding is strongly dependent on the relative 4s and 3d orbital sizes and relative ordering of states for the atom. For the Ca atom the ground state of the atom is $4s^2$ and Ca_3 shows little 3d involvement in the bonding. For equilateral triangle geometries the Ca_3 molecular state arising from three $4s^2$ Ca atoms is bound by about 0.5 eV. The bonding here arises from $4s \rightarrow 4p$ promotion. The lowest state of Sc_3 arises from three ground state $4s^2 3d^1$ Sc atoms. The 4s bonding here is similiar to Ca_3, but the 3d orbitals are also bonding leading to a doublet ground state and an equilibrium Sc-Sc distance of 5.75 a_0 compared to an equilibrium Ca-Ca distance of 7.5 a_0 for Ca_3. The best 3d bond orbitals (no molecular nodal planes) are derived from the $3d\sigma$ and $3d\pi''$ atomic orbitals, while the $3d\pi'$ orbitals have one molecular nodal plane but are bonding between pairs of atoms, and the $3d\delta$ orbitals are essentially non bonding. The lowest state of Sc_3 is:

$$^2A_2'' = 4sa_1'^2\ 4se'^4\ 3da_1'^2\ 3da_2''^1$$

although there is a low-lying $^2E'$ state arising by excitation of an electron from 4se$'$ to 3da$_2''$. The $^2A_2''$ state is consistent with the ESR spectrum which shows equivalent Sc atoms, a doublet spin state, and very little 4s character in the radical orbital.

Moving from Sc_3 toward Cr_3, the $4s^2 3d^n \rightarrow 4s^1 3d^{n+1}$ atomic excitation energy decreases monotonically leading to more $4s \rightarrow 3d$ promotion. Here only the positive ions have been studied. The lowest states of Sc_3^+ and Ti_3^+ have 4s-4p populations near four, while for V_3^+ the 4s-4p population is approaching two. The lower 4s population is crucial in allowing the continued formation of 3d bonds as the 3d orbitals become smaller relative to the 4s orbitals. The additional 3d electrons go into the $3d\pi'$ and $3d\delta$ levels. Here there is a balance between additional 3d bonding associated with placing electrons in bonding orbitals and minimal disruption of exchange interactions for the atom which is initially high spin coupled. For Mn_3, the ground state of the atom is $4s^2 3d^5$ and once again a high 4s-4p population (analogous to Sc_3) is expected, leading to a larger metal-metal bond length. Because of the more contracted 3d orbitals for Mn as compared to Sc, the Mn_3 molecule is not expected to form 3d bonds and the ground state of Mn_3 is predicted to be high-spin. This is found to be the case for Mn_5.

For Cu_3 the ground state arises from three Cu atoms in the $4s^1 3d^{10}$ state. For Cu the 4s orbitals are much larger than the 3d orbitals and there is a closed 3d shell leading to mainly 4s bonding. The lowest state is found to have a near equilateral triangle geometry arising from a $4sa_1'^2 4se'^1$ configuration in D_{3h} symmetry. The low-lying excited states of Cu_3 arise by excitation of a single electron from 4se$'$ to virtual orbitals. Most of these upper states are found to be Rydberg like.

ACKNOWLEDGMENTS

S.P. Walch was supported by a NASA grant(NCC2-296). We would like to acknowledge helpful discussions with Dr. Paul S. Bagus. We would also like to thank Drs. P. J. Hay and W. R. Wadt for making their effective core potentials available to us prior to publication.

REFERENCES

1. L.B. Knight,Jr., R.W. Woodward, R.J. VanZee, and W. Weltner,Jr., J. Chem. Phys., **79**, 5820 (1983).
2. K. Kernisant, G.A. Thompson, and D.M. Lindsay, J. Chem. Phys., **82**, 4739 (1985).
3. J.A. Howard, K.F. Preston, R. Sutcliffe, and B. Mile, J. Phys. Chem., **87**, 536(1983).
4. M. Moskovits, D.P. DiLella, and W. Limm, J. Chem. Phys., **80**, 626(1984).
5. D.P. DiLella, K.V. Taylor, and M. Moskovits, J. Phys. Chem., **87**, 524(1983)
6. M.D. Morse, J.B. Hopkins, P.R.R. Langridge-Smith, and R.E. Smalley, J. Chem. Phys., **79**, 5316(1983)
7. H. Poppa, Vacuum, **34**, 1081(1985); H.Poppa, Ultramicroscopy, **11**, 105(1983).
8. P.G. Hezot, Catal.-Rev. Sci. Eng., **20**, 121(1979).
9. R.D. Moorhead and H. Poppa, Proc. IX Int. Vac. Congress Madrid(1983), Ext. Abstr., p11.
10. G.C. Bond,"Metal support and additive effects in catalysis", B. Imelik et al. Editors., p.1. Elsevier, Amsterdam (1982).
11. W. Weltner,Jr., and R.J. VanZee, Ann. Rev. Phys. Chem., **35**, 291 (1984).
12. M.E. Geusic, M.D. Morse, and R.E. Smalley, J. Chem. Phys., **82** ,590(1985).
13. E.K. Parks, K. Liu, S.C. Richsmeier, L.G. Pobo, and S.J. Riley, J. Chem. Phys., **82**, 5470(1985), and references quoted therein; J. Chem. phys., **82**, 590(1985).
14. C. Backmann, J. Demuynck, and A. Viellard, Faraday Symp. Chem. Soc., **14**, 170 (1980); J. Demuynck, M. Rohmer, A. Strich, and A. Viellard, J. Chem. Phys. **75**, 3443 (1981)
15. H. Tatawaki, E. Migashi, and T. Nakamura, J. Chem. Phys., **76**, 5073 (1982); E. Miyoshi, H. Tatewaki, and T. Nakamura, J. Chem. Phys., **78**, 815 (1982); E. Miyoshi, H. Tatewaki, and T. Nakamura, Int. J. Quant. Chem., **23**, 1201(1983)
16. G.H. Jeung, M. Pelissier, and J.C. Barthelat, Chem Phys Lett., **97** ,369(1983).
17. H. Basch, M.D. Newton, and J.W. Moskowitz, J. Chem. Phys., **73** ,4492(1980).
18. S.C. Richtsmeier, J.L. Gole, and D.A. Dixon, Proc. Natl. Acad. Sci. USA, **77**, 5611 (1980); S.C. Richtsmeier, D.A. Dixon, and J.L. Gole, J. Phys. Chem. **86**, 3937 (1982)
19. P. E. M. Siegbahn, A. Heiberg, B. O. Roos, and B. Levy, Physica Scripta, **21**, 323 (1980); B. O. Roos, P. R. Taylor, P. E. M. Siegbahn, Chem. Phys, **48**, 157 (1980); P. E. M. Siegbahn, J. Almlof, A. Heiberg, and B. O. Roos, J. Chem. Phys., **74**, 2381 (1981).
20. P.E.M. Siegbahn, Int. J. Quantum Chem., **23**, 1869(1983).
21. S.P. Walch and C. W. Bauschlicher, Jr., submitted to J. Chem. Phys.

22. S.P. Walch and B. C. Laskowski, submitted to J. Chem. Phys.
23. S.P. Walch and C.W. Bauschlicher,Jr., in "Comparison of Ab Initio Quantum Chemistry with Experiment", edited by R.J. Bartlett, Reidel Publishing, p. 17.
24. C.E. Moore, Atomic Energy Levels, Circular 467 (National Bureau of Standards, Washington, 1949)
25. S.P. Walch, H. Partridge, and C.W. Bauschlicher,Jr., unpublished results
26. C.W. Bauschlicher,Jr., P.S. Bagus, and B.N. Cox, J. Chem. Phys., **77**, 4032(1982).
27. S.P. Walch and C.W. Bauschlicher,Jr., Chem. Phys. Lett., **94**, 290(1983).
28. S.P. Walch and C.W. Bauschlicher,Jr., J Chem. Phys., **79**, 3590(1983).
29. C. Kittel, Introduction to Solid State Physics. 3rd ed. (Wiley, New York, 1967).
30. C. Baumann, R.J. VanZee, S. Bhat, and W. Weltner,Jr., J. Chem. Phys., **74**, 6977 (1981); J. Chem. Phys. **78**, 190 (1983)
31. P.J. Hay, and W.R. Wadt, J. Chem. Phys., **82**, 299(1985).
32. O. Gropen, U. Wahlgren, and L. Pettersson, Chem. Phys., **66**, 459(1982).
33. L.G.M. Pettersson, U. Walgren, and O. Gropen, Chem. Phys., **80**, 7(1983).
34. A.J.H. Wachters, J. Chem. Phys., **58**, 4452 (1973).
35. P.J. Hay, J. Chem. Phys., **66**, 4377 (1977).
36. L.G.M. Pettersson, P.E.M. Siegbahn, and S. Ismail, Chem. Phys., **82**, 355(1983).
37. S.P. Walch, and C.W. Bauschlicher,Jr., J. Chem. Phys., **79**, 3590(1983).
38. C.W. Bauschlicher,Jr., S.P. Walch, and P.E.M. Siegbahn, J. Chem. Phys., **76**, 6015(1982).
39. A.D. McLean, and B. Liu, Chem. Phys. Lett., **101**, 144(1983).
40. S.P. Walch, C.W. Bauschlicher,Jr., B.O. Roos, and C.J. Nelin, Chem. Phys. Lett, **103**, 175(1983).
41. J. Almlof, MOLECULE, a gaussian integral program.
42. P.E.M. Seigbahn, C.W. Bauschlicher,Jr., B. Roos, A. Hieberg, P. R. Taylor, and J. Almof, SWEDEN, a vectorized SCF MCSCF direct CI.
43. B.C. Laskowski, S.R. Langhoff, and C.W. Bauschlicher,Jr., to be published.
44. R. Ahlrichs, S. Scharf, and C. Ehrhardt, J. Chem. Phys., **82**, 890(1985).
45. R. Martin, J. Chem. Phys., **78**, 5840(1983).
46. P. Scharf, S. Brode, and R. Ahlrichs, Chem. Phys. Lett., **113**, 447 (1985).
47 H. Werner and R.L. Martin, Chem. Phys. Lett., **113**, 451 (1985).

Electronic Structure and Reactions of Transition Metal Complexes Using Effective Core Potentials

P. Jeffrey Hay and Celeste McMichael Rohlfing
Theoretical Division, MS-J569
Los Alamos National Laboratory
Los Alamos, New Mexico USA 87545

ABSTRACT. Effective core potentials are employed to study the geometries and binding energies of the species MCO and $M(CO)_4$ for M=Ni, Pd and Pt, and addition reactions of H_2 to $W(d^6)$ and $Pt(d^{10})$ centers. Møller-Plesset perturbation theory is assessed as a means of incorporating electron correlation effects in transition metal species.

1. INTRODUCTION

A theoretical understanding of the structures, chemistry, and photochemistry of transition metal complexes will require the reliable calculation of geometries and relative energies of the reactants, products, and transition states. In this report we summarize some of the recent advances employing effective core potentials (ECP's) for carrying out electronic structure calculations on transition metal species and we present representative examples in the above areas.

The ECP's employed in these calculations rigorously replace the core electrons by an effective one-electron potential that is required to produce a nodeless pseudo-orbital ϕ, which has the same orbital energy (ϵ_i) as the Hartree-Fock orbital (ϕ_i), and which has the identical shape as ϕ_i in the valence region of the atom. In addition, for the heavier elements (Z > 36), the relativistic mass-velocity and Darwin terms are implicitly incorporated into the relativistic effective core potentials (RECP), which are derived from all-electron relativistic Hartree-Fock calculations.

Recently, we have published a series of ECP's in analytic form for integral evaluation and the corresponding basis sets for the valence orbitals.[1-3] These include the transition metals Sc-Hg and the main group elements Na-Bi. A third series of ECPs was generated for the elements K-Hg including the outer "core" orbitals. In cases such as the K atom, where the "core" 3s and 3p orbitals can overlap significantly with electrons on other atoms, the core approximation breaks down and these electrons must be explicitly included along with the other valence electrons.

A. Veillard (ed.), Quantum Chemistry: The Challenge of Transition Metals and Coordination Chemistry, 135–141.

2. STRUCTURES AND ENERGIES OF METAL CARBONYLS

We have recently completed[4] a systematic study of the transition metal monocarbonyls MCO and tetracarbonyls $M(CO)_4$ for M=Ni, Pd and Pt employing ECP's on the metal atoms. A 4-31G basis was used on C and O, and a [1s, 1p, 2d] contracted basis was used to describe the nd and (n+1)s,p orbitals of the metal. Geometry optimizations were carried out at the Hartree-Fock (HF) level, and also at the second-order Møller-Plesset perturbation theory (MP2) level to assess the effects of electron correlation. Other recent investigations on NiCO include ECP calculations by Rives and Fenske[5] and all-electron calculations by Ha and Nguyen[6] and by Blomberg et al.[7] All-electron studies of $Ni(CO)_4$ with geometry optimization include Faegri and Almlof[8] and Blomberg et al.[7] A more thorough review of the literature, including the Pd and Pt species, is given in Ref. 4.

Table I gives the geometries and dissociation energies of the $^1\Sigma^+$ state of the monocarbonyls. We shall examine the SCF results first. The r(M-C) bond lengths fall in the predicted order, Ni < Pd $\leq$ Pt. The r(C-O) distances are slightly longer than that in free CO, indicating that some backbonding to the CO π^* is occurring. The Pd and Pt compounds have similar r(C-O) separations, while in the Ni complex this distance is longer. This is in accord with stronger π back-donation in NiCO than in PdCO and PtCO as suggested by the metal atom ionization potentials. SCF dissociation energies relative to ^1S metal plus free CO yield the same relative stabilities as experiment, i.e., NiCO > PtCO $\geq$ PdCO. The MP2 results parallel exactly all of the trends found at the SCF level. Inclusion of correlation strengthens the M-C interaction through increased backbonding, thus shortening r(M-C) and lengthening r(C-O). The increased importance of π back-donation at the MP2 level is shown by the substantial lengthening of the monocarbonyl r(C-O) relative to MP2-optimized free CO. The magnitude of this r(C-O) stretch is much larger than the corresponding lengthening observed at the SCF level. The SCF-to-MP2 changes in the monocarbonyl r(C-O)'s are also larger than the corresponding change in free CO.

For the $^3\Delta$ states no bonding is apparent at the SCF level, while MP2 optimization yields weakly bound states for NiCO and PtCO with M-C bond lengths of 2.2 and 2.4 Å, respectively. Thus the ground state of linear MCO is predicted to be $^1\Sigma^+$.

Table II lists the bond lengths and dissociation energies calculated for the $^1A_1$ states of $Ni(CO)_4$, $Pd(CO)_4$, and $Pt(CO)_4$. First, the trend in the SCF r(M-C) bond length, Ni < Pd < Pt, is as expected, and follows that for the $^1\Sigma^+$ monocarbonyls, with the Pd-C and Pt-C bond lengths being similar. The relative stabilities as predicted by the SCF dissociation energies are $Ni(CO)_4 > Pt(CO)_4 > Pd(CO)_4$, in agreement with experiment.

MP2 geometry optimizations were performed for $Ni(CO)_4$ and $Pt(CO)_4$, and it was found that the r(C-O) bond length was nearly the same for both molecules. Hence only the Pd-C distance was optimized at the MP2 level for $Pd(CO)_4$. Relative to the SCF results, the MP2 r(M-C) bond lengths are shorter due to increased backbonding, and the r(C-O) bond lengths are longer. MP2 increases the latter by 0.05 Å to a distance slightly longer than that in MP2-optimized free CO. The metal-carbonyl interaction is still weaker than that in $^1\Sigma^+$ MCO, since the average bond strengths in the tetracarbonyls are calculated to be 2.41, 1.25, and 1.46 eV for Ni, Pd, and Pt, respectively. This is not an unexpected result since there exists a competition among four equal ligands in the tetracarbonyls.

The agreement between experiment and the results in Table II is satisfactory. The relative stabilities of the three tetracarbonyls are predicted correctly. The MP2 bond lengths and strengths for $Ni(CO)_4$ agree fairly well with the experimental values, considering the basis set size and method of correlation. In conclusion, SCF and MP2 yield parallel trends. Correlation effects are very similar between the mono- and tetracarbonyls. The Pd species are found to be the least stable of the series Ni, Pd, and Pt.

TABLE I. Results for the $^1\Sigma^+$ state of MCO and the $^1\Sigma^+$ state of CO. Bond lengths in Å and dissociation energies in eV.

		R(M-C)	R(C-O)	D_e(to ^{1}S M)	D_e(to ^{3}D M)
NiCO	SCF	1.844	1.134	1.30	-3.44
	MP2	1.711	1.208	4.05	2.52
	expt.[a]	--	--	2.97 ±0.65	1.23 ±0.65
PdCO	SCF	2.056	1.130	0.57	0.44
	MP2	1.882	1.185	1.62	2.48
PtCO	SCF	2.073	1.129	0.86	-0.18
	MP2	1.977	1.184	1.93	1.62
CO	SCF	--	1.128	--	--
	MP2	--	1.172	--	--
	expt.[b]	--	1.128	--	--

[a] Reference 9.

[b] K. P. Huber and G. Herzberg, Constants of Diatomic Molecules (Van Nostrand Reinhold Co., New York, 1979.)

TABLE II. Results for $^1A_1$ state of $M(CO)_4$. Bond lengths in Å and dissociation energies in eV.

M		R(M-C)	R(C-O)	D_e (to ^1S M)
Ni	SCF	1.971	1.129	3.10
	MP2	1.873	1.181	9.64
	expt.[a)]	1.82-1.84	1.12-1.15	6.96-7.81
Pd	SCF	2.169	1.128	1.64
	MP2	2.032	(1.178)	4.98
Pt	SCF	2.202	1.128	2.22
	MP2	2.100	1.178	5.82

[a)] See Reference 4 for a summary of the experimental results.

3. ADDITION REACTIONS OF H_2 TO METAL CENTERS

The well-established oxidative addition reaction of H_2 to metal centers

$$ML_n + H_2 \rightarrow L_nMH_2$$

has often been postulated in catalytic cycles. The reaction has been observed for d^8 and d^{10} complexes. The model reaction for ML_n=$Pt(PH_3)_2$ has been studied by numerous investigators[10-12] using Hartree-Fock and correlated wavefunctions and employing relativistic ECP's on the Pt atom. Earlier calculations[10,11] showed the stable existence of the cis - L_2MH_2 species, the initial species formed in this reaction. Although the trans form appears to be the more stable, particularly with bulky PR_3 groups, the cis form has been detected recently in solution.[13]

Table III summarizes the results of the various studies to date including the present results with the recent[1] ECP for Pt. The previous studies[10-12], which did not use "norm-conserving" ECP's, differ in the basis sets employed and (in the case of Ref. 10) in the geometry constraints assumed in the calculations.

Recently Kubas et al.[14] have prepared 5-coordinate compounds $M(PR_3)_2(CO)_3$, where M=Mo and W, that reversibly add hydrogen. Preliminary x-ray diffraction studies indicated that the H_2 bond remained intact, in contrast to the usual situation discussed above for d^8 and d^{10} complexes. <u>Ab initio</u> studies[15] were carried out on various geometries of the model compound $W(PH_3)_2(CO)_3H_2$(see Fig. 1), including 6-coordinate side-on bonded and end-on bonded H_2 forms and the 7-coordinate dihydride form. The side-on bonded forms were found to be stable relative to the $WL_5 + H_2$ fragments by 17 kcal/mole. The lowest energy form had the H-H axis parallel to the P-W-P axis but with very little barrier (0.3 kcal/mole) to rotation about the $W-H_2$ axis (Table IV). The W-H distance was calculated to be 2.15 Å at the Hartree-Fock level and the H-H separation was found to be 0.79 Å, only 0.05 Å longer than in H_2 itself. The published structure[14] showed inequivalent W-H bonds of 1.76 and 2.12 Å and an H_2 distance of 0.75 ± 0.16 Å from low temperature x-ray diffraction. Room temperature neutron diffraction studies suggested R(W-H) = 1.75 Å and R(H-H) = 0.86 Å, although there were problems arising from disorder in the crystal. More recent unpublished low temperature neutron diffraction studies show equivalent W-H bonds 1.89 ± 0.01 Å in length, an H_2 separation of 0.82 ± 0.01 Å, and the H_2 lying parallel to the P-W-P axis in accord with the theoretical results.

Calculations on the unobserved 7-coordinate dihydride,[15] corresponding to the oxidative addition product, find it 17 kcal/mole higher than the side-on bonded form--or nearly equal to the energy of the separate fragments. When PH_3 groups are substituted for the π-accepting CO groups, we find the oxidative addition is 2 kcal/mole exothermic, compared with 17 kcal/mole endothermic when three CO ligands are present. These results suggest that ligand substitution could be used to favor either the dihydrogen or dihydride forms. Experimentally, 7-coordinate $Mo(PR_3)_5(H)_2$ complexes have been observed to support this hypothesis.

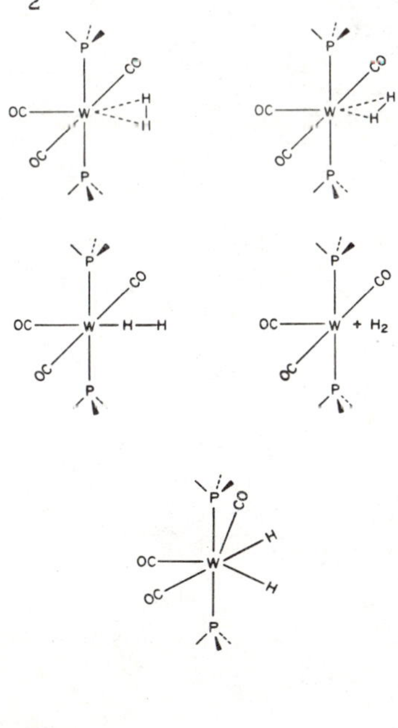

Figure 1. Various geometries of the model compound $W(PH_3)_2(CO)_3 + H_2$.

TABLE III. Reaction energies (E_{react}), activation energies (E_{act}), and Pt-H distances for the transition state in the $Pt(PH_3)_2 + H_2$ reaction.

	E_{react}	E_{act}	Transition State R(Pt-H)
	(kcal/mole)	(kcal/mole)	(Å)
	Hartree-Fock Wavefunctions		
Noell, Hay[10]	-16	17	1.81
Kitaura et al.[11]	-37	5.2	2.07
Low, Goddard[12]	-6.7	4.0	2.20
Hay	-4.2	8.4	1.70
	Correlated Wavefunctions		
Noell, Hay[10]	-5.0	16.6	[1.76]
Kitaura et al.[11]	-27	7.1	[2.07]
Low, Goddard[12]	-15.9	2.3	2.42
Hay--MP2	7.5	0.8	[1.70]
Hay--MP3	-8.6	4.0	[1.70]

TABLE IV. Relative energies of various forms of $W(L_1)_3(L_2)_2(H_2)$ complexes.

	Relative energy, (kcal/mole)	R(W-H), (Å)
L_1 = CO, L_2 = PH_3		
Side-on complex	0.0	2.15
Side-on complex	0.3	2.18
End-on complex	11	2.44
Fragments	17	∞
7-Coordinate complex	17	1.89
L_1 = L_2 = PH_3		
Side-on complex	0	1.84
7-Coordinate complex	-2	1.84

4. REFERENCES

1. P.J. Hay and W. R. Wadt, J. Chem. Phys. **82**, 270 (1985).
2. W. R. Wadt and P. J. Hay, J. Chem. Phys. **82**, 284 (1985).
3. P. J. Hay and W.R. Wadt, J. Chem. Phys. **82**, 299 (1985).
4. C. M. Rohlfing and P. J. Hay, J. Chem. Phys., in press.
5. A. B. Rives and R. F. Fenske, J. Chem. Phy. **65**, 1293 (1981).
6. T. K. Ha and M. T. Nguyen, J. Molec. Struct. **109**, 331 (1984).
7. M. R. A. Blomberg, U. B. Brandemark, P. E. M. Siegbahn, K. B. Mathisen, and G. Karlstrom, J. Chem. Phys. **89**, 2171 (1985).
8. K. Faegri, Jr. and J. Almlof, J. Chem. Phys. Lett. **107**, 121 (1984).
9. A. E. Stevens, C. S. Feigerle, and W. C. Lineberger, J. Amer. Chem. Soc. **104**, 5026 (1982).
10. J. O. Noell and P. J. Hay, J. Am. Chem. Soc. **104**, 4578 (1982).
11. K. Kitaura, S. Obara and K. Morokuma, J. Am. Chem. Soc. **103**, 2891 (1981).
12. J. J. Low and W. A. Goddard III, J. Am. Chem. Soc. **102**, 6933 (1984).
13. R. S. Paonessa and W. G. Trogler, J. Am. Chem. Soc. **104**, 1138 (1982).
14. R. Kubas, R. Ryan, B. Swanson, P. Vergamini, and H. Wasserman, J. Am. Chem. Soc. **106**, 451 (1984).
15. P. J. Hay, Chem. Phys. Lett. **103**, 466 (1984).

CORRELATION EFFECTS IN THE GROUND AND IONIZED STATES OF TRANSITION METAL COMPLEXES

Ian H. Hillier
Chemistry Department
University of Manchester
Manchester M13 9PL
United Kingdom

ABSTRACT. Correlation effects in the ground states of bis(π-allyl)nickel and $Ni(CN)_4^{2-}$ have been studied using configuration interaction calculations. In both species a redistribution of electron density occurs increasing the density on the metal atom compared to that given by a RHF wavefunction. The valence ionization energies of $Ni(CN)_4^{2-}$, $Co(CN)_6^{3-}$ and $Fe(CN)_6^{4-}$ have been measured using X-ray emission spectroscopy. These data and the photoelectron spectrum of bis(π-allyl)nickel have been compared with the ionization energies calculated using the extended-two-particle-hole Tamm-Dancoff method, which includes both relaxation and correlation effects. For all species studied, excellent agreement between theory and experiment is found, which, for bis(π-allyl)nickel, is absent for ΔSCF-CI calculations.

INTRODUCTION

Photoelectron (PE) spectroscopy has proved an invaluable aid in the study of molecular electronic structure, particularly when combined with theoretical calculations of molecular ionization energies (IE) [1]. For molecules consisting of first and second row atoms IEs obtained from *ab initio* calculations [2], suitably scaled [3], often provide a satisfactory assignment of the experimental spectra. More accurate values may be calculated by considering not only orbital relaxation effects (ΔSCF method), but also the differential correlation between the unionized and ionized species [4]. For transition metal complexes, the situation is more complicated since Koopmans' theorem is often totally inadequate due to the considerably larger orbital relaxation which may arise from metal, compared to ligand, ionization [4,5]. For a range of transition metal complexes, ΔSCF calculations, which include such differential orbital relaxation, have been extremely useful in assignment of their PE spectra. However, to date, the question of the role of correlation effects in the prediction of the ionization energies of transition metal complexes has been largely unanswered.

A. Veillard (ed.), Quantum Chemistry: The Challenge of Transition Metals and Coordination Chemistry, 143–157.

From the experimental viewpoint, it is often necessary to study a series of related molecules in order to obtain a credible assignment of the PE spectrum of transition metal complexes and even then doubts as to the assignment frequently remain. It is possible, in principle, to obtain further information on the energies and more importantly the nature of the ionized states via the use of X-ray emission (XRE) spectroscopy, in which the emission spectrum accompanying valence→core transitions is observed. The use of core electron ionization energies, obtained from ESCA measurements, allows the XRE spectra involving transitions to core holes on different atoms in the molecule to be aligned. From such aligned spectra, information on the atomic character of the ionized states can be obtained, due to the operation of dipole selection rules, and to the dominant contribution of one-center terms to the corresponding transition moment. Thus, in the field of transition metal complexes, it is feasible, at least in principle to use this technique to obtain the relative energies of those ionized states arising from metal and ligand electron ionization. Such information is not so readily available from PE spectroscopy.

Two of the first transition metal complexes studied by <u>ab initio</u> methods were bis(π-allyl)nickel [$Ni(C_3H_5)_2$] [6,7] and the tetracyanonickelate anion [$Ni(CN)_4^{2-}$] [5]. Veillard and coworkers recognized the importance of including orbital relaxation effects in the calculation of the valence IEs of these molecules. Since these original calculations, additional experimental studies on the PE spectrum of bis(π-allyl)nickel have suggested that ΔSCF calculations lead to an incorrect ordering of the ionic states, and furthermore we have recently obtained XRE spectra of the transition metal cyanides, $Co(CN)_6^{3-}$, $Ni(CN)_4^{2-}$ and $Fe(CN)_6^{4-}$ which lead to the relative energies of metal and ligand valence electron ionization. In this paper we briefly present our experimental data, and discuss the calculation of the valence IEs of these cyanides, and of bis(π-allyl)nickel including both relaxation and correlation effects.

EXPERIMENTAL DETAILS AND RESULTS

XRE spectra of $K_2Ni(CN)_4$, $K_3Co(CN)_6$ and $K_4Fe(CN)_6$ were obtained using the photoelectron converter technique with the apparatus previously described [8]. The exciting radiation was obtained from the storage ring at the SERC Daresbury Laboratory, using the soft X-ray line, 20 metres from the tangent point. This line incorporates a chromium coated mirror in the horizontal plane to provide a high energy cut-off which can be varied from 4 to 20 Å by changing the angle of incidence of the radiation with the mirror. The X-ray photons were energy analysed using Ne(2p) as the converter gas for the C Kα, and C_6H_6(C1s) for the N Kα and the Ni, Co and Fe L$\alpha_{1,2}$, emissions, which arise from transitions from the valence orbitals to the C1s, N1s and Ni, Co and Fe $2p_{3/2}$ core holes respectively. For each molecule, the three XRE spectra were aligned on a common energy scale using measurements of the corresponding core electron IEs obtained using a conventional ESCA spectrometer (Table I).

TABLE I. MEASURED VALENCE IEs (eV) OF $Ni(CN)_4^{2-}$, $Co(CN)_6^{3-}$ AND $Fe(CN)_6^{4-}$

	$K_2Ni(CN)_4$	$K_3Co(CN)_6$	$K_4Fe(CN)_6$
Cls IE	283.7	283.3	284.8
C Kα XRE peak	278.5	278.3	278.6
Valence IE	5.2	5.0	6.2
Nls IE	396.9	396.6	398.1
N Kα XRE peak	392.1	392.1	391.8
Valence IE	4.8	4.5	6.3
$M2p_{3/2}$ IE	855.6	781.3	708.6
M $L\alpha_{1,2}$ XRE peak	851.4	778.2	706.8
Valence IE	4.2	3.1	1.8

For $K_4Fe(CN)_6$ our XRE spectra are in good agreement with those obtained previously by more conventional methods [9]. For $K_2Ni(CN)_4$, the XRE spectra thus obtained are shown in Figure 1.

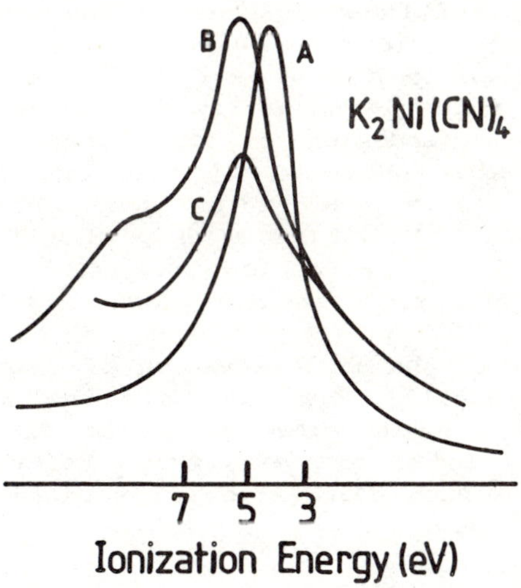

Figure 1. X-ray emission spectra of $K_2Ni(CN)_4$.A, Ni $L\alpha_{1,2}$; B,C Kα; C, N Kα.
The spectra are aligned using the corresponding core IEs to yield the ionization energy scale given.

The measured IEs corresponding to the N Kα, C Kα and M L$\alpha_{1,2}$ peaks are given in Table I, where it can be seen that for all three anions, the lowest IEs correspond to metal 3d ionization, closely followed by those ionic states arising from ligand ionization. It is to be noted that the separation of the metal L$\alpha_{1,2}$ line from the ligand Kα lines increases along the series $Ni(CN)_4^{2-}$, $Co(CN)_6^{3-}$, $Fe(CN)_6^{4-}$.

We now describe the results of our calculations of the bonding and IEs of bis(π-allyl)nickel and the three cyanide complexes.

THEORETICAL CONSIDERATIONS

There are two, well-proven, approaches to the accurate calculation of the ionization energies of molecules lacking a transition metal atom. In the first, SCF calculations are carried out on both the ground and various ionic states, correlation effects being included by carrying out configuration interaction (CI) calculations on each state. Ionization energies are then obtained as the energy difference between the ground and ionic state (ΔSCF CI method) [4]. To avoid carrying out individual calculations on both the unionized, and various ionic states, several authors have used the single-particle propagator or Green's function techniques to calculate ionization energies directly taking account of both relaxation and correlation effects. Both approaches have been successful in yielding ionization energies accurate to ~0.5 eV for a range of organic molecules [10,11], and accurate to ~0.2 eV for smaller molecules for which extended basis sets can be used [12].

However, the extension of these methods to realistic transition metal complexes which may be studied experimentally, can present formidable technical difficulties. Thus, a double zeta representation of the valence orbitals, which is generally deemed the minimum size needed for a meaningful discussion of correlation effects, may generate more than 100 basis functions. The 4-index transformation for such a basis size is a non-trivial computational problem. Furthermore, the size of the CI expansion generated from the large number of valence electrons in the complex, and from a basis of this size, may be upwards of 5×10^5, again presenting computational problems.

Over the last few years, the implementation of quantum chemistry codes on vector processors [13], combined with the development of direct CI methods [14], has increased the ease of carrying out calculations on transition metal complexes which include correlation effects. We now describe such calculations for bis(π-allyl)nickel, $Ni(CN)_4^{2-}$, $Co(CN)_6^{3-}$ and $Fe(CN)_6^{4-}$.

COMPUTATIONAL DETAILS

The calculations were carried out using contracted Gaussian functions. For carbon and nitrogen the (9s5p) basis of Huzinaga [15] was contracted to (3s2p) [16]. The metal bases were constructed from the (12s6p4d) functions of Roos et al. [17]. The two most diffuse s functions were replaced by those having exponents of 0.32 and 0.08. An additional p function having an exponent of 0.32, and a diffuse d function were added [18]. These functions were contracted (6s3p2d), giving a close to double zeta representation of the valence region. Calculations were carried out at the experimental geometries.

For bis(π-allyl)nickel, restricted Hartree Fock (RHF) calculations were carried out on the ^1A_g ground state of the neutral molecule, and lowest ^2A_g, ^2B_g, ^2A_u and ^2B_u states of the ion. Correlation effects were considered in all 5 states by carrying out CI calculations. In general, the CI expansion consisted of single and double excitations from a single root, the RHF configuration. All valence electrons, except the carbon 2s were correlated in this way using all the virtual orbitals.

The second method used to calculate the valence IEs of bis(π-allyl)nickel and of $Ni(CN)_4^{2-}$, $Co(CN)_6^{3-}$ and $Fe(CN)_6^{4-}$ is a Green's function method. The IEs and their relative intensities (pole strengths) appear as the poles and the residues of the one-particle Green's function and can thus be calculated directly, instead of as the difference of the energies resulting from two independent calculations for the unionized ground state and for the various ionized states, as in the ΔSCF-CI method. The method includes both correlation and relaxation effects. Two distinct approximations to the one-particle Green's function are generally employed. These are discussed in detail in references 19-22. In the outer valence region where a one-particle picture holds for the physical description of the ionization process (i.e where ionization can be reasonably well described by the ejection of an electron from a ground state molecular orbital) the outer-valence Green's function (OVGF) may be used [19,20]. If there is very strong relaxation in the sense that the SCF orbitals for the ionized system are a strong mixture of the ground state SCF orbitals, plus the virtual orbitals, as is the case for bis(π-allyl)nickel, $Ni(CN)_4^{2-}$, $Co(CN)_6^{3-}$ and $Fe(CN)_6^{4-}$ for metal localized orbitals, or if in a configurational expansion of the ionic wavefunction the single hole configurations mix strongly with configurations involving ionization plus excitation, as is the case in the inner valence region, then the extended two-particle-hole Tamm Dancoff method (extended 2ph-TDA) should be used [21,22]. This latter method is applicable in the entire valence region. Typical for the inner valence region is the appearance of satellite lines which borrow their intensity from the ionization processes which lead to simple hole states, and the possible disappearance of main lines with the intensity being distributed over many lines. This effect has been termed the breakdown of the molecular orbital model of ionization [23]. Both the OVGF and the extended 2ph-TDA methods are accurate to third order in the electron-electron interaction in the outer valence region. However, in the inner valence region, when the breakdown of the orbital model of ionization occurs, the structure of the ionization spectrum is only described qualitatively by the extended

2ph-TDA calculations. This is due to a) the complicated configurational structure of the states and to the missing higher excitations and to b) basis set inadequacies in the neighbourhood of a continuum of states.

For bis(π-allyl)nickel, the extended 2ph-TDA calculation included all the filled valence orbitals including the carbon 2s, and 20 virtual orbitals. For $Ni(CN)_4^{2-}$, $Co(CN)_6^{3-}$ and $Fe(CN)_6^{4-}$ these calculations included all the filled valence MOs which correlate with the 1π, 4σ and 5σ MOs of CN^- and with the metal d orbitals, together with the lowest 22 virtual orbitals.

COMPUTATIONAL RESULTS

THE GROUND STATES OF BIS(π-ALLYL)NICKEL, $Ni(CN)_4^{2-}$, $Co(CN)_6^{3-}$ AND $Fe(CN)_6^{4-}$.

RHF calculations of the singlet ground states of bis(π-allyl)nickel and $Ni(CN)_4^{2-}$ yielded the same description of the bonding as that obtained previously by Veillard and coworkers [5,7], although the exact ordering of the closely-spaced MOs is somewhat basis set dependent. For bis(π-allyl)nickel, the highest filled MO, $7a_u$ involves no metal character, and derives from the allyl a_2 π-MOs. The other MO involving these ligand orbitals, $6b_g$, is the second highest filled orbital, and has the greatest contribution to the metal-ligand bonding, via interaction with the nickel $3d_{xz}$ orbital (we here use the conventional axis labelling in C_{2h}, rather than that used in reference 7). The antisymmetric combination of the ligand b_1 π-MOs, $11b_u$, is non-bonding, whilst the symmetric combination interacts with the metal $3d_{xy}$ orbital leading to the $9a_g$ and $13a_g$ MOs. The $10a_g$, $5b_g$ and $11a_g$ MOs are mainly non-bonding, consisting of the metal $3d_{z^2}$, $3d_{yz}$ and $3d_{x^2-y^2}$ orbitals respectively. The remaining valence orbitals involve mainly the σ-framework of the ligands. The energies of these valence orbitals are used in Table II, where it can be seen that Koopmans' theorem predicts the mainly metal ionizations to occur at considerably higher energy than the ligand π ionizations.

TABLE II. IONIZATION ENERGIES (eV) OF THE VALENCE ORBITALS OF BIS(π-ALLYL)NICKEL

Orbital	Character	Koopmans' Theorem	ΔSCF	ΔSCF-CI	extended 2ph-TDA	expt[a]
$7a_u$	π(L)	7.5	6.8	6.7	6.4	7.7(1)
$6b_g$	$3d_{xz}$; π(L)	9.0	5.6	6.6	7.7	8.1(2)
$11b_u$	π(L)	11.8	11.0	10.8	10.3	10.3(5)
$13a_g$	$3d_{xy}$; π(L)	11.7	5.5	6.4	7.6	8.1(2)
$12a_g$	σ(L)	14.0			13.5	12.7(7)
$5b_g$	$3d_{yz}$	14.0			8.5	
$11a_g$	$3d_{x^2-y^2}$	14.2			8.2	8.5(3)
$6a_u$	σ(L)	14.6			13.3	
$10b_u$	σ(L)	14.6			13.4	12.7(7)
$4b_g$	σ(L)	15.0			13.7	
$10a_g$	$3d_{z^2}$	15.3			8.8	9.4(4)
$9a_g$	$3d_{xy}$; π(L)	16.4			11.5	11.5(6)
$5a_u$	σ(L)	16.5			14.9	
$3b_g$	σ(L)	17.3			15.1	14.2(8)
$9b_u$	σ(L)	18.0			16.2	
$8a_g$	σ(L)	19.0			16.5	15.6(9)

a The band number is given in parentheses.

The results of the RHF calculations on $Ni(CN)_4^{2-}$, $Co(CN)_6^{3-}$ and $Fe(CN)_6^{4-}$ are summarized in Tables III, IV and V.

TABLE III. CALCULATED VALENCE IONIZATION ENERGIES (eV) OF $Ni(CN)_4^{2-}$

Orbital	Character	Ionization Energy	
		Koopmans' Theorem	extended 2ph-TDA
$1a_{2g}$	π	3.9	3.1
$8e_u$	π	3.9	2.5
$2e_g$	π	4.2	4.4
$1b_{2u}$	π	4.3	3.6
$2b_{2g}$	π	4.8	2.9
$3a_{2u}$	π	4.9	4.3
$5b_{1g}$	5σ	5.7	4.0
$7e_u$	5σ	5.8	4.2
$9a_{1g}$	5σ	5.9	2.7
$8a_{1g}$	d_{z^2}	7.3	3.4
$1e_g$	d_{xz}, d_{yz}	8.4	2.2
$4b_{1g}$	4σ	8.5	6.5
$6e_u$	4σ	8.9	6.7
$1b_{2g}$	d_{xy}	10.2	5.4
$7a_{1g}$	4σ	11.7	8.1

TABLE IV. CALCULATED VALENCE IONIZATION ENERGIES (eV) OF $Co(CN)_6^{3-}$

Orbital	Character	Ionization Energy	
		Koopmans' Theorem	extended 2ph-TDA
$8t_{1u}$	π	0.7	−0.8
$1t_{1g}$	π	1.1	0.4
$1t_{2u}$	π	1.5	0.8
$2t_{2g}$	π	2.0	2.1
$5e_g$	5σ	3.0	0.8
$7t_{1u}$	5σ	3.1	1.5
$8a_{1g}$	5σ	3.6	1.5
$4e_g$	4σ	5.5	3.1
$6t_{1u}$	4σ	5.8	3.5
$1t_{2g}$	3d	7.2	−2.1
$7a_{1g}$	4σ	9.1	6.0

TABLE V. CALCULATED VALENCE IONIZATION ENERGIES (eV) OF $Fe(CN)_6^{4-}$.

Orbital	Character	Ionization Energy	
		Koopmans' Theorem	extended 2ph-TDA
$2t_{2g}$	3d	-4.2	-9.2
$8t_{1u}$	π	-3.8	-5.4
$1t_{1g}$	π	-3.4	-4.1
$1t_{2u}$	π	-3.0	-3.8
$5e_g$	5σ	-2.5	-4.7
$7t_{1u}$	5σ	-1.6	-3.0
$1t_{2g}$	π	-1.1	-2.7
$8a_{1g}$	5σ	-0.8	-2.8
$4e_g$	4σ	0.7	-1.3
$6t_{1u}$	4σ	1.1	-0.9
$7a_{1g}$	4σ	3.7	0.8

In the nickel and cobalt species, the highest filled MOs correlate with the 1π and 5σ orbitals of CN^-, and the more strongly bound orbitals with the 4σ orbitals of CN^- and with the metal d orbitals. By comparison with our reported XRE spectra of these molecules, we see immediate confirmation of the well-known inadequacies of Koopmans' theorem for these molecules. In the case of $Fe(CN)_6^{4-}$, the highest filled MO, $2t_{2g}$ is predominantly 3d in character, although here the metal-ligand separation predicted by Koopmans' theorem is much smaller than the experimental value.

We now summarize the results of the CI calculation on the ground state of bis(π-allyl)nickel. It was found that the RHF configuration contributed 83% to this CI wavefunction, a value considerably less than that usually found for molecules lacking a transition metal atom (>95%). The major correlating configurations were found to be the single and double excitations $6b_g \rightarrow 7b_g$ and $6b_g^2 \rightarrow 7b_g^2$. The $7b_g$ MO is the metal-ligand antibonding counterpart of the $6b_g$ bonding MO. Thus, the important $6b_g^2 \rightarrow 7b_g^2$ configuration yields the left-right correlation of the metal-ligand bonding electrons, a correlation effect absent in the isolated metal atom and allyl ligands. Both the $6b_g$ and $7b_g$ MOs are composed of Ni 3d and allyl π-orbitals. However, the Ni 3d character is greater in the $7b_g$ (43%) than in the $6b_g$ (38%) so that this excitation corresponds to an increase in the population of the 3d(b_g) atomic orbitals. Although these two $6b_g \rightarrow 7b_g$ excitations are the most important, they contribute less than 2% of the total wavefunction. The effect of all the correlating configurations additional to the RHF configuration which make up 18% of the total wavefunction can be seen from a population analysis of the natural orbitals of this CI wavefunction. The major effects of correlation on the molecular charge distribution are

(i) an increase in the 3d(b_g) populations and an associated decrease in the π-populations of the terminal carbon atoms of the allyl ligands, corresponding to correlation of the metal-ligand bonding electrons.

(ii) a smaller decrease in the population of the 3d(a_g) non-bonding electrons and an increase in the π-population of the central carbon atoms of the allyl ligands. This may be associated with in-out type correlation of the non-bonding 3d electrons.

The net result of these two effects is an increase (0.07e) in the metal 3d electron density and a decrease (0.03e) in the ligand π density.

THE CALCULATED IONIZATION ENERGIES OF BIS(π-ALLYL)NICKEL

The valence ionization energies, calculated by the various methods, are summarized in Table II. As in previous studies [6,7], Koopmans' theorem predicts the mainly metal 3d MOs ($10a_g$, $11a_g$, $5b_g$) to be considerably more tightly bound than the ligand π MOs ($7a_u$, $11b_u$). However, when direct SCF calculations are carried out on the ion states, the considerably greater reorganization energy associated with metal, compared to ligand, ionization, results in the ground state of the ion being predicted to be ^2A_g, arising from metal 3d ionization, in agreement with the calculations of Veillard and coworkers. The results of our CI calculations on the lowest four ion states of different symmetry show that considerably greater correlation energy is associated with the ^2A_u and ^2B_u states arising from ligand ionization, than from the ^2A_g and ^2B_g states which correspond to metal ionization. For the ^2A_u and ^2B_u states, the most important correlating configurations are similar to those in the ^1A_g ground state, and involve the $7b_g$ MO, whose characters are similar in these ion states and in the ^1A_g ground state. Thus, the important correlation effects which we have identified in the ground state, are also present in the ^2A_u and ^2B_u states, and indeed appear to be slightly more important in the latter two.

Examination of both the metal orbital populations, and the metal-ligand overlap populations in the RHF ^2A_u and ^2B_u states reveals similar values to those found in the ^1A_g ground state. Thus, similar correlation energies associated with the metal and metal-ligand bonding electrons in these three states are to be expected, as found by our CI calculations.

However, in the RHF ^2A_g and ^2B_g ion states, there is substantial electron reorganization accompanying metal electron ionization, leading to ligand → metal electron migration. As a result, the total metal 3d populations in the ^2A_g and ^2B_g ions are close to 9 electrons (8.9 in both cases). The $3d_{xz}$ (b_g) population is 1.8 in the ^2A_g state and 1.4 in the ^2B_g state, so that there is no strongly metal-ligand bonding electron pair in either state, as witnessed by the decrease in the bond overlap populations in these two states, compared with that in the ^1A_g ground state. There is thus a reduction of correlation energy in the ^2A_g and ^2B_g ion states compared to that found in the ^2B_u and ^2A_u ion states and in the ^1A_g ground state associated mainly in the loss of the correlation energy associated with the metal-ligand bonding electrons in the $6b_g$ MO. The differential correlation energy associated with the (^2A_u, ^2B_u) and (^2A_g, ^2B_g) ion states is reflected in the ΔSCF-CI ionization energies shown in Table II. Thus, whereas the ^2A_g - ^2A_u separation is 1.3 eV at the ΔSCF level, the introduction of electron correlation reduces this difference to 0.3 eV. However, at both the ΔSCF and ΔSCF-CI level, the ground ionic state is predicted to be ^2A_g.

The ionization energies calculated by the extended 2ph-TDA approximation (Table II), are, as expected, closer to the ΔSCF-CI, than to the ΔSCF results. However, the electron reorganization and correlation given by this method now places the ^2A_u as the ground ionic state, separated by 1.2 eV from the ^2A_g state. This method also allows all the valence ionization energies to be calculated, rather than just the first of each symmetry given by our ΔSCF CI calculations.

We now discuss the valence PE spectrum of bis(π-allyl)nickel in the light of these calculations.

THE PHOTOELECTRON SPECTRUM OF BIS(π-ALLYL)NICKEL

The PE spectrum of bis(π-allyl)nickel (Figure 2) shows nine bands

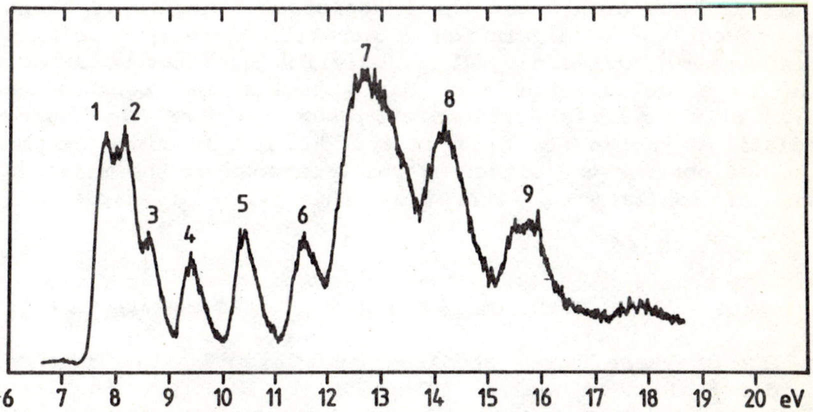

Figure 2. Photoelectron spectrum of bis(π-allyl)nickel [24].

below 17 eV, and was originally assigned by Batich [24] on the basis of methyl-substituent effects and using both He(I) and He(II) radiation, as follows: The first, third and fourth bands are due to nickel 3d ionizations. The second is due to 3d plus a ligand π-orbital, and the fifth and six are due to ligand π-orbitals. On the basis of a comparison of the PE spectra of a series of nickel, palladium and platinum bis(π-allyl) derivatives an alternative assignment has been presented [25]. Here the first and fifth bands arise from the $7a_u$ and $11b_u$ ligand MOs respectively, bands two and three from the mainly metal $9a_g$, $10a_g$, $11a_g$ and $5b_g$ MOs, and bands four and six from the $6b_g$ and $13a_g$ orbitals respectively having significant ligand π-character. Bands seven to nine are assigned to orbitals of allyl σ-character. A Green's function approach based upon the semi-empirical INDO method yields [26] an assignment in complete agreement with the above, although it is of interest to note that the INDO method predicts considerably smaller deviations from Koopmans' theorem than are given by _ab initio_ methods. Our _ab initio_ Green's function calculation yields an assignment of the PE spectrum close to that deduced by Bohm _et al_ [26]. In particular the ground ionic state is predicted to be ^2A_u. The major difference is an interchange of the $9a_g$ and $13a_g$ MOs. However, both MOs have similar atomic orbital characters so that any distinction between them must be somewhat arbitrary. In addition the ordering of the closely spaced states arising from the $6b_g$ and $10a_g$ MOs is inverted in our calculation.

The valence ionization spectrum has been calculated by the extended 2ph-TDA method in the energy range up to about 30 eV. In the energy range up to 11 eV, the lines have a large relative intensity, but frequently the intensity is borrowed from many ground state orbitals, particularly in the case of a_g and b_g symmetries. This is a reflection of the strong electron reorganization in the electron ejection from orbitals with a strong nickel d character. Already at about 11 eV satellite lines start to appear and they soon become dense in energy. It is thus clear that the PE spectrum above 10 eV cannot be explained without the consideration of satellite lines. A detailed discussion is however not meaningful as only the qualitative aspects are amenable to a calculation for this molecule and because the calculation is only strictly applicable in the limit of high energy exciting radiation. Above about 17 eV the density of lines becomes very large and we observe the effect of the breakdown of the molecular orbital model of ionization. There are thus no more simple hole states.

THE VALENCE IONIZED STATES OF $Ni(CN)_4^{2-}$, $Co(CN)_6^{3-}$ AND $Fe(CN)_6^{4-}$.

The valence IEs of these three species, calculated by the extended 2ph-TDA method are given in Tables III to V. For $Ni(CN)_4^{2-}$ the first IE at 2.2 eV, and the one at 4.4 eV involve strong mixtures of both Koopmans' theorem configurations arising from the metal $1e_g$ MO and the ligand $2e_g$ MO, reflecting the electron reorganization occurring upon ionization. A similar situation arises for the b_{2g} and a_{1g}

ionizations, where significant mixing between the various Koopmans' theorem configurations occurs. In contrast, the ionic states of other symmetries have a single dominant Koopmans' theorem component arising from ligand ionization. When comparing our calculated IEs with the values obtained from the experimental XRE spectrum (Table I), it should be noted that due to the neglect of the counter-ion, our calculated IEs will be smaller than the experimental values. In our comparison between theory and experiment, we are thus concerned with relative IEs. The single peak in the Ni $L\alpha_{1,2}$ spectrum clearly arises from those ionic states which correlate with $1b_{2g}$, $1e_g$, $8a_{1g}$, $9a_{1g}$ and $2e_g$ ionizations spanning a calculated energy range of 3.2 eV (Table III). All of these states have significant contributions from configurations arising from ionization from orbitals having substantial Ni3d character. The maximum in the Ni $L\alpha_{1,2}$ emission may be estimated to correspond to an ionization energy of ~3.3 eV, since most intensity is likely to be associated with the degenerate ^2E_g states correlating with $1e_g$ and $2e_g$ ionization. The major peak in both the C and N $K\alpha$ spectra clearly arises from states associated with ligand 1π and 5σ MOs, orbitals having significant 2p character. The corresponding calculated IEs span an energy range 2.5-4.4 eV placing these peaks very close to the estimated Ni $L\alpha_{1,2}$ maximum at 3.3 eV. Indeed the spectra (Figure 1) show substantial overlap of the three major emission bands, in agreement with the results of our calculations. The pronounced shoulder observed in the C $K\alpha$ emission to ~3 eV higher IE of the main peak is assigned to transitons from those MOs which correlate with the 4σ orbitals of CN^-. This observed energy separation is in good agreement with our calculation (Table III).

The IEs computed for $Co(CN)_6^{3-}$ (Table IV) show a similar trend to those for $Ni(CN)_4^{2-}$. Upon the inclusion of relaxation and correlation effects, the IEs of those MOs of mainly ligand 1π and 5σ character decrease by about 1-2 eV, those of ligand 4σ character decrease by 2-3 eV, whilst there is a much larger decrease in the metal localized IE, here about 9 eV. The 2ph-TDA results predict the first IE of $Co(CN)_6^{3-}$ arises from the metal localized $1t_{2g}$ MO, with a pronounced gap of greater than 1eV between it, and the first ligand ionization. In the case of $Fe(CN)_6^{4-}$, the 2ph-TDA calculations yield a splitting of the metal and ligand IEs of more than 3 eV, compared to the value of less than 1eV at the Koopmans' theorem level of approximation. Thus, along the series $Ni(CN)_4^{2-}$, $Co(CN)_6^{3-}$, $Fe(CN)_6^{4-}$, the calculated separation of the metal and ligand IEs is < 1 eV, 1-2 eV and more than 3 eV. These values are to be compared with our experimental estimates of 0.5 eV, 1.5 eV and 4.5 eV. (Table I).

CONCLUSIONS

The configuration interaction calculations we have here reported for bis(π-allyl)nickel have shown that the important correlation effects, absent in the isolated metal and ligand entities lead to an increase in metal electron density. We have found a similar result for

$Ni(CN)_4^{2-}$ where the population of the metal $d_{x^2-y^2}$ orbital (b_{1g}) is increased (by 0.06e) upon the inclusion of correlation effects. Thus, in both molecules, electron correlation leads to a less polar situation. A similar decrease in ionicity has been shown to occur in ferrocene [27] and leads to a decrease in the metal-ligand bond length. It is thus probable that calculations beyond the Hartree-Fock description are particularly important, even for a semi-quantitative description of the electronic structure of transition metal complexes.

The correct assignment of the ionized states of transiton metal complexes is a continuing challenge to both theoreticians and experimentalists. We have here shown that XRE spectroscopy can be extremely useful in distinguishing between metal and ligand IEs and in this respect provides a critical test of the valence IEs of transition metal complexes computed by different methods.

The 2ph-TDA method is particularly attractive for computing the whole manifold of valence IEs of transition metal complexes. In addition to avoiding the number of 4-index transformations needed in ΔSCF-CI methods, the necessity of obtaining a number of roots of very large matrices is also removed. The calculations of IEs of bis(π-allyl)nickel, $Ni(CN)_4^{2-}$, $Co(CN)_6^{3-}$ and $Fe(CN)_6^{4-}$ which we have reported are the first to employ an ab initio Green's function technique on transition metal complexes. The agreement we obtain between such theory and experiment is extremely encouraging. Thus, for bis(π-allyl)nickel, our assignment of the PE spectrum agrees with that obtained experimentally except for some ionizations of similar atomic character which are difficult to distinguish experimentally. For the transiton metal cyanides which we have studied, the calculated separation of the metal and ligand ionizations are in excellent agreement with the results from XRE spectra.

To obtain the correct ordering of the closely spaced levels arising from metal and ligand ionization requires a balanced calculation of relaxation and correlation effects. The ΔSCF method allows for the calculation of all the relaxation effects, which lower the metal relative to the ligand IEs. For bis(π-allyl)nickel, correlation effects increase the metal relative to the ligand IEs. Due to the difficulty of recovering such differential correlation effects, the ΔSCF-CI results lead to an incorrect ordering of the metal and ligand ionized states. The success of the 2ph-TDA calculations must be ascribed to the more balanced treatment of these relaxation and correlation effects.

ACKNOWLEDGEMENTS

The work described herein was carried out in collaboration with M.F. Guest, A.A. MacDowell, D. Moncrieff, W. von Niessen, V.R. Saunders, D. Taylor, D.S. Urch and M. Vincent. Support from the Science and Engineering Research Council (U.K.) is gratefully acknowledged.

REFERENCES

[1] Schwartz, M.E. in 'Modern Theoretical Chemistry', Schaefer, H.F., ed; Plenum Press: New York, 1977; Vol.4, p.358.

[2] Richards, W.G., Int.J.Mass Spectrom.Ion.Phys. 1969, 2, 419.

[3] Brundle, C.R.; Robin, M.B.; Basch, H., J.Chem.Phys. 1970, 53, 2196.

[4] Hillier, I.H., Pure and Appl.Chem. 1979, 51, 2183.

[5] Veillard, A.; Demuynck, J. in "Modern Theoretical Chemistry', Schaefer, H.F., ed; Plenum Press: New York, 1977; Vol.4, p.187.

[6] Rohmer, M-M.; Veillard, A., J.Chem.Soc.(Chem.Comm.) 1973, 250.

[7] Rohmer, M-M.; Demuynck, J.; Veillard, A. Theoret.Chim.Acta.(Berl.) 1974, 36, 93.

[8] MacDowell, A.A.; Hillier, I.H.; West, J.B. J.Phys. 1983, E16, 487.

[9] Bergknut, L.; Andermann, G.; Haycock, D.; Kasrai, M.; Urch, D.S. J.Chem.Soc., Faraday Trans.2, 1981, 77, 1879.

[10] von Niessen, W.; Cederbaum, L.S.; Kraemer, W.P.; Diercksen, G.H.F. J.Am.Chem.Soc. 1978, 98, 2066.

[11] von Niessen, W.; Kraemer, W.P.; Cederbaum, L.S. J.Electron Spectrosc. 1976, 8, 179; Schirmer, J.; Domcke, W.; Cederbaum, L.S.; von Niessen, W. J.Phys. 1978, B11, 1901.

[12] von Niessen, W.; Cederbaum, L.S.; Diercksen, G.H.F. J.Chem. Phys. 1977, 67, 4124; and references cited there.

[13] Saunders, V.R.; Guest, M.F. Comp.Phys.Comm. 1982, 26, 389.

[14] Saunders, V.R.; van Lenthe, J.H. Mol.Phys. 1983, 48, 923.

[15] Huzinaga, S. J.Chem.Phys. 1965, 42, 1293.

[16] Dunning, T.H. J.Chem.Phys. 1970, 53, 2823.

[17] Roos, B.; Veillard, A.; Vinot. G. Theoret.Chim.Acta.(Berl.) 1971, 20, 1.

[18] Hay, P.J. J.Chem.Phys. 1977, 66, 4377.

[19] Cederbaum, L.S.; Domcke, W. Adv.Chem.Phys. 1977, 36, 205.

[20] Cederbaum, L.S. Theoret.Chim.Acta.(Berl.) 1973, 31, 239; J.Phys. 1975, B8, 290.

[21] Schirmer, J.; Cederbaum, L.S. J.Phys. 1978, B11, 1889.

[22] Schirmer, J.; Cederbaum, L.S.; Walter, O. Phys.Rev. 1983, 28, 1237.

[23] von Niessen, W.; Schirmer, J.; Cederbaum, L.S. Computer Phys.Rep. 1984, 1, 57; and references cited there.

[24] Batich, C.D. J.Am.Chem.Soc. 1976, 98, 7585.

[25] Bohm, M.C.; Gleiter, R.; Batich, C.D. Helv.Chim.Acta. 1980, 63, 990.

[26] Bohm, M.C.; Gleiter, R. Theoret.Chim.Acta.(Berl.) 1980, 57, 315.

[27] Taylor, T.E.; Hall, M.B. Chem.Phys.Lett. 1985, 114, 338.

Analysis of σ-bonding, π-(back)bonding and the synergic effect in $Cr(CO)_6$. Comparison of Hartree-Fock and Xα results for metal-CO bonding.

E.J. Baerends and A. Rozendaal
Chemistry Department
Vrije Universiteit
De Boelelaan 1083
1081 HV Amsterdam
The Netherlands

Abstract.

LCAO-Xα calculations with large STO basis sets are used to establish the relative contributions of σ-bonding, π-(back)bonding and notably the synergic effect to the total metal-CO bond strength in $Cr(CO)_6$. It appears that the π-bond is about twice as strong as the σ-bond. The synergic effect makes a contribution of the same order of magnitude as the σ- and π-bonds. The Cr-C and C-O bond distances are optimized, and the A_{1g} vibration frequencies are calculated and compared to both Hartree-Fock calculations and experiment. Contrary to the Hartree-Fock model, which gives generally too low metal-ligand bond energies and too long bonds, the Xα model leads to too large metal-ligand bond energies, too short bond lengths and too high metal-CO vibration energies. The reduction of the C-O stretch frequency upon coordination is calculated very accurately.

I. Introduction.

The metal-CO bond in transition-metal carbonyl complexes is one of the best known and most extensively studied coordinative bonds. The interest stems partly from the fact that CO may act as both a σ-base, through the 5σ carbon lone-pair orbital, and as a π-acid through the 2π orbital. After a considerable amount of semi-empirical work, several of the early ab-inito studies of transition-metal complexes dealt with carbonyl complexes [1,2]. These systems have also been studied by Xα methods, both with the multiple-scattering formalism [3] and with LCAO calculations [4]. All of these studies focused on an analysis of the molecular orbitals, mostly by Mulliken population analysis, in order to obtain a picture of the coordinative bond. Bond energies were seldom quoted: they were indeed disappointingly far

A. Veillard (ed.), Quantum Chemistry: The Challenge of Transition Metals and Coordination Chemistry, 159–177.

off the experimental values, $Ni(CO)_4$ for instance not even being bound with respect to the Ni atom and CO molecules in their ground states at the Hartree-Fock level [2,5].
Today there is a fair understanding of the problems associated with Hartree-Fock calculations on transition-metal complexes. It has been recognized that in Hartree-Fock calculations on the transition-metal atoms the energetic separation between the $d^{n+1}\ s^1$ and $d^{n+2}\ s^0$ configurations on one hand and the $d^n\ s^2$ configuration on the other hand (to which the atomic ground state mostly belongs) is strongly overestimated. Ni(d^{10}; ^1S) is for instance computed at 4.19 eV above the experimental ground state Ni(d^9s^1; ^3D), whereas the experimental excitation energy is 1.74 eV. The Ni (d^8s^2; ^3F) state is experimentally 0.03 eV above the ^3D groundstate but Hartree-Fock finds it 1.26 eV*below* the ^3D (d^9s^1) groundstate. It is clear that Hartree-Fock strongly favours 4s occupation, the energy difference between the lowest states of the d^8s^2 and d^{10} configurations being 1.71 eV experimentally and 5.45 eV in Hartree-Fock. However, the $d^n\ s^2$ and $d^{n+1}\ s^1$ configurations are well known to be very unsuitable for bonding to ligands, the diffuse filled 4s shell leading to strong exchange repulsion with donor orbitals on the ligands such as the CO 5σ. It is well known indeed that transition metal atoms in complexes have the "valence state" configuration $d^{n+2}\ s^0$ rather than $d^n\ s^2$ (e.g. Ni is d^{10} in $Ni(CO)_4$, not d^9s^1 or d^8s^2). Unfortunately the error in the HF energy for Ni d^{10} is so large that when an effectively d^{10} Ni atom is present in an SCF calculation on $Ni(CO)_4$ the energy is far too high, and therefore the bond energy with respect to CO molecules and Ni (d^9s^1; ^3D) far too low. A related problem is the poor representation of the transition metal-ligand bond lengths in SCF calculations. These tend to be far too long as has been extensively documented for metallocenes [6-9] and for carbonylcomplexes [9-10]. One way to understand this is again the incorrect preference of the Hartree-Fock model for 4s-occupation instead of 3d occupation, leading to too much exchange repulsion. As has been pointed out by Lüthi [9], however, there are other aspects of the metal-ligand bond that play a role here, such as the different ranges of metal-ligand distance which are optimal for σ-bonding and π-bonding in metal-CO coordination. The π-bond develops optimal strength at distances which are too short for the bulky 5σ (the C lone pair) which at such distance already has considerable repulsion with the 3d electrons. The correlation effects diminish this repulsion and allow shortening of the metal-CO bond. This picture is very much in line with the difference between π- and σ-bonds in diatomic molecules we recently observed in a momentum-space study [12] of the chemical bond in the first-row diatomics Li_2-F_2. The same difference in distance dependence for σ- and π-bonding in the metal-CO bond as alluded to by Lüthi has indeed been observed in a momentum-space study of $Cr(CO)_6$ [13]. The two problems cited here, the bond energy and the bond length, have been extensively studied on the model systems Ni-CO [14-18] and Cu_2 [19,20].

Apart from the efforts to obtain quantitative agreement between calculation and experiment, there has also been a growing interest in the qualitative picture of the metal-ligand bond, notably the relative importance of σ (donative) bonding and π-(back)bonding, and the role of exchange repulsion and electrostatic interaction (cf. the methods and applications by Morokuma and coworkers [21-23]).
For the metal-CO bond are particularly relevant the investigation by Ziegler on the bonding of CO and other ligands (N_2, PF_3) to a $Ni(CO)_3$ fragment [24], and by Bauschlicher and Bagus on $Ni(CO)_4$ and $Fe(CO)_5$ [5]. Similar investigations on model systems for CO adsorption have been carried out by Post and Baerends [25, 27] and by Bagus and coworkers [28-30].

In this paper we reinvestigate the bonding in $Cr(CO)_6$ using the Xα model with a two-fold purpose. In the first place it is interesting to see to which extent the problems associated with the application of the Hartree-Fock model on transition-metal complexes are also present in the Xα model. Although the theoretical foundation of Xα, and local density methods in general, is rather weak, they have shown surprisingly good agreement with experiment in many cases. As the local density approximation has been exploited in computational schemes [31-35] that are considerably more efficient than current ab initio Hartree-Fock programs, it is useful to establish the limits of validity of the local density approaches. There are indeed indications that, in contrast to the under-binding found in Hartree-Fock, the Xα model leads to over-binding in the case of metal-ligand bonds [36].

We have for instance found too strong Cu-CO bonds for CO bound at hollow sites to Cu clusters [27], and also the Li-CO bond appears to be too strong [25,37]. It is relevant to mention here that, although the atomic multiplet problem has not been investigated in nearly as much detail in the HFS model as in the HF model, there are indications that the situation is rather different in the two models. For instance we have computed the $Ni(d^{10}; {}^1S)$ state at 1.45 eV above the ${}^3D(d^9s^1)$ ground state (exp. 1.74 eV, HF 4.19 eV).

In the second place we use a partitioning of the total bond energy into various components such as electrostatic interaction + exchange repulsion (=steric repulsion), σ-bonding and π-(back)bonding in order to obtain insight in the relative importance of these well-known bonding mechanisms, to establish the magnitude of the synergic effect, and to compare to the bonding of CO to clusters of metal atoms.

II. Method; results for CO and Cr.

The one-electron equation to be solved is

$$f\,\varphi_i = \varepsilon_i\,\varphi_i\,,$$

$$f = -1/2\,\nabla^2 + V_N + V_C + V_{X\alpha}\,, \tag{1}$$

where V_N and V_C represent the Coulomb potentials due to the nuclei and the electrons repectively, and $V_{X\alpha}$ is Slater's exchange potential:

$$V_{X\alpha}(1) = -3\alpha\,[3/8\pi\,\rho\,(1)]^{1/3} \tag{2}$$

The scaling constant α is always chosen as 0.70 [38]. We use the scheme developed by Baerends et al. [32] in which the molecular orbitals φ_i are expanded in a basis set of Slater-type orbitals (exponential functions) and the Fock matrix elements are calculated with a numerical integration method. In order to reduce computation time and yet get an accurate representation of the Coulomb and exchange potentials the charge density is expanded in an auxiliary set of exponential functions centred on the nuclei:

$$\rho\,(\mathbf{r}) \rightarrow \tilde{\rho}\,(\mathbf{r}) = \sum_i a_i\,f_i\,(\mathbf{r}). \tag{3}$$

The expansion coefficients a_i are determined in a least-squares fitting procedure, i.e. minimizing the deviation

$$D = \int (\rho - \tilde{\rho})^2\,d\mathbf{r}, \tag{4}$$

with the constraint of charge conservation.

Table 1. STO basissets and fitfuntionsets used in the $Cr(CO)_6$ calculations. The same fitfunctions have been used in the DZ and extended basis calculations.

Basissets			Fitfunctionsets		
Cr	C	O	Cr	C	O
double-zeta:					
1s 18.70	1s 5.40	7.36	1s 37.40	1s 10.88	1s 14.72
2s 8.45	2s 1.24	1.70	2s 27.15	2s 10.00	2s 14.94
2p 9.90	2s 1.98	2.82	3s 21.40	2s 6.68	2s 9.08
3s 3.25	2p 0.96	1.30	4s 19.70	3s 6.70	3s 10.46
3s 5.00	2p 2.20	3.06	4s 12.75	3s 4.42	3s 8.16
3p 2.85			5s 11.40	3s 3.38	3s 6.16
3p 4.65			5s 8.40	3s 2.56	3s 4.60
3d 1.24			5s 6.50	3s 1.64	3s 3.44
3d 2.70			6s 6.00	4s 1.50	3s 2.24
3d 5.70			6s 4.25	2p 7.60	2p 10.42
4s 1.75			7s 4.00	2p 5.03	2p 6.90
4s 1.00			7s 2.75	3p 4.89	3p 6.71
4p 2.00			7s 2.00	3p 3.46	3p 4.74
extended:			5g 12.00	3p 2.44	3p 3.35
DZ +	DZ 2s +	DZ 2s +	5g 8.50	4p 2.28	4p 3.13
4f 1.00	2p 2.94	2p 4.08	5g 6.02	4p 1.68	4p 2.30
4f 2.00	2p 1.48	2p 2.08	5g 4.27	3d 7.90	3d 9.36
	2p 0.82	2p 1.12	5g 3.02	3d 5.58	3d 6.61
	3d 3.00	3d 2.50	5g 2.14	3d 3.94	3d 4.67
	3d 1.50	3d 1.20		4d 3.68	4d 4.36
				4d 2.70	4d 3.21
				5d 2.48	5d 2.94
				4f 4.70	4f 5.06
				4f 3.46	4f 3.72
				4f 2.55	4f 2.74
				5g 5.00	5f 2.51
				5g 3.79	5g 4.00
				5g 2.87	5g 3.03
					5g 2.30

In table 1 the basis sets and fit function sets are specified. The extended basis set is triple-zeta in the Cr 3d and C and O 2p AOs and double-zeta in the other valence AOs as well as in the 4f and 3d polarization functions on Cr and C and O respectively. In table 2 we compare a number of calculated properties of CO to experiment and to converged basis results obtained in completely numerical calculations [39]. We refer to ref. [40] for a detailed discussion of basis set effects in CO, in particular the importance of polarization functions (compare the DZ and extended basisset results in table 2).

Fig. 1 illustrates the need for polarization functions in order to get close to converged basis results. Saturating the s,p basis from double-zeta (DZ) through triple-zeta (TZ) to quadruple-zeta (QZ) does not nearly have as much effect as adding a single 3d polarization

Table 2. Spectroscopic constants of CO from various Hartree-Fock-Slater calculations. (α=0.70) compared to experiment.
DZ: double zeta basis (table 1)
EXT: extended basis (table 1)
CONV: converged basis, a completely numerical (basis-free) calculation [39].

	DZ	EXT	CONV	EXP.	
R_e	2.173	2.131	2.128	2.132	bohr
f_{CO}	16.1	19.0	19.1	19.0	mdynÅ
ω_e	1997.	2170.	2174.	2170.	cm^{-1}
$\omega_e x_e$	14.4	13.5	-	13.3	"
B_e	1.93	1.93	-	1.93	"
α_e	.021	.018	-	.0175	"
D_e	10.1	11.9	12.1	11.2	eV

function (cf. the DZD curve). Given the importance of polarization functions in free CO we felt obliged to retain them in the calculations on the carbonyl complex. As also the Cr basis is very large, the results are believed to be close to the basisset limit.

The fit function set is also very extensive. In fig. 2 we compare the density difference for CO, i.e. $\rho(CO) - \rho(C) - \rho(O)$, with the difference $\rho(CO) - \tilde{\rho}(CO)$. Clearly the auxiliary set is perfectly able to fit the density on the scale of the density variations induced by the chemical bond formation, which are themselves in most regions already small compared to the total density.
As for the chromium atom we note that the ground state is a high spin ^7S corresponding to the $3d^5 4s^1$ configuration, whereas in the $Cr(CO)_6$ complex the atomic configuration is essentially $(dt_{2g})^6 (de_g)^0 = (d_{xz})^2 (d_{yz})^2 (d_{xy})^2 (d_{x^2-y^2})^0 (d_{z^2})^0$. For our bond energy analysis we will mostly use the closed shell $Cr(t_{2g})^6$ atom as our metal atom fragment which interacts with 6 CO molecules, or with a $(CO)_6$ cage. In table 3 the relative energy of the Cr atom ground state (^7S) with respect to the $(t_{2g})^6$ "valence state" is given. It is quite high, 9.4 eV. It is well known that the Xα model overestimates the effect of spin polarization, so this large energy difference is maybe not very realistic. It may be more useful when e.g. comparing Xα and Hartree-Fock to take the $(t_{2g})^6$ atom as a common reference point. We will return to this point when discussing the metal-CO bond energy. In table 3 the energy of the $(de_g)^4 (4s)^2$ configuration is also given, as this configuration will be used in the study of some factors in the bond energy (*vide infra*).

Table 3. Relative energies of Cr atom "valence states" with respect to the ^7S ground state (Hartree-Fock-Slater calculations with extended basis of table 1 and α=0.70).

$Cr(d^5s^1;\ ^7S)$	0.00 eV
$Cr(dt_{2g})^6$	9.423 eV
$Cr(de_g)^4(4s)^2$	13.325 eV

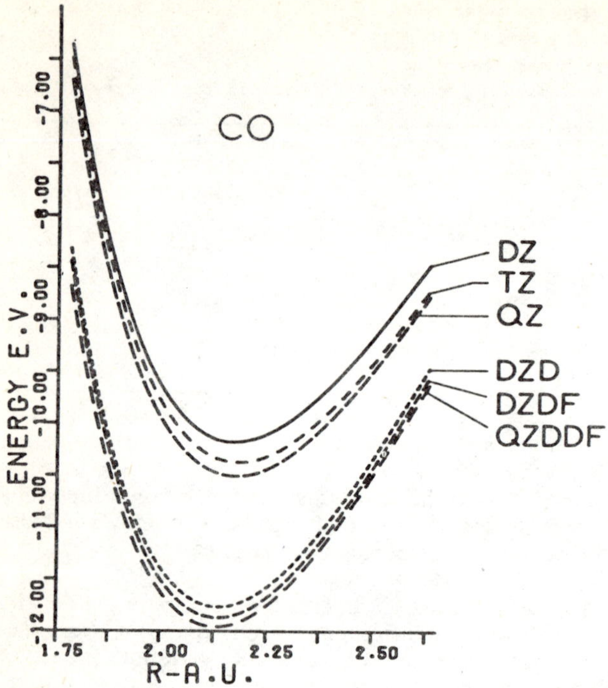

Fig. 1. The bond energy of CO $\Delta E = E(CO) - E(C) - E(O)$ plotted as function of R for various basissets. The quality of the valence (2s,2p) basisset is double-zeta (DZ), triple-zeta (TZ) or quadruple-zeta (QZ). D and F denoted 3d and 5f polarization functions.

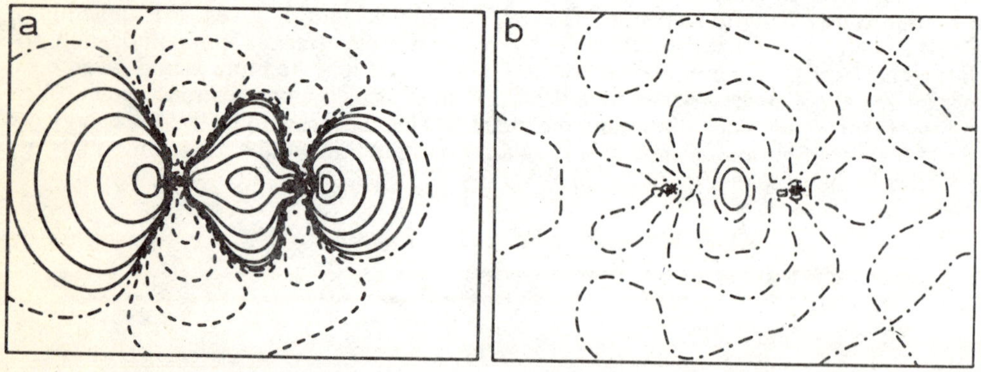

Fig. 2. Electron density differences for CO in TZDF basis. a) molecule minus C and O atoms. b) true density ρ minus fit density $\tilde{\rho}$. Contours at: 0., ± 0.001, ± 0.003, ± 0.01, ± 0.03, ± 0.10, ± 0.15, ± 0.20 e/bohr^3. Drawn lines: positive; dashed lines: negative; dash-dot lines: zero contour. (C atom to the left).

In the analysis of the metal-CO interaction we use a decomposition of the total bond energy between molecular fragments which is in the same spirit as the ones used by Morokuma and coworkers [21], Whangbo and Wolfe and coworkers [41], Bernardi, Bottoni et al. [42], Stone and Erskine [43], Ziegler [24], Bagus et al. [29] and many others. Our method is based on the transition-state method for bond energies by Ziegler and Rauk [44] , and has been described and applied before in studies of CO interacting with small clusters of metal atoms [27]. The first term in the decomposition of the bonding energy is the steric repulsion, which is defined as the difference between the energies of the separate fragments and the energy E^0 corresponding to the determinantal wavefunction

$$\Psi^0 = | \varphi_1^A \varphi_2^A \dots \varphi_N^A \psi_1^B \psi_2^B \dots \psi_M^B | \tag{5}$$

constructed from the (non-orthogonal!) occupied SCF orbitals φ_i^A and ψ_j^B of the fragments A and B:

$$\Delta E_{steric} \equiv \Delta E^0 = E^0 - E^A - E^B . \tag{6}$$

Because of the non-orthogonality of φ_i^A and ψ_j^B one cannot use the Condon-Slater rules for the matrixelement $< \Psi^0 | H | \Psi^0 > = E^0$, and also the one-electron density ρ^0 corresponding to Ψ^0 is not simply the sum of ρ^A and ρ^B. If we apply a linear transformation to the set of orbitals in (5), such that the resulting orbitals are orthogonal (e.g. a Schmidt or Löwdin transformation) the wavefunction Ψ^0 is not changed, and the energy and density may be evaluated in the usual way. It is sometimes useful to consider separately the electrostatic interaction energy, which is defined as

$$\Delta E_{elstat} = \int \rho^A(1)\, \rho^B(2) / r_{12}\, dr_1\, dr_2 + \int V_N^A(1)\, \rho^B(1)\, dr_1 + \int V_N^B(1)\, \rho^A(1)\, dr_1 , \tag{7}$$

where V_N^A (V_N^B) is the nuclear potential of fragment A (resp. B). The steric repulsion ΔE^0 is considered to be the sum of the electrostatic interaction energy and the exchange repulsion ΔE_{XREP}, which is defined by the relation:

$$\Delta E^0 = \Delta E_{elstat} + \Delta E_{XREP} . \tag{8}$$

The exchange repulsion is particularly large when orbitals of the two fragments strongly overlap (it is also called overlap repulsion), as may be expected from the large effect of the orthogonalization in that case. The role of exchange repulsion, and the difference between ρ^0 and $\rho^A + \rho^B$ is discussed in ref. [45].

The next step in the interaction is the energy lowering from the E^0-level due to admixing of virtual orbitals on A or on B or on both. It is often broken up in parts, such as polarization of A (only mixing of occupied orbitals on A with virtuals on A allowed), charge transfer from A to B (only mixing of occupied orbitals on A with virtuals on B allowed), etc. Such calculations are easily carried out in practice by setting appropriate blocks of off-diagonal interaction matrix elements in the Fock matrix equal to zero [29,46]. It is well known that the determination of charge-transfer and polarization contributions to the electronic interaction energy is not without ambiguity. For instance, the contribution of the polarization of A will depend on whether we allow A to polarize in the field of the unmodified fragment B or in the field due to the B orbitals orthogonalized on A, and in the latter case it makes a difference whether one Schmidt orthogonalizes the B orbitals onto the occupied A space, or Löwdin orthogonalizes the occupied A and B orbitals, or uses some other orthogonalization. It should be noted that the total density ρ^0, and ΔE^0, are invariant to transformations between different sets of orthogonalized orbitals, but of course the individual orbitals are not. This means that the orbitals which are identified, after the orthogonalization, as A-type or B-type, depend on the actual orthogonalization procedure. The same ambiguity exists in identifying virtual orbitals as A-type and B-type. Moreover, charge-transfer and polarization type of orbital mixings influence each other, so the combined effect will differ from the sum of the separate effects. Finally, the extent of the virtual spaces on A and B is rather arbitrary, one complete virtual set on one fragment would in principle suffice, destroying the whole analysis. In spite of these difficulties the distinction between charge-transfer and polarization has often been made, and has added considerably to our understanding of molecular bonding.

We will in the present paper not use this distinction, but we will break up the electronic interaction energy in contributions labelled with the irreducible representations Γ of the pertinent symmetry group:

$$\Delta E = E_{AB} - E_A - E_B = \Delta E^0 + \Delta E_{int} ,$$
$$\Delta E_{int} = \sum_{\Gamma} \Delta E^{\Gamma} . \tag{9}$$

It is intuitively clear that such a division can be made, as only A- and B-orbitals of the same symmetry species Γ will interact. Explicitly, when writing:

$$\Delta E_{int} = 1/2 \sum_{\mu,\nu} P_{\mu\nu} \, (H^{core}_{\mu\nu} + F_{\mu\nu}) - 1/2 \sum_{\mu,\nu} P^0_{\mu\nu} \, (H^{core}_{\mu\nu} + F^0_{\mu\nu}) , \tag{10}$$

where H^{core} and F are the matrices of the purely one-electron operators (kinetic energy, nuclear attraction) and full Fock operator respectively, and the zero superscript refers to Ψ^0, it is seen that the division in irreducible representations immediately follows from the symmetry blocking of the F, H and P matrices. We note that this division of the total energy ΔE_{int} is unambiguous. We may also switch on the interactions in one symmetry species Γ only, but we will then not obtain the same ΔE^{Γ} as with all interactions present, as there will be a mutual influence. This is precisely what is meant by the synergic effect in coordinative bonding: σ-donation and π-backdonation may reinforce each other, so that the combined bonding effect is larger than the sum of each of the bonding mechanisms taken alone. If σ-bonding and π-backbonding occur in different irreducible representations we may study their relative strengths, as we have done before for CO interacting with metal clusters [25-27] (cf. also [24]), as well as the synergic effect. This will be done in this paper for $Cr(CO)_6$.

We wish to stress that, although common parlance hardly distinguishes between σ-bonding and σ-donation, there is a difference: σ-donation supposedly refers to a charge-transfer type of contribution to the bonding due to orbitals of σ symmetry with respect to the metal-ligand bond, whereas σ-bonding comprises all interactions (including polarization) due to those fragment orbitals.

III. Analysis of the metal-CO bonding.

In table 4 the irreducible representations of the O_h group of $Cr(CO)_6$ are listed together with the MO's of CO and AO's of Cr which occur, and the type of orbital-interaction that may be operative. It is to be noted that almost all interactions may be classified by symmetry: $5\sigma \rightarrow 4s$ donation in A_{1g}, $5\sigma \rightarrow 3d$ donation in E_g, $3d \rightarrow 2\pi$ backbonding in T_{2g}. Only T_{2u} contains two interactions, $5\sigma \rightarrow 4p_\sigma$ σ-donation, and possibly some $1\pi \rightarrow 4p_\pi$ backdonation. The equilibrium bond lengths are calculated to be 3.53 bohr and 2.16 bohr for Cr-C and C-O respectively (table 11). We will first discuss the bonding at this geometry, and consider geometry variations later.

Table 4. Fragment orbitals (CO molecular orbitals and Cr atomic orbitals) and types of interaction occurring in each irreducible representation of O_h.

Irreducible Representation (O_h)	CO MO's	Cr AO's	Type of interaction
A_{1g}	σ	4s	$5\sigma \rightarrow 4s$ bonding
E_g	σ	3d	$5\sigma \rightarrow 3de_g$ bonding
T_{2u}	σ	4p	$5\sigma \rightarrow 4p_\sigma$ bonding
	π		($1\pi \rightarrow 4p_\pi$ bonding)
T_{2g}	π	3d	$3d \rightarrow 2\pi$ backbonding
T_{1g}	π	-	
T_{1u}	π	-	
E_u	(δ)	-	

Table 5. Energy decomposition for the formation of a $(CO)_6$ cage (at the same geometry as in $Cr(CO)_6$) from 6 CO molecules. Energies in electronvolts.

	$(CO)_6$ from 6 CO
$\Delta E_{el.stat.}$	-1.911
ΔE_{XREP}	+4.743
ΔE^0	2.832
$\Delta E(A_{1g})$	0.067
$\Delta E(E_g)$	-0.518
$\Delta E(T_{2u})$	-1.728
$\Delta E(T_{2g})$	-0.048
$\Delta E(T_{1u})$	-0.034
$\Delta E(T_{1g})$	-0.012
$\Delta E_{el.int.}$	-2.379
ΔE_{total}	+0.453 eV

Let us first consider the formation of the $(CO)_6$ cage. The data in table 5 show that there is some exchange repulsion (+4.7 eV) which together with the electrostatic attraction of -1.9 eV leads to a fairly small steric repulsion ΔE^0 of 2.8 eV. From a symmetry analysis of ΔE^0 (not given here) it appears that the exchange repulsion is primarily due to the 5σ's of neighbouring CO's, as is evident from the overlaps in table 6, which show that at R=3.53 bohr the 5σ-5σ overlap is by far the largest one (amongst the occupied-occupied overlaps). When we next allow the $(CO)_6$ calculation to proceed to self-consistency the electronic interactions, which are given per symmetry species in table 5, lead to a lowering by -2.4 eV, leaving a net repulsive effect of 0.45 eV. By far the largest contribution to the electronic interaction comes from the T_{2u} symmetry, where both σ and π orbitals of the CO molecules are present (see table 4). The important interaction is mixing of the occupied 5σ's with the unoccupied neighbouring 2π's, as is already indicated by the relatively large (5σ,2π) overlap in table 6.
The total interaction is fairly small: the repulsion of 0.45 eV implies a repulsive contribution of 0.075 eV per Cr-CO bond, which is to be compared to the experimental bond energy of 1.2 eV [26].

Table 6. Overlaps for a number of selected CO MO's and Cr AO's at four Cr-C distances (3.53 bohr is the calculated equilibrium Cr-C distance).

	Cr-C distances (bohr)			
(CO,Cr)	3.06	3.53	4.71	6.28
overlaps				
(5σ,3s)	.22	.15	.05	.01
(5σ,4s)	.44	.42	.33	.19
(5σ,3p)	.20	.14	.05	.01
(5σ,4p)	.46	.50	.43	.21
(5σ,$3d_\sigma$)	.23	.23	.17	.08
(2π,3p)	.10	.07	.02	.01
(2π,4p)	.41	.35	.22	.09
(2π,$3d_\pi$	.26	.21	.12	.05
(CO,CO) (cis)				
overlaps				
(5σ,4σ)	.07	.04	.01	.00
(5σ,5σ)	.24	.15	.05	.01
(2π,1π)	.05	.03	.01	.00
(2π,2π)	.15	.11	.04	.01
(2π,4σ)	.06	.04	.02	.01
(2π,5σ)	.18	.14	.07	.02
(CO,CO) (trans)				
overlaps				
(5σ,5σ)	.09	.04	.01	.00
(2π,2π)	.06	.04	.01	.00

Next we compare the steric repulsion of the six CO molecules with the Cr atom in either the $(t_{2g})^6$ or the $(e_g)^4(4s)^2$ configuration (table 7). The first thing to note from table 7 is the extremely large exchange repulsion of the CO's with Cr $(e_g)^4(4s)^2$, whereas the purely electrostatic interaction is strongly attractive. Both effects have their origin in the diffuseness of the 4s AO. As may be deduced from table 6, a normalized A_{1g} combination of 5σ orbitals has an overlap 0.806 with the Cr 4s AO. This large overlap causes a huge exchange repulsion. There is also of course exchange repulsion between the 5σ's and the occupied $3d_{x^2-y^2}$ and $3d_{z^2}$ (de_g) orbitals. We find this exchange repulsion to be significant, but still more than three times less than that between 5σ's and 4s. This fits in with an overlap of 0.47 between an E_g combination of 5σ's and a $3de_g$. This overlap is significant, but considerably less than the 0.81 between 4s and 5σ in the A_{1g} symmetry. The diffuseness of 4s also accounts for the large electrostatic attraction: the 4s has a significant amplitude at the (six!) Carbon and Oxygen nuclei, which are not sufficiently screened by their own electrons and exert a strong attraction on the electronic 4s charge. The total exchange repulsion however dominates by far, the net effect being a repulsion of 40.3 eV.
It is obvious that a change to the configuration $(t_{2g})^6$ is advantageous in many ways: the exchange repulsion of 5σ with both 4s and $3de_g$ disappears, the now empty 4s and $3de_g$ may act as acceptor orbitals for 5σ, and the Cr electrons now occupy the $3dt_{2g}$ AO's which can interact with the CO LUMO, 2π. Looking at the energies in table 7 we observe indeed a large reduction in exchange repulsion. The remaining exchange repulsion is mostly due to 5σ's overlapping the Cr 3s and 3p AO's (there is also a 4.7 eV contribution from the 12 5σ-5σ overlaps, cf. table 5). There is also an understable drop in the electrostatic attraction, the total effect being a reduction in ΔE^0 from 40.3 for $(e_g)^4(4s)^2$ to 4.1 eV for $(t_{2g})^6$. Given these large effects it is perfectly understandable that the energetic cost of 9.4 eV to promote Cr from the $d^5\,s^1$ ground configuration to the "valence configuration" $(t_{2g})^6$ is compensated by the more favourable interaction with $(CO)_6$, both for the repulsion just discussed and for the electronic interactions to be discussed below. It is gratifying that very similar results for ΔE^0 (called Frozen Orbital repulsion) have been found by Bauschlicher and Bagus [5] for Ni $(CO)_4$ and Fe $(CO)_5$. For Ni $(CO)_4$ they noted a change from 27.5 eV to 4.7 eV in going from Ni $(3d^8\,4s^2)$ to Ni $(3d^{10})$, and for Fe $(CO)_5$ they found 36.5 eV for Fe $(3d^6\,4s^2)$ against 6.1 eV for Fe $(3d^8)$.

Table 7. *Steric repulsion (ΔE^0) of $Cr(e_g)^4(4s)^2$ and $Cr(t_{2g})^6$ valence states with six CO's at R(Cr-C) = 3.53 bohr (Energies in eV).*

	$Cr(eg)^4(4s)^2$	$Cr(t_{2g})^6$
$\Delta E_{el.stat.}$	-53.1	-22.0
ΔE_{XREP}	+93.4	+26.1
ΔE^0	+40.3	+ 4.1

We wish to stress that the configuration change we have examined here is not particular to the Cr atom or to carbonyl complexes. It is occurring almost always in transition metal-ligand interactions. We have given a similar analysis as the one for Cr and CO here for Co interacting with two cyclopentadienyl rings in $Co(Cp)_2$ [36]. A completely analogous configuration change often occurs in clusters of metal atoms interacting with CO [25-27]. Always the most important driving force of those configuration changes is the reduction of the exchange repulsion of occupied metal orbitals with the ligand lone pair orbital. The

Table 8. *σ-bonding only in $Cr(CO)_6$ with various sets of virtual fragment orbitals admitted. Energies (in eV) are with respect to 6 CO molecules and $Cr(t_{2g})^6$ as zero.*

	Virtual orbitals admitted		
	$3de_g$ only	$3de_g$+4s+4p	All vituals of Cr s,p de_g characters, and of CO σ character
$\Delta E(A_{1g})$	-0.771	-0.925	-1.342
$\Delta E(E_g)$	-3.022	-2.989	-3.222
$\Delta E(T_{2u})$	0.030	-0.121	-0.796
$\Delta E(T_{2g})$	+0.238	+0.229	+0.165
$\Delta E(T_{1g})$	-0.020	-0.024	-0.031
$\Delta E(T_{1u})$	-0.006	-0.007	-0.009
$\Delta E_{el.int.}$	-3.550	-3.836	-5.235
ΔE^0	+4.124	+4.124	+4.124
ΔE_{total}	+0.574	0.288	-1.111

Table 9. *π-bonding only in $Cr(CO)_6$ (energies in eV).*

	Virtual orbitals admitted	
	CO 2π only	All CO π + Cr t_{2g}
$\Delta E(A_{1g})$	+1.661	1.360
$\Delta E(E_g)$	-0.687	-0.461
$\Delta E(T_{2u})$	-1.197	-1.444
$\Delta E(T_{2g})$	-7.062	-10.492
$\Delta E(T_{1g})$	-0.020	-0.026
$\Delta E(T_{1u})$	-0.025	-0.110
$\Delta E_{el.int.}$	-7.330	-11.173
ΔE^0	4.124	4.124
ΔE_{total}	-3.206	-7.049

electrons leave orbitals which have overlap with the 5σ, and move to metal orbitals which can interact with the CO 2π orbitals. In case there is a single metal atom the electrons go into the metal 3d if available (transition metals) or else into p_π orbitals, e.g. $2p_\pi$ in Li [25], $3p_\pi$ in Al [26,28], $4p_\pi$ in Cu [52]. In the case of metal clusters there is, also with transition metals, the additional possibility of the electrons going into sp (conduction electron) states with the right symmetry to interact with the 2π [21,24,25]. The Ni-CO system, which has recently been the focus of much computational work [14-18] may be considered from the same point of view.

At this point, with only the steric repulsion present, the total energy is 4.1 eV above the energy of six CO's and a valence state Cr $(t_{2g})^6$ atom, and 13.5 eV above the initial situation with the Cr atom in the ^7S ground state. We now consider the effect of σ-bonding, by admitting into the orbital space first only the Cr $3de_g$ orbitals, next also the Cr 4s and 4p orbitals, and finally all virtual Cr AO's of $s(A_{1g}$ in $O_h)$, $p(T_{2u})$ and $d_{x^2-y^2}$, d_{z^2} (E_g) symmetry. It is to be noted that we employ a transformation of the primitive STO basis to the SCF AO's and MO's of Cr and CO respectively. When all virtual (and occupied) orbitals of the fragments are used this is of course completely equivalent to the use of the full primitive set. It is possible, however, to restrict the basis to only one or a few virtual fragment orbitals (plus all occupied ones). This is implicitly done in qualitative discussions, when e.g. the metal-CO interaction is described in terms of donation out of 5σ into 3d (e_g) and backdonation out of 3d (t_{2g}) into 2π. We study the various bonding contributions by applying the appropriate restrictions of the orbital space. The first column in table 8 shows the energetic effect of only allowing $5\sigma \rightarrow 3de_g$ donation. With only the $3d_{x^2-y^2}$ and $3d_{z^2}$ AO's of Cr present no variational change of shape in the orbitals is possible. As expected the largest contribution is in the E_g symmetry (3.0 eV). The charge shift from 5σ to $3de_g$ also affects the energies in the other symmetries due to the concomitant change in the potential. These effects are small but not negligible, being e.g. -0.8 eV in A_{1g} symmetry and +0.2 eV in T_{2g} symmetry. The total effect is -3.55 eV, which is not sufficient to offset the steric repulsion of 4.1 eV. In the second column of table 8 the effect of adding the Cr 4s and 4p AO's is shown. The energetic contribution of $5\sigma \rightarrow 4p$ donation is obviously very small, in the order of 0.3 eV. This is somewhat surprising, as this donation is the basis of the 18-electron rule. Our result is however in perfect agreement with the conclusion of Bauschlicher and Bagus [5] that the contribution of the metal 4s and 4p AO's to the bonding in $Ni(CO)_4$ and $Fe(CO)_5$ is only ~0.3 eV (N.B. Our procedure is slightly different than the one of ref. [5]. As we omit or admit the unoccupied metal AO's resulting from the $Cr(t_{2g})^6$ SCF calculation, and not primitive basis functions, the reference energy of the metal atom does not change). Finally, in the third column, the effect is shown of including all virtual orbitals of the fragment (Cr and CO) SCF calculations of s,p and de_g symmetry (Cr) and σ symmetry (CO). The previous calculations could be called minimal linear-combination-of-fragment-orbital (LCFO) calculations (cf. minimal LCAO). Now the orbitals are allowed the full variational freedom enabled by the large primitive STO set, although still $3d \rightarrow 2\pi$ π-(back)bonding is excluded by omitting virtual orbitals in other symmetries than A_{1g}, E_g and T_{2u}. As shown in column three of table 8, the σ-bonding changes by 36%, to -5.2 eV. This change comprises the σ-donation into higher virtual orbitals of the metal, the σ-backdonation from metal to CO 6σ (and higher CO σ virtual orbitals), and that part of the polarization of the metal and CO which may occur in the restricted orbital space admitted here. We do not separate these effects further.

We now examine the π-bonding. In column 1 of table 9 the energies are shown when only the CO 2π orbital (occurring in T_{2g}, T_{2u} and T_{1g} symmetries) is admitted in the variation space. The effect, in particular in the T_{2g} symmetry, is quite striking: more than 7 eV binding. The π-bonding is obviously much stronger than the σ-bonding. This conclusion is strengthened by the result of admitting full variational freedom for the π-bond by admitting all CO π orbitals and all virtual Cr AO's of T_{2g} symmetry (column 2): the π-bond increases to 11.2 eV, out of which 10.5 eV is due to the T_{2g} symmetry. The π-bond appears to be about double as strong as the σ-bond! The contribution in the T_{2g} symmetry accounts for almost but not all of the π-bonding. The contributions in A_{1g}, E_g, and T_{2u} symmetry are

much smaller, but not negligible. For T_{2u} the contribution is partly due to the $(CO)_6$ cage, as the 5σ's are allowed to interact with the neighbouring 2π's. This interaction resulted in a T_{2u} energy of -1.7 eV in $(CO)_6$ (table 5).
The A_{1g} and E_g energies are possibly due to Coulombic potential effects. The flow of electrons from the metal atom $3dt_{2g}$ to CO 2π is expected to lead to a more unfavourable electronic potential for the CO σ electrons, hence the repulsive A_{1g} energy. The σ orbitals of CO also occur in E_g, but there the loss of t_{2g} electrons from the metal is expected to lead to a more favourable potential for the e_g electrons, hence a (small) negative energy in E_g. These explanations are however tentative.
The π-bonded $Cr(CO)_6$ system is 7 eV more stable than Cr $(t_{2g})^6$ and six CO's. When the 5.2 eV of the σ-bonding of table 8 would be added, there would only be 12.2 - 9.4 = 2.8 eV bonding with respect to Cr atom in the ^7S ground state. It is clear that a synergic effect, a mutual strengthening of σ- and π-bond, is called for.

The synergic effect is quantitatively investigated in table 10. In column 1 the energies are displayed if only the Cr $3de_g$ σ-acceptor orbital and the CO 2π π-acceptor orbital are allowed in the virtual space. The synergic effect is quite striking: the total electronic interaction energy of 18.1 eV is much larger than the sum of the first columns of tables 8 and 9, 10.88 eV. Both the σ- and π-bonds are strengthened, about in the same 1:2 ratio in which they contribute to the total bond: the E_g energy of -3.0 eV in column 1 of table 8 increases by -2.8 to -5.8 eV, the T_{2g} energy of -7.1 eV in column 1 of table 9 increases by -5.4 eV to -12.5 eV. A completely analogous picture holds when column 2 of table 10, where all virtuals on Cr and CO are simultaneously admitted, is compared to columns 3 and 2 of tables 8 and 9 respectively. The E_g energy (σ-bonding) increases from -3.2 eV in table 8 to -5.7 eV, and the T_{2g} energy (π-bonding) of -10.5 eV in table 9 increases to -16.5 eV. It is interesting that the synergic effect is about equally large at the level of the minimal LCFO basis as when the full virtual space is used. This follows by comparing the minimal LCFO results, -18.1 eV for synergic bonding and -10.9 eV for the sum of σ- and π-bonding separately, with the ones obtained with the full basis, i.e. -24.0 eV for synergic bonding and -16.4 eV for the sum of σ- and π-bonding separately. In both cases the synergic effect is ca 7.5 eV, which is of the same order of magnitude as the σ- and π-bonding energies.

It is interesting to compare our HFS results with the Hartree-Fock calculations by Bauschlicher and Bagus [5] on Ni $(CO)_4$ and Fe $(CO)_5$. Due to differences in the way of analysing the various energy components the comparison can only be semi-quantitative. In spite of this there is one feature that stands out rather clearly. Taking the NSP MBS basis (table VI, ref. [5]) we may say, somewhat loosely, that Bauschlicher and Bagus find for Ni $(CO)_4$ ca 0.3 eV σ-donation energy, 5.1 eV π-backdonation energy, and 2.6 eV polarization energy (both metal and CO) in Ni $(CO)_4$, and 2.4 eV σ-donation energy, 6.7 eV π-backdonation energy, 4.2 eV polarization energy for Fe $(CO)_5$. This is to be compared to the 5.8 eV σ-bonding and 16.5 eV π-bonding in the present calculation, where it is to be noted that these numbers contain both donation (charge-transfer) and polarization effects. Our σ-bonding appears to be in line with the Hartree-Fock result. The increase from 0.3 to 2.3 eV going from Ni $(CO)_4$ to Fe $(CO)_5$ is due to the presence of the unoccupied $3d_{z^2}$ in Fe $(CO)_5$. The additional unoccupied $3d_{x^2-y^2}$ in Cr $(CO)_6$ will again increase the σ-donation, maybe by a roughly similar amount. May our σ-bonding therefore be reasonable, the π-bonding is obviously considerably larger than could be expected on the basis of the Hartree-Fock results: HFS 16.5 eV π-bonding in Cr $(CO)_6$ against HF 6.5 eV π-donation energy (+3.6 eV polarization) in Fe $(CO)_5$. Although some increase from Fe $(CO)_5$ to Cr $(CO)_6$ seems reasonable, this difference appears to be too large. The origin of the difference is not completely clear. The first possibility is that the Xα model for some reason overestimates π-bonding of a metal atom with CO. This fits in with the observation that bonds of a single metal atom (Li, Al, Cu) with CO appear stronger in the Xα model than in HF [37,28]. On the other hand, the metal-CO bonding is too weak in HF, CI calculations lead to considerable strengthening of the π-bond in particular [16,48,49,50]. There may be a relation here to the wrong covalency often found in HF, i.e. metal-ligand orbital mixing in the wrong proportion [8]. We are currently investigating this point comparing calculated unpaired spin distributions in radicals with ESR measurements.

In this paper we will make two other comparisons to experimental data. First the calculated metal-CO bond energy will be compared to the experimental one, and next the calculated and experimental A_{1g} vibrations will be compared.

IV. Comparison to experiment: metal-CO bond energy and metal-CO and C-O vibrations.

According to table 10 the total bonding energy is 19.881 eV with respect to Cr $(t_{2g})^6$ and six CO's with a C-O distance of 2.160 bohr. Correcting for the excitation energy of 9.423 eV from Cr (^7S) ground state to Cr $(t_{2g})^6$ (table 3), and the stretching of the C-O distance from 2.132 to 2.160 bohr (0.015 eV), this means a bonding energy of 10.368 eV with respect to the fragments in their ground states. This would lead to a bond energy of 39.9 kcal/mol per Cr-CO bond. This number has to be corrected for the basis-set-superposition error (BSSE). We have obtained an upper limit to the BSSE by calculating the energy lowering of $(CO)_6$ when a ghost nucleus with the Cr basis functions is added, and similarly the energy lowering of Cr $(t_{2g})^6$ atom when the $(CO)_6$ basis functions are added. This procedure overestimates the BSSE, it would be more appropriate, as pointed out by Bauschlicher and Bagus, to use only the virtual orbitals of the other fragment. Our BSSE correction leads to a bond energy of 34.6 kcal/mol.. Comparing this to the experimental bond energy of 25.7 kcal/mol [47] this is too high, whereas Hartree-Fock results are in general too low. Ni $(CO)_4$ and Fe $(CO)_5$ are

Table 10. Synergic effect in $Cr(CO)_6$ (energies in eV)
Column 1: simultaneous σ-donation into 3de_g and π-backdonation into CO 2π in a minimal LCFO description.
Column 2: all interactions (charge transfer + polarization) simultaneously present.

	Virtual orbitals admitted	
	Cr 3de_g, CO 2π	All virtuals
$\Delta E(A_{1g})$	+1.478	+0.47
$\Delta E(E_g)$	-5.794	-5.757
$\Delta E(T_{2u})$	-1.219	-2.028
$\Delta E(T_{2g})$	-12.510	-16.514
$\Delta E(T_{1g})$	-0.044	-0.054
$\Delta E(T_{1u})$	-0.040	-0.122
$\Delta E_{el.int.}$	-18.128	-24.005
ΔE^0	4.124	4.124
ΔE_{total}	-14.004	-19.881

even found to be unbound in HF by 1.97 eV and 2.36 eV respectively with respect to the groundstate Ni ($3d^8\,4s^2$; ^3F) and Fe ($3d^6\,4s^2$; ^5D) and $(CO)_4$, resp. $(CO)_5$ [5]. This is approximately 11.4 and 10.9 kcal/mol per metal-CO bond. $Cr(CO)_6$ however is found to be bound in HF [48] by 17.2 kcal/mol per Cr-CO bond, which is still too low compared to the experimental bond energy. This value is increased to 33.5 kcal/mol after CI (not corrected for BSSE). None of these calculated values is very close to the experimental bond energy. The problem with Hartree-Fock is well understood: the $3d^{n+2}\,4s^0$ configuration of the atom is calculated to lie far too high in energy, and as this configuration is effectively present in the complexes the energies of the complexes are too high, i.e. the bond energy with respect to the metal atom in its ground state (with $3d^n\,4s^2$ or $3d^{n+1}\,4s^1$ configuration) is too low. The $X\alpha$ model makes an error in the opposite direction (bond energy too high instead of too low), as in fact the quoted CI calculation does [48]. This may be related to an underestimation of s→d transfer in $X\alpha$ (cf the Ni atom mentioned in the introduction). This point requires more detailed investigation.

A full geometry optimization has been carried out employing both a double-zeta (DZ) and an extended basis set (see table 11). Energies have been calculated at 24 geometries (various C-O distances (d) as well as Cr-C distances (R), retaining O_h symmetry. They have been fitted with a generalized Morse potential:

$$V(R,d) = \sum_{n=0}^{3} \sum_{m=0}^{3} C_{nm}\, e^{-n\alpha(R-R_e)} \,.\, e^{-m\beta(d-d_e)} . \qquad (11)$$

An excellent fit is obtained with $\alpha=\beta=1.0$ (R and d in bohr), so the exponents are simple integers. The coefficients C_{nm} are determined by a least-squares fit. The harmonic force constants for the A_{1g} symmetry coordinates R and d, as well as the interaction constant, as calculated from (11) are given in table 11. The vibration frequencies calculated with a normal coordinate analysis [51] are also given in table 11.

For the extended basis set calculations a lenthening of 0.028 bohr in the C-O distance relative to free CO is calculated, slightly more than the experimental 0.022 bohr. DZ calculations predict a similar increase of the C-O distance (0.032 bohr), but the C-O distance itself is too long in the DZ basis by 0.04-0.05 bohr. The extended basis gives very good agreement of the calculated C-O distance with experiment (cf. also ref. [40]).

The Cr-C distance does not differ much between DZ and EXT basis (3.522 and 3.531 bohr),

Table 11. Calculated spectroscopic constants for $Cr(CO)_6$ in double zeta and extended basissets compared to experiment.
† with BSSE correction.

	DZ	EXT	EXP	
R_{CO} (d)	2.205	2.160	2.154	bohr
R_{CrC} (R)	3.522	3.531	3.616	
f_{dd}	15.27	17.70	18.11	mdyn/ Å
f_{Rd}	0.22	0.20	0.38	"
f_{RR}	3.22	3.19	2.44	"
ω_{C-O}	1997	2139	2139	cm^{-1}
ω_{Cr-C}	430	430	(379)	"
D_e (Cr-CO)	43.8	39.9 34.6†	25.7	kcal/mol

and is too short compared to experiment (3.616 bohr) in contrast to Hartree-Fock metal-ligand distances which are generally too long. This would seem to fit in with our tentative conclusion that the π-bonding is too strong. Due to the short range of π-bonding, CO would be pulled in too closely by a too strong π-bond. The calculated C-O vibration frequency, however, does not support this conclusion. Instead of being too low, as would be expected to result from too strong π-bonding, it is in perfect agreement with experiment for the EXT basis calculation. As also the harmonic vibration frequency of free CO is virtually identical in the EXT calculation and experiment, the experimental lowering by 31 cm^{-1} is precisely predicted by the calculation. It is remarkable that the DZ basis, which gives a reasonable Cr-C bond distance, Cr-CO bonding energy, and C-O bond lengthening, fails for the lowering of the C-O stretching frequency, giving no shift at all from free to coordinated CO.

(The DZ calculation appeared to be numerically less stable, for reasons we do not understand. As the frequency shift is small, and the calculation of ω_e sensitive to the details of the curve-fitting, we can not rule out the possibility that this behaviour of DZ is an artifact of the calculations).

The Cr-C vibration frequency, which is equal in DZ and EXT calculations, is a bit high in comparison to experiment (430 cm^{-1} versus 379 cm^{-1}; the experimental value is not the harmonic vibration frequency [53]). This would be in line with a too strong bond.

V. Summary.

We have investigated the metal-carbonyl bond in $Cr(CO)_6$ using the Xα model. Taking a Cr $(3d\ t_{2g})^6$ atom as the reference we find the π-bond to be considerably stronger than the σ-bond (about a factor two). The synergic effect has been quantitatively investigated by computing σ- and π bonds both when present alone and when acting cooperatively. The synergic effect has been found to be quite strong, ca 7.5 eV, of the same order of magnitude as the individual σ- and π-bonding in the complex.

A linear-combination-of-fragment-orbitals (LCFO) approach, in which the fragment valence molecular and atomic orbitals are used, is found to give a qualitatively correct picture of e.g. relative importance of σ- and π-bonding, and of the synergic effect. This supports the ubiquitous use of this approach in the qualitative MO discussions of coordination chemistry. For a quantitative description, however, the higher lying virtuals of the fragments are absolutely necessary. This is an indication for the importance of polarization effects.

Comparing the bonding in the Xα model with the one given by HF we note a great similarity in the qualitative picture. In particular when we take the $3d^{n+2}\,4s^0$ atomic configurations as a reference, our present calculation fits in with the Hartree-Fock calculations on $Ni(CO)_4$ and $Fe(CO)_5$ [5]. The repulsion of CO 5σ with the atomic 4s orbital is evident in both models, and the σ-bonding increases regularly from Ni to Cr due to the increasing availability of 3d AO's as σ-acceptor orbitals. Both models also show very little 4s,4p involvement in the bonding. There is, however, a difference in the amount of π-bonding: the Xα model predicts considerably stronger π-bonding.

The values of the bond energy when referred to the atomic ground states ($3d^{n+1}\,4s^1$ or $3d^n\,4s^2$ configurations) are plagued by the difficulties associated with the atomic multiplet problem, which even lead in HF calculations on $Ni(CO)_4$ and $Fe(CO)_5$ to antibonding. In the case of $Cr(CO)_6$ a HF calculation gives 17.2 kcal/mol per Cr-CO bond (33.5 after CI, without BSSE correction), our LCAO-Xα calculation gives 34.9 kcal/mol, whereas the experimental number is 25.7 kcal/mol.

The calculation of the equilibrium distances and A_{1g} vibration frequencies gives good agreement with experiment for the lowering of the C-O vibration frequency and the lenthening of the C-O bond. The Cr-C bond distance however is too short (3.53 bohr *versus* 3.62 experimentally) and the Cr-CO vibration frequency too high (ω_e = 430. cm^{-1}, *versus* 379. cm^{-1} for the observed (not corrected for anharmonicity) frequency). This indicates, as does the bond energy, overestimation of the metal-CO bond strength in the Xα model.

In a recent investigation of $Ni(CO)_4$ with a Gaussian-based LCAO-Xα program [34], Jörg and Rösch [54] came to virtually identical conclusions regarding the comparison between the Hartree-Fock and Xα models for the metal-ligand bonding.

Acknowledgement.

This investigation was supported in part by the Netherlands Organization for Chemical Research (SON) with financial aid from the Netherlands Organization for the Advancement of Pure Research (ZWO).

References.

1. I.H. Hillier, V.R. Saunders, Mol. Phys. 22 (1971) 1025.
2. J. Demuynck, A. Veillard, Theoret. Chim. Acta 28 (1973) 241.
3. J.B. Johnson, W.G. Klemperer, J. Am. Chem. Soc. 99 (1977) 7132.
4. E.J. Baerends, P. Ros, Mol. Phys. 30 (1975) 1735.
5. C.W. Bauschlicher Jr., P.S. Bagus, J. Chem. Phys. 81 (1984) 5889.
6. H.P. Lüthi, J.H. Ammeter, J. Almlöf, K. Korsell, Chem. Phys. Lett. 69 (1980) 540.
7. H.P. Lüthi, J.H. Ammeter, J. Almlöf, K. Faegri Jr., J. Chem. Phys. 77 (1982) 2002.
8. T.E. Taylor, M.B. Hall, Chem. Phys. Lett. 114 (1985) 338.
9. H.P. Lüthi, dissertation, Zürich, 1984.
10. J. Demuynck, A. Strich, A. Veillard, Nouv. J. Chim. 1 (1977) 217.
11. D. Spangler, J.J. Wendoloski, M. Dupuis, M.M.L. Chen, H.F. Schaefer, J. Am. Chem. Soc. 103 (1981) 3985.
12. A. Rozendaal, E.J. Baerends, Chem. Phys. 95 (1985) 57.
13. A. Rozendaal, dissertation, Amsterdam, 1985.
14. K. Hermann, P.S. Bagus, Phys. Rev. B16 (1977) 4195.
15. S.P. Walch, W.A. Goddard, J. Am. Chem. Soc. 98 (1976) 7908.
16. A.B. Rives, K. Fenske, J. Chem. Phys. 75 (1981) 1293.
17. C.M. Kao, R.P. Messmer, Phys. Rev. B31 (1985) 4835.
18. B.I. Dunlap. H.L. Yu, P.R. Antoniewicz, Phys. Rev. A25 (1982) 7.
19. C.W. Bauschlicher, S.P. Walch, P.E.M. Siegbahn, J. Chem. Phys. 78 (1983) 3347; J. Chem. Phys. 76 (1982) 6015; C.W. Bauschlicher, Chem. Phys. Lett. 97 (1983) 204.
20. R.L. Martin, J. Chem. Phys. 78 (1983) 5840.
21. K. Morokuma, Acc. Chem. Res. 10 (1977) 294.
22. J.O. Noell, K. Morokuma, Inorg. Chem. 18 (1979) 2774.
23. K. Kitaura, S. Sakaki, K. Morokuma, Inorg. Chem. 20 (1981) 2292.
24. T. Ziegler, A. Rauk, Inorg. Chem. 18 (1979) 1755.
25. D. Post, E.J. Baerends, Surf. Sci. 109 (1981) 167.
26. D. Post, E.J. Baerends, Surf. Sci. 116 (1982) 177.
27. D. Post, E.J. Baerends, J. Chem. Phys. 78 (1983) 5663.

28. P.S. Bagus, C.J. Nelin, C.W. Bauschlicher, Phys. Rev. B28 (1983) 4378.
29. P.S. Bagus, K. Hermann, C.W. Bauschlicher, J. Chem. Phys. 80 (1984) 4378.
30. P.S. Bagus, K. Hermann, C.W. Bauschlicher, J. Chem. Phys. 81 (1984) 1966.
31. K.H. Johnson, F.C. Smith Jr., Phys. Rev. B5 (1982) 831.
32. E.J. Baerends, D.E. Ellis, P. Ros, Chem. Phys. 2 (1973) 41.
33. H. Sambe, R.H. Felton, J. Chem. Phys. 62 (1975) 1122.
34. B.I. Dunlap, J.W.D. Connolly, J.R. Sabin, J. Chem. Phys. 71 (1979) 3396.
35. O. Gunnarsson, J. Harris, R.O. Jones, J. Chem. Phys. 67 (1977) 3970.
36. C. Famiglietti, E.J. Baerends, Chem. Phys, 62 (1981) 407.
37. J. Koutecky, personal communication.
38. E.J. Baerends, P. Ros, Chem. Phys. 2 (1973) 52.
39. L. Laaksonen, D. Sundholm, P. Pyykkö, Int. J. Quantum Chem. 1985, in press.
40. E.J. Baerends, P. Vernooijs, A. Rozendaal, P.M. Boerrigter, M. Krijn, D. Feil, D. Sundholm, J. Molec. Struct. (THEOCHEM), 1985, in press.
41. S. Wolfe, D.J. Mitchell, M.-H. Whangbo, J. Am. Chem. Soc. 100 (1978) 1936.
42. F. Bernardi, A. Bottoni, A. Mangini, G. Tonachini, J. Molec. Struct. (THEOCHEM), 86 (1981) 163.
43. A.J. Stone, R.W. Erskine, J. Am. Chem. Soc. 102 (1980) 7185.
44. T. Ziegler, A. Rauk, Theor. Chim. Acta 46 (1977) 1.
45. H. Fujimoto, Y. Osamura, T. Minato, J. Am. Chem. Soc. 100 (1978) 2954.
46. K. Kitaura, K. Morokuma, Inr. J. Quantum Chem. 10 (1976) 325.
47. J.A. Connor, Topics in Current Chemistry, 71 (1977) 71.
48. D. Moncrieff, P.C. Ford, I.H. Hillier, V.R. Saunders, J.C.S. Chem. Comm. 1983, 1108.
49. A. Dedieu, S. Nakamura,,personal communication.
50. P.S. Bagus, B.O. Roos, J. Chem. Phys. 75 (1981) 5971.
51. J.G. Snijders, program NORMANAL, Vrije Universiteit, Amsterdam.
52. a) J.A. Howard, B. Mile, J.R. Morton, K.F. Preston, R. Sutcliffe, Chem. Phys. Lett. 117 (1985) 115.
    b) P. Kasai, P.M. Jones, J. Am. Chem. Soc. 107 (1985) 813.
53. L.H. Jones, R.S. McDowell, M. Goldblatt, Inorg. Chem. 8 (1969) 2349.
54. H. Jörg, N. Rösch, Chem. Phys. Lett., in press.

APPLICATIONS OF THE LCGTO-Xα METHOD TO TRANSITION METAL CARBONYLS

N. Rösch and H. Jörg
Lehrstuhl für Theoretische Chemie
Technische Universität München
8046 Garching, Fed. Rep. Germany

B.I. Dunlap
Naval Research Laboratory (Code 6129)
Theoretical Chemistry Section
Washington, D.C. 20375, USA

ABSTRACT. LCGTO-Xα calculations have been performed on the compounds $Ni(CO)_4$ and $Fe(CO)_5$. The agreement with experimental data for the metal to carbon bond lengths and for the force constants is satisfactory. The deviations are significantly smaller than those encountered in Hartree-Fock calculations. A systematic trend to overestimate the metal to carbon bond strength is observed and attributed to the local density approximation.

1. INTRODUCTION

The chemistry of transition metals is governed by the interplay of two atomic levels quite similar in energy, but very different in spatial extension. The valence (n)s orbital, usually doubly occupied in the atomic ground state, is radially diffuse and spherically symmetric, whereas the (n-1)d orbitals, occupied with the remaining electrons, are contracted and have two angular nodes. An accurate description of the electronic structure of transition metal compounds, especially those of the d-electron rich metals, therefore requires methods which properly treat these dissimilar atomic orbitals in a balanced way.

It is now well known that the Hartree-Fock method fails to do so /1-3/. The s-d excitation energies in transition metal atoms are frequently overestimated by several eV /4/. This is due to neglected correlation, which becomes important when metal electrons are redistributed during bond formation.

These shortcomings can be rectified by using MCSCF and CI methods. In the last two years, several transition metal compounds have been investigated /3,5/. However, the computational effort necessary to go beyond one-determinant wavefunctions limits the size of systems that can be treated conveniently.

For these reasons it may be illuminating to investigate the capabi-

A. Veillard (ed.), Quantum Chemistry: The Challenge of Transition Metals and Coordination Chemistry, 179–187.

lities of other non-empirical approaches, such as the density functional theory. The LCGTO-Xα technique /6/, featuring the local density (LD) approximation in the form of a computationally simple exchange-correlation potential, uses Gaussian basis set expansion techniques and in this respect resembles current HF and CI procedures. This facilitates a methodological comparison between the LD and HF based models that will be attempted in the following for two of the most familiar carbonyl compounds, $Ni(CO)_4$ and $Fe(CO)_5$.

2. COMPUTATIONAL DETAILS

The present calculations have been performed using the LCGTO-Xα program /6/ with an exchange scaling factor of 0.7. The orbital basis sets and contraction schemes used for $Ni(CO)_4$ were identical with those of our previous study /7/. For $Fe(CO)_5$, the basis sets were constructed in an analogous manner. For iron, the (14s/9p/5d) basis set of Wachters /8/ has been augmented by one s-function to fill the gap between 3s and 4s functions (exponent 0.298), by two p-functions (exponents 0.275 and 0.08 /3/), and by one d-function (exponent 0.1133 /9/). The basis sets for carbon and oxygen were identical to those used for $Ni(CO)_4$. They consisted of the (9s/5p) basis sets of van Duijneveldt /10/ augmented with an appropriate d function following Huzinaga /11/ (exponents: 0.600 for carbon, 1.154 for oxygen). As for $Ni(CO)_4$, the iron basis set was contracted to (11s/8p/4d) and the ligand basis sets to (6s/4p/1d). These basis sets are of the same quality as those used in the CI study on $Fe(CO)_5$ /3/ with two minor differences. Here, no f polarization functions were employed on the metal and the contractions were somewhat more flexible.

The auxiliary basis sets were derived from the orbital basis sets in standard fashion /7,12/. In contrast to the calculations on $Ni(CO)_4$, no polarization functions were included in the fitting basis sets of iron to minimize any bias between axial and equatorial ligands. Due to the reduced symmetry in $Fe(CO)_5$ (D_{3h}) the number of fitting functions is higher (85 functions) than the corresponding number in $Ni(CO)_4$ (T_d symmetry, 58 functions).

3. GEOMETRY OPTIMIZATION

3.1. Nickel tetracarbonyl

For convenience, we will first briefly review the results of our recent investigation on $Ni(CO)_4$ /7/. In this study the Ni-C and the C-O bond length were optimized indepently by calculating 25 points on the two-dimensional potential energy surface representing the totally symmetric stretching motions.

The nickel to carbon bond was found to be too short by 0.020 Å (see table 1). This deviation is quite satisfactory in comparison with the deviation of 0.096 Å found in HF calculations /2/. The remarkable relativistic bond contraction (0.026 Å) estimated in ref. /2/ is not suitable

TABLE 1
Theoretical and experimental data for $Ni(CO)_4$ and $Fe(CO)_5$

a) $Ni(CO)_4$

Method	d_{Ni-C} (Å)	d_{C-O} (Å)	$f_{NiC,NiC}$ (N/m)	$f_{CO,CO}$ (N/m)	$f_{NiC,CO}$ (N/m)	D_o (eV)
LCGTO-Xα[a)]	1.805	1.137	300.	1800.	24.	8.57
HF	1.921[b)]	-	-	-	-	-1.97[c)]
CI[d)]	1.884	-	-	-	-	4.14
Experiment	1.825[e)]	1.122[e)]	235.5[e)]	1823.3[e)]	23.5[e)]	6.12[f)]

b) $Fe(CO)_5$

Method	d_{Fe-C}(ax) (Å)	d_{Fe-C}(eq) (Å)	$f_{ax,ax}$ (N/m)	$f_{eq,eq}$ (N/m)	D_o (eV)
LCGTO-Xα	1.774	1.798	404.	278.	10.04
HF[g)]	2.047	1.874	50.	150.	-
CI[g)]	1.798	1.836	330.	280.	-
Experiment	1.807[h)]	1.827[h)]	257.[i)]	264.[i)]	6.10[f)]

a) Ref. /7/. b) Ref. /2/. c) Ref. /14/. d) Ref. /5/.
e) Ref. /16/. f) Ref. /17/. g) Ref. /3/.
h) Geometrically inconsistent r_a values, ref. /26/. i) Ref. /26/.

for Xα data since it reflects an artifact of the very shallow HF potential energy curve. A correction by about 0.01 Å seems to be more appropriate /13/.

The overestimation of the metal to carbon bond strength becomes more obvious when the Ni-C force constant (28% higher than experiment) and the dissociation energy with respect to CO molecules (40% higher than experiment) are considered. However, this is closer to experiment than HF results which predict the molecule to be unstable with respect to the ground state of the nickel atom /14/.

The harmonic frequencies for the a_1 vibrations derived from the force constants in table 1 are $\omega_1 = 2150\ cm^{-1}$ and $\omega_2 = 418\ cm^{-1}$, in satisfactory agreement with the experimental values of 2154.1 cm^{-1} and 370.8 cm^{-1}.

3.2 Iron pentacarbonyl

The equilibrium geometry for $Fe(CO)_5$ was calculated assuming D_{3h} symmetry keeping the CO distances fixed at their experimental value of 1.152 Å.

This apprixmation should be appropriate in studies of the Fe-C bond, given the weak coupling between Ni-C and the C-O bond that we found in $Ni(CO)_4$ /7/. The calculations were performed for four different axial and five equatorial Fe-C distances. The resulting grid of twenty points turned out to be sufficient for a fit stable with respect to the degree of the fitting polynomial within 0.001 Å for the bond lengths. As in $Ni(CO)_4$, the equilibrium geometry shows slightly too short metal-carbon bonds (see table 1) compared to the gas phase electron diffraction data /15/. Unlike for $Ni(CO)_4$, the experimental distances used here are geometrically inconsistent r_a values. (The effects of vibrational averaging are not taken into account /16/.) To our knowledge, a force field based refinement is still lacking for $Fe(CO)_5$. From the data given for $Ni(CO)_4$ /16/ one should expect the geometrically consistent distances in $Fe(CO)_5$ to be shorter by about 0.01 Å, thus reducing the error of our calculation. The calculated bonding energy D_o of 10.01 eV is significantly higher than the experimental value of 6.10 eV /17/.

As can be seen from table 1, especially the axial force constant is too large in comparison with experiment. Since in HF calculations the axial force constant is much too low, we may conclude that the deficiencies of the different theoretical approaches are predominantly reflected in the axial bonds, whereas a good description of the equatorial bonds seems easier to achieve.

The coupling force constant $f_{ax,eq}$ was found to be negligible. Thus the vibrational frequencies were determined separately from the two force constants $f_{ax,ax}$ and $f_{eq,eq}$, yielding 495 cm^{-1} and 410 cm^{-1}. The experimental values are 443 cm^{-1} and 413 cm^{-1}, respectively /26/. The coupling between the metal-carbon and the carbon-oxygen vibration was not included in the present treatment, nevertheless the vibrational frequencies are estimated (based on our $Ni(CO)_4$ work) to be within 20 cm^{-1} of a coupled Xα calculation. However, such uncertainties will have little influence on the subsequent discussion, since the overestimation of the metal to carbon bond strength remains unquestionable.

Inspection of table 1 shows that the quality of the LCGTO-Xα results for both molecules is definitely superior to those obtained from HF calculations, although the computational effort required differs by an order of magnitude.

4. DISCUSSION

4.1. Comparison to other calculations

A critical basis set dependence of the nickel-carbon bond length in $Ni(CO)_4$ has been noted in HF studies /2/. Therefore it seemed desirable to perform similar investigations within the Xα model (see table 2). The carbon-oxygen distance was kept fixed at the value of 1.137 Å as optimized for the standard basis set. Firstly, the single d-type orbital polarization function on carbon and oxygen in the standard set described above was replaced by a set of two functions for each center (exponents for carbon: 1.335, 0.228; oxygen: 2.704, 0.535 /11/). Similar to the findings of Faegri and Almlöf /2/ this modification results in an elongation of the nickel-carbon distance, but only by an extremely small amount.

Next, the standard basis was augmented by a set of f-type orbital polarization functions on nickel. The exponent $\zeta = 0.56$ was chosen in the spirit of Huzinaga /11/ to maximize the radial overlap with the Ni 4s atomic orbital. The bond length decreased by 0.008 Å. A rather marginal shortening was observed when one more d exponent was released from the contraction. Similarly to the bond lengths, only minor variations with the basis sets were noticed for the dissociation energy (see table 2). From these findings we conclude that the large basis set effects found in the HF calculations on $Ni(CO)_4$ should be attributed to the rather shallow HF potential energy curve rather than to any serious basis set deficiencies.

In contrast to HF data one finds a systematic trend in LCGTO-Xα calculations (or, more common, in local density methods /18,19/) to yield too strong bonds. This feature is not a consequence of the Xα approximation to the exchange-correlation potential, since the use of a more sophisticated potential /27/ yields a further bond contraction by about 0.004 Å. An interpretation of this overbinding may be offered following a recent study of Jones and Gunnarsson /19/. As mentioned above, the formation of the metal-ligand bonds in transition metal carbonyls leads to an intraatomic charge transfer from (n)s to (n-1)d orbitals via the repulsion of the (n)s orbital by the CO 5σ orbitals and an metal-ligand charge transfer via backbonding into CO π^* orbitals. Both these features tend to strengthen the metal-carbon bond but at the same time induce an overestimation of the exchange energy in the LD approximation. The former is due to an incorrect description of the (n-1)d core-valence exchange, the latter brings orbitals with additional nodes into play whose exchange contribution may not be described adequately in a local model /19/.

TABLE 2
Ni-C bond length and change in binding energy obtained for various basis sets

Basis set	d_{Ni-C} (Å)	ΔD_o (eV)[a]
Standard set [b]	1.805	-
Two d-functions on ligand atoms[c]	1.809	-0.12
One f-function on Ni[c]	1.797	+0.08
Contraction (11s/8p/4d) on Ni[c]	1.802	+0.05
No Ni 4s and 4p exponents[c]	1.796	-0.30[d]

[a] Change in the binding energy D_o = 8.57 eV with respect to Ni ^3D.
[b] See section 2.
[c] See section 4.
[d] Change in the binding energy with respect to Ni ^1S.

An improvement of the quantitative results obtained with density functional methods may be achieved by the inclusion of non-local terms into the exchange-correlation energy /20/. Treating such terms will undoubtedly increase the computational effort. To our knowledge, applications to large-scale molecular calculations are still lacking.

4.2. Discussion of one-electron orbitals

$Ni(CO)_4$ and $Fe(CO)_5$ are both experimentally and theoretically well characterized. The subsequent discussion is therefore intended to elaborate on some methodological aspects. Vertical ionization energies have been calculated by Baerends et al. /21/ using the DVM-Xα method, which is a well-established procedure for solving the LDF equations. The LCGTO-Xα results, obtained by ΔSCF calculations, are rather similar, as indicated by the three uppermost potentials given for each compound in table 3. Transition state calculations on $Ni(CO)_4$ resulted in marginal (<0.1 eV) differences. Some qualitative features of the calculated ionization potentials nevertheless should be mentioned.

The levels derived from CO 1π orbitals are bracketed by those derived from the CO 5σ orbitals in both the compounds. The 5σ splitting is twice as high in $Fe(CO)_5$ than in $Ni(CO)_4$, (2.04 eV in $Fe(CO)_5$, 0.93 eV in $Ni(CO)_4$) which may be attributed to the unoccupied d_{z^2} orbital on iron. This allows additional ligand to metal σ-type donation from the totally symmetric ligand orbitals.

TABLE 3
Selected ionization potentials for $Ni(CO)_4$ and $Fe(CO)_5$

a) $Ni(CO)_4$

Level	Assignment	This work	DVM-Xα[a)]	Exp.[b)]
$9t_2$	d_{xy}, d_{xz}, d_{yz}	8.86	9.3	8.9
2e	$d_{z^2}, d_{(x^2-y^2)}$	10.30	10.3	9.8
$8t_2$	CO 5σ	13.53	14.1	14.1

b) $Fe(CO)_5$

Level	Assignment	This work	DVM-Xα[a)]	Exp.[c)]
10e'	$d_{(x^2-y^2)}, d_{xy}$	8.39	9.0	8.6
3e"	d_{xz}, d_{yz}	9.60	10.2	9.9
$8a_2''$	CO 5σ(ax), π(eq)	12.59	13.4	13.9

a) Ref. /21/. b) Ref. /23/. c) Ref. /24/.

Both Xα calculations underestimate the splitting of the orbitals derived from CO 4σ orbitals in $Ni(CO)_4$. The experimental splitting is 1.5 eV, whereas the calculations yield 0.7 eV (DVM-Xα) and 0.57 eV (LCGTO-Xα). A possible explanation of this discrepancy with experiment may be attributed to the LD approximation. It has been found /19/ that in atoms s-p excitation energies are underestimated due to the different nodal structures of s and p orbitals. The energy splitting between the $6t_2$ and the $7a_1$ orbital in $Ni(CO)_4$ may be formally described as an excitation $7a_1 \rightarrow 6t_2$ of the positive ion $Ni(CO)_4^+$. The nodal structure of these orbitals is, however, equivalent to that of atomic s and p orbitals, respectively. Further one notes that LD calculations result in systematically too low ionization potentials when other than the uppermost orbitals are concerned. However, the results are better than those obtained from HF calculations /22/, which should not be surprising due to the importance of correlation effects noted above.

Let us finally mention the results of a Mulliken population analysis. A comparison with data obtained by other calculations shows (see table 4) that above all such an analysis reflects the differences in the basis sets used. In the case of the Ni 4p population, for example, Baerends et al. /21,24/ noted that an additional diffuse p function on nickel increases the corresponding population by 0.90 electrons. As a consequence, large differences in the resulting atomic charge of nickel are not surprising.

In an attempt to avoid such ambiguities arising from large two-center overlaps in the LCAO basis the diffuse s and p exponents on nickel, representing 4s and 4p orbitals, have been omitted in an additional calculation. The resulting basis set is very similar to one of those used by Bauschlicher and Bagus in their HF study /14/ to show that ligand to metal donation contributes little to the bonding. The effect of this truncation on the binding energy is in fact rather small (<5%)

TABLE 4
Population analysis for the nickel atom in $Ni(CO)_4$

Method	Population Ni 3d	Ni 4s	Ni 4p	Q(Ni)
LCGTO-Xα	8.71	0.36	1.55	-0.61
HF[a)]	9.13	0.29	0.77	-0.18
CI[b)]	8.82	0.23	0.57	+0.38
DVM-Xα[c)]	8.13	-0.01	0.88	+1.01
LCGTO-Xα trunc.[d)]	9.12	0.01	0.11	+0.76
HF trunc.[e)]	9.14	0.00	-0.01	+0.88

a) Ref. /14/. b) Ref. /25/. c) Ref. /21/.
d) See section 4. e) Ref. /14/, see section 4.

in our LCGTO-Xα calculations (see table 2), similar to the results in ref. /14/. The bond length was shortened by 0.009 Å.

However, the basis set truncation has enormous effects on the nickel 4s and 4p population (see table 4). Furthermore, the populations obtained with HF and LCGTO-Xα using the truncated sets are nearly identical although the description of the bond is quite different. Consequently, a methodological comparison should not be based on a Mulliken population analysis.

5. CONCLUSIONS

The present study clearly indicates the capabilities of the LCGTO-Xα method to calculate the electronic structure and properties of transition metal carbonyls. The results turn out to be better than those of HF calculations, almost reaching the quality of recent CI studies. To further improve the accuracy of the calculations a computationally manageable non-local correction to the exchange-correlation potential seems to be highly desirable.

ACKNOWLEDGEMENT

This work has been supported in part by the Deutsche Forschungsgemeinschaft.

REFERENCES

1. H.P. Lüthi, J. Ammeter, J. Almlöf, and K. Faegri Jr., J. Chem. Phys. 77 (1982) 2002.
2. K. Faegri Jr. and J. Almlöf, Chem. Phys. Lett. 107 (1984) 121.
3. H.P. Lüthi, P.E.M. Siegbahn, and J. Almlöf, J. Phys. Chem. 89 (1985) 2156.
4. C.M. Rohlfing and R.L. Martin, Chem. Phys. Lett. 115 (1985) 104.
5. H.P. Lüthi, P.E.M. Siegbahn, J. Almlöf, K. Faegri Jr., and A. Heiberg, Chem. Phys. Lett. 111 (1984) 1.
6. B.I. Dunlap, J.W.D. Connolly, and J.R. Sabin, J. Chem. Phys. 71 (1979) 3396; 71 (1979) 4993.
7. H. Jörg and N. Rösch, Chem. Phys. Lett., to be published.
8. A.J.H. Wachters, J. Chem. Phys. 52 (1970) 1033.
9. P.J. Hay, J. Chem. Phys. 66 (1977) 4377.
10. F.B. van Duijneveldt, IBM Research Report RJ 945 (1971).
11. S. Huzinaga (Ed.), Gaussian basis sets for molecular calculations (Elsevier, Amsterdam, 1984).
12. H. Jörg, N. Rösch, J.R. Sabin, and B.I. Dunlap, Chem. Phys. Lett. 114 (1985) 529.
13. P. Pyykkö, private communication.
14. C.W. Bauschlicher Jr. and P.S. Bagus, J. Chem. Phys. 81 (1984) 5889.
15. B. Beagley and D.G. Schmidling, J. Mol. Struc. 22 (1974) 466.
16. L. Hedberg, T. Iijima, and K. Hedberg, J. Chem. Phys. 70 (1979) 3224.

17. W.E. Dasent, Inorganic Energetics, Cambridge University Press, Cambridge, 1982.
18. H. Jörg and N. Rösch, Surface Sci., to be published.
19. R.O. Jones and O. Gunnarsson, Phys. Rev. Lett. 55 (1985) 107; Phys. Rev. B 31 (1985) 7588.
20. A. Savin, U. Wedig, H. Preuss, and H. Stoll, Phys. Rev. Lett. 53 (1984) 2087.
21. E.J. Baerends and P. Ros, Mol. Phys. 30 (1975) 1735.
22. J. Demuynck and A. Veillard, Theoret. Chim. Acta 28 (1973) 241.
23. I.H. Hillier, M.F. Guest, B.R. Higginson, and D.R. Lloyd, Mol. Phys. 27 (1974) 215.
24. E.J. Baerends, Ch. Oudshoorn, and A. Oskam, J. Electron Spectr. Relat. Phenom. 6 (1975) 259.
25. M.R.A. Blomberg, U.B. Brandemark, P.E.M. Siegbahn, K.B. Mathisen, and G. Karlström, J. Phys. Chem. 89 (1985) 2171.
26. L.H. Jones, R.S. McDowell, M. Goldblatt, and B.I. Swanson, J. Chem. Phys. 57 (1972) 2050.
27. J.P. Perdew and A. Zunger, Phys. Rev. B 23 (1981) 5048.

QUANTITATIVE RESULTS AND QUALITATIVE ANALYSIS BY THE HARTREE-FOCK-SLATER TRANSITION STATE METHOD

Tom Ziegler, Vincenzo Tschinke and Louis Versluis
Department of Chemistry, University of Calgary
Calgary, Alberta, CANADA

I. INTRODUCTION

The challenge from transition metal chemistry has been met over the last twenty years with increasing success by theoretical chemists using a variety of methods ranging from calculations on the back of an envelope to computations on mainframe computers. The qualitative "back of the envelope" type of calculations have clearly had a profound influence not only on the field of transition metal chemistry as a whole, but also on practitioners of more extensive calculation schemes by forcing them to provide a qualitative analysis of the (at best) quantitative results, and we present here a scheme, the generalized transition state method[1], by which quantitative (or at least extensive) calculations can be subjected to a qualitative analysis.

The extensive calculation scheme used here is not of the usual ab initio Hartree-Fock type but rather based on the Hartree-Fock-Slater method[2] (or its density functional extensions[3],) as implemented by Baerends et al.[4]

II. ENERGY FUNCTIONALS AND THE GENERALIZED TRANSITION STATE METHOD

The total electronic energy for a molecular system with the electron density $\rho(1)=\rho^{\alpha}(1)+\rho^{\beta}(1)$ and the one electron density matrix $\rho(1,1')$ can be written as:

$$E = -\tfrac{1}{2}\int \left[\nabla_1^2\, \rho(1,1')\right]_{1\to 1'} dx_1 + \sum_A \int \rho(1)\frac{Z_A}{|\vec{R}_A - \vec{r}_1|}\, dx_1 + \tfrac{1}{2}\int\int \rho(1)\rho(2)\, 1/r_{12}\, dx_1 dx_2 + \int \rho^{\alpha}(1)[\varepsilon_x^{\alpha}(1) + \varepsilon_c^{\alpha}(1)]dx_1$$

A. Veillard (ed.), Quantum Chemistry: The Challenge of Transition Metals and Coordination Chemistry, 189–198.

$$+ \int \rho^{\beta}(1)\ [\varepsilon_{x}^{\beta}(1)+\varepsilon_{c}^{\beta}(1)]dx_1 \qquad (1)$$

In ab initio calculations on the Hartree-Fock level correlation between electrons of different spins is neglected by setting $\varepsilon_c^{\alpha}, \varepsilon_c^{\beta}$ *to zero, and correlation between electrons of the same spin is included by approximating* $\varepsilon_x^{\alpha}, \varepsilon_x^{\beta}$ *with the well known HF-exchange potentials* $\varepsilon_{xhf}^{\alpha}, \varepsilon_{xhf}^{\beta}$. *Experience from the last decade has shown that the simple HF-model for calculations on transition metal complexes involving energy differences is somewhat inadequate. The correlation potentials* ε_c *as well as the exchange potentials* ε_x *can however in principle be described to any desired degree of accuracy by resorting to ab initio configuration interaction calculations (HF-CI). Such calculations are except for the smallest of systems computationally quite demanding and it is for real size transition metal complexes only possible to obtain an approximate (limited CI) description of* ε_c *and* ε_x. *New technics and computer technology might well change this situation and it will thus in addition to a quantitative study of chemical systems be possible to examine accurate exchange and correlation potentials in detail.*

In particular physicists have over the years adopted an approach somewhat different from HF-CI in which ε_c *and* ε_x *by appealing physical arguments are given a preconceived form*[3]. *This approach is often referred to as density functional theory since* ε_c *and* ε_x *are expressed in terms of* ρ^{α} *and* ρ^{β}. *The simplest and certainly the most widely applied density functional theory in chemistry is the Hartree-Fock-Slater method*[2] *(HFS or Xα) where* ε_c *is set to zero and* ε_x *approximated by the local potential*

$$\varepsilon_x^{\gamma}(1) = V_x^{\gamma}(1) \equiv -3\alpha_{ex}[\frac{3}{4\pi}\rho^{\gamma}(1)]^{\frac{1}{3}}\ ;\ \gamma = \alpha,\beta \qquad (2)$$

Other more complex functionals have also come to fore recently[3,5].

The expression for E based on the HFS-approximations $(\varepsilon_c^{\gamma}=0, \varepsilon_x^{\gamma}=V_x^{\gamma})$ *with which we will be dealing in the next sections is most often referred to as the statistical energy and can be used to evaluate energy changes as they occur in connection with ionizations, excitations and chemical processes by simply subtracting the energies of the final and initial states. It is however in the ab initio type implementation of the HFS method by Baerends et al.*[4]*, which we will adopt here, computationally expedient and somewhat illuminating to write down a direct expression for an energy difference. It is readily shown*[1] *that the enthalphy for the chemical reaction A+B→AB within the HFS-approximation is given by:*

$$\Delta E = E_{elst} + E_{exch} + \int \Delta\rho(1)\hat{f}[\rho_A(1)+\rho_B(1)+\tfrac{1}{2}\Delta\rho(1)]dx_1, \quad (3)$$

where

$$E_{elst} = \sum_{\nu}^{A}\sum_{\mu}^{B} \frac{Z_\nu Z_\mu}{R_{\nu\mu}} + \int \frac{\rho_A(\vec{r}_1)\ \rho_B(\vec{r}_2)}{r_{12}}\, d\vec{r}_1\, d\vec{r}_2 - \sum_{\nu}^{A}\int \frac{Z_\nu \rho_B(\vec{r}_1)}{|\vec{R}_\nu - \vec{r}_1|}\, d\vec{r}_1 - \sum_{\mu}^{B}\int \frac{Z_\mu \rho_A(\vec{r}_1)}{|\vec{R}_\mu - \vec{r}_1|}\, d\vec{r}_1, \tag{4}$$

is the electrostatic interaction between fragments (atoms or molecules) A and B at the positions they take up in the combined molecule AB, and

$$E_{exch} = \frac{3}{4}\sum_{\gamma}\left[\int(\rho_A^\gamma+\rho_B^\gamma)V_x^\gamma(\rho_A^\gamma+\rho_B^\gamma)d\vec{r}_1 - \int\rho_A^\gamma V_x^\gamma(\rho_A^\gamma)d\vec{r}_1 - \int\rho_B^\gamma V_x^\gamma(\rho_B^\gamma)d\vec{r}_1\right], \tag{5}$$

the corresponding exchange term. The one electron operator $\hat{f}(1)$, in fact the one electron HFS operator, is finally given by

$$\hat{f}(\rho(1)) = -\tfrac{1}{2}\nabla_1^2 - \sum_{\nu}\frac{Z_\nu}{|\vec{R}_\nu - \vec{r}_1|} + \int\frac{\rho(\vec{r}_2)}{r_{12}}d\vec{r}_2 + \sum_{\gamma}V_x^\gamma(\rho^\gamma(\vec{r}_1)). \tag{6}$$

The calculation of energy differences by (3) is often referred to as the generalized transition state procedure since the operator $\hat{f}$ in (3) is defined with respect to the density $\rho = \rho_A + \rho_B + \frac{1}{2}\Delta\rho$, intermediate between the initial density of the two fragments $\rho_A + \rho_B$ and the final density of the combined molecule $\rho_{AB} = \rho_A + \rho_B + \Delta\rho$. The expression (3) can also be used to calculate ionization energies[3b,6] and excitation energies[6]. In those cases E_{elst} and E_{exch} are zero and $\Delta\rho$ represents the difference in density between the initial and final state.

III. APPLICATIONS

Periodic Trends in the Nucleophilicity of Metal Centers

The ability of pentacarbonyls 1 and tetracarbonyls 5 as well as related low valent complexes to act as nucleophiles is of crucial importance in several key-step catalytic processes (e.g. the S_{N_2}-type oxidative addition). We have[7] proposed the nucleophilicity of 1 and 5 as a function of the metal center by calculating the protonation energy ΔE_p for the processes 1→2 and 5→6. The protonation processes are further in order to foster some insight considered in a sequence of steps. The framework of the carbonyls are in the first step deformed from the conformations 1 and 5 of the free carbonyls to the geometries 3 and 7 they have in the combined hydrides 2 and 6. The contribution to ΔE_p from this step is termed ΔE_{prep}. The proton is placed in the second step at the position it will have in the hydride complex without allowing for any rearrangement of the electron density at the carbonyl. The contribution to ΔE_P from the second step is thus purely electrostatic (E_{elst} of Eq. (4) and Eq. (5)) and referred to as $\Delta E°$. We allow in the third step the electron density to relax by $\Delta\rho$ to that of the metal hydrides 3 and 6, that is, we carry out a full SCF-calculation. The contribution from this step to ΔE_P is $\Delta E_{el} = \int \Delta\rho \, \hat{f}_T dx_1$, with $\hat{f}_T$ defined in Eq. (6). Since further $\Delta\rho$ can be blocked[1] into density changes of the orbitals in each of the irreducible representations corresponding to the point groups of 2 and 6 as $\Delta\rho = \Sigma_\Gamma \Delta\rho_\Gamma$, one can decompose ΔE_{el} as $\Delta E_{el} = \Sigma_\Gamma \Delta E_\Gamma$, where ΔE_Γ is the contribution to ΔE_{el} from symmetry representation Γ. We finally have as a separate term the contribution from relativistic[8,16] effects ΔE_R.

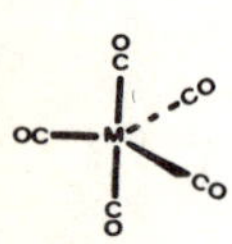

1

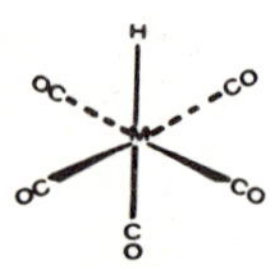

2

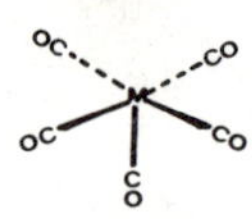

3

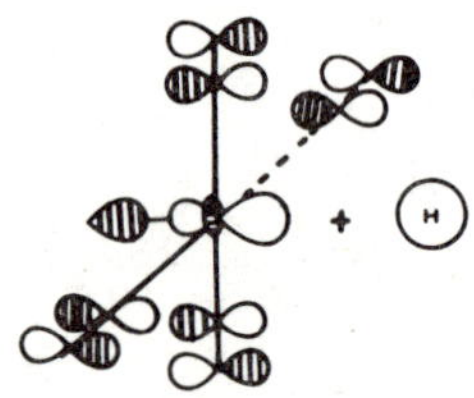

a_1 + 1s

4

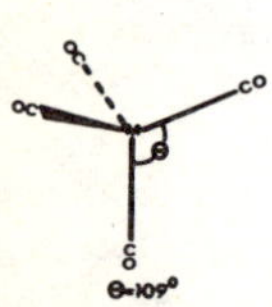

5

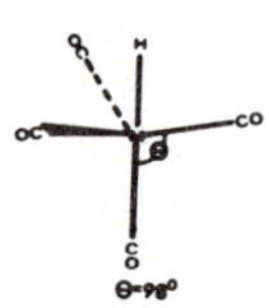

6

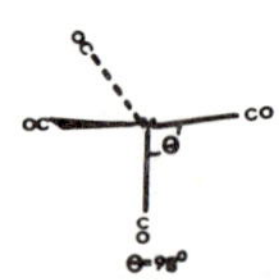

7

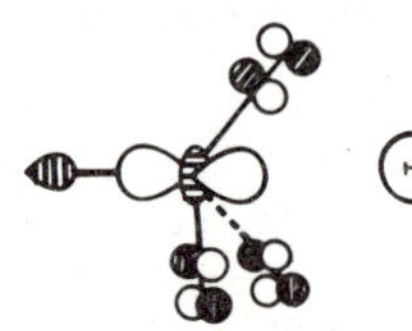

a_1 + 1s

8

The calculated protonation energies for 1 and 5 are given in Table 1, and Table 2, decomposed into the various contributions, $E_P = \Delta E_{prep} + \Delta E° + \Sigma_\Gamma \Delta E_\Gamma + \Delta E_R$. It follows from the tables that the protonation energy within a homologous series (e.g. $Mn(CO)_5^-$, $Tc(CO)_5^-$, $Re(CO)_5^-$) increases down a triad and that the trend setting term in ΔE_P is ΔE_{A_1} corresponding to the donor acceptor interactions 4 and 8. The donor acceptor interaction energy ΔE_{A_1} depends on the overlap between the $1s_H$ acceptor orbital and the d_{z^2}-type donor orbital on the carbonyl fragment, 4 or 8, and the trend is explained by noting that this overlap increases as the d-orbitals become more diffuse down a triad. We predict in general for a homologous series that the nucleophilicity will increase down a triad.

A comparison of protonation energies within a family of isoelectronic carbonyls made up of metals from the same transition series ($Cr(CO)_5^{2-}$, $Mn(CO)_5^-$, $Fe(CO)_5$, $Co(CO)_5^+$) and ($Fe(CO)_4^{2-}$, $Co(CO)_4^-$, $Ni(CO)_4$, $Cu(CO)_4^+$) is not all that informative since the different charges on the carbonyls will have a strong (uninteresting) influence on the *variation in ΔE_P through ΔE°. It is however interesting to compare ΔE_{A_1} which in our opinion is a good measure of the nucleophilicity. It is seen that ΔE_{A_1} decreases along a series with increasing nuclear charge (Cr to Co; Fe to Cu), and this trend can be explained by noting that the overlaps in 4 and 8 will decrease as the 3d orbitals contract with increasing nuclear charge along the transition series. We predict in general the nucleophilicity to decrease along a transition series with increasing nuclear charge. The calculated protonation energy of 816 kJ mol^{-1} is in fair agreement with the experimental value[9] of 853 kJ mol^{-1}.*

Table 1 A decomposition of calculated protonation energies (in kJ mol^{-1}) for $M(CO)_5^-$ (M = Mn, Tc, Re) and $M(CO)_5$ (M = Fe, Ru, Os) as well as $Cr(CO)_5^{2-}$, $Co(CO)_5^+$.

$M(CO)_5^{n-}$	ΔE_{prep}	ΔE°	ΔE_{elec}	ΔE_{A_1}	$\Delta E_{elec}-\Delta E_{A_1}$	ΔE_R	ΔE_P
$Mn(CO)_5^-$	-30.0	403.8	843.5[b]	754.2	89.3	-	1217.3[a]
$Tc(CO)_5^-$	-30.7	356.1	934.9	839.3	95.6	-	1260.3
$Re(CO)_5^-$	-31.8	360.2	955.9	882.9	73.0	34.4	1318.7
$Fe(CO)_5$	-27.1	46.8	796.3	687.0	109.3	-	816.0
$Ru(CO)_5$	-26.3	33.1	833.8	713.5	120.3	-	840.6
$Os(CO)_5$	-31.9	24.7	860.9	771.6	89.3	28.9	882.6
$Cr(CO)_5^{2-}$	-29.2	941.7	890.7	819.6	71.2	-	1803.2
$Co(CO)_5^+$	-27.8	-496.6	702.7	581.1	121.6	-	178.3

a The total protonation energy ΔE_P is given by: $\Delta E_P = \Delta E_{prep} + \Delta E^\circ + \Delta E_{elec} + \Delta E_R$

b $\Delta E_{elec} = \Delta E_{A_1} + \Delta E_{A_2} + \Delta E_{B_1} + \Delta E_{B_2} + \Delta E_E$

Table 2 A decomposition of calculated protonation energies (in kJ mol^{-1}) for $M(CO)_4$ (M = Ni, Pd, Pt) and $M(CO)_4^-$ (M = Co, Rh, Ir) as well as $Fe(CO)_4^{2-}$, $Cu(CO)_4^+$, $Co(CO)_3(PH_3)^-$.

$M(CO)_3L^{n-}$	ΔE_{prep}	ΔE°	ΔE_{elec}	ΔE_{A_1}	$\Delta E_{A_2} + \Delta E_E$	ΔE_R	ΔE_P
$Co(CO)_4^-$	-30.2	475.3	747.1[b]	638.4	108.7	-	1192.2[a]
$Rh(CO)_4^-$	-35.1	462.4	777.6	671.5	106.1	-	1204.9
$Ir(CO)_4^-$	-29.8	441.4	812.3	702.2	110.1	23.1	1247.0
$Ni(CO)_4$	-32.3	-46.0	569.1	550.7	108.4	-	580.8
$Pd(CO)_4$	-39.2	-56.5	697.9	608.9	89.0	-	602.2
$Pt(CO)_4$	-27.3	-71.9	721.7	632.1	89.6	23.9	646.4
$Fe(CO)_4^{2-}$	-29.1	967.2	799.4	704.3	95.1	-	1735.5
$Cu(CO)_4^+$	-26.3	-599.3	594.3	498.7	95.6	-	-31.3
$Co(CO)_3(PH_3)^-$	-34.8	481.1	775.2	670.2	105.1	-	1221.5

a The total protonation energy ΔE_P is given by: $\Delta E_P = \Delta E_{prep} + \Delta E^\circ + \Delta E_{elec} + \Delta E_R$.

b $\Delta E_{elec} = \Delta E_{A_1} + \Delta E_{A_2} + \Delta E_E$.

Multiple Metal-Metal Bonds

Binuclear metal systems such as M_2L_6 (M = Cr, Mo, W; L = Cl, OR, NR_2, H) of conformation 9 and $M_2(PR_3)_4Cl_4$(M = Mn, Tc, Re) of conformation 10, all with a triple metal-metal bond, as well as the quadruple bonded systems $M_2(PR_3)_4Cl_4$(M = Cr, Mo, W) and $M_2(CO_2R)_4$ (R = H, CH_3, CF_3; M = Cr, Mo, W) of conformation 10 or 11 have been studied[10] extensively over the last decade in order to further our understanding of multiple metal metal bonds first accounted for by Cotton[11] in terms of σ, π, and δ bonding orbitals as shown for $M_2(PR_3)_4Cl_4$ in 12.

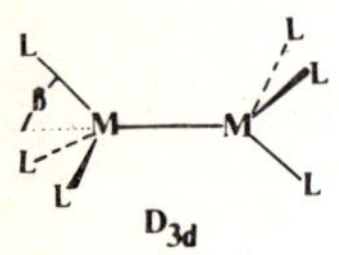

9

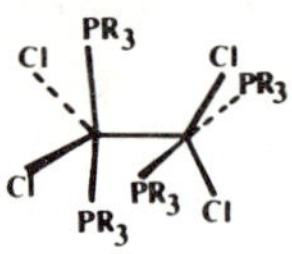

10

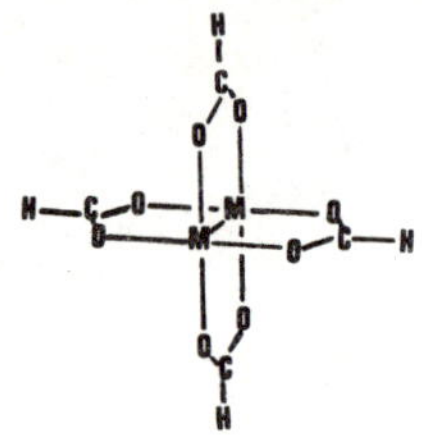

11

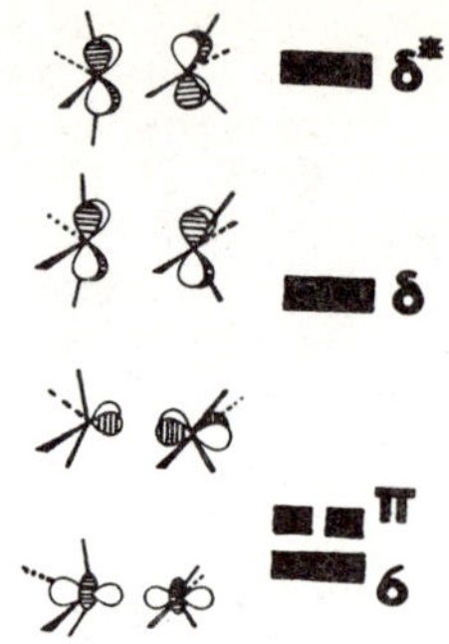

12

We have in a series of HFS-calculations[12,6b], in which some of the key-geometrical parameters including the metal-metal bond distance R_{MM} were optimized, calculated the metal-metal bond strengths decomposed as $\Delta E_{MM} = \Delta E^{\circ} + \Delta E_{\sigma} + \Delta E_{\pi} + \Delta E_{\delta} + \Delta E_{R}$, see Table III. The term ΔE° (steric interaction term) corresponds to bringing together two high-spin ML_n fragments of opposite spin-polarization to the positions they will have in the combined M_2L_{2n} complex without allowing the density to relax, whereas ΔE_{σ}, ΔE_{π} and ΔE_{δ} are the contributions to ΔE_{MM} from the σ, π, and δ bonds when the density is allowed to relax to that of M_2L_{2n}. It follows from Table III that the sigma bond is somewhat stronger than the phi-bond since $\Delta E_{\sigma} > \frac{1}{2}\Delta E_{\pi}$, and that the delta-bond in $M_2(PH_3)_4Cl_4$ (M = Cr, Mo, W) is relatively weak with $\Delta E_{\delta} \ll \Delta E_{\sigma}$. Relativistic effects ΔE_R are further seen to be quite important for the M_2H_6 systems making the W-W bond stronger than the Mo-Mo bond.

We have further probed[6b] the nature of the multiple metal-metal bond in the case of the quadruple bonded systems $M_2(CO_2H)_4$ (M = Cr, Mo, W) by calculating the change in the metal-metal bond distance R°_{MM} as an electron is ionized out of the σ, π and δ orbitals, respectively. We calculate, Figure 1, in accordance with the relative strengths of the delta- and phi-bonds a modest elongation (.04 Å) of R_{MM} when an electron is ionized out of the δ-orbital and a substantial elongation (.12 Å) on ionization out of the π-orbital. The ionization of an electron out of the σ-orbital does on the other hand not result in any elongation. A rationalization for this surprising result is given in Ref. 6b.

Migratory Aptitude of Hydride and Methyl

The intramolecular migration of a methyl group to a cis-CO ligand as in (7) has been observed in several instances[13].

Table 3. Optimized Metal-Metal Bond Distances R_{MM}(Å), and Decomposition of Calculated Metal-Metal Bonding Energy ΔE_{MM} (kJ mol^{-1}).

Complex	R_{MM}	ΔE_{MM}	$\Delta E°$	ΔE_{σ}	ΔE_{π}	ΔE_{δ}	ΔE_{R}
$Mn_2(PH_3)_4Cl_4$	1.92	294.8	-376.5	246.3	423.8	1.2	-
$Tc_2(PH_3)_4Cl_4$	2.16	598.7	-319.6	331.4	584.3	2.6	-
$Re_2(PH_3)_4Cl_4$	2.27	562.1	-256.0	303.6	503.9	1.1	9.2
$Cr_2(PH_3)_4Cl_4$	1.89	152.9	-498.5	-222.6	420.7	8.1	-
$Mo_2(PH_3)_4Cl_4$	2.18	523.5	-389.5	288.6	591.6	32.8	-
$W_2(PH_3)_4Cl_4$	2.28	428.3	-430.5	285.8	540.2	25.2	7.2
Cr_2H_6	1.92	300.6	-461.8	314.5	447.9	-	-
Mo_2H_6	2.22	447.1	-374.5	334.1	487.5	-	-
W_2H_6	2.36	535.2	-379.3	318.5	482.4	-	113.6

[a] The total bonding energy is given as $\Delta E_{MM} = \Delta E° + \Delta E_{\sigma} + \Delta E_{\pi} + \Delta E_{\delta} + E_R$.

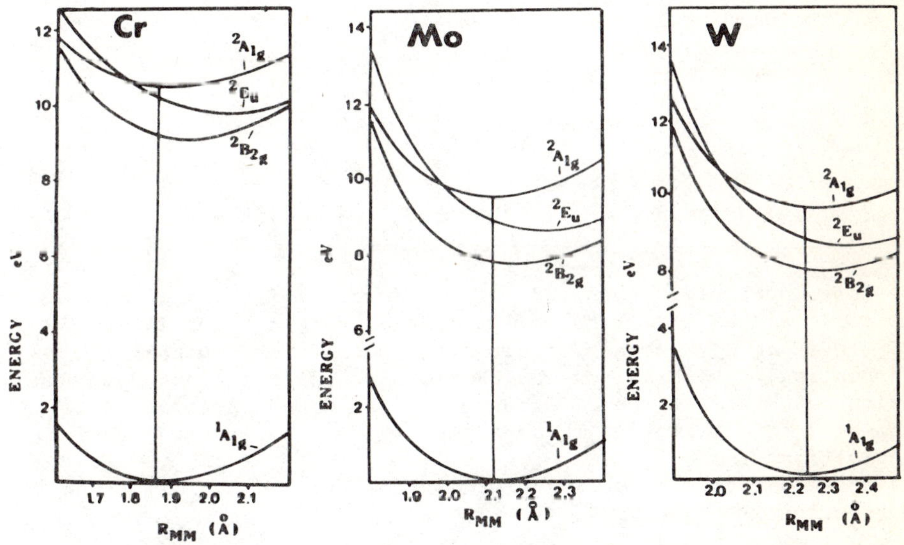

Figure 1 Energies (eV) of the $^1A_{1g}$ ground state for $M_2(CO_2H)_4$(M=Cr,Mo,W) and the three states $^2B_{2g}$ ($\delta\pi^4\sigma^2$), $^2E_u(\delta^2\pi^3\sigma^2)$ and $^2A_{1g}(\delta^2\pi^4\sigma)$ of $M_2(CO_2H)_4^+$ (M=Cr,Mo,W) as a function of the metal-metal bond distance R_{MM} (Å).

$$\begin{array}{c} CH_3 \\ | \\ (CO)_4Mn - CO \end{array} \rightarrow (CO)_4Mn - C\begin{array}{l} \diagup CH_3 \\ \diagdown\!\!\diagdown O \end{array} \qquad (7).$$

The corresponding shift reaction (8) with hydride instead of methyl has on the other hand proved to be rather elusive[14].

$$\begin{array}{c} H \\ | \\ (CO)_4Mn - CO \end{array} \rightarrow (CO)_4Mn - C\begin{array}{l} \diagup H \\ \diagdown\!\!\diagdown O \end{array} \qquad (8).$$

The apparent poor migratory aptitude of hydride towards CO in comparison to methyl seems surprising in view of the fact that hydride appears to be at a par with methyl as a migratory group towards other ligands such as CS[15] and CH_2[16].

We find the reaction (8) with an enthalpy of $\Delta H_8 = 159$ kJ mol^{-1} to be more endothermic than the corresponding methyl migration (7) with $\Delta H_7 = 75$ kJ mol^{-1}, and attribute the order $\Delta H_8 > \Delta H_7$ to the Mn-H bond in $HMn(CO)_5$ being stronger than the Mn-CH_3 bond in $CH_3Mn(CO)_5$.

We find further the model reactions:

$$\begin{array}{c} CH_3 \\ | \\ (CO)_4Mn - CS \end{array} \rightarrow (CO)_4Mn - C\begin{array}{l} \diagup CH_3 \\ \diagdown\!\!\diagdown S \end{array} \qquad (9)$$

and

$$\begin{array}{c} H \\ | \\ (CO)_4Mn - CS \end{array} \rightarrow (CO)_4Mn - C\begin{array}{l} \diagup H \\ \diagdown\!\!\diagdown S \end{array} \qquad (10)$$

to be less endothermic than (7) and (8) respectively with $\Delta H_9 =$ 20 kJ mol^{-1} and $\Delta H_{10} = 71$ kJ mol^{-1}. The migration to CS is easier than to CO since the π^*_{CS}-acceptor orbital is lower in energy than the corresponding π^*_{CO}-orbital on CO. Exploratory calculations along the reaction paths for (7) to (10) revealed only modest activation barriers (1 kJ mol^{-1} to 8 kJ mol^{-1}) in agreement with a previous theoretical investigation[14].

The reactions (7) to (10) are key-steps in the overall carbonylation reaction

$$RMn(CO)_4CX + CO \rightarrow RC(X)Mn(CO)_5 \qquad (11),$$

where the incoming CO ligand takes up the vacated site after the migration of R[13]. The degree to which the 16-electron complex $RC(X)Mn(CO)_4$ rearranges once it is formed with conformation 13a is thus of importance for the stereochemistry of the overall carbonylation *reaction (11). Kinetic labeling measurements*[13] *in solution indicate that CO is coordinated to the site vacated by R without isomerization of* $RC(X)Mn(CO)_4$ *in conformation* 13a. *We find that* 13a *can rearrange to the dihapto structure* 13d *via* 13b *and* 13c, *and that the dihapto structure* 13d *is 79* $kJ\ mol^{-1}$ $(R\text{-}CH_3,\ X = O)$, *68* $kJ\ mol^{-1}$ $(R = H,\ X = O)$, *130* $kJ\ mol^{-1}$ $(R = H,\ X = S)$ *and 136* $kJ\ mol^{-1}$ $(R = CH_3,\ X = S)$, *respectively, more stable than* 13a. *We explain the inability of* $RC(X)Mn(CO)_4$ *with conformation* 13a *to isomerize in solution by the formation of the solvated complex* 13e.

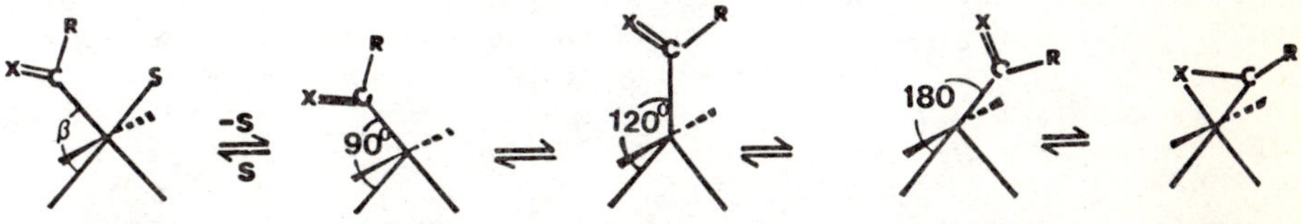

13e **13a** **13b** **13c** **13d**

IV. EXTENSIONS OF THE HFS-METHOD

The HFS-method, where ε_c *of Eq. 1 is zero and* ε_x *given by Eq. 2, is known to overestimate bond energies in several instances as it is shown in Table 4 for the first CO-dissociation energy of* $Ni(CO)_4$ *and* $Cr(CO)_6$. *Including correlation between electrons of different spins by adopting for* ε_c *the self interaction corrected functional due to Stoll*[18] *et al. does not improve the calculated dissociation energies, LSD of Table 4. If, however, one add to* ε_x *of Eq. 2 the lowest order gradient correction term* ($X_{\alpha\beta\gamma}$) *derived by Becke*[19] *a dramatic improvement in the calculated dissociation energies is obtained,* $X_{\alpha\beta\gamma}$ *of Table 4. We are currently*[20] *investigating this new functional by Becke.*

Table 4. Calculated first CO-dissociation energy ($kcal\ mol^{-1}$) *for* $Ni(CO)_4$ *and* $Cr(CO)_6$.

	HFS	LSD[a]	$X_{\alpha\beta\gamma}$[b]	Exp.[c]
$Ni(CO)_4$	45	46	18	22
$Cr(CO)_6$	65	66	31	36

[a]*Functional by Stoll et al.*[18] [b]*Functional due to Becke*[19]
[c]*Ref. 21.*

ACKNOWLEDGEMENT

We would like to thank Professor A. Becke for numerous useful discussions and Professor E.J. Baerends for providing us with a copy of his HFS-program. This investigation was supported by the Natural Sciences and Engineering Research Council of Canada (NSERC).

REFERENCES

1. (a) Ziegler, T.; Rauk, A. Theor. Chim. Acta 1977, 46, 1.
   (b) Ziegler, T. in 'Relativistic Effects in Atoms, Molecules and Solids', G. Malli (ed.), Plenum (1982).
2. Slater, J.C. Adv. Quantum Chem. 1972, 6, 1.
3. Dahl, J.P.; Avery, J. 'Local Density Approximations in Quantum Chemistry and Solid State Physics', Plenum Press, 1984.
4. (a) Baerends, E.J.; Ellis, D.E.; Ros, P. Chem. Phys. 1973, 2, 41.
   (b) Baerends, E.J.; Snijders, J.G.; deLange, C.A.; Jonkers, G. Ref. 3, pp.: 414-485.
5. Dreizler, R.; da Providencia, J. 'Density Functional Methods in Physics', Plenum Press (to be published).
6. (a) Ziegler, T. Ref. 3, pp.: 273-285.
   (b) Ziegler, T. J. Am. Chem. Soc. 1985, 107.
7. Ziegler, T. Organometallics 1985, 4, 675.
8. (a) Snijders, J.G.; Baerends, E.J.; Ros, P. Mol. Phys. 1979, 38, 1909.
   (b) Ziegler, T.; Snijders, J.G.; Baerends, E.J. J. Chem. Phys. 1981, 74, 1271.
9. Beauchamp, J.L.; Stevens, A.E.; Corderman, R.R. Pure Appl. Chem. 1979, 51, 967.
10. Cotton, F.A.; Walton, R.A. 'Multiple Bonds Between Metal Atoms', Wiley, New York, 1982.
11. Cotton, F.A. Inorg. Chem. 1965, 4, 334.
12. (a) Ziegler, T. J. Am. Chem. Soc. 1983, 105, 7543.
    (b) Ziegler, T. J. Am. Chem. Soc. 1984, 106, 5901.
13. Flood, T.C.; Jensen, J.E.; Statler, J.A. J. Am. Chem. Soc. 1981, 103, 4411.
14. Berke, H.; Hoffmann, R. J. Am. Chem. Soc. 1978, 100, 7224.
15. Clark, G.B.; Collins, T.J.; Marsden, K.; Roper, W.R. J. Organomet. Chem. 1978, C23, 157.
16. Collman, H.P.; Hegedus, L.S. 'Principles and Applications of Organotransition Metal Chemistry', University Science Books, Mill Valley, Ca., 1980.
17. Ziegler, T.; Versluis, L.; Tschinke, V. J. Am. Chem. Soc. submitted.
18. Stoll, H.; Pavlidov, C.M.E.; Preuss, H. Theoret. Chim. Acta 1978, 49, 143.
19. Becke, A. Int. J. Quantum Chem. submitted.
20. Tschinke, V.; Ziegler, T. International Conference on Quantum Chemistry, Montreal 1985.

Xα ELECTRONIC STRUCTURE OF TRANSITION METAL HALIDES: CALCULATION OF LIGAND-FIELD MULTIPLET STATES

A. GOURSOT and C. DAUL

UA CNRS n°431, Ecole de Chimie
68093 Mulhouse
France

Institut Chimie Inorganique
FRIBOURG
SUISSE

Abstract : Multiple scattering Xα wavefunctions are used in connection with irreducible tensor operators to calculate the many-electron CI matrices, needed to predict the energies of ligand-field multiplet states. A non-muffin-tinned form of the Xα charge distribution (charge-partitioning procedure of Case and Karplus) is used to evaluate the Coulomb an exchange integrals in terms of the Griffith's parameters. The energies of the ligand-field excited states of CrF_6^{3-} and $CrCl_6^{3-}$ are thus predicted and compared with experiment.

1. Introduction

Recent studies have shown that one-electron and also two-electron properties can be obtained with useful accuracy from the MS-Xα wavefunctions (1-4). The interest of such calculations lies in their use of the self-consistent MS-Xα molecular distribution within the atomic spheres rather than that of the spherically-averaged, muffin-tin charge distribution. Their fundamental approximation is the partitioning of the intersphere charge among the basis functions centered on the atoms (1). This technique allows to evaluate the proper two-electron integrals needed to calculate the electrostatic interactions between the d^n configurations of transition metal complexes (5). It is applied here to the d^3 complexes CrF_6^{3-} and $CrCl_6^{3-}$, whose ligand-field spectra have been extensively studied.

2. Method

2.1. Intersphere charge-partitioning

We will only recall briefly the general features and approximations of the charge partitioning procedure (1-4) applied to the SCF-MS-Xα wavefunctions, whose characteristics have already been established in details (8,9).

Each SCF-Xα MO is expressed, in the atomic and outersphere regions, in terms of single-center expansions of products of radial functions and real spherical harmonics. In the intersphere region, the MO has a

A. Veillard (ed.), Quantum Chemistry: The Challenge of Transition Metals and Coordination Chemistry, 199–207.

multicenter form, consisting of products of spherical Bessel functions and real spherical harmonics (9). The charge partitioning of the intersphere charge of each MO is performed by extending the radial functions $P^{\alpha}_{i,\ell}(\vec{r})$ beyond the atomic sphere radii, according to an algorithm written by Case and Karplus (1). Under the NDDO approximation (neglect of diatomic differential overlap), the molecular orbitals can be expressed simply in terms of atomic contributions :

$$\phi_i(\vec{r}) = \sum_{\ell,m} C^{\alpha}_{i,\ell m} P^{\alpha}_{i,\ell}(\vec{r}_\alpha) Y_{\ell,m}(\Omega_\alpha) \text{ in the atomic sphere } \alpha$$

where $C^{\alpha}_{i,\ell m}$ are the expansion coefficients of the atomic basis functions in the MO

- $P^{\alpha}_{i,\ell}(\vec{r}_\alpha)$ are the radial parts, characteristic of the one-electron energy $\varepsilon i(\phi i)$
- $Y_{\ell,m}(\Omega_\alpha)$ are the real spherical harmonics

For two-electron properties, the expansions of the radial functions are best performed by treating each atom individually and self-consistently (4).

2.2. Evaluation of two-electron electrostatic integrals

In order to calculate the electrostatic integrals, the operator $1/|\vec{r}_1-\vec{r}_2|$ must be expanded in a form that separates the radial and angular parts. When both electrons are located on the same atomic sphere α, the Laplace expansion is used, while the Steinborn-Filter formulation (5) is employed when r_1 is located on α but r_2 refers to an electron of a different sphere β (4).

A general two-electron integral can be written as

$$< ij|k\ell > = \int \phi_i(\vec{r}_1)\, \phi_j(\vec{r}_1) \frac{1}{|\vec{r}_1-\vec{r}_2|} \phi_k(\vec{r}_2)\, \phi_\ell(\vec{r}_2)\, dr_1\, dr_2$$

$$\sum_{\substack{\ell_i m_i\\ \ell_j m_j\\ \ell_k m_k\\ \ell_\ell m_\ell\\ \alpha\beta}} C^{\alpha}_{i\ell_i m_i} C^{\alpha}_{j\ell_j m_j} C^{\beta}_{k\ell_k m_k} C^{\beta}_{\ell\ell_\ell m_\ell} < X^{\alpha}_i X^{\alpha}_j \,|\, X^{\beta}_k X^{\beta}_\ell >$$

where $X^{\alpha}_i = P^{\alpha}_{i\ell_i}(r_\alpha)\, Y_{\ell_i m_i}(\Omega_\alpha)$ and so on ...

Under the NDDO approximation, the three-and four-center integrals vanish so that $\vec{r}_1$ identifies with $\vec{r}_\alpha$ and $\vec{r}_2$ with $\vec{r}_\beta$. The X^{α}_i can thus be expressed in the atomic coordinate systems for which their fonctional forms are known. The calculation of $< X^{\alpha}_i X^{\alpha}_j \,|\, X^{\beta}_k X^{\beta}_\ell >$ requires to consider 3 cases : $\alpha \neq \beta$ ($\beta \neq$ outersphere) and $\alpha = \beta$ (β = outersphere).

In all cases, the angular parts of the integrals are expressed as Gaunt coefficients and the radial integrals are evaluated numerically.

2.3. Evaluation of the electrostatic matrix elements connecting the d^n states

In the case of cubic complexes, it has been shown that any electrostatic matrix element may be expressed as linear combinations of ten reduced two-electron matrix elements, by using irreducible tensor operator (7,10). These ten parameters, called a, b, c, d, e, f, g, h, i, j, according to Griffith's notation (6), are characteristic of a particular complex. As usual, the values of the electrostatic integrals have been reduced by 10% in order to account for the electron correlation (11).

The evaluation of the diagonal elements of the CI matrix necessitates the determination of an additional parameter $\Delta_1 = n\varepsilon(e) + m\varepsilon(t_2)$, where $\varepsilon(e)$ and $\varepsilon(t_2)$ are respectively the one-electron energies of orbitals e and t_2, in the configuration $t_2^m e^n$. Δ_1 may be expressed as a function of the averaged energy of this configuration :

$$\Delta_1 = E_{av}(t_2^m\, e^n) - \frac{n(n-1)}{2}\, g(e,e) - \frac{m(m-1)}{2} g(t_2,t_2) - mng(e,t_2)$$

where $g(t_2,t_2) = 1/5\ a + 4/5\ b - 2/5\ j$; $g(e,e) = e - 5/3\ f$ and $g(e,t_2) = d + c - 1/2\ (g + h)$, as expressed in terms of the Griffith's parameters. Indeed, $g(t_2,t_2)$, $g(e,e)$ and $g(e,t_2)$ represent the repulsion between two electrons occupying orbital t_2, orbital e, and both orbitals t_2 and e.

Δ_1 is easily evaluated by using MS-Xα transition-state calculations, which yield the energy separation Δ_c between the barycentres of two configurations :

$$\Delta_c = E_{av}(e^{n-x} t_2^{m+x}) - E_{av}(e^n t_2^m)$$

The Δ_c values are evaluated by fixing the ground-state as origin of the energies. It must be noted that the Δ_1 quantities are generally not linear functions of n and m, which necessitates different transition -state calculations for each excited-state configuration.

3. MS-Xα calculation parameters

The experimental Cr-F bond distance of 1.93 Å (in Na_2KCrF_6) was chosen for CrF_6^{3-} (12). The Cr-Cl bond distance was fixed to 2.45 Å , weighted average value between the Cr-Cl distance in $CsNaInCl_6$ and in $CrCl_3$. Both complexes have O_h symmetry. A 20% overlapping ratio has been chosen for the metal and halogen sphere radii (R_{Cr} = 2.3116 and R_F = 2.0684 a.u. in CrF_6^{3-} ; R_{Cr} = 2.6245 and R_{Cl} = 2.9313 a.u. in $CrCl_6^{3-}$).

An external tangent outer-sphere is used in each case. It also serves as a Watson sphere (13), on which a +3 charge is distributed to simulate the environment of the anion. The values of the exchange parameters (α_{Cr} = 0.71352, α_F = 0.73732, α_{Cl} = 0.72325) are those optimized by Schwarz (14). A weighted average of these atomic values is chosen for the α value of the interatomic and extramolecular regions. Partial waves up to $\ell = 4$ are included in the multiple scattering expansion in the Cr sphere and the extramolecular region and up to $\ell = 1$ in the halogen spheres.

4. Results and discussion

The calculated values of the electrostatic integrals are presented in tables I and II.
The MS-Xα values of the Δ_c parameters for the t_{2g}^2eg, $t_{2g}e_g^2$ and e_g^3 configurations are 15.80, 16.05, 16.20 kK for CrF_6^{3-} and 11.70, 11.80, 11.93 kK for $CrCl_6^{3-}$, respectively.

TABLE I : Xα calculated values of the Griffith's parameters for $CrCl_6^{3-}$

Two electron integrals (b)	Major contributions (a) (cm^{-1}) Cr-Cr	Cr-F	F-F	Xα calculated values (cm^{-1})
a = $(\xi^2;\xi^2)$	156929	7994	482	165405
b = $(\eta^2;\xi^2)$	144684	7797	306	152787
c = $(\theta\varepsilon;\xi^2)$	2783	357	190	3330
d = $(\varepsilon^2;\xi^2)$	124250	22790	638	147678
e = $(\varepsilon^2;\varepsilon^2)$	118912	19249	3205	141366
f = $(\theta\varepsilon;\theta\varepsilon)$	5583	315	507	6405
g = $(\theta\xi;\theta\xi)$	3795	15	42	3852
h = $(\varepsilon\eta;\theta\eta)$	1380	4	-14	1370
i = $(\theta\eta;\zeta\xi)$	-1470	-11	0	-1481
j = $(\eta\zeta;\eta\zeta)$	6122	6	9	6137

(a) the (OUT-OUT), (Metal-OUT) and (ligand-OUT) contributions are always less than 5 cm^{-1}
(b) (ξ,η,ζ) and (ε,θ) are components of t_2 and e, respectively

The Cr^{3+} halides appeared to be suitable compounds to perform the analysis of their ligand-field states, since their CT bands, especially for CrF_6^{3-}, do not hide the major part of their ligand-field spectra. Moreover, these compounds have been extensively studied both experimentally and theoretically. The ground-state of these t_{2g}^3 compounds is $^4A_{2g}$ and the most intense d-d bands are thus quartet-to quartet transitions. However, in most cases, quartet-to-doublet spin-forbidden transitions are also observed.

TABLE II : Xα calculated values of the Griffith's parameters for CrF_6^{3-}

Two electron integrals(a)	Major contributions(b) (cm^{-1}) Cr-Cr	Cr-Cl	Cl-Cl	Xα calculated values (cm^{-1})
a	150418	6396	315	157129
b	138824	6281	214	145319
c	2469	287	153	2909
d	111234	24845	601	136680
e	99830	19465	3769	123064
f	4620	220	412	5252
g	3359	13	31	3403
h	1230	5	-11	1224
i	-1350	-13	0	-1363
j	5797	6	6	5809

(a) same definition as in Table I

(b) the (OUT-OUT), (Metal-OUT) and (ligand-OUT) contributions are always less than 5 cm^{-1}

One reason for the observation of these bands is the possible mixing of doublet and quartet states of adequate symmetry through the spin-orbit interaction. If the spin-orbit interaction modifies the transition probabilities for the spin-forbidden lines, the spin-orbit coupling in the Cr^{3+} ion is so small ($\zeta_{3d} \simeq 200$ cm^{-1}) that the splittings of the terms are generally not observable. We have thus neglected, in the present calculation, the effect of the spin-orbit coupling.

Experimental data and ligand-field analyses are available for different CrF_6^{3-} systems, but we will focus here on the studies devoted to K_2NaCrF_6 or Cr^{3+} in K_2NaGaF_6, since these crystals are known to possess a cubic symmetry. Their detailed absorption and emission spectra have been reported (15, 16, 17), so as assignments based on ligand-field (15, 16) or MO calculations (18, 19, 20). From these works, it comes out that the lowest complex band at 15.0 - 17.0 kK is formed by the three overlapping absorptions $^4A_{2g} \rightarrow {}^2E_g$, $^4A_{2g} \rightarrow {}^2T_{1g}$ and $^4A_{2g} \rightarrow {}^4T_{2g}$. All the studies agree that $(t_{2g}^3)\,^2E_g$ is the lowest excited state. The relative location of $^2T_{1g}$ and $^4T_{2g}$ is subject of controversy. The theoretical studies generally report $^4T_{2g}$ as the highest component of the lowest band, while a recent analysis of the vibronic structure of the $^4A_{2g} \rightarrow {}^2E_g$ transition leads to the conclusion that $^4T_{2g}$ and $^2T_{1g}$ lie about 550 and 1600 cm^{-1} above ^2E_g respectively.

TABLE III : calculated and experimental transition energies (kK) for CrF_6^{3-}

$(t_2^3)^{(a)}\ ^4A_{2g} \longrightarrow$	calc. values	Exp. values *(b)*	*(c)*
$t_2^3\ ^2E_g$	15.3	15.2	15.1
$t_2^3\ ^2T_{1g}$	16.3	15.6	16.6
$t_2^2(^3T_1)e\ ^4T_{2g}$	17.6	16.4	15.6
$t_2^3\ ^2T_{2g}$	23.2	22.5	
$t_2^2(^3T_1)e\ ^4T_{1g}$	25.3	23.5	
$t_2^2(^1E)e\ ^2A_{1g}$	29.7	30.5$^{(d)}$	
$t_2e^2(^3A_1)\ ^4T_{1g}$	37.8	37.4$^{(e)}$	
$t_2^2\ (^1T_2)e\ ^2T_{2g}$	31.1	31.5$^{(e)}$	
$t_2^2\ (^1T_2)e\ ^2T_{1g}$	32.6	33.1$^{(e)}$	
$t_2^2\ (^3T_1)e\ ^2T_{1g}$	37.2	37.7$^{(e)}$	
$t_2^2\ (^1T_2)e\ ^2T_{2g}$	43.4	43.4$^{(e)}$	

(a) the subscript g has been omitted for simplicity
(b) in K_2NaCrF_6 at 77°K - from ref. 15 and 16 if not otherwise specified.
(c) from ref. 17 in $K_2NaGaF_6 : Cr^{3+}$
(d) very weak feature at 77°K
(e) observations in optically pumped ruby-from ref. 20

A detailed analysis of the absorption spectrum of Cr^{3+} in $CsNaInCl_6$ concludes to the following ordering of the three lowest excited states :

$$^4T_{2g} < {}^2E_g < {}^2T_{1g} \qquad (21,22)$$

Tables III and IV present the calculated ligand-field states of CrF_6^{3-} and $CrCl_6^{3-}$ compared with available experimental peak positions. For the chlorine complex, this comparison is more limited since the ligand-field absorptions higher than 25 kK are hidden by CT bands.

TABLE IV : calculated and experimental transition energies (kK) for $CrCl_6^{3-}$

$(t_2^3)^{(a)}\ {}^4A_{2g}$	calc. values	Exp. values$^{(b)}$
$t_2^2(^3T_1)e\ {}^4T_{2g}$	14.2	13.5, 13.2$^{(c)}$
$t_2^3\ {}^2E_g$	14.4	14.4
$t_2^3\ {}^2T_{1g}$	15.2	15.0
$t_2^2(^3T_1)e\ {}^4T_{1g}$	20.6	18.2, 18.7$^{(c)}$
$t_2^3\ {}^2T_{2g}$	21.3	19.9
$t_2^2(^1A_1)e\ {}^2A_{1g}$	25.5	25.0

(a) the subscript g has been omitted for simplicity
(b) from ref.21, if not otherwise specified
(c) from ref.23

Examination of tables III and IV shows that the calculated multiplet energies of the studied Cr^{3+} compounds reproduce with a very useful accuracy the observed spectra. The agreement with experiment is even more interesting if one bears in mind that no optimization has been performed, concerning the atomic sphere radii, the corrections for correlation energy or the Cr-halogen bond distances. These results could probably be improved by using well-fitted metal-halogen bond lengths, especially for $CrCl_6^{3-}$ and also by usingCI admixtures of charge transfer states.

Nevertheless, in their present state, these results show that the method described above, using only the MS-Xα wavefunctions of the ground-state and, in case of cubic symmetry, the evaluation of ten interelectronic repulsion integrals, could be very useful to predict reliable multiplet energies.

References

1. D.A. Case and M. Karplus, *Chem. Phys. Lett.*, **39**, 33 (1976)

2. M. Cook and M. Karplus, *J. Chem. Phys.*, **72**, 7 (1980)

3. D.A. Case, M. Cook and M. Karplus, *J. Chem. Phys.*, **73**, 3294 (1980)

4. M. Cook and M. Karplus, *Chem. Phys. Lett.*, **84**, 565 (1981)

5. E.O. Steinborn and E. Filter, *Intern. Quantum Chem.*, **9S**, 435 (1975)

6. J.S. Griffith *"The theory of transition-metal ions"*, Cambridge, University Press (1961)

7. C. Daul, *Thèse d'agrégation*, Fribourg (1981)

8. J.C. Slater, *"The theory of the self-consistent field for molecules and solids"* vol. **4**, Mc Graw-Hill New-York (1974)

9. K.H. Johnson, *Adv. Quantum Chem.*, **7**, 143 (1973)

10. C. Daul and P. Day, *Mol. Phys.*, **34**, 1707 (1977)

11. J. Ferguson, *Progress in Inorganic Chemistry*, **12**, 159 (1970)

12. K. Knox and D.W. Mitchell, *J. Inorg. Nucl. Chem.*, **21**, 253 (1961)

13. R.E. Watson, *Phys. Rev.*, **111**, 1108 (1958)

14. K. Schwarz, *Phys. Rev.B*, **5**, 2466 (1972)

15. J. Ferguson, H.J. Guggenheim and D.L. Wood, *J. Chem. Phys.*, **54**, 504 (1971)

16. D.L. Wood, J. Ferguson, K. Knox and J.F. Dillon Jr., *J. Chem. Phys.* **39**, 890 (1963)

17. L. Dubicki, J. Ferguson and B. van Oosterhout, *J. Phys. C*, **13**, 2791 (1980)

18. D.R. Armstrong, R. Fortune and P.G. Perkins, *J. Chem. Soc. Dalton Trans.*, 753 (1976)

19. L. Pueyo and J.W. Richardson, *J. Chem. Phys.*, **67**, 3577 (1977)

20. L. Pueyo and J.W. Richardson, *J. Chem. Phys.*, **67**, 3583 (1977)

21. P. Shaw, *D. Phil. Thesis*, University of Oxford (1975)

22. H.U. Güdel, private communication

23. W.E. Hatfield, R.C. Fay, C.E. Pfluger and T.S. Piper, *J. Am. Chem. Soc.*, **85**, 265 (1983)

COMPLEMENTARY SPHERICAL ELECTRON DENSITY MODEL FOR CO-ORDINATION AND ORGANOMETALLIC COMPOUNDS

D. M. P. Mingos
Inorganic Chemistry Laboratory
University of Oxford
South Parks Road
Oxford OX1 3QR
United Kingdom.

ABSTRACT. If a molecule is viewed initially as a central atom surrounded by a spherical shell of electron density representing the ligand sphere then the atomic orbitals of the central atom and those of the spherical shell can both be expressed in terms of spherical harmonics and the quantum mechanical problem is accurately resolved. If the spherical shell is concentrated into N regions of electron density corresponding to the ligand positions in an ML_N complex then the linear combination of atomic orbitals can be expressed in terms of a spherical harmonic expansion with each linear combinations defined by l and m quantum numbers. Not all ligand polyhedra generate linear combinations whose quantum numbers match the d, s and p valence orbitals of the central atom. Co-ordination polyhedra which are solutions to the packing and covering problems on a spherical surface generate S, P and D linear combinations in a sequential fashion which emulate those of the central atom. Therefore, the inert gas rule can be reformulated in terms of complementary sets of ligand and metal functions which when taken together have a net angular momentum of zero and an electron distribution which approximates as closely as possible to spherical. This quantum mechanical restatement of the inert gas rule has specific electronic and stereochemical implications. The scope of this Complementary Spherical Electron Density Model is illustrated by reference not only to co-ordination compounds but also metal cluster compounds.

A. Veillard (ed.), Quantum Chemistry: The Challenge of Transition Metals and Coordination Chemistry, 209–224.

1. INTRODUCTION

During the last ten years semi-empirical molecular orbital calculations have been widely used particularly by Hoffmann and his coworkers [1] to describe the stereochemical preferences of co-ordination and organometallic complexes of the transition metals. Two cornerstones of this approach have been the construction of Walsh diagrams to trace the influence of angular deformations on the energies of one-electron molecular orbitals and the use of perturbation theory to describe in a semi-quantitative fashion the interactions between molecular fragments. Considering the approximate nature of the extended Hückel model used for the majority of the calculations and the total neglect of electron-electron repulsion, nuclear-nuclear repulsion and electron correlation effects [2] the success of this model in accounting for ground state geometries and rotational barriers in compounds [3] has been remarkably good.

An examination of the ground state structures of organometallic compounds of the transition metals reveals a particularly simple pattern which can be represented in a matrix form such as that illustrated in Figure 1. Complexes which conform to the inert gas rule adopt co-ordination polyhedra which are solutions to either the covering or packing problem on a spherical surface [4]. These covering and packing problems have been solved by mathematicians interested in seed pollination problems and can be summarised as follows:

No. of vertices	Best covering polyhedron	Best packing polyhedron
3	triangle	triangle
4	tetrahedron	tetrahedron
5	trigonal bipyramid	square-pyramid
6	octahedron	octahedron
7	pentagonal bipyramid	capped octahedron
8	dodecahedron	square-antiprism
9	tricapped trigonal prism	capped square-antiprism

The packing solutions also form the basis of the Valence Shell Electron Pair Repulsion Model developed by Sidgwick, Powell, Nyholm and Gillespie [5,6]. However, since electron repulsion effects are neglected in the extended Huckel method the packing of electron pairs on a spherical surface cannot explain the apparent success of this method. Furthermore the observation that 16, 14 and 12 electron organometallic complexes generally have stereochemistries which are related to the spherical polyhedra by the loss of one, two and three vertices respectively is in disagreement with the Valence Shell Electron Pair Model since it implies that electron pair holes are stereochemically active in transition metal compounds [7]. For example in Figure 1 the metal carbonyls $Cr(CO)_6$, $Cr(CO)_5$,

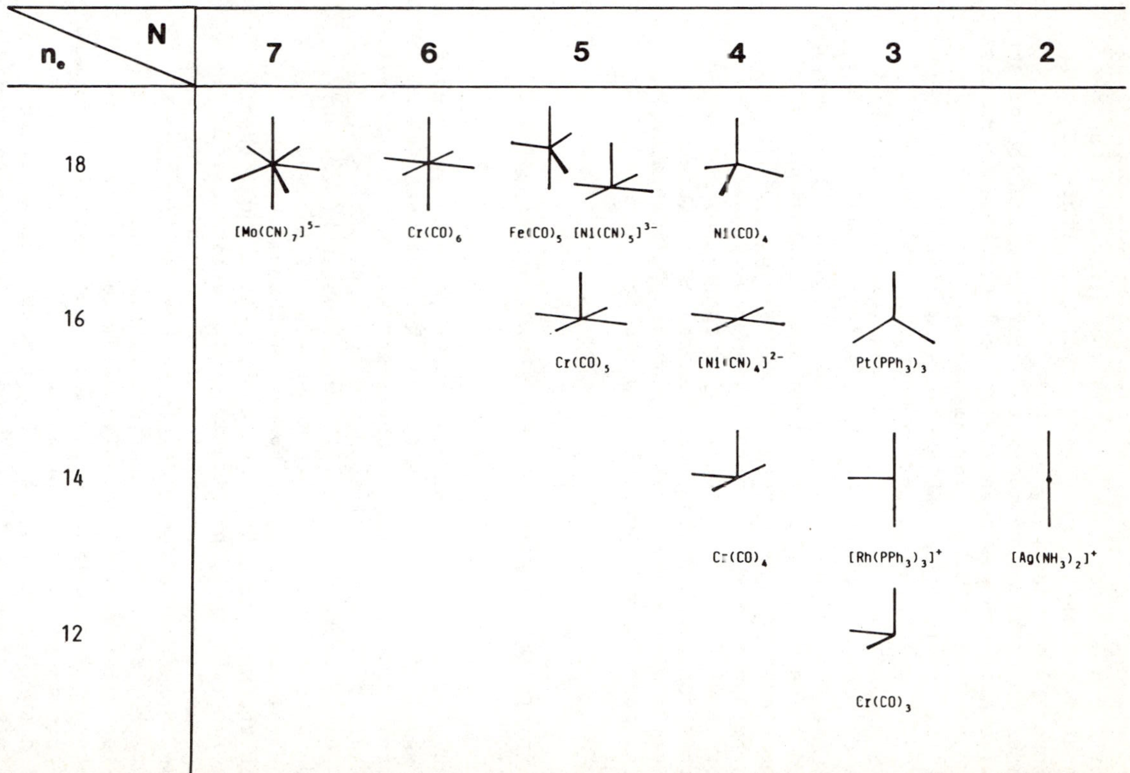

Figure 1. Matrix representation of the preferred stereochemistries for covalent transition metal complexes. N is the co-ordination number, n_e is the total number of valence electrons.

$Cr(CO)_4$ and $Cr(CO)_3$ have structures which can be related to the octahedron. This generalisation is only valid for complexes where the metal d shell is incompletely filled. For d^{10} metals the pattern reverts to that established for main group molecules and the ligands adopt the best covering or best packing polyhedra on ligand dissociation in agreement with the V.S.E.P.R. conclusions. For example, for the d^{10} metal complexes shown in Figure 1 the following stereochemistries are observed - $Ni(CO)_4$, (tetrahedron), $Pt(PPh_3)_3$ (triangular) and $Ag(NH_3)_2^+$ (linear) .

In this paper an alternative model for inorganic stereochemistries is developed which although based on the molecular orbital formalism does not require the use of Walsh diagrams and yet accounts for the adoption of the co-ordination polyhedra summarised in Figure 1. This model views molecules as interpenetrating spheres of electron density and unifies the inert gas and molecular orbital formalisms. It depends critically on the idea that the linear combinations of ligand lone pair orbitals in a co-ordination compound can be expressed in terms of a spherical harmonic expansion. This idea can be traced back to a paper by Verkade et al [8], but has been expressed in a more general fashion by Stone [9] as part of a more general analysis of cluster compounds. The importance of this concept has also been recognised by Quinn who has also described a simple and general method for illustrating spherical, vector and tensor spherical harmonic functions as projections [10].

2 LIGAND LINEAR COMBINATIONS AS A SPHERICAL HARMONIC EXPANSION

If a molecule is viewed initially as a central atom surrounded by a spherical shell of electron density, which represents the ligand sphere, then the derivation of the inert gas rule is trivial. The atomic orbitals of the central atom and the spherical shell are both expressed in terms of spherical harmonics $Y_{l,m}(\theta, \phi)$ and are governed by the same l and m quantum numbers. If the wave functions of the valence orbitals of the central atom overlap effectively with the wave functions of the spherical shell then a matching set of n (n=4 or 9) bonding molecular orbitals are generated. For example, for a transition metal atom with 3d, 4s and 4p valence atomic wave functions interacts exclusively with the nine $S\sigma$, $P\sigma$, and $D\sigma$ wave functions with identical l and m quantum numbers. If the resultant nine molecular orbitals are fully occupied then the resultant total electron density is spherically symmetric and approximates to that of an inert gas atom.

The radial distribution functions for the valence orbitals of a transition metal shown in Figure 2 indicate that the nd orbitals are substantially more contracted than the (n+1)s and (n+1)p orbitals and therefore good overlap between the metal and the ligand sphere is only achieved for all the valence orbitals when the ligand

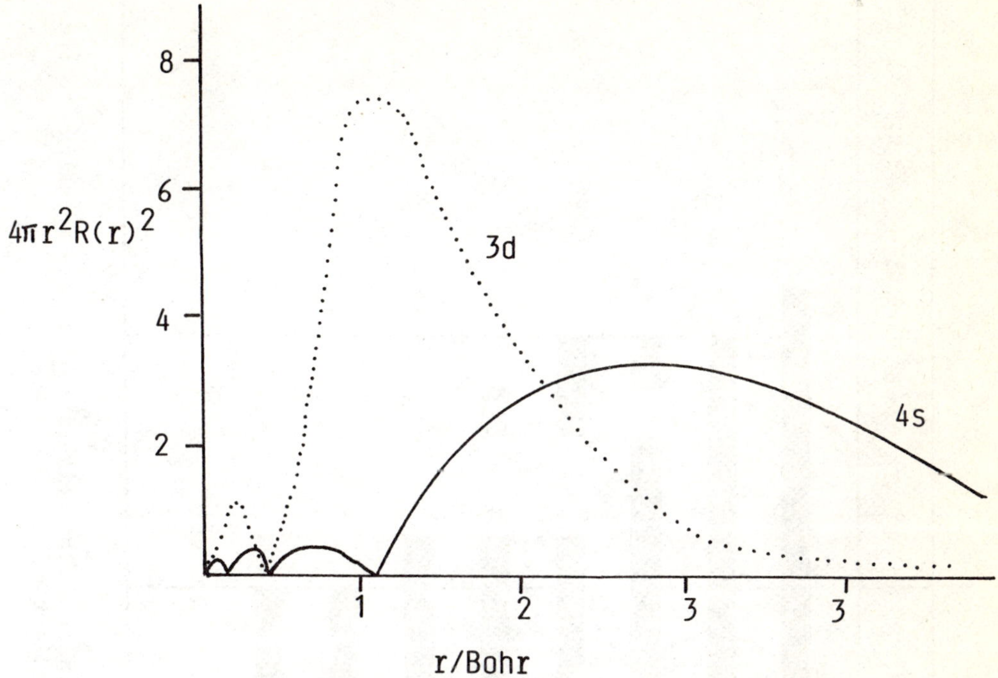

Figure 2. Radial distribution functions of the 5s and 4d atomic orbitals of a ruthenium atom inits ground state.

sphere has a small radius. In real co-ordination compounds this is realised either by having ligands with small covalent radii (e.g. hydride) or ligands which have short metal-ligand internuclear distances by virtue of multiple bonding effects (e.g. CO). Therefore, adherence to the inert gas formalism is closely related to having ligands which overlap effectively with the nd, (n+1)s and (n+1)p metal valence orbitals.

If the spherical ligand shell is now concentrated into N regions of electron density corresponding to the ligand positions of an MH_N complex then the linear combinations of hydrogen 1s orbitals, σ_i, can be expressed in terms of the following spherical harmonic expansion :

$$\Phi(l,m) = \sum_i c_i \sigma_i$$
$$= N' \sum_i Y_{lm}(\theta_i \phi_i) \sigma_i$$

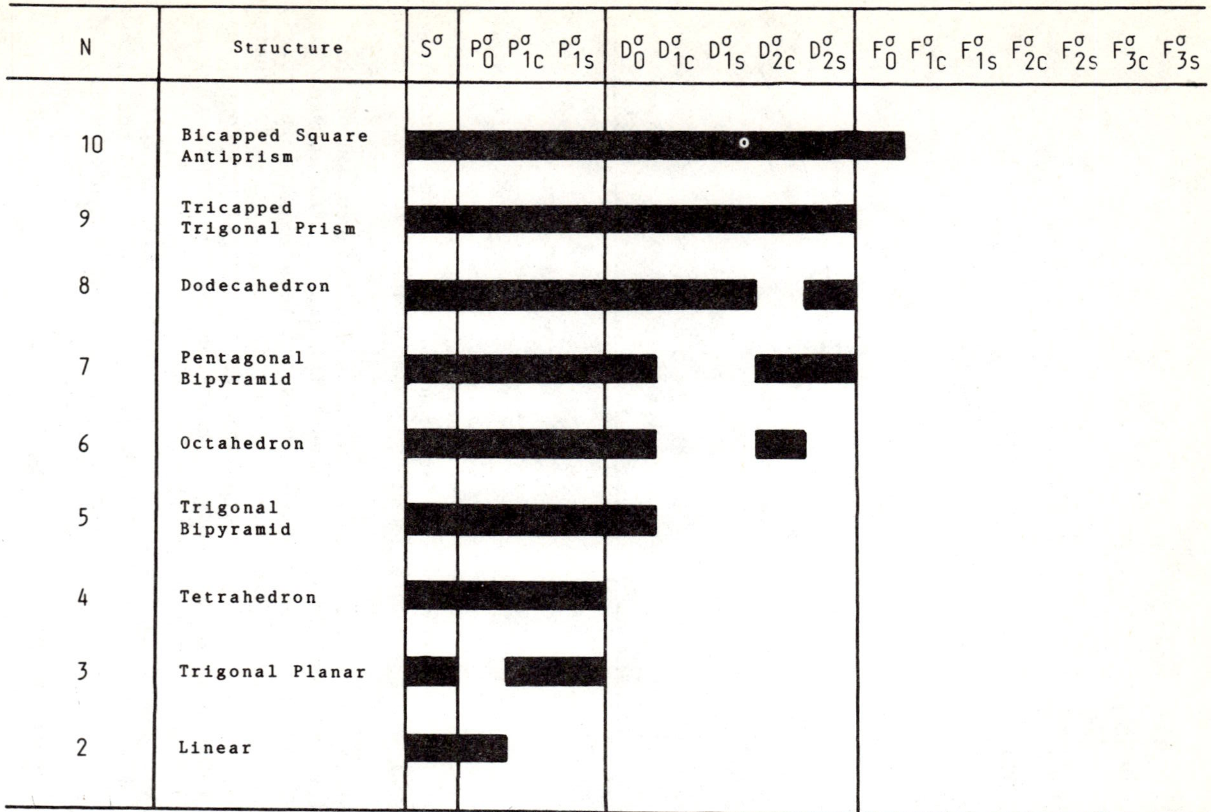

Figure 3. Summary of ligand linear combinations for polyhedra which are solutions to the covering problem on a spherical surface.

$$= \quad L_m^{\sigma} \qquad m = 0, 1c, 1s, \ldots\ldots \quad L = S, P, D, \ldots$$

The quantum numbers are no longer strictly valid because of the descent in symmetry, but they do accurately represent the nodal characteristics of the ligand linear combinations, which determine the overlap preferences between ligand and central atom orbitals with the same symmetries.

In the extended Hückel formalism the central atom wave functions are expressed in terms of spherical harmonics and the metal-ligand overlap integrals are evaluated as a function of the angular co-ordinates utilising the spherical harmonic characteristics of the wave functions. In this model the pseudo- spherical symmetry of the co-ordination polyhedron is utilised to define the bonding characteristics of the ligand linear combinations. For example, the D_o linear combination of ligand wave functions has a significant overlap integral with a d_o function of the central atom, but a near zero overlap integral with the metal d_{1s}, d_{1c}, d_{2s}, d_{2c} wave functions. As the number of ligand atoms increases this approximation improves because the spherical ideal is more closely approached.

Therefore, since θ_i and ϕ_i in the spherical harmonic expansion given above represent the locations of the ligand atoms on the spherical surface this mode of expressing the ligand linear combinations has important stereochemical implications. In particular only ligand polyhedra which represent solutions to the packing and covering problems generate S^{σ}, P^{σ} and D^{σ} functions in a sequential fashion (see Figure 3 for example). Other general classes of polyhedra, e.g. three-connected polyhedra and bipyramids (see Figures 4 and 5 for example) utilise F^{σ} functions in preference to D^{σ} for the higher co-ordination polyhedra because they have atoms lying on the nodal planes of the D^{σ} functions. For example, for 8 co-ordination the dodecahedron and the square-antiprism generate S^{σ}, P^{σ} and four D^{σ} linear combinations. In contrast the cube and the hexagonal bipyramid generate S^{σ}, P^{σ}, three D^{σ} functions and one F^{σ} function. Therefore, although the dodecahedron and square antiprism are compatible with the attainment of the inert gas configuration about the central atom the cube and the hexagonal bipyramid are not because they contribute wave functions with f pseudo-symmetries.

An alternative way of appreciating this stereochemical distinction is based on a 'United Atom' approach [11]. A dodecahedral or square antiprismatic 18 electron transition metal complex MH_8 has wave functions which correlate with the ground state of the corresponding inert gas atom, whereas the cubic and hexagonal bipyramidal complexes correlate with the inert gas atom in the excited state, $d^8s^2p^6f^2$.

A hydrido- co-ordination compound achieves a pseudo-spherical eighteen electron configuration by adopting one of the best covering polyhedra and utilising a complementary set of metal d functions. As the co-ordination number of the complex is reduced from nine then vacancies in the D^{σ} shell of linear combinations occur (see

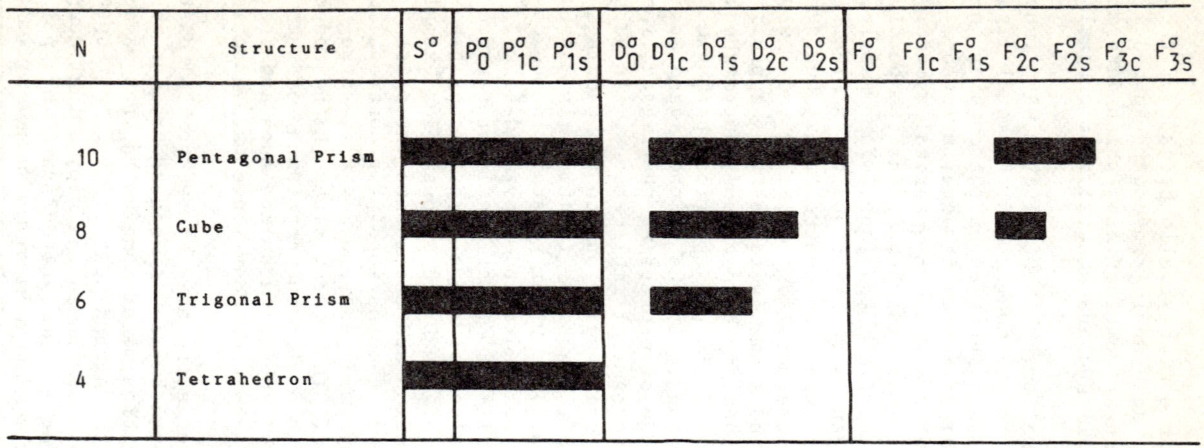

Figure 4. Summary of ligand linear combinations for three-connected co-ordination polyhedra.

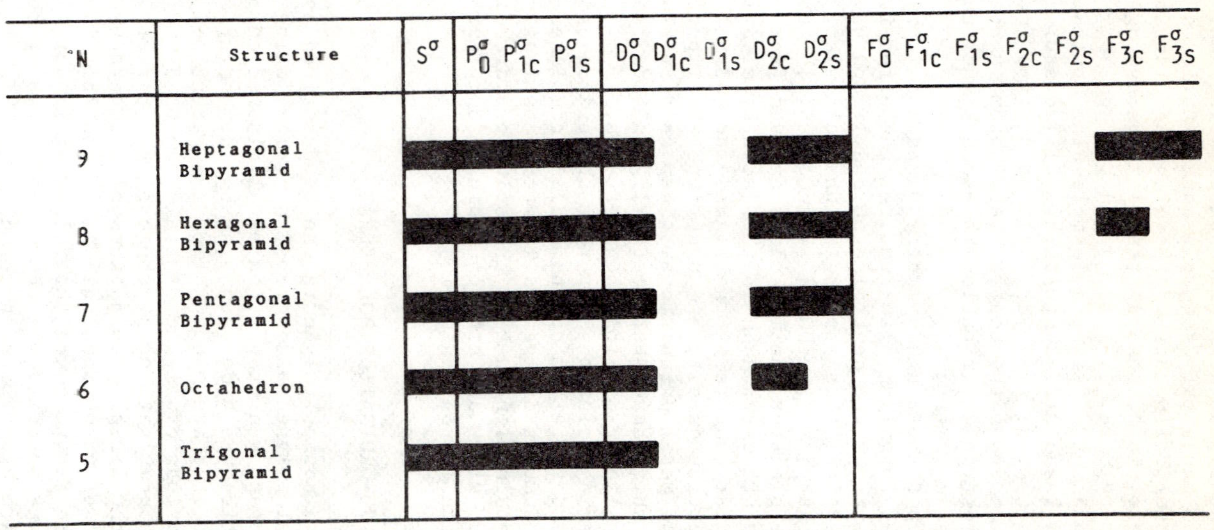

Figure 5. Summary of ligand linear combinations for bipyramidal co-ordination polyhedra.

Figure 3 for example) which have the effect of creating holes in what originally was a near spherical electron distribution. However, if the metal atom has a set of complementary filled d orbitals then the near spherical nature of the electron distribution is maintained. For example, in octahedral $[RuH_6]^{2-}$ the hydrogen linear combinations generate $S^{\sigma}(a_{1g})$, $P^{\sigma}(t_{1u})$ and D^{σ}_{o} and D^{σ}_{2c} (e_g) functions which form a set of six stable molecular orbitals with the metal s, p and d_{z^2} and $d_{x^2-y^2}$ orbitals. Filled orbitals localised on the central atom, d_{1s}(xz), d_{1c}(yz) and d_{2c}(xy) complete the inert gas configuration about the central metal atom in the following senses - the electron distribution is pseudo-spherical and the net angular momentum is zero.

The arguments developed above are symmetry based and in reality adherence to the inert gas formalism also requires the effective matching of the radial parts of the central atom and ligand wave functions. In order to explore this aspect, X-alpha molecular orbital calculations on $[RuH_6]^{2-}$, $[Ru(CO)_6]^{2+}$ and $[RuCl_6]^{2-}$ have been completed [12]. The electron density plots derived from these calculations, which are discu ssed in the lecture, clearly demonstrate the complementary nature of the ligand and central metal atom regions of electron density and underline the importance of the radial parts of the metal d functions. They only overlap effectively with ligands which have short metal-ligand bond lengths by virtue of their small size or multiple bond effects.

The packing and covering solutions for four and six co-ordinate polyhedra are identical and therefore organometallic complexes with these co-ordination numbers are generally stereochemically rigid. For the other co-ordination numbers the solutions are different and therefore five, seven, eight and nine co-ordinate complexes are generally stereochemically non-rigid [13].

3. NON-SPHERICAL CO-ORDINATION POLYHEDRA

Lower symmetry co-ordination polyhedra can be described as fragments of the spherical polyhedra described above. Following the precedent set for borane and transition metal cluster compounds these fragments are described as _nido_- and _arachno_- depending on whether one or two vertices are missing from the parent polyhedron [14]. The linear combinations of ligand orbitals can be described as sub-sets of the linear combinations of those of the parent polyhedra. For example, the linear combinations of a square-pyramid can be derived from those of an octahedron. An octahedral ligand set is defined as shown in Figure 6 and the following linear combinations are derived by substituting the co-ordinates of the ligands into spherical harmonic expansion formulae.

$$S^{\sigma} = 1/\sqrt{6}\,\sigma_1 + 1/\sqrt{6}\,\sigma_2 + 1/\sqrt{6}\,\sigma_3 + 1/\sqrt{6}\,\sigma_4 + 1/\sqrt{6}\,\sigma_5 + 1/\sqrt{6}\,\sigma_6$$

$$P^{\sigma}_{o} = 1/\sqrt{2}\,\sigma_1 - 1/\sqrt{2}\,\sigma_6$$

$$P^{\sigma}_{1c} = 1/\sqrt{2}\,\sigma_2 - 1/\sqrt{2}\,\sigma_4$$

$$P^{\sigma}_{1s} = 1/\sqrt{2}\,\sigma_3 - 1/\sqrt{2}\,\sigma_5$$

$$D^{\sigma}_{o} = 1/\sqrt{3}\,\sigma_1 - 1/\sqrt{12}\,\sigma_2 - 1/\sqrt{12}\,\sigma_3 - 1/\sqrt{12}\,\sigma_4 - 1/\sqrt{12}\,\sigma_5 + 1/\sqrt{3}\,\sigma_6$$

$$D^{\sigma}_{2c} = 1/2\,\sigma_2 - 1/2\,\sigma_3 + 1/2\,\sigma_4 - 1/2\,\sigma_5$$

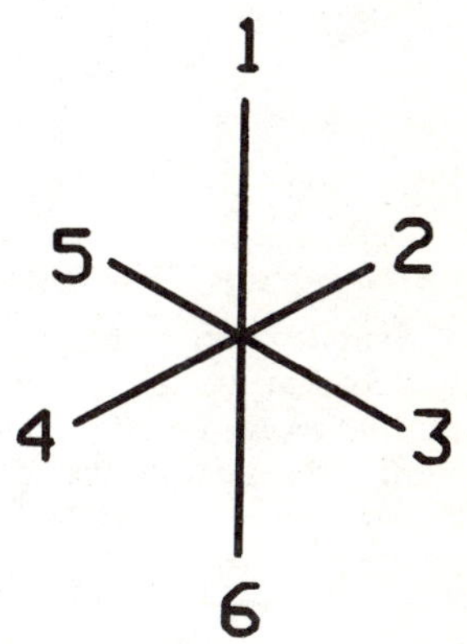

Figure 6. Numbering of ligand positions in an octahedron.

The linear combination of the octahedral symmetry adapted linear combinations $1/\sqrt{6}S^{\sigma} + 1/\sqrt{2}P^{\sigma}_{o} + 1/\sqrt{3}D^{\sigma}_{o}$ is localised exclusively at position 1 on the sphere (see Figure 6) and the remaining 5 linear combinations are delocalised over the five remaining locations and may be constructed in a way that makes them equivalent to the symmetry adapted linear combinations for a square pyramid.

In a sixteen electron complex the total electron density contributed by the metal and ligands generally deviates from pseudo-spherical. Furthermore, the 'united atom' approach would suggest that the most stable stereochemistry will result when the electron pair hole has the maximum amount of p orbital character, since the orbital energies for the central atom follow the order $nd > (n+1)s \gg (n+1)p$. For example, if an eighteen electron tetrahedral complex, e.g. $Ni(CO)_4$, loses a ligand the stereochemical choices for the resulting $Ni(CO)_3$ fragment are planar, D_{3h}, pyramidal C_{3v} and T-shaped, C_{2v}. In the planar geometry the electron pair hole can be localised exclusively in a p orbital which is orthogonal to the trigonal plane. The pyramidal fragment is a sub-group of the tetrahedron (nido- tetrahedral) and therefore the electron pair hole will be localised predominantly in the region of the missing vertex and have approximately 1/4s character and 3/4p character. The T-shaped fragment is a sub-group of the five co-ordinate polyhedra (arachno- square pyramid or trigonal bipyramid) and has approximately 1/5d, 1/5s and 3/5p character. Clearly the planar, D_{3h}, geometry maximises the p character of the electron pair hole. Similarly loss of two ligands from $Ni(CO)_4$ results in a preference for a linear geometry where two orthogonal p orbitals can accomodate the electron pair holes.

D_{3h} C_{3v} C_{2v}

Loss of a ligand from a five co-ordinate d^8 complex, e.g. $[Ni(CN)_5]^{3-}$, gives rise to structures based either on square-planar (D_{4h}) or trigonal bipyramid with the elimination of an axial or equatorial vertex. The square planar geometry results in the localisation of the electron pair hole in a p orbital perpendicular to the co-ordination plane and is therefore preferred. For the corresponding three coordinate ML_3 d^8 complex it is not possible to generate a structure with two non-bonding p orbitals and it can be shown that the observed T-shaped geometry (arachno- square pyramid or trigonal bipyramid) maximises the localisation of p-orbital character of the electron pair hole.

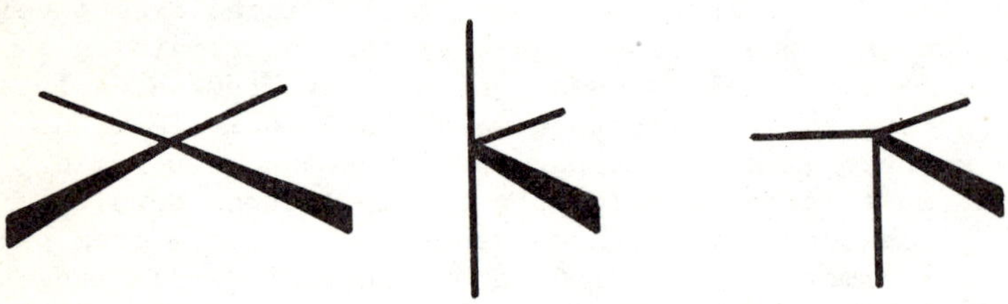

D_{4h} C_{2v} C_{3v}

Similar considerations apply to sixteen, fourteen and twelve electron d^6 complexes which adopt geometries based on the octahedron (see Figure 1). The isolobal analogies [15] between main group and transition metal $M(CO)_n$ fragments depend critically on the presence of these empty orbitals which point towards the missing vertices of the octahedron.

For the earlier transition metals with $d^o - d^4$ configurations the electron affinities of the metal d orbitals is much smaller and localisation of electron pair holes in d orbitals which are orthogonal to the metal orbitals begins to compete with the localisation of electron pair holes in hybrid orbitals with the maximum amount of p character. Consequently d^o-d^4 octahedral complexes with electron pair holes in the t_{2g} set of orbitals are common. Furthermore $d^o - d^2$ pentagonal bipyramidal complexes with electron pair holes in the d_{1c} and d_{1s} (xz, yz) orbitals are observed. For example, $[ZrF_7]^{3-}$ (d^o), $[V(CN)_7]^{4-}$ (d^2) and the eighteeen electron d^4 complex $[OsH_4(PMe_2Ph)_3]$ all have structures based on the pentagonal bipyramid. Similarly $d^o - d^2$ eight co-ordinate structures based on the square-antiprism, the dodecahedron or the square-antiprism are observed.

4. CLUSTER COMPOUNDS

The bonding in the cluster compounds of gold $[Au(AuPR_3)_m]^{x+}$ is dominated by radial interactions between the peripheral gold atoms on the surface of the cluster and the central interstitial gold atom [16]. These bonding interactions occur primarily through the 6s valence orbitals of the surface atoms and the 6s and 6p valence orbitals of the central atom. For gold the bonding contributions from the metal 5d orbitals are not large because of their high ionisation enrgies. The combinations of 6s orbitals of the surface atoms follows that described above for co-ordination compounds and can also be described by the L^{σ}_m quantum number designators derived from the spherical harmonic expansion formulae. However, because the 6s-6s overlaps between adjacent gold atoms are significant the energy spread of the L^{σ}_m energy levels is greater than that found for co-ordination compounds. Only the S^{σ} and P^{σ}_o, P^{σ}_{1c} and P^{σ}_{1s} linear combinations are bonding between the metal atoms when m is greater than 7. The D^{σ} and F^{σ} linear combinations are strongly metal-metal antibonding. The S^{σ} and P^{σ} functions overlap effectively with the 6s and 6p orbitals of the central gold atom giving rise to a set of four stable bonding molecular orbitals. The 5d orbitals of the central atom interact only weakly with the higher lying D^{σ} linear combinations of the surface atoms, but because of the nearly core like nature of the former they are occupied by 10 valence electrons. It follows that spherical clusters of this type are characterised by a total of 12m + 18 valence electrons. These clusters achieve an inert gas configuration about the central gold atom in a manner reminiscent to that described above for co-ordination compounds, but with the following important difference. The surface gold atoms contribute only S^{σ} and P^{σ} linear combinations even when the co-ordination number exceeds four and the central atom contributes a complete set of five d functions, which because of the high effective nuclear charge of gold are approximately core like. Some examples of these spherical gold cluster compounds are illustrated in Figure 7.

The analogy between these clusters and co-ordination compounds can be taken a step further to discuss the structural features of a second class of gold cluster compounds with 12m+16 valence electrons. For 16 electron co-ordination compounds a structure which localised the electron pair hole in a p orbital orthogonal to the co-ordination plane was preferred. This is also the case with gold cluster compounds which are observed to have a toroidal topology when the electron count is 12m+16. Some examples of these clusters are illustrated at the bottom of Figure 7.

This complementary spherical electron density model can be extended to high nuclearity clusters with several interstitial atoms. For those clusters where the radial bonding interactions predominate the electron count is $12n_s + \Delta_i$, where n_s is the number of surface atoms and Δ_i is the electron count characteristic of the interstitial group of atoms. The electronic structures of hugh clusters with face centred cubic, e.g. $[Ni_{38}Pt_6(CO)_{48}H_{6-n}]^{n-}$, hexagonal close packed, e.g. $[Pt_{26}(CO)_{32}]^{2-}$, and icosahedral, e.g. $[Pt_{19}(CO)_{22}]^{4-}$, packing arrangements have been rationalised in this fashion [17].

5. SUMMARY

The simple ideas which form the basis of the <u>complementary spherical electron density model</u> which have been developed above are widely applicable and provide a consistent way of describing the bonding not only in co-ordination and organometallic compounds, but also high nuclearity cluster compounds.

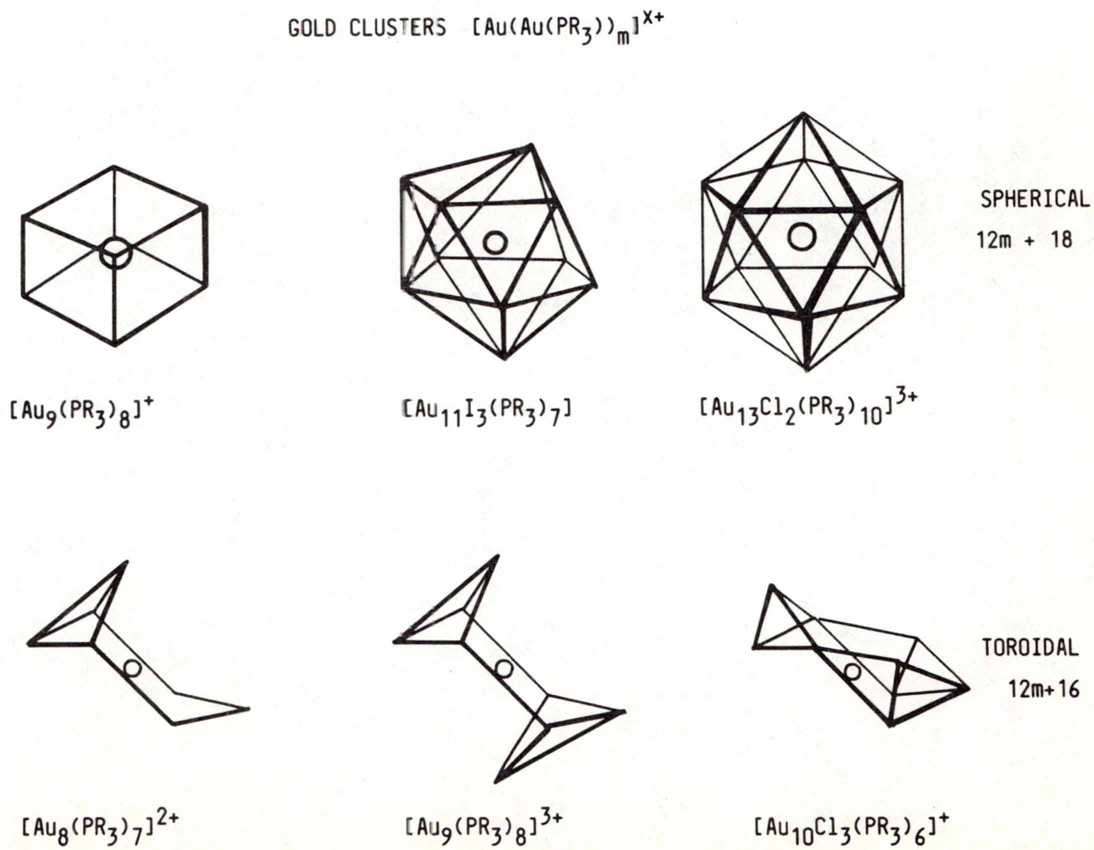

Figure 7. Some examples of spherical and toroidal gold cluster compounds.

References

1. R. Hoffmann, Science, 211, 995 (1981).
2. M.B. Hall, J. Amer. Chem. Soc., 100, 6333 (1978); Inorg. Chem., 17, 2261 (1978).
3. D.M.P. Mingos in "Comprehensive Organometallic Chemistry", Ed. G. Wilkinson, F.G.A. Stone and E.W. Abel, 3, 1 (1982).
4. H. Fejes-Toth, "Regular Figures", Pergamon Press, Oxford, 1964.
5. N.V. Sidgwick and H.M. Powell, Proc. Roy. Soc. London, A176, 153 (1940).
6. R.J. Gillespie and R.S. Nyholm, Quart. Rev., 11, 339 (1957).
7. T.A. Albright, Tetrahedron Reports, 38, 1339 (1982).
8. D.K. Hoffmann, K. Ruedenberg and J.G. Verkade, Struct. Bond., 33, 57 (1977).
9. A.J. Stone, Mol. Phys., 41, 1339 (1980).
10. C.M. Quinn, J.G. McKieman and D.B. Raymond, J. Chem. Educ., 61, 569, 572 (1984).
11. D.M.P. Mingos and J.C. Hawes, Struct. Bond., in press.
12. J.C. Hawes and D.M.P. Mingos, unpublished results.
13. E.L. Muetterties, Acc. Chem. Res., 3, 266 (1970).
14. R.E. Williams, Inorg. Chem., 10, 210 (1971).
15. M. Elian, M.M.L. Chen, D.M.P. Mingos and R. Hoffmann, Inorg. Chem., 15, 1148 (1976).
16. K.P. Hall and D.M.P. Mingos, Prog. Inorg. Chem., 32, 329 (1984); D.M.P. Mingos, Polyhedron, 3, 1289 (1984).
17. D.M.P. Mingos, J. Chem. Soc. Chem. Commun., in press.

THEORETICAL ANALYSIS OF BONDING PROPERTIES IN TRANSITION METAL COMPLEXES. LIGAND SUBSTITUTION EFFECTS.

L.G. Vanquickenborne, I. Hyla-Kryspin and M. Hendrickx
University of Leuven, Department of Chemistry
Celestijnenlaan 200F, 3030 Leuven, Belgium

ABSTRACT. Ligand substitution is analyzed at the Hartree-Fock level of approximation, by a comparison of $Co(CN)_6^{3-}$ and $Co(CN)_5(OH)^{3-}$. The general features of the metal-ligand interactions are found to be as expected from ligand field theory : the CN^- ligand is a strong σ-donor and a weak π-acceptor, whereas OH^- is a weaker σ-donor and a π-donor.

An analysis of the excited states shows that the intraconfigurational energy splittings are not primarily due to the open-shell repulsions, as assumed in ligand field theory. On the contrary, the higher the total energy of a given state, the lower its open shell repulsion. Still, the ligand field description of the photochemical behavior of these complexes is confirmed by the ab initio results.

1. INTRODUCTION

A vast amount of experimental data on transition metal complexes has been rationalized on the basis of ligand field theory (LFT). It is well known that this theory provides an adequate qualitative description of a variety of spectral, magnetic, structural, thermodynamic and kinetic phenomena in coordination chemistry. Yet, ligand field theory is characterized by a set of drastic approximations and obvious internal inconsistencies. In the past few years, we have been trying to analyze the ligand field scheme from an ab initio point of view. More specifically, a number of SCF-calculations have been carried out on a series of transition metal ions, and on a series of transition metal complexes (1,2). In the present paper, we discuss certain aspects of large basis set SCF calculations on two hexacoordinated Co(III)-complexes, $Co(CN)_6^{3-}$ and $Co(CN)_5(OH)^{3-}$. A more detailed account of this work is given in reference 3. A (15s, 11p, 6d/11s, 8p, 4d) basis set was used for the Co-atom; the exponents and the contraction scheme are detailed in previous papers (2,4). The Huzinaga-Dunning (9s, 5p/5s, 3p) bases were used for the ligand atoms. The interatomic distances were either taken directly from experiment, or extrapolated from related experimental data R(Co-C) = 1.89 Å, R(C-N) = 1.15 Å, R(Co-O) = 1.85 Å, R(O-H) = 0.97 Å.

A. Veillard (ed.), Quantum Chemistry: The Challenge of Transition Metals and Coordination Chemistry, 225–233.

2. ANALYSIS OF THE GROUND STATES

For the $Co(CN)_6^{3-}$ ground state ($^1A_{1g}$), the total energy was found at − 1934.8834 hartree, which is approximately 0.5 hartree lower than the best calculation reported by Sano et al. (5,6). The total energy of the $Co(CN)_5(OH)^{3-}$ ground state was found at − 1917.9725 hartree. The C_{4v}-orbitals can be connected either to an O_h-parent orbital, or to an OH^--parent orbital : the parentage is preserved for more than 80 % in each case. Since ligand field theory classifies OH^- as a stronger π-donor than CN^-, one would expect the octahedral metal dπ($1t_{2g}$)-orbitals to be split by the C_{4v}-field into a lower-lying $1b_2(d_{xy})$ and two higher-lying $6e(d_{xz},d_{yz})$ orbitals. From the SCF-treatment the relevant energy splitting is calculated at 1953 cm^{-1}, but $1b_2$ is above 6e. In the next Section it will be shown however, that the OH^--ligand still turns out to be the stronger π-donor, if the π-parameters are determined from the calculated excited state energies.

The notions of σ and π donor or acceptor ligands can be given an alternative meaning by concentrating on density shifts. Table I

TABLE I : Population analysis of the metal orbitals for the two ground states; q is the total charge on Co.

	3dσ	3dπ	4s	4p	q
$Co(CN)_6^{3-}$	1.63	5.72	0.45	− 0.04	1.24
$Co(CN)_5(OH)^{3-}$	1.37	5.73	0.37	− 0.07	1.59

shows the results of a standard Mulliken population analysis. According to this Table, the population ratio of the 4s:4p:3dσ orbitals is approximately 1:0:4. If the sp^3d^2 hybridization were valid, this ratio would be 1:3:2. This is obviously not the case and therefore, the often invoked rationalization of the 18-electron rule may have to be reconsidered. Clearly, the SCF results are closer to the conventional ligand field theory, where the bonding is supposed to take place essentially in the d-orbitals.

Table I further obviously suggests that CN^- is a strong σ-donor and a weak π-acceptor, while OH^- seems to have a smaller σ-donor strength, leaving the mono-substituted complex with the larger positive charge (q = + 1.59). If the ligand populations are analyzed in a similar way, one finds that the CN-populations are very nearly identical in the two complexes. This result is in marked contrast with the Xα-calculations of Goursot et al. (7) where the equatorial C_{4v}-ligands are found to be π-acceptor ligands, quite similar to the O_h-ligands, but where the axial CN^- ligand is calculated to be a π-donor. Here again, the Hartree-Fock results are closer to the conventional picture where the inert ligands are largely unaffected by substitution.

The general pattern is confirmed by density difference maps $\Delta\rho = \rho(\text{complex}) - \rho(\text{metal-ligand system})$. For $Co(CN)_6^{3-}$, Δρ refers to the difference between $\rho(^1A_{1g})$ and on the other hand $\rho(Co^{3+};t_{2g}^6;{}^1A_{1g})$

plus the density corresponding to six $CN^{-}(^{1}\Sigma^{+})$ ligands, all non-interacting, but kept at the internuclear distances of the complex. This map shows the σ-density shift from the ligands to the metal and the π-back bonding shift from the metal to the ligands. It is rather similar to the map derived by Sano et al. (5). Apart from the σ- and π-shifts, one observes a density build-up between Co and C, which is traditionally linked to covalent bonding.

In order to analyze the density maps, it is interesting to calculate subsets of the density difference, corresponding to specific irreducible representations. Figure 1 shows the density shifts $\Delta\rho_{\sigma}$ taking place in the e_g-representation of O_h. Figure 1 is much simpler than a total density difference map and it shows the shifts populating a the dσ-orbitals, illustrating quite clearly the σ-donor properties of a CN^{-}-ligand. Similarly the density shifts $\Delta\rho_{\pi}$ in the t_{2g}-representation describe the π-bonding and π-back-bonding phenomenon. Within the conventional picture, the essential features of the metal-ligand bonding should be described by the density shifts taking place in these two representations. Figure 2 shows the sum of the t_{2g} and the e_g shifts. And indeed, the general outlines of Figure 2 come rather close to the total density difference map Δρ. One rather important exception is the covalent plateau between Co en C, which is only partly present in Figure 2 : the outer part of this plateau is apparently associated with the 4s- and 4p-population of the Co-atom.

Figures 3 and 4 describe the density difference Δρ for the $Co(CN)_5(OH)^{3-}$ complex, in the xy and x(-z) planes respectively. Both Figures are quite similar to each other and to the corresponding $Co(CN)_6^{3-}$ plot. This fact confirms the results of the population analysis and seems to support the idea of independent bonds, which is at the basis of the additivity postulate of ligand field theory. A similar plot can also be made for the Co-OH bond in the x(+z) plane, showing the σ-donor properties of the OH^{-}-ligand. Here again, in the Co-O bonding zone, a covalent plateau is observed, although its height is only about half the height of the Co-C plateau.

3. ANALYSIS OF THE EXCITED STATES

Table II shows the results of the SCF-calculation and a comparison with experiment. The four O_h excited states correspond to the t^5e^1 configuration. On the average, these states are calculated too low : the interconfigurational gap, described by the $t_{2g} \rightarrow e_g$ transition, or alternatively by the ligand field parameter 10 Dq is too small. The SCF value of 10 Dq is found at 27534 cm^{-1}, whereas the experimental 10 Dq(Co-CN) = 34890 cm^{-1}.

If the SCF-state energies are used in conjunction with the first order ligand field expressions, one obtains the following theoretical LF-parameters : $\Delta\sigma = \sigma(CN) - \sigma(OH) = 6450\ cm^{-1}$, $\Delta\pi = \pi(CN) - \pi(OH) = -1650\ cm^{-1}$ as compared to the semi-empirical values of 2540 and $-4150\ cm^{-1}$ respectively. Therefore, the qualitative predictions on the relative bond properties (σ, π-donor or acceptor character) of both ligands are identical in both schemes, but obviously the SCF-calcula-

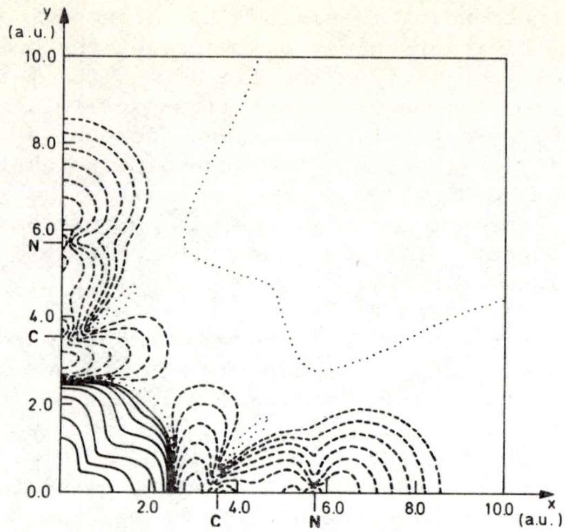

Figure 1 : Density differences $\Delta\rho_\sigma$ in the $Co(CN)_6^{3-}$ ground state (x,y plane), due to electronic shifts within the e_g-representation. Full contours correspond to an electron density increase, dashed contours to an electron density decrease; at the dotted lines $\Delta\rho = 0$; the values of the $\Delta\rho$-contours are $\pm$ 0.000625, $\pm$ 0.00125, $\pm$ 0.0025, $\pm$ 0.005, $\pm$ 0.01, $\pm$ 0.02, $\pm$ 0.04 and $\pm$ 0.08 a.u.$^{-3}$.

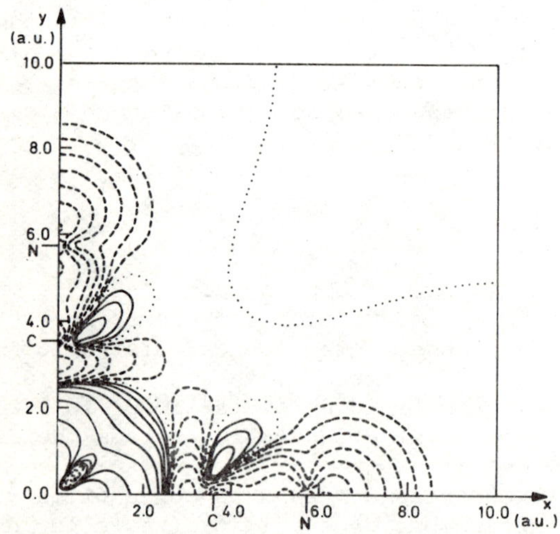

Figure 2 : $\Delta\rho_\sigma + \Delta\rho_\pi$, describing the density shifts upon bond formation in $Co(CN)_6^{3-}$, corresponding to the classical bond description in terms of σ and π interactions.The values of the $\Delta\rho$-contours are as in Figure 1.

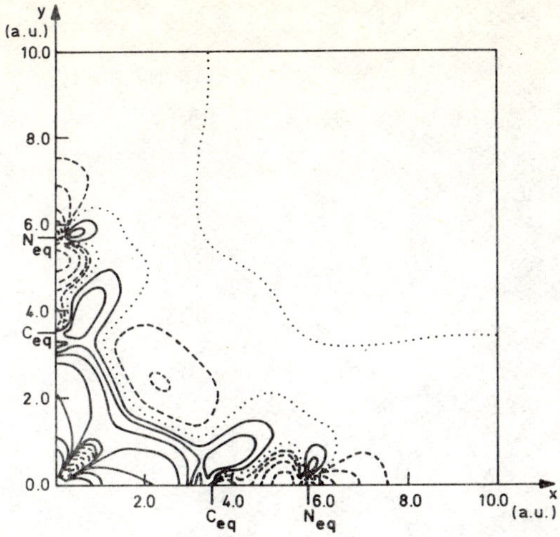

Figure 3 : Total density difference plot $\Delta\rho = \rho(\text{complex}) - \rho(\text{metal-ligand system})$ for the $^1A_1$ ground state of $Co(CN)_5(OH)^{3-}$. The plot describes the electronic density shifts in the xy-plane upon bond formation; the subscript eq refers to the equatorial bonds. The separated metal-ligand system is considered to be made up of Co^{3+} $(t_{2g};{}^1A_1)$ plus five $CN^-({}^1\Sigma^+)$ and one OH^- $({}^1\Sigma^+)$ ligand, all non-interacting but kept at the internuclear distnaces of the complex. The values of the $\Delta\rho$ contours are $\pm$ 0.0025, $\pm$ 0.005, $\pm$ 0.01, $\pm$ 0.02, $\pm$ 0.04, $\pm$ 0.08 and $\pm$ 0.16 a.u.$^{-3}$.

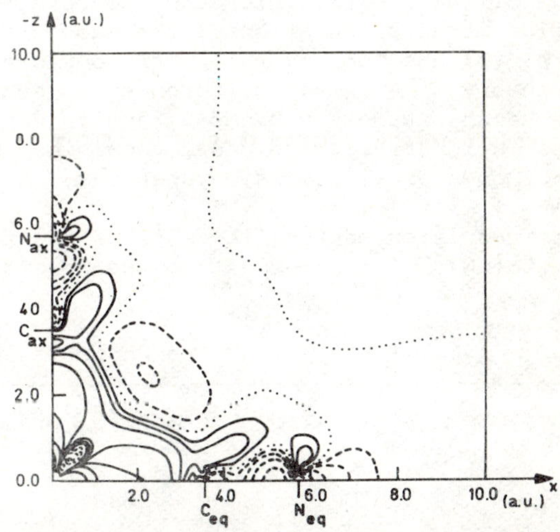

Figure 4 : Total density difference plot similar to Figure 3, but in the x(-z) plane, describing the bond between the Co-ion and the axial CN^--ligand.

TABLE II : Transition energies (in cm^{-1}) of the ligand field spectra

	$Co(CN)_6^{3-}$			$Co(CN)_5(OH)^{3-}$	
State	SCF	experiment	State	SCF	experiment
$^1A_{1g}$	0	0	$^1A_1$	0	0
$^3T_{1g}$	16372	25200	^3E_a	10074	
			$^3A_2$	15903	
$^3T_{2g}$	24779		$^3B_2$	16294	
			^3E_b	19527	
$^1T_{1g}$	23813	32500	^1E_a	20598	26400
			$^1A_2$	23523	
$^1T_{2g}$	39558	39200	$^1B_2$	32730	
			^1E_b	33564	34600

tions do not reproduce the very strong π-donor properties usually associated with the OH^--ligand in ligand field theory.

The intraconfigurational energy splitting between the four states of the t^5e^1 configuration is calculated too high : 23186 cm^{-1} versus the experimental value of 14000 cm^{-1}. This seems to be a rather general phenomenon in transition metal calculations : the same effect has also been observed for NiF_6^{4-} (8) and for several other complexes (2,5).

In ligand field theory, the energy difference between these four states depends on the Racah parameters B and C, and is supposed to be due to the differences in inter-electronic d-d repulsion energy. Table III shows the SCF open shell repulsion energies, as compared to

Table III : Open shell repulsion energy (in cm^{-1}) of the four t^5e^1 states of $Co(CN)_6^{3-}$ (with respect to the lowest excited $^3T_{1g}$)

State	SCF	frozen orbitals
$^3T_{1g}$	0	0
$^3T_{2g}$	- 16570	8342
$^1T_{1g}$	- 20093	7373
$^1T_{2g}$	- 62846	23598

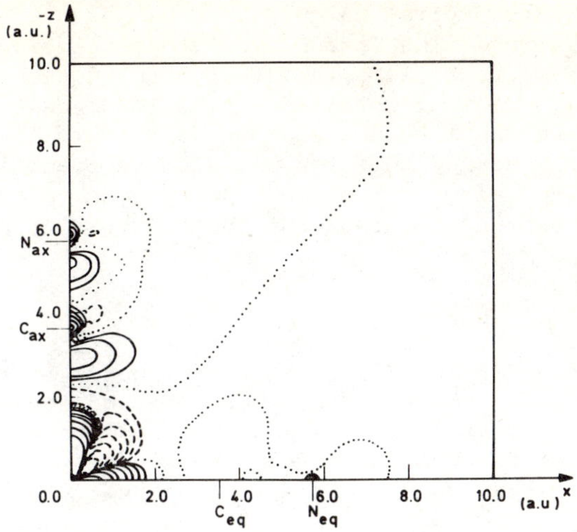

Figure 5 : Total density difference plot $\rho(^3E_a) - \rho(^1A_1)$ of the substituted complex in the x(-z) plane. The contour conventions are as in Fig. 3.

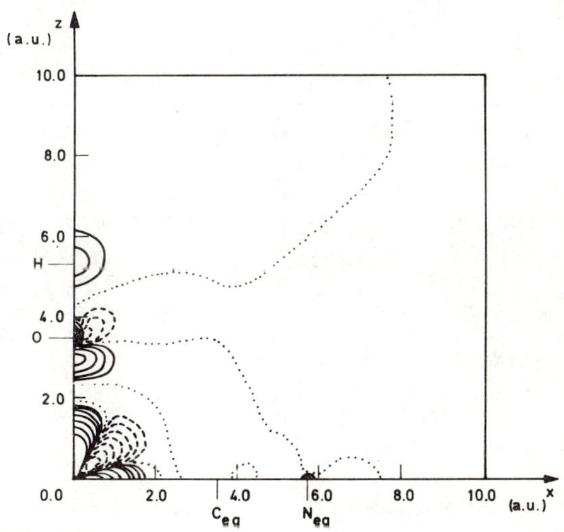

Figure 6 : Total density difference plot $\rho(^3E_a) - \rho(^1A_1)$ of the substituted complex in the x(+z) plane. The contour conventions are as in Fig. 3.

the results of a frozen orbital calculation, based on the orbitals of the configuration average. The frozen orbital calculation, being related to the ligand field approach, faithfully reproduces the increasing d-d repulsion with increasing total energy. The SCF-calculation however reveals that the open shell repulsion decreases from $^3T_{1g}$ to $^1T_{2g}$: the relaxation from the average orbitals to the state-specific orbitals reverses the relative value of the d-d repulsion energies, thereby questioning the conceptual framework of the classical theories (1,2).

4. REMARKS ON THE PHOTOCHEMISTRY

The first excited state of $Co(CN)_6^{3-}$, $^3T_{1g}$ corresponds to an excitation of the general type pq → p^2-q^2, where p and q stand for x, y and z. This $\pi \rightarrow \sigma$ excitation thus effectively corresponds to a rotation of electron density from the region between the ligands into the region of the bond axes. On the basis of SCF-calculations, it is possible to draw a density difference map $\rho(^3T_{1g}) - \rho(^1A_{1g})$ showing the density shifts upon excitation. The $d\pi \rightarrow d\sigma$ shifts are seen to be accompanied by a depletion of the internuclear region and a corresponding accumulation of charge on the ligands, suggesting a decreased σ-donation in the excited state, and a corresponding weakening of the metal-ligand bond. The latter prediction is in agreement with ligand field theory (9).

The first excited state of $Co(CN)_5(OH)^{3-}$, ^3E_a essentially corresponds to the excitation $(d_{xz}, d_{yz}) \rightarrow d_{z^2}$. A density difference plot in the xy plane shows the expected density increase on the metal ion, but the metal-ligand region remains largely unaffected. In the xz (or yz) plane on the other hand, Figure 5 and 6 show that the axial bond regions are modified considerably, especially by a significant decrease of ligand to metal σ-donation. In a similar vein, the next excited state, $^3A_2$, corresponding to the xy → x^2-y^2 excitation also shows the expected density shifts. In this case, the axial bond region remains virtually unchanged (with respect to the ground state), but the equatorial bonds show rather important density shifts, suggesting a significant bond labilization. As a whole, the SCF description of the photochemistry agrees quite closely with the ligand field rationalizations (9).

REFERENCES

1. Vanquickenborne, L.G.; Haspeslagh, L. Inorg. Chem. 1982, 21, 2448
2. Vanquickenborne, L.G.; Haspeslagh, L.; Hendrickx, M.; Verhulst, J. Inorg. Chem. 1984, 23, 1677
3. Vanquickenborne, L.G.; Hendrickx, M.; Hyla-Kryspin, I. Inorg. Chem. 1985, 24,
4. Vanquickenborne, L.G.; Verhulst, J. J. Am. Chem. Soc., 1983, 105, 1769
5. Sano, M.; Hatano, Y.; Kashiwaga, H.; Yamatera, H. Bull. Chem.Soc. Jpn 1981, 54, 1523

6. Sano, M.; Yamatera, H.; Hatano, Y. Chem. Phys. Lett. 1979, 60, 257
7. Goursot, A.; Pénigault, E. Chem. Phys. 1981, 61, 83
8. Wachters, A.J.H.; Nieuwpoort, W.C. Phys. Rev. B : Solid State 1972, 5, 1972

TOPOLOGY OF POTENTIAL ENERGY SURFACES : THE COMPLEMENTARITY OF MO AND VB APPROACHES.

ALAIN SEVIN
Laboratoire de Chimie Organique Théorique
Bâtiment F. 4 Place Jussieu, 75230 Paris Cedex 05

ABSTRACT. The complementarity of MO-and-State and VB correlation diagrams is illustrated by the study of very different reactions. Extended basis ab initio SCF+CI calculations are compared with diabatic and adiabatic VB results for the system $Na^{x}+HF \longrightarrow NaF + H$. The efficiency of the VB approach, as well for a qualitative understanding of the reaction mechanism as for obtaining accurate data is outlined. The second example deals with the study of oxygen transfer from $Fe^{(IV)}$-oxo porphyrins to ethylene yielding ethyleneoxide. The radical character of the reaction is emphasized on the ground of qualitative VB considerations. The inherent complementarity of both approaches is discussed throughout.

The ideas of orbital correlations are essential for the understanding of electronic mechanisms and underlying notions such as symmetry conservation[1], topicity[2], natural orbital correlations[3] have extensively been used in the study of chemical reactions. In most or-

A. Veillard (ed.), Quantum Chemistry: The Challenge of Transition Metals and Coordination Chemistry, 235–252.

ganic processes, state correlation diagrams may be built, using for both reactants and products benchmarks provided for by known electronic properties. When this is not possible, reliable sets of data may be obtained from elaborate model calculations. This is no longer the case dealing with reactions occurring at transition metal centers for which data largely remain conjectural. Further complexity arises from the fact that oxido-reduction often takes place, corresponding, in state-to-state-correlation language, to the crossing of covalent and ionic diabatic potential energy surfaces (PESs).

In what follows, we will focus our attention on two typical reactions involving rupture and formation of bonds :

$$(L)_nM + AB \longrightarrow (L)_nMAB \qquad (1)$$

$$(L)_nM + AB \longrightarrow (L)_nMA + B \qquad (2)$$

$(L)_n$ stands for the ligands and AB is a covalent molecule where A and B are more electronegative than M. It is therefore assumed that, to some variable extent, the MA/MB bonds have the $M^{\delta+}$—$A^{\delta-}$ polarity, and that both processes are oxidative in nature : i.e they correspond to the transfer of electronic density from the metal to A and B. Figure 1 schematically depicts the important features of reaction (1).

In the left part (up), the MOs that are of concern are outlined : σ and $\sigma^{\times}$ for the AB (or A_2) moiety, and two frontier levels of M labelled "d" for convenience. All other MOs are assumed to be "spectators".The natural MO correlations are straightforward : σ and $\sigma^{\times}$ are correlated with the couple of bonding MOs of the product while the d set is correlated with its nonbonding (or antibonding) counterpart. The process at least preserves a local symmetry element and an avoided MO

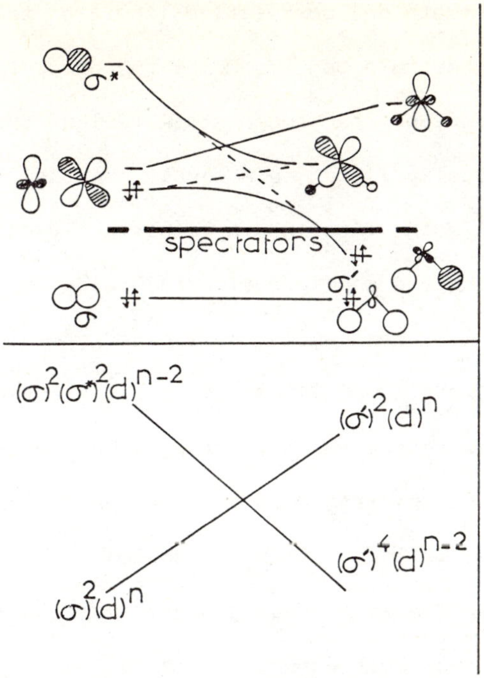

Figure 1. a)left. MO and state correlation diagram for oxidative addition. The relative contributions have been emphasized on purpose. b)right. VB correlation diagram for the same reaction. The A_2 bond is assumed to be mainly covalent.

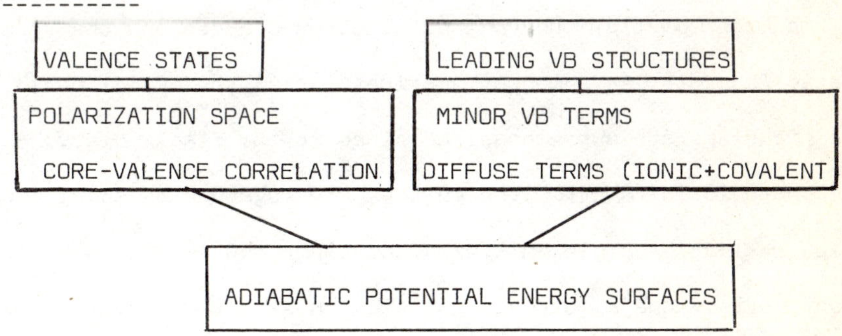

Figure 2. Comparison of MO-and States and VB correlations.

crossing results between the antisymmetrical correlation lines, so that finally, an initial d level is linked to a bonding MO of the product. State correlations (left, bottom) derive. The ground state (GS) of the starting system $(\sigma)^2(d)^n$ is diabatically linked to a doubly excited state of the product $(\sigma')^2(d)^n$, while the final GS is diabatically correlated with a charge transfer having the configuration $(\sigma)^2(d)^{n-2}(\sigma^{\times})^2$ or the formal Lewis structure M^{++}/H_2^{--}. Here again, an avoided crossing takes place and a non-monotonous adiabatic PES links both GSs. A potential energy barrier thus arises, reflecting the more or less endothermic synchronous transfer of two electrons from d to $\sigma^{\times}$.[4] Indeed, in real system, a great number of states will have to be taken into account and it is clear that they will interfere with these diabatic correlations, rendering the whole process much less abrupt, and allowing for a continuous distorsion of the electronic density. It remains that the initial diabatic correlations play a leading role int the overall process and constitute the backbone of any further investigation. These MO-and-State correlations deserve comment. First of all, they only describe the major part of the valence electron changes and the resulting adiabatic PESs are, in some way, caricatures of the actual ones. Electronic transfers are overemphasized since in the final moiety, A and B are not pure ionic ligands although they are regarded as such. Bond polarities reflect a complex mixing of covalent and ionic characters. The covalent part of the electronic correlation is not adequately described in this model. Moreover, the correlation effects due to the rest of the system are not treated and obviously remain to be estimated as soon as a correct value of the activation energy is required.

The related VB correlations (right part) have the merit of condensing into a single scheme the aforementioned trends. Here again, one has to note that the diabatic PES corresponding to any given VB structure only partly depicts the real system. For example the A_2 linkage is a mixing of ionic $A^+A^- + A^-A^+$ and covalent A°B° terms whose relative weights have to be variationally optimized, the procedure being analogous to a 2x2 CI. (vide infra the description of C_2H_4). We must insist on the fact that although state and VB correlation diagrams bear strong resemblance, the corresponding wavefunctions differ in nature. Each VB PES corresponds to a given location of the valence electrons on the atomic fragments, while an electronic PES corresponds to a given electronic configuration built on the valence MOs. Both descriptions cannot be reduced to an unique one but rather appear to be complementary as resumed in figure 2. This will be illustrated by studying two very different reactions. In a first step, we will compare both approaches at quantitative level in a rather simple system for which accurate calculations have been carried out, and, in a second step, we will examine at qualitative level, a reaction for which both the size of the ligands and the reaction path complexity prevent all-electron calculations from being achieved at the present time.

$Na^{\times} + HF \longrightarrow NaF + H$.

The thermodynamic values of table 1 clearly show that either in the GS or in the excited state, formation of NaF is the only reactive channel. In what follows, we will restrict ourselves to this reaction, whose optimal reaction coordinate (RC), is $C_{\infty v}$. Model calculations were carried out using two independent methods : i) open-shell

SCF with a 6-31Gxx+Rydberg+polarization basis, followed by Møller-Plesset CI on a set of 80 to 120 reference configurations, yielding around to $3\text{-}4\times10^6$ perturbative terms (CIPSI series of programs).[5] ii) a VB method developped in our group by J.M.Lefour, J.P.Flament and P.C. Hiberty,[6] in which antisymmetrized determinantal wavefunctions are built on fragment MOs of Na, F, H (same basis as in i). We thus get diabatic VB functions, Φ_i, whose variationally determined linear combination $\Psi = \Sigma_i\, c_i \Phi_i$, yields adiabatic VB functions. Using a set of 15 to 20 reference structures taking into account the correlation of the three "active" electrons, we obtain results that can be compared to the preceding ones with a very good agreement.

Let us first compare the MO and VB diagrams. The MO correlation diagram (figure 3, left), is very simple. The bonding σ_{FH} MO is correlated with the bonding σ_{FNa} MO, since in both the F character is very dominant. σ^{x}_{FH} having a dominant H character is correlated with the MO of the leaving hydrogen, while 3s and $3p_z$ of Na tend to retain their nature. Two avoided crossings result and finally 3s is linked to H. This behavior is reflected in the diabatic state correlations (left bottom). The $(\sigma_{FH})^2(3s)^1$ GS of Na+FH is correlated with an ionic state of the products, while the GS of the products is correlated with a charge transfer of the reactants : Na^+/FH^-. The correlation of the lowest excited state of same symmetry has also been indicated. Avoided state crossings result, and a non-monotonous PES adiabatically takes the system from reactants to products. A parent diagram is straightforwardly obtained using the leading VB structures. The GS having one

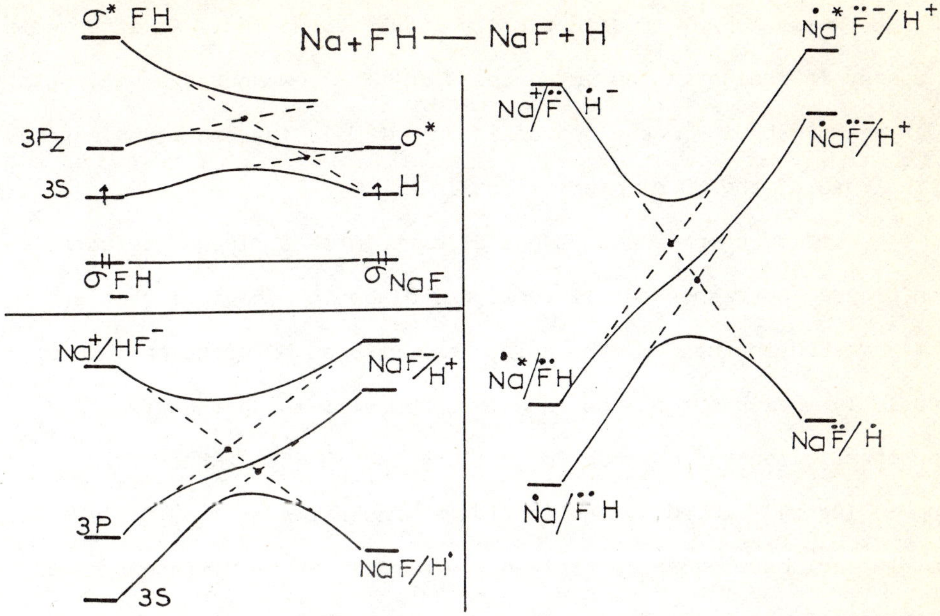

Figure 3. a) left. MO-and-state correlation diagrams. The core levels have been omitted, and the scheme is restricted to three active electrons. The 3p levels are restricted to a single component. The $3p_{x,y}$ MOs, orthogonal to the preceding ones remain unchanged along the reaction path. b) right. VB correlation diagram. Only the dominant contributions of FH and NaF are shown.

NaH + F (3.85)

4p (3.75)

4s(3.20)

3p (2.10)

NaH +F (1.36)

Na + FH (0.00)

Table 1. Thermodynamics data. Energies are in eV.

electron located on Na diabatically yields an ionic final structure, NaF^-/H^+, while NaF/H° is diabatically linked to Na^+/FH^-. At intermediate energy is indicated the VB correlation of a low energy excited state (for example having one electron in $3p_z$ of Na). The same topology as in the case of the MO diagram is obtained.

Let us compare both types of correlations. The state correlation diagram corresponds to intentional diabatic behaviors that are partly masked in the SCF step, since the related MO crossings are progressively worked out as the reaction proceeds. In that perspective, the potential energy maximum found along the GS PES, rather results from a "legacy" leaved by the intrinsic properties of the initial system. The adiabatic VB potential energy barrier is of different nature. It results from the diabatic correlations of VB terms representing limiting situations. For example, it is clear that FH is only partly ionic, $\overset{\circ\circ}{F}$—H, the covalent structures $\overset{\circ}{F}$—$\overset{\circ}{H}$ (valence and diffuse) having a noticeable weight in the optimal wavefunction. By restricting ourselves to the leading VB terms, we only correlate the dominant weights of r reactants and products. Obviously, in both cases, the description is not complete and additional work is necessary in order to include the diffuse and Rydberg spaces. But a general remark can be made about these complementary approaches : in the SCF+CI formalism, fragment contributions are spread over many occupied and empty MOs so that obtaining correct electronic correlation imposes a very large number of excitations. In the VB formalism, which widely reposes on chemical intuition, electronic localization is imposed by construction of the determinantal wavefunctions, hence, the same type of correlation is ensured through a

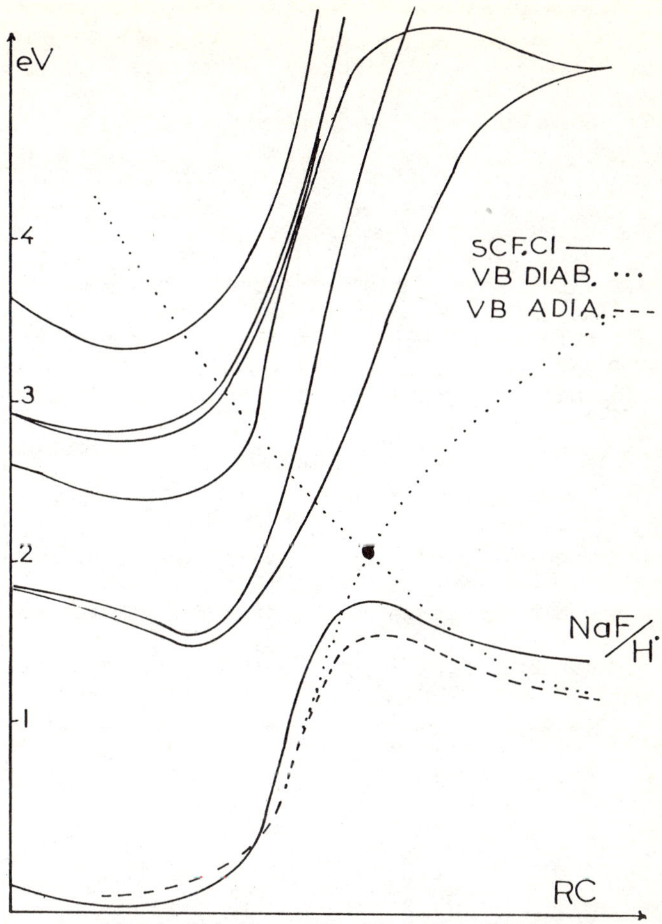

Figure 4. Calculated PESs. The VB curves have been reported to a common origin in order to compare both activation energies. The RC is complex : in a first step; the formation of an attractive complex occurs without significant FH relaxation. Then, FH stretching begins, leading to NaF+H very suddenly (harpooning mechanism). Then H is no longer bonded to the NaF molecule. (last part of the RC). The VB transition state is found at the same geometry.

smaller number of terms.

The calculated PESs of figure 4 illustrate this discussion. Very comparable activation energies, corresponding to quite identical transition geometries, are found by both methods along the GS PES. Exciplexes are obtained in the low energy excited states, whose potential energy wells lie in the vicinity of the GS maximum. We have seen in figure 3 that this region corresponds to the crossing of three diabatic curves, we thus can infere that their mixing will provide us with an efficient mechanism for reaching the Naf+H° situation, starting from the lowest excited state (^2P) of Na.

To conclude this part, let us recall that complementary use of MO and VB methodologies is very effective for understanding the actual nature of diabatic PESs which, in turn, rule the behavior of more elaborate adiabatic PESs. The adiabatic VB method affords accurate results at low cost, either in complexity or computing time.

OXYGEN TRANSFER FROM $Fe^{(IV)}$-OXO PORPHYRINS TO ALKENES.

$Fe^{(IV)}$-oxo porphyrins and their reactions are of paramount importance in human metabolism and have been the subject of many experimental and theoretical investigations.[7] In the present report we will only focus our attention on the primary step of oxygen transfer, yielding epoxides, according to the sheme :

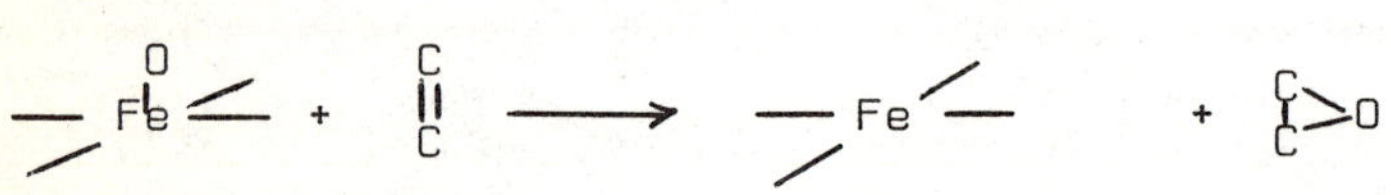

Experimentally, it is well established that the reactive species is $Fe^{(IV)}$-O , d^4, and that the presence of a porphyrin ring is not always necessary. Other isoelectronic transition metals, such as Mn are also effective oxygen carriers.[8] On the ground of stereochemical studies, it has been claimed that these species bear an important radical character and rather react as such.[9]

The size of the system is a heavy handicap for an all-electron study of reactivity, and our purpose is to show that very simple VB considerations bring about a reaction mechanism which fits very well most experimental findings. The central idea consists in considering the $Fe^{(IV)}$-O unit as a chromophore, in terms of localized atomic contributions. Accordingly, the EHT MOs of figure 5 have been divided into three sets. The bonding and antibonding ligand MOs have been confined into a box which imposes the nature of the central field (C_{4v}), but is assumed not to interfere too much with the central chromophore. At low energy, we have a set composed of four MOs, mainly located on O, combined with minor in-phase metal contributions. The frontier set is composed of four metal MOs : d_{xy}, then the degenerate set d_{xz},d_{yz} and, at slightly higher energy, d_{z^2}, the latter three MOs having minor out-of-phase oxygen contributions. The GS d^4 configuration is triplet, as indicated. It is woth noticing that the degenerate MOs located in the xz,yz planes bear some resemblance with the classical π and π^* MOs of O_2, as outlined in our drawing.

Static properties of the $Fe^{(IV)}$- chromophore.

After the EHT MOs of figure 5, the oxygen atom bears an important ionic character, corresponding to the limiting Lewis structure $[L]^{2-}Fe^{4+}O^{2-}$.

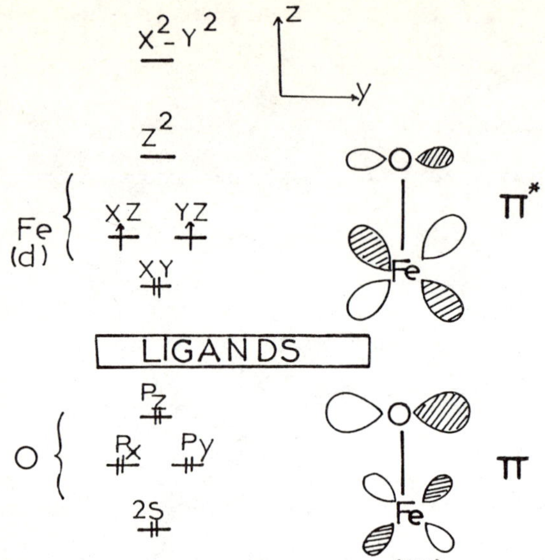

Figure 5. Fe and O contributions to the $Fe^{(IV)}$-O chromophore.

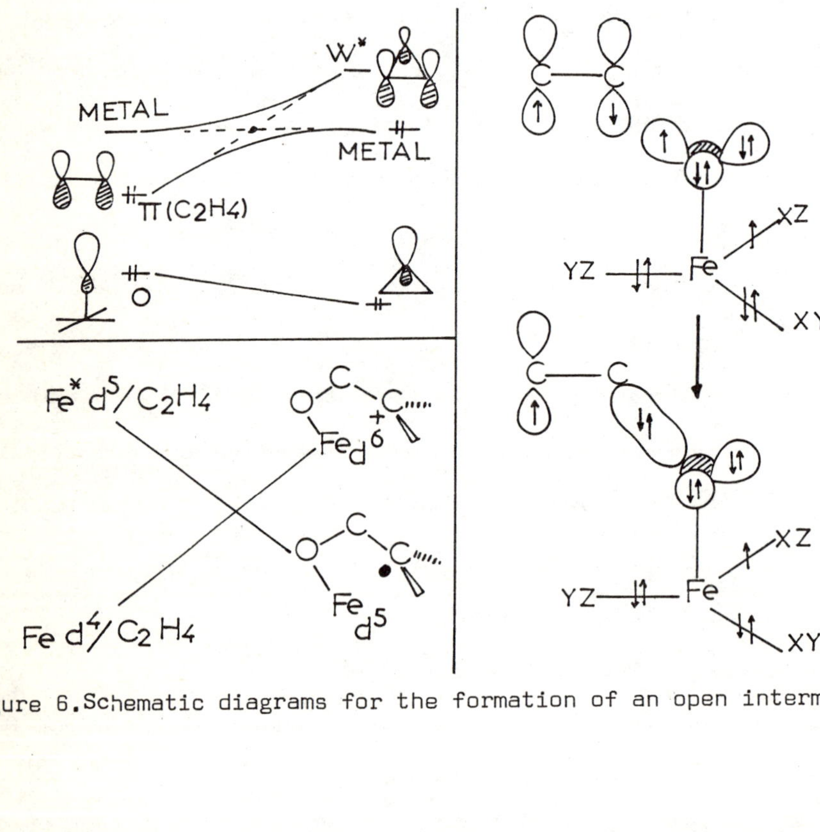

Figure 6. Schematic diagrams for the formation of an open intermediate.

This finding is in contradiction with the aforementioned radical-like reactivity. The reason is that the EHT electronic distribution only attributes the open shell character to the metal atom, while in reality it is delocalized on both atoms. Going through the analogy with O_2, the oxygen extremity of the chromophore possesses four paired electrons directed along the z axis, thus leaving six electrons shared by Fe and O, in the xz,yz planes. Through appropriate axis rotation, we can define two orthogonal MO sets, labelled π_x, π_y, $\pi^{\times}_x$, $\pi^{\times}_y$ whose analytical form is easily extracted from EHT coefficients. To a very good approximation,and apart from normalization factors, these π-type MOs can be written :

$\pi_x = 1.\ (2p_x) + \alpha(\ 3d_{xz})$; $\pi^{\times}_x = \alpha(2p_x) - 1.\ (3d_{xz})$, and similarly for the y components. A realistic value of α is 0.5. (To take the same value of α in π and $\pi^{\times}$ amounts to neglecting the overlap in the normalization factors). It is now easy to build up determinantal wavefunctions corresponding to the electrónic configurations :

$|\ldots.\ core\ ..(\pi_x \ldots \pi^{\times}_y)^6|$ and to work out separately the Δ(6) determinant of the π system. Projection of a conveniently antisymmetrized determinant on Fe (3d) and O (2p) AOs yield the VB structures of table 2, whose weights are only functions of α. Following Harcourt's treatment of the three electron bond[10], it is instructive to begin with the case of three electrons in one plane, corresponding to the doublet determinants $\Delta(3) = |\pi_x \bar{\pi}_x \pi^{\times}_x|$; $\Delta(3)^{\times} = |\pi^{\times}_x \bar{\pi}^{\times}_x \pi_x|$ .We see that in Δ(3) we have $\overset{\circ}{Fe}-\overset{\circ\circ}{O}$ (67%) and $\overset{\circ\circ}{Fe}-\overset{\circ}{O}$ (33%); the relative percentage being inverted in $\Delta(3)^{\times}$. In Δ(6), further delocalization is offered to the system and the VB structures having a covalent oxygen, i.e bearing at least

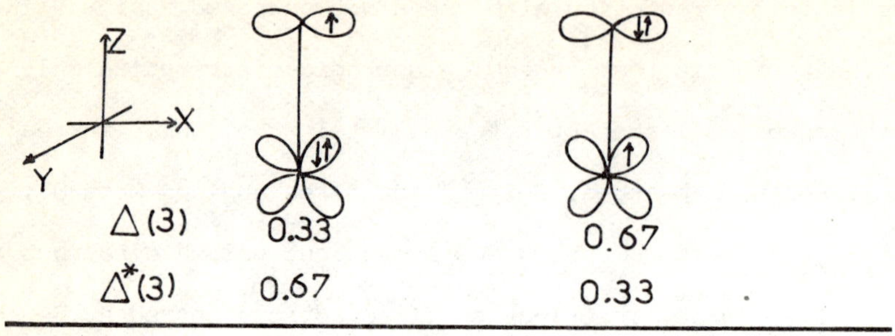

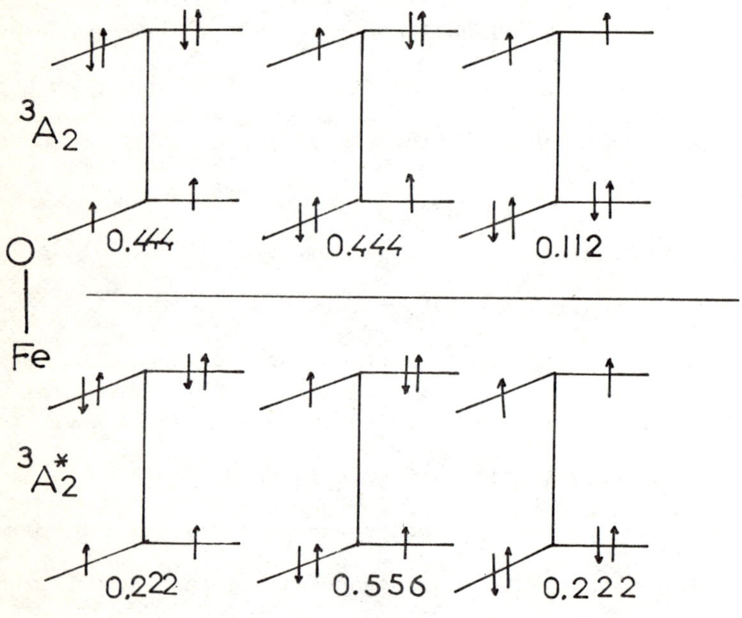

Table 2. Weights of the various VB structures, normalized to unity, calculated for α=0.5. The determinantal forms of Δ(6) and $\Delta^{\times}$(6) are :

$$^3A_2 = N \left| \pi_x \bar{\pi}_x \pi_y \bar{\pi}_y (\pi_y^{\times} \bar{\pi}_x^{\times} - \pi_x^{\times} \bar{\pi}_y^{\times}) \right|$$

$$^3A_2^{\times} = N \left| \pi_x \bar{\pi}_x \pi_y^{\times} \bar{\pi}_y^{\times} (\pi_y \bar{\pi}_x^{\times} - \pi_x^{\times} \bar{\pi}_y) - \pi_y \bar{\pi}_y \pi_x^{\times} \bar{\pi}_x^{\times} (\pi_x \bar{\pi}_y^{\times} - \pi_y^{\times} \bar{\pi}_x) \right|$$

where N stands for a normalization factor which does not change the normalized weights.

one unpaired electron, amount to 55% in the $^3A_2$ GS and up to 77% in the related $^3A_2{}^*$ excited state. We thus conceive that using a two-configurational wavefunction of the form $\Psi = (1-\lambda)\ {}^3A_2 + \lambda\ {}^3A_2{}^*$ allows for a continuous variation of the oxygen radical character from 55% to 77%, as a function of increasing λ. Athough qualitative, these findings clearly point out the important radical character of the $Fe^{(IV)}$-O chromophore, and, moreover we see that the adiabatic mixing of $^3A_2$ and $^3A_2{}^*$ acts as a driving force in radical-like reactivity.

Reactivity study.

The VB approach efficiency will be illustrated by comparison with a classical MO analysis. First of all, let us examine a limiting reaction path which has the only merit of underlying the major electronic events occurring during the process. In the upper part of figure 6, the important MO interactions arising in an isoceles geometry (C_{2v}) are schematically drawn. The oxygen lone pair, pointing out Fe-O, along the z axis, and the π MO of C_2H_4 give rise to a strong repulsive four-electron interaction.The consequence is that the $2p_z$ level of O is correlated with a nonbonding lone pair of C_2H_4O, while the destabilized π MO tends to reach an antibonding Walsh's orbital (W^*). This intended correlation is intercepted by a metal level of suitable symmetry and, as an end result, the metal is reduced by two units ($d^n \rightarrow d^{n+2}$). In the case of $Fe^{(IV)}$-O, the reaction is spin and symmetry allowed, but it is obvious that the reactants approach, which is very endothermic, will prevent the reaction from occurring in this geometry. Model studies show that in the optimal reaction path, only one carbon atom of C_2H_4 approaches the O atom, leading to an intermediate Fe-O-C-C open species.

We are now faced with the electronic structure of this intermediate. In the lower part of figure 6, the initial system is diabatically linked to a d^6 intermediate (according to the preceding MO discussion), bearing a positive charge on the terminal carbon atom $Fe^{(II)}$-O-C-C^+. The parent d^5 species, bearing an unpaired electron on the same carbon atom : $Fe^{(III)}$-O-C-C°, is correlated with an excited d^5 species of the starting moiety (rather than with a charge transfer excitation of much higher energy). We thus see that an adiabatic PES links the reactants and the product, which, in turn is likely to evolve. This one-electron mechanism : $d^4 \rightarrow d^5$ however remainscomplex since we do not clearly see the origin of the "reactive" PES.

The VB scheme depicted in the right part of figure 6 explains the latter fact. On the one hand, it is well established that the optimal multireference wavefunction of C_2H_4, which is a linear combination $\Psi \simeq \{1-\lambda\}|\ldots \pi\bar{\pi}| \ -\lambda\ |\ldots\ \pi^{x}\bar{\pi}^{x}|$ corresponds to ≃80% of covalent VB structure : $\overset{\uparrow}{C}$—$\overset{\downarrow}{C}$. On the other hand, it has been shown that approaching a radical species strongly polarizes these spins[11], according to the scheme : R↑....C↓——— ↑C . Using these premises, we select in table 2, the covalent VB structure which is the most likely to allow for the formation of a covalent C-O bond, all other spins being frozen during the process (strong spin polarization). Qualitative conclusions may be drawn from this very simple model . i) The reactive VB structure is d^5 and since its weight is only 55% in the GS, activation energy will be necessary in order to increase it. This will be done upon excitation : $^3A_2 \rightarrow {}^3A_2{}^{x}$. ii) The mixing of these VB PESs yields an adiabatic surface which drives the system to a d^5 radical moiety in agreement with experiment. iii)The intermediate radical species is likely to live for some time. On the one hand, it cannot close in the yz plane giving a metallacycle, for we have three electrons preventing the formation of a bond in that plane. On the other hand, the other unpaired electron have the wrong spin and thus also cannot form a bond. Therefore, prior to any further evolution of the system, one needs activation energy, in order to properly arrange the various spins. In other words, the intermediate species lies in a potential well, whose deepness obviously cannot be estimated at this stage of our study.

CONCLUSION

The complementary use of MO and VB methodologies appears to be very effective for the study of reaction mechanisms involving important electronic density changes, such as in ionic or oxido-reductive processes. We have shown that although the topology of correlation diagrams is similar in both cases, the actual nature of the diabatic PESs is intrinsically different. Converging descriptions are finally obtained, either at SCF+CI or adiabatic VB level. We have also shown that a 20x20 VB calculation affords results that are comparable to a CI involving more than 3×10^{6} Møller-Plesset perturbative terms, using the same extended basis.

REFERENCES

1- R.B.Woodward and R.Hoffmann, The conservation of Orbital symmetry, Verlag Chemie, Academic Press, N.Y., 1970.

2-L.Salem, Electrons in Chemical Reactions : First Principles, J.Wiley and sons Ed.,N.Y.,1982.

3-A.Devaquet, A.Sevin and B.Bigot, J.Am.Chem.Soc., 1978, 100, 2009; A.Sevin and P.Chaquin, Nouveau Journal de Chimie, 1983 7,353.

4-A.Veillard and A.Dedieu, ibidem, 1983, 7, 6863.

5-B.Huron, J.P.Malrieu and P.Rancurel, J.Chem.Phys., 1973, 58, 5745.

6-To be published.

7-Review in V.Ulrich, Topics Curr.Chem., 1979, 83, 68.

8-O.Bortoloni and B.Meunier, J.Chem.Soc., Perkin Trans., II, 1984, 1967.

9-J.P.Collman, J.I.Brauman, B.Meunier, T.Hayashi,T.Kododek, and S.A.Raybuck, J.Am.Chem.Soc., 1985, 107, 2000.

10-R.D.Harcourt, Qualitative VB Description of Electron Rich Molecules; Pauling Three Electron Bonds and Increased Valence Theory, Springer Verlag Ed., Berlin, 1982.

11-S.Nagase, T.Takatsuka and T.Fueno, J.Am.Chem.Soc., 1976, 98 3838; V.Bonacic-Koutecký, J.Koutecký and L.Salem, ibidem, 1977, 99,84; A.Sevin, H.T.Yu and E.Evleth, Journal of Molecular Structure Theochem, 1983, 104, 163.

APPLIED MO THEORY: ORGANOMETALLIC STRUCTURE AND REACTIVITY PROBLEMS

P.Hofmann
Anorganisch-chemisches Institut
Technische Universität München
Lichtenbergstr. 4
D-8046 Garching, FRG

ABSTRACT. Molecular orbital investigations at the Extended Hückel level will be discussed for two independent cases of organometallic structure and reactivity problems, each representing a typical chapter of "applied" MO theory, closely related to experimental work. The examples chosen are: 1. the activation of carbon monoxide by biscyclopentadienyl derivatives of group 4 transition metals, and, 2. ligand coupling and cleavage processes in biscarbene and olefin transition metal complexes. Only specific examples and selected aspects of each topic will be detailed; the primary intention of this report is to illustrate the utility and capability of a one electron valence MO treatment to describe and to understand structures and reactivity patterns of even rather complex organometallic systems and to stress the importance and value of a mutual interplay between MO theory and experiment.

1. INTRODUCTION

Organotransition metal chemistry with its continuously increasing wealth of complex molecular structures, reactivity patterns, metal ligand bonding modes and dynamic processes, often without precedent and in many cases unforeseen, represents an enormous and permanent challenge for both qualitative and quantitative electronic structure theory. Molecular orbital calculations or equivalent approaches at various levels of sophistication, along with symmetry and perturbation theory based arguments frequently can provide an appropriate framework to understand or even to predict aspects of interest to the experimental chemist. It is certainly fair to state that qualitative concepts of electronic structure and semiempirical quantitative calculations on one side and ab initio electronic structure calculations on the other hand both had and have their share in shaping our present picture of bonding in organometallics, and that these various levels of methodology, if employed properly (that is to say with their limitations in mind), have their very specific merits and therefore their own right of existence. With respect to many problems in the field of experimental organometallic chemistry, one has to simply admit, however, that ab initio quantum theory of "chemically interesting" (e.g. moderately sized) transition metal containing molecules for obvious reasons has not yet reached the same stage of applicability and reliability as have ab initio treatments of small molecules composed of light atoms. Nevertheless, qualitatively reliable and trans-

A. Veillard (ed.), Quantum Chemistry: The Challenge of Transition Metals and Coordination Chemistry, 253–276.

ferable concepts of electronic structure, bonding and reactivity, hopefully method independent, have to be and can be worked out here, utilizing appropriate simpler molecular orbital methods with proper expertise and care. The analysis of known experimental facts may then provide the understanding necessary to predict chemical trends correctly. A reliable prediction of structural and reactivity trends in many cases is exactly what experimental chemistry needs as a frame in which to plan. Moreover - in the author's experience - often a qualitative picture on the one electron theory level, extracted from whatever level of theoretical sophistication, may be a prerequisite for discussing the results of theoretical calculations with experimentalists.

Our own involvement in the field of organometallic MO theory has been mainly based upon Extended Hückel theory as developed by R.Hoffmann and his group /1/, together with orbital interaction rules, frontier orbital theory and symmetry considerations, to tackle problems of structure, dynamics or reactivity of "real" molecules of interest in experimental organometallic (or inorganic) chemistry.

Two representative examples of this type of applied MO theory will be dicussed in the following sections.

2. CO ACTIVATION BY BISCYCLOPENTADIENYL DERIVATIVES OF GROUP 4 TRANSITION METALS

2.1. Experimental background - theoretical problems

The activation of carbon monoxide by transition metal compounds at present is being actively investigated by many groups. In particular the carbonylation chemistry of biscyclopentadienyl dialkyls (diaryls) and of related derivatives of group 4 transition metals has shown very interesting aspects /2/. CO insertion processes into M-R bonds of such compounds Cp_2MR_2 ($Cp = h^5$-C_5H_5, M = Ti, Zr, Hf) or their Cp*analogs ($Cp^* = C_5$-Me_5) are extremely facile, sometimes leading to isolable monoacyl complexes which display a rather unusual geometric feature: spectroscopic and structural investigations have shown that they contain h^2acyl groups RCO with both carbon and oxygen bound to the metal center. Although two possible h^2 acyl isomers could in principle exist, **1** (O-outside) and **2** (O-inside), all isolated compounds, e.g. $Cp_2Zr(COCH_3)(CH_3)$ /3/, exclusively show structure **2**. Spectroscopic observations in the course of carbonylation reactions of various Cp_2ZrR_2 molecules have indicated, however, that initially stereoisomers **1** (O-outside) are formed, which cannot be isolated but rearrange spontaneously and irreversibely to the apparently thermodynamically more stable structures **2** already at very low temperature /4/.The initial formation of type **1** structures of course indicates that CO inserts in a regiospecific way into the M-R bond. Where does this regiospecificity come from? What is the mechanism of the rearrangement from h^2 (O-outside) to h^2(O-inside) acyl coordination? Why is the latter bonding mode preferred and of lower energy? What about the electronic structure of these h^2 acyl complexes and its consequences for the known or expected chemical reactivity of such species?

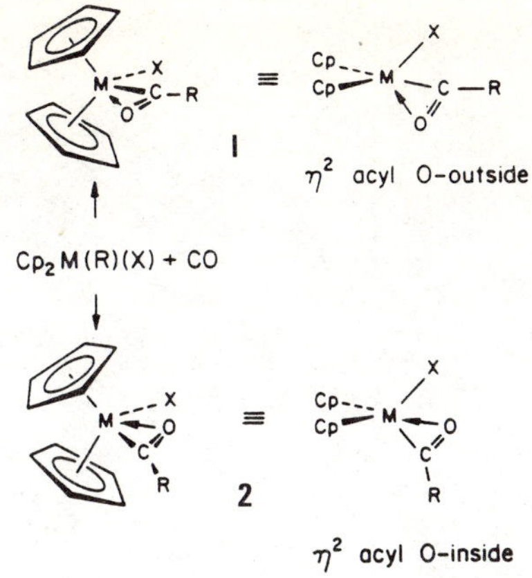

A particularly fascinating reaction is the direct carbonylation of some Cp* derivatives **3**, which leads to enediolate complexes **4**. A metal bound organic ligand system is formed by reductive coupling of two inserted CO molecules in a one-pot reaction /5/.

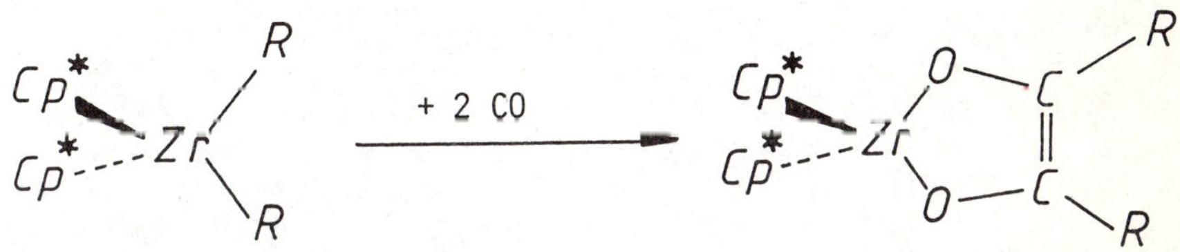

3 **4**

The mechanistic details of this very remarkable transformation are still unknown. An intramolecular pathway has been suggested, in which double carbonylation of a Cp_2MR_2 molecule first leads to a bis-acyl complex **5**, not isolated, which then undergoes ligand (C-C) coupling at the metal via some sort of bis-oxycarbene intermediate **6** /5/ to form products **4**.

Not only the electronic and geometric stucture of postulated bisacyl complexes of group 4 biscyclopentadienyls is of theoretical interest here. In elucidating the mechanism of acyl coupling (intra - vs. intermolecular, least motion or synchronous vs. more complicated geometric pathways) theory should be able to provide some information to guide proper experiments. Moreover, enediolate complexes like **4** are unusual compounds themselves and deserve a closer inspection with regard to their own electronic structure. It is remarkable, for instance, that

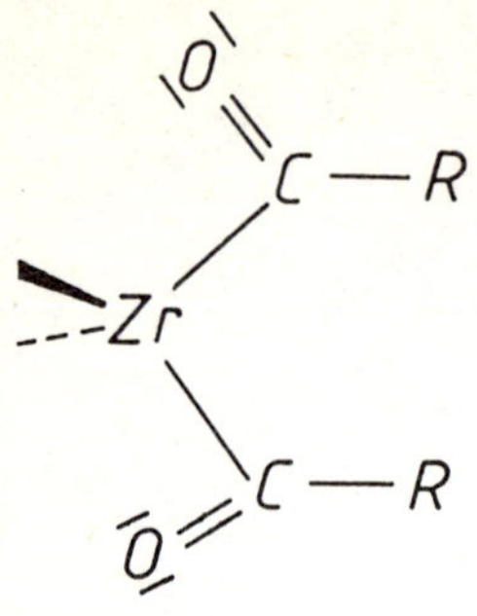

5

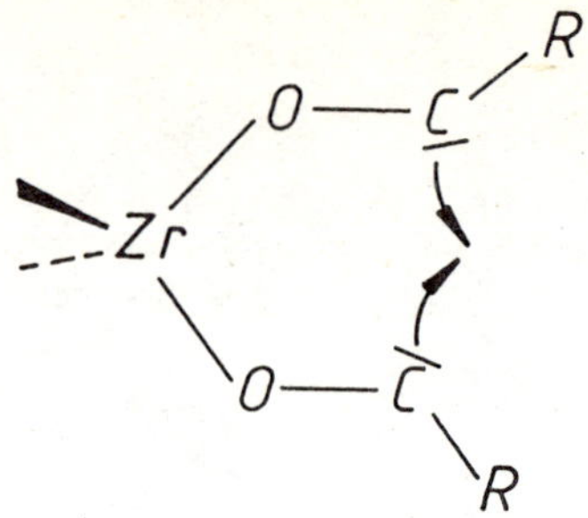

6

the known representatives of complexes **4**, e.g. with R = Me or vinyl /5/, are deeply colored compounds, which seems rather unusual for d^0 systems with formally two OR groups attached to Zr (IV).

We have studied the theoretical aspects of the structural and reactivity problems mentioned above using EH calculations, /6,7/ combining them with some experimental work related to the theoretical questions or results.

2.2. CO insertion in $Cp_2Zr(CH_3)_2$

$Cp_2Zr(CH_3)_2$ has been taken as a model system to study the energetic differentiation between a lateral attack and a central approach (pathways A and B in **7**, fig. 1) of CO. The same compound has been studied experimentally /4a/, so a direct comparison is possible. The approximate nature of the Extended Hückel method does not allow to reliably calculate the complete insertion processes with bond-breaking and bond-formation. The important initial stages of the reaction, where frontier orbital interactions should play the dominant role, can be examined, however. The calculations were performed by varying the carbon(CO)-to-metal distance, L, from 4.5 to 3.0 Å, with optimization of the relative angular positions of CO and the methyl groups, thereby allowing for the best ligand arrangement within the xy plane, as CO approaches. Figure 1, top, shows that the lateral pathway A, accompanied by an appropriate relaxation of the methyl positions, is energetically favored. This is in agreement with an early prediction of Lauher and Hoffmann /8/ based upon the spatial extension of the LUMO of $d^0Cp_2ML_2$, of y^2 -type, **8**, which finds better overlap with the HOMO of CO, **9**, when the approach of the Lewis base follows pathway A. Both curves in fig. 1 go through a small barrier at about L = 3.5 Å. Path A is preferred by about 5 kcal, certainly enough to induce regiospecificity for the primary CO attack. The small absolute barrier height of again about 5 kcal is in accord with experimental oberservations: CO insertion into $Cp_2Zr(CH_3)_2$ occurs at temperatures of - 130°C in solution /4/. So both qualitative frontier orbital considerations and explicit calculations allow to understand the initial formation of isomers **1** with O-outside orientation of the h^2 acyl group in the course of CO insertion.

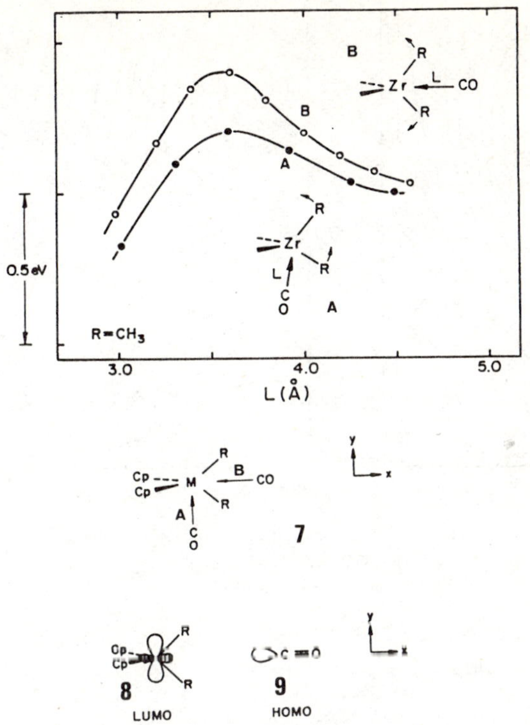

Fig.1. Total energy profiles for the two CO approaches toward $Cp_2Zr(CH_3)_2$ (top) and frontier orbitals of both reacting species (bottom).

2.3. Acyl ligand coordination

To investigate more closely the details of acyl ligand to metal coordination and to compare the stabilities of O-outside isomers and O-inside isomers, **1** and **2**, we have calculated the two potential energy surfaces, shown in fig. 2 for $Cp_2Zr(CH_3CO)(CH_3)$. Variables are the angular parameters α (Zr-C-CH) and ß (C-Zr-C).

The lowest potential minimum is found for **2**, O-inside, at $\alpha = 170^o$ and ß = 106°, corresponding to a clearcut h^2coordination mode of CH_3-CO.

This structure is approximately 5 kcal/mol more stable than the alternative O-outside h^2 geometry **1** found on the O-outside potential surface at $\alpha = 170^o$ and ß = 85°, thus reflecting nicely the experimental findings for structurally characterized stable h^2 acyl complexes.

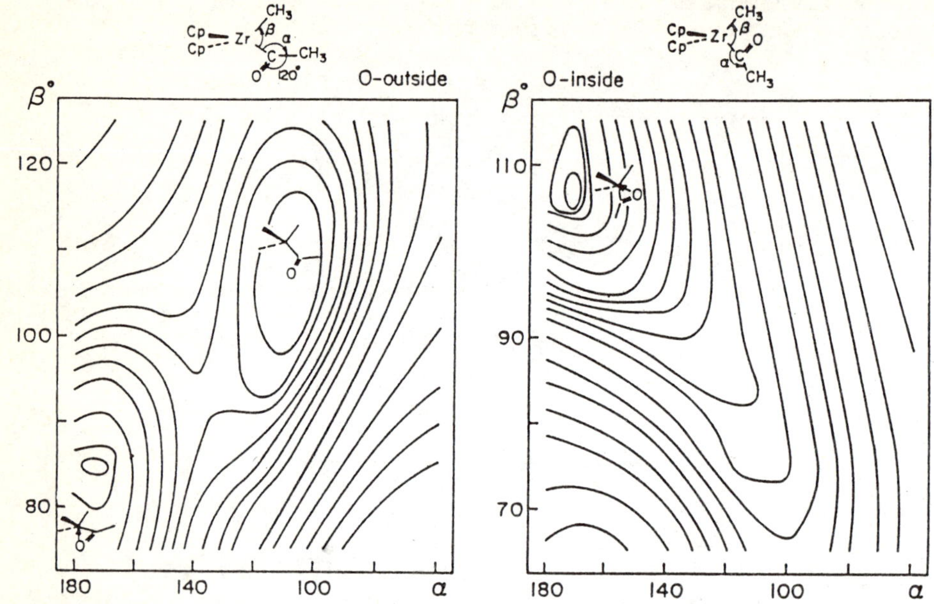

Fig.2. Potential energy surfaces for variation of the two angles, α ($Zr-C-CH_3$) and β (C-Zr-C) in O-inside (right) and O-outside (left) geometries of $Cp_2Zr(CH_3CO)(CH_3)$.

The energetic preference of **2** can be traced back in the calculations to a better overlap situation between the LUMO of a $Cp_2Zr(CH_3)^+$ or in general a d^0-Cp_2ML fragment, d_π /8,9/, and the HOMO, n_- of the CH_3CO^- ligand. This different overlap situation between both frontier orbitals is displayed qualitatively in **10b** and **11b**.

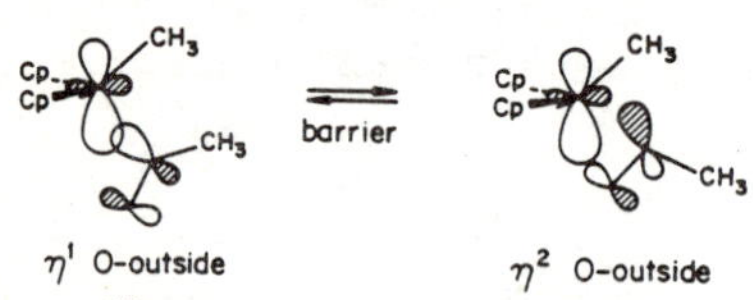

10a **10b**

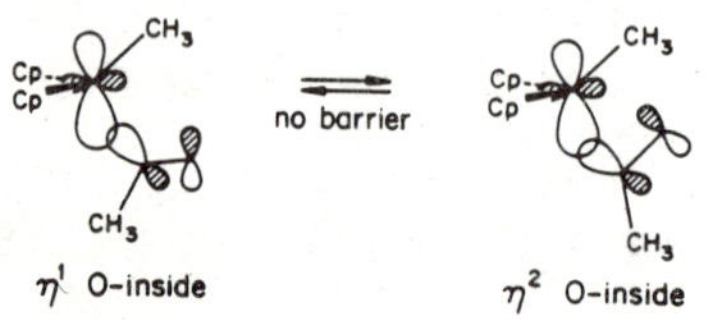

11a **11b**

The metal fragment to acyl ligand total overlap is better in **11b**, leading to a lower lying HOMO of $Cp_2Zr(CH_3CO)(CH_3)$ and a stronger Zr-C(acyl) bond for the O-inside h^2 ligand arrangement.

The most striking feature of the potential energy surface for the O-outside acyl coordination in fig. 2, rather unexpected, is the computed second energy minimum at $\alpha = 110^o$ and $\beta = 105^o$. This minimum, about 5 kcal/mol above **1**, is clearly of the h^1 type, representing a structure with a 16 valence electron count at Zr, in contrast to **1** and **2**. It is surprising that such a geometry should appear as a local minimum at all, because its existence implies an energy barrier for attaching the acyl oxygen to the metal and for going from an O-outside h^1 to an O-outside h^2 structure. This seems sowewhat unusual, in particular as the same feature is not present for the inverse O-inside acyl coordination. Actually a more detailed look upon the orbital structure of the h^1 O-outside minimum and its comparison with an analogous h^1 O-inside geometry (not an energy minimum) uncovers a simple symmetry based origin of the computational findings. The dominant interaction, again between d_π and n_- of the metal fragment and of acyl, and seen in the HOMO, is depicted in **10a** and **11a** for both possible h^1 bonding situations. The notable difference lies in the nodal properties. It is clear from **11a** and **11b**, that the h^1 O-inside structure relaxes to the h^2 geometry without destroying the phase match between both interacting orbitals. The h^1 to h^2 transformation can occur smoothly, without energy barrier, downhill, and leaving the h^2 structure as the only minimum. On the contrary, an h^1 to h^2 interconversion with O-outside transforms **10a** into **10b**, involving a phase discontinuity on the way. The origin of the barrier between h^1 and h^2 O-outside acyl coordination therefore lies in a strongly avoided crossing between HOMO and LUMO, as evident from an appropriate Walsh-diagram / 6/. The HOMO on each side, **10a** and **10b**, tends to correlate to the LUMO on the other side (antibonding counterparts of **10a** and **10b**), this causes a HOMO energy profile and accordingly a potential energy surface with a double minimum, isolating h^1 structures of $Cp_2M(COCH_3)X$ complexes as local minima for O-outside coordination.

Our detailed calculations also show that these h^1 O-outside structures play an important role in isomerizing **1**, the initial products of regiospecific CO insertion, to the finally isolable, more stable h^2 O-inside compounds **2**. This rearrangement cannot occur by a direct rotat - tion of h^2 bound acyl groups (about 2 eV barrier) but involves h^1 acyl coordination and Zr-C single bond rotation. The transition state has its h^1 acyl group rotated out of the plane by about 90° ("perpendicular").

In fig. 3 the calculated overall energy profile for the **1** to **2** rearrangement (with appropriate geometry optimization of each relevant structure) is shown. The energetic results are in surprisingly good agreement with known experimental data, given the approximate nature of the calculations. The experimentally observed activation free energy for isomerizing $Cp_2Zr(COR)(R)$ compounds ranges form 11.4 to 15.6 kcal/mol, /4/, the calculations find 13.5 kcal/mol.

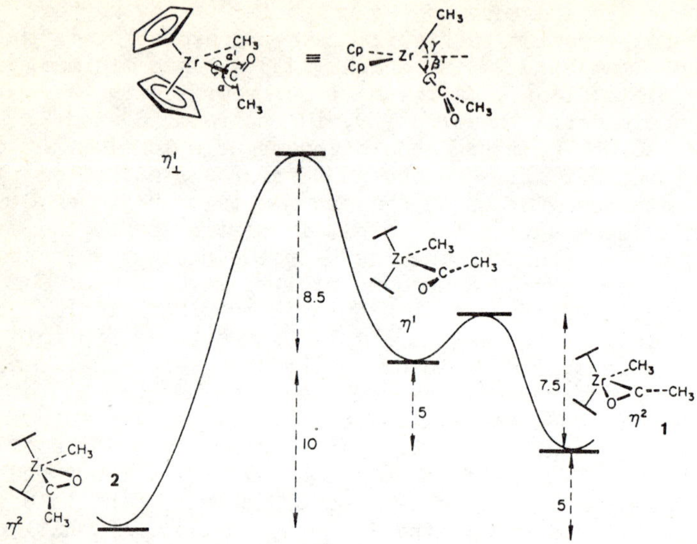

Fig.3. Computed energy profile for $Cp_2Zr(COCH_3)(CH_3)$ along the isomerization pathway from the h^2 O-outside structure **1**, the kinetic product of CO insertion into $Cp_2Zr(CH_3)_2$, to the thermodynamically more stable h^2 O-inside structure. Energies in kcal/mol.

Our calculations have also shown, that the oftentimes cited "carbene type" reactivity of h^2 bound acyl groups, frequently invoked to explain the chemical behavior of group 4 acyl complexes, is a mere consequence of the electronic properties of the ground state h^2 acyl structures themselves.

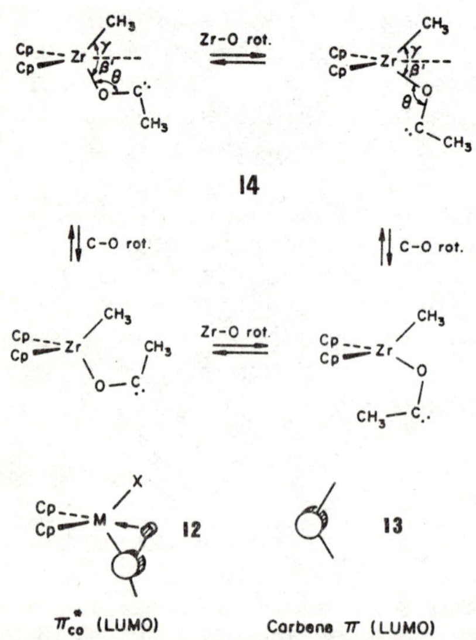

They all share the common feature of a low lying LUMO **12**, mainly of acyl π^* character, heavily localized on the carbonyl carbon and quite analogous to the LUMO of carbenes, **13**. In addition coordination of oxygen to the Lewis acidic metal increases the positive charge of the acyl carbon and all this together accounts well for the observed "carbene type" chemistry of such h^2 acyl complexes, which probably would be more appropriately described as "carbenium type".

As far as still annother mode of acyl coordination is concerned, namely that of a true oxycarbene structure as shwon in **14** for some particular conformations,calculations find them to be local minima, with Zr-O-C angles close to 180°. Any preequilibria of acyl complexes with such species as conceivable entries towards subsequent "carbene type" reactions have to be disregarded, however, as the oxycarbene isomers of e.g. $Cp_2Zr(COCH_3)(CH_3)$ turn out ot have energies about 2 eV higher than their Zr-C bound acyl congeners. This is of relevance for the above mentioned mechanism of reductive CO coupling to enediolate ligands, which we will turn to now.

2.4. Electronic structure and qroundstate geometry of mononuclear enediolate complexes of Zr

Model calculations for enediolate complexes of type 4, $Cp_2Zr(C_2O_2R_2)$, R = Me, which are formed as carbonylation products of the corresponding dialkyls in some cases, lead to the interesting prediction, that the ZrO_2C_2 - chelate rings should not be planar as shown in **15** and as normally anticipated, but that they should posess folded ring geometries as in **16**, i.e. the enediolate ligands should be coordinated with a σ^2,π tendency rather than in a pure σ^2 fashion. Thus a close analogy to the structures of s-cis diene complexes /10/, **17**, is suggested by MO calculations.

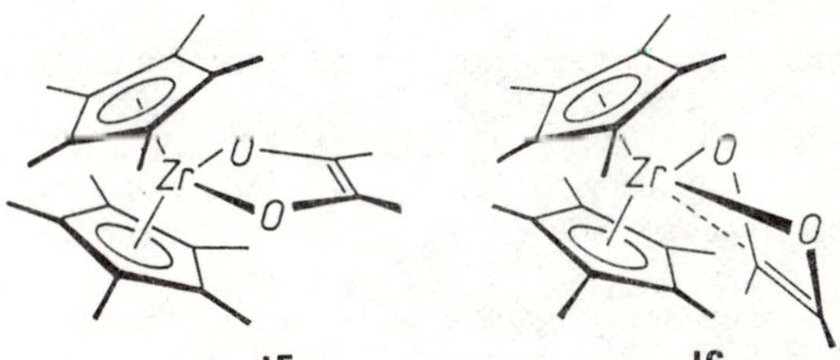

15 **16**

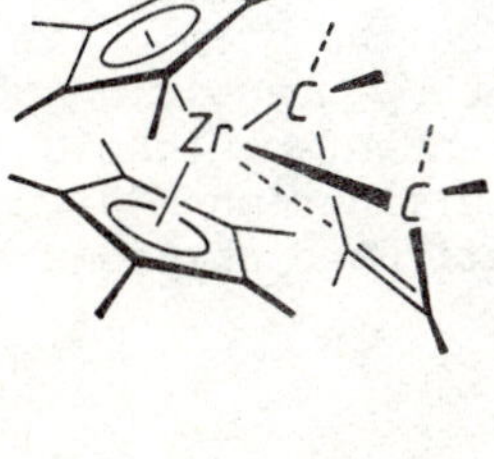

17

Only for nonplanar chelate rings the HOMO, which essentially is the π-orbital of the double bond, can interact in a stabilizing way with the metal localized LUMO of y^2 character. This situation is displayed in fig. 4 and is analogous to the interpretation of bonding in related dithiolene and diene complexes /8,10b/.

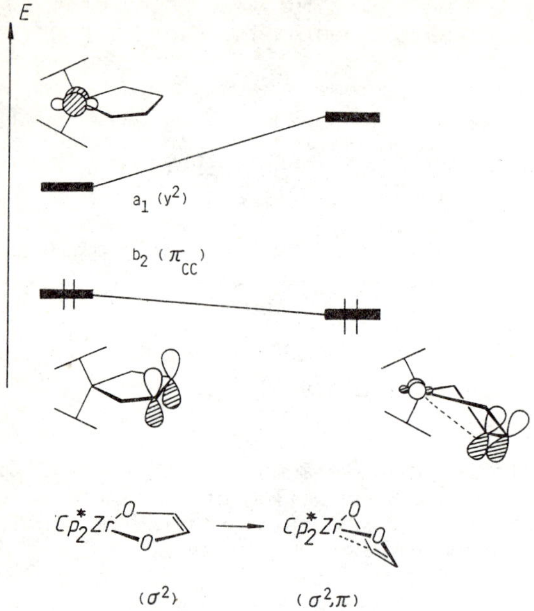

Fig. 4. Frontier orbital energy variation (qualitative) upon ring folding of $Cp_2Zr(C_2O_2R_2)$ enediolate complexes.

Indeed an X-ray structure determination of the enediolate complex from direct carbonylation of $Cp*_2Zr(CH_3)_2$ /5a/, performed by us /11/ recently, revealed a molecular structure with a ZrO_2C_2 ring slightly folded around the O-O axis and with an obvious tendency towards a σ^2, π coordination mode, despite of the sterically demanding Cp* ligands.

The MO calculations give only very small energetic preferences for nonplanar enediolate coordination geometries (in accord with extremly facile ring flipping in solution), thus suggesting that it should be possible to sterically enforce the electronically less favorable planar structure by large R groups at the ring carbons. An enforced planar groundstate structure should then in turn lead to a bathochromic shift of the long wavelength (low extinction) absorption band in the electronic spectrum, compared to a nonplanar system. This is a simple conclusion from fig.4 which suggests to assign the deep color of monomeric $d^0Cp_2^*$-M-enediolate complexes to a HOMO-LUMO transition of only weak intensity in the planar structure. An independent synthesis of $Cp*_2Zr(O_2C_2R_2)$, R = t-Bu, and its X-ray structure determination /11/ corroborate the MO based expectations: the t-Bu substituted compound has a planar ring system, shown in fig.5 at right, the folded structure of the methyl derivative is displayed on the left for comparison. The lowest energy absorption is shifted by 75 nm on going from the purple CH_3 substituted to the dark blue t-Bu substituted compound.

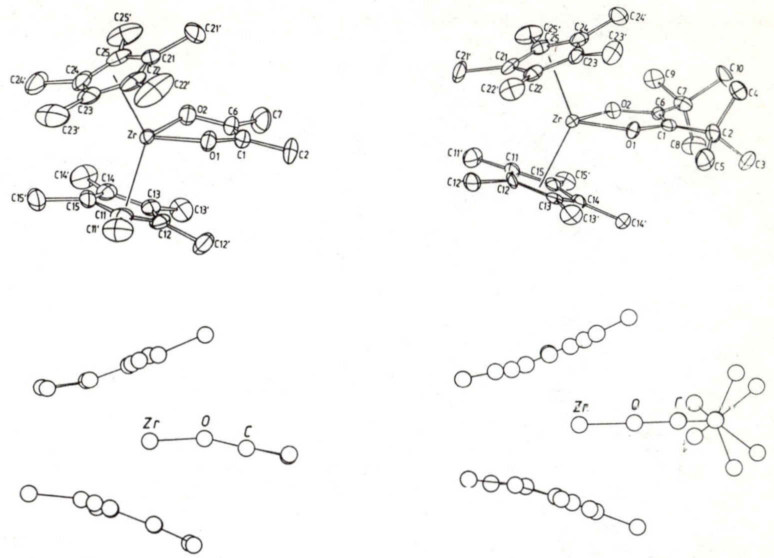

Fig.5. Molecular structures of $Cp^*_2Zr(O_2C_2R_2)$ enediolate complexes; R = CH_3 (left) and R = t-Bu (right).

2.5. The mechanism of enediolate ligand formation

Experimentally we have been able recently to show, that the reductive coupling of two CO molecules in the course of enediolate ligand formation (3 + 2 CO → 4, see above) is an intramolecular process /12/. In an isotope labeling crossover experiment a 1:1 mixture of $Cp^*_2Zr(CH_3)_2$ and $Cp^*_2Zr(CD_3)_2$ was carbonylated and only enediolate complexes were obtained which contained either two CH_3 or two CD_3 groups.
Based upon this knowledge we considered it worthwhile to take a closer look at the postulated metal centered reaction mechanism in which initial double carbonylation is supposed to generate bis-acyl complexes, e.g. $Cp^*_2Zr(COR)_2$, **18** , which subsequently rearrange and couple to form the enediolate system of **4**, possibly via an intermediate bis-oxycarbene stage of type **6**. The simplest pathway then of course would be a concerted in-plane motion of the acyl ligands of **18** (see arrows) retaining C_{2v} symmetry.

Three problems arise immediately, however, from qualitative MO considerations and from model calculations: i) The groundstate structure of a bis-acyl complex of Cp^*_2Zr (or Cp_2Zr) should have one h^2 O-inside

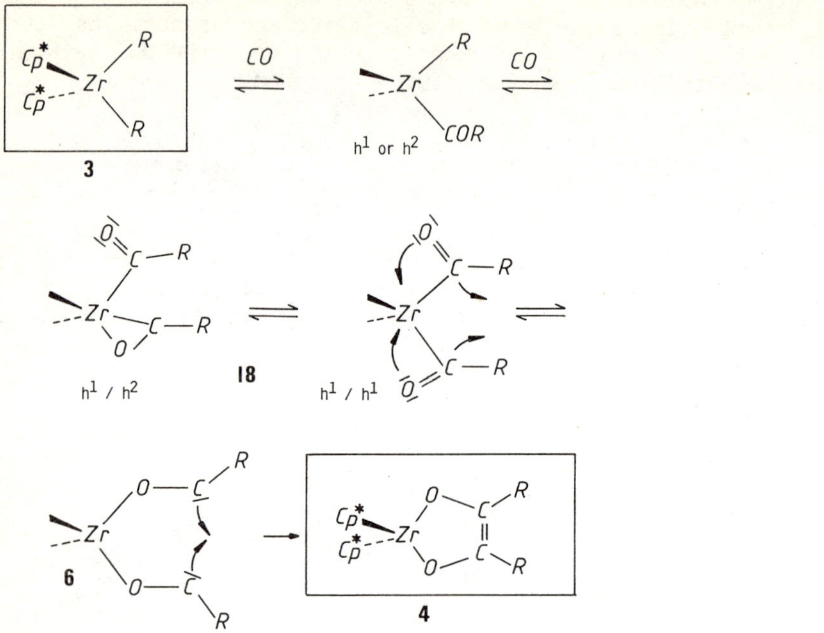

and one h^1 acyl ligand, and, although alternative h^2,h^1 or bis-h^1 geometries are close in energy, a bis-h^2 structure with two O-outside ligands is a 20 electron system and should not be involved; ii) zirconoxycarbene structures are high energy species (see 2.3.); iii) a least motion C_{2v} pathway is symmetry forbidden.

The latter conclusion evolves from an appropriate correlation diagram for a bis-acyl (h^1,h^1)complex to enediolate complex transformation, and fig. 6, top left, displays the relevant levels of both sides and their crossings. As also shown in fig. 6, retaining only C_s symmetry (either by independent in-plane movements of both acyl groups or by disrotatory out-of-plane coupling (top right, bottom left), does not remove the symmetry restrictions, only a conrotatory (C_2) pathway could avoid the correlation of occupied with unoccupied levels (bottom, right).

In model calculations we have tested explicitly the conrotatory and the least motion (C_{2v}) pathway, using two variables, namely an out-of-plane rotation angle θ of acyl coordination and various sets of distances and angles as reaction coordinate r_i, the stepwise synchronous change of which carries **18** over to planar **4** in 10 steps. This is a quite crude approach but should give at least a trend. Suffice it here to say, that even symmetry allowed conrotatory coupling model pathways, starting out from a bis-h^2 structure of the bis-acyl complex (as in fig. 7) require barriers of nearly 2 eV. The bis-h^2 system, however, lies still about 1 eV higher already than the h^2,h^1 groundstate of a $Cp_2Zr(COCH_3)_2$. So the energy barrier separating a bis-acyl complex with an 18 electron groundstate geometry from the enediolate system on a C_2 conrotatory coupling path seems prohibitively high.

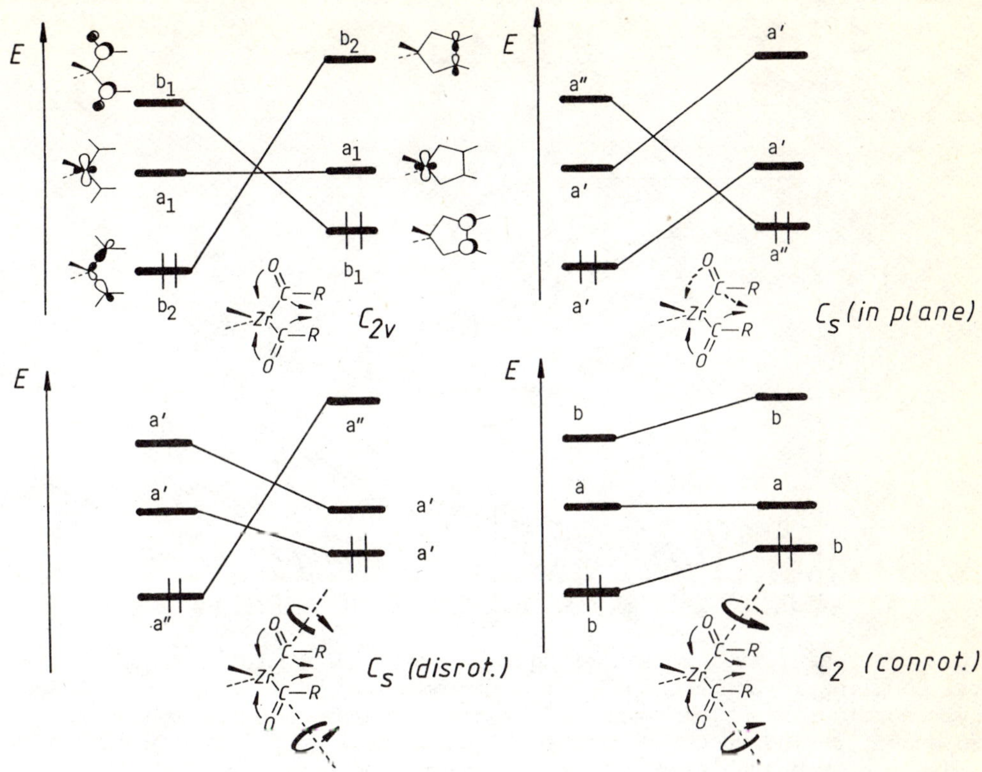

Fig.6. Qualitative frontier orbital correlations for different intramolecular acyl ligand coupling modes, transforming a bis-(h[1]acyl) complex into its enediolate product.

Naturally these computed energies in our model calculations have to be taken with care, but the point here is, that the results suggest that other mechanisms rather than direct acyl coupling should be taken into serious consideration and tested by further calculations and in experiments.

In summary, molecular orbital calculations have been and certainly will continue to be of valuable help in understanding this specific field of group 4 transition metal and, not discussed here, of related actinide chemistry /6,7,13/.

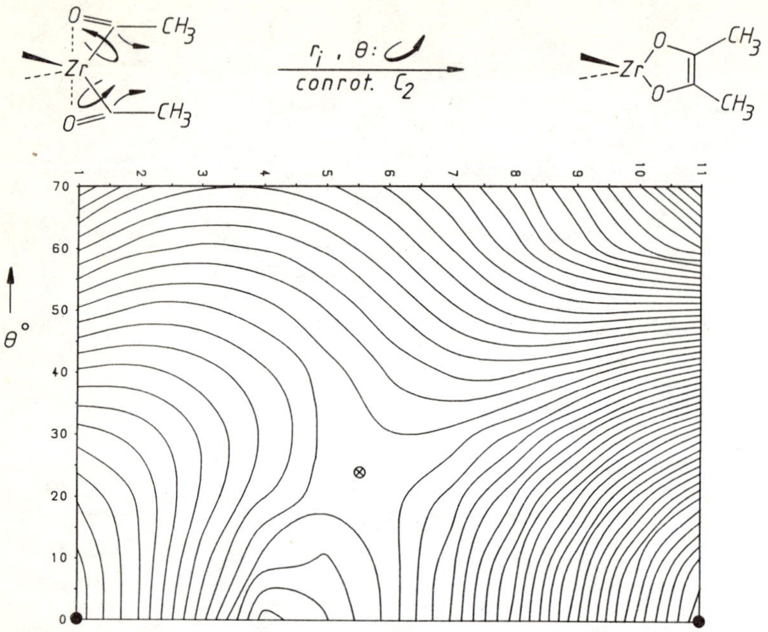

Fig.7. Representative potential energy model surface for intramolecular acyl coupling in $Cp_2Zr(COCH_3)_2$ .The variables are the rotation angle of acyl coordination θ , and a set of geometric parameters r_i , angles and distances, varied synchronously in 10 steps from a bis-acyl (bis-h^2) toward the more stable enediolate structure (both planar, C_{2v} ,filled circle at bottom left and right). Crossed circle: approximate transition state for conrotatory pathway. The energy contours are 0.16 eV apart.

3. LIGAND COUPLING AND CLEAVAGE PROCESSES IN BISCARBENE AND OLEFIN TRANISTION METAL COMPLEXES

3.1. Experimental background

The formation and breaking of carbon carbon bonds within the coordination sphere of transition metals is one of the most actively studied areas in organotransition metal chemistry. Particularly fascinating are processes in which either two carbene ligands in a biscarben transition metal complex are coupled to yield olefin complexes, or, in the reverse sense, where olefinic or olefin type ligands undergo double bond cleavage to generate two ligands with one carbon atom. Some experimental evidence has been collected in the past for the occurence of such reactions at metal centers.They also have been found to exist for ketene ligands,which are formed by carbene carbonyl coupling or which are cleaved at their C-C double bond, yielding a carbene and a carbonyl ligand /14/. Cases of carbyne to carbonyl coupling are also known, leading to ketenyl ligands

/15/. The need for at least a qualitative theoretical understanding of such reactions is obvious. In the present context only one specific problem will be discussed, namely carbene carbene coupling and olefin cleavage at a d^6 $M(CO)_4$ fragment. The particular point of inerest here will be the influence of electron releasing olefin or carbene substituents, because experimental work indicates, that electron rich olefins can be easily cleaved inter alia at d^6 transition metal centers yielding stable octahedral bis-carbene complexes, e.g. in the reaction of d^6 metal hexacarbonyls $M(CO)_6$, (M = Cr,Mo,W) with such alkenes /16/. Normal olefines do not seem to undergo the same reaction.

3.1. The model system

We will focus here on the model system represented in **19** and **20**, trying to understand the basic electronic features of this potential interconversion between the octahedral (18 electron) bis-carbene complex **19** and its isomeric olefin analog (16 electrons), **20**, for R = H as well as for R = NR_2.

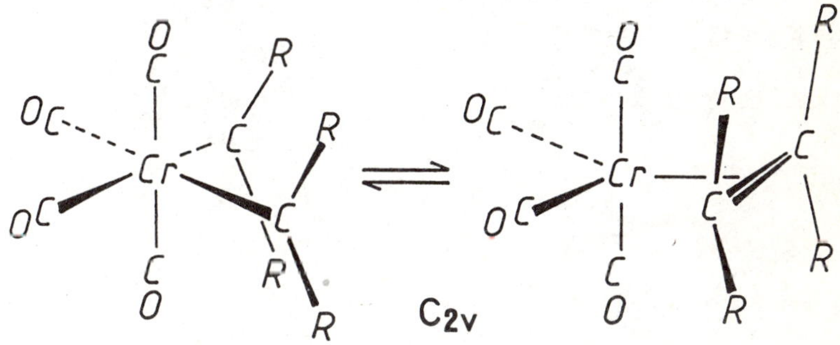

19 **20**

The simplest (least motion) pathway for a coupling/decoupling reaction retains C_{2v} symmetry, and it is most practical to start out our analysis from there.

3.2. The $(CO)_4Cr(CH_2)_2$ and $(CO)_4Cr(C_2H_4)$ systems

Hoffmann et al. /17/ have very elegantly analyzed the electronic situation for the coupling of two methylene groups bound to a "naked" transition metal center in the case of $W(CH_2)_2$ and have already discussed the effect of varying electron counts. We can use their approach as a proper and convenient basis for the $(CO)_4Cr(CH_2)_2$ system. A correlation diagram for the "naked metal" case, analogous to the one calculated by Hoffmann et al. is shown in fig. 8.

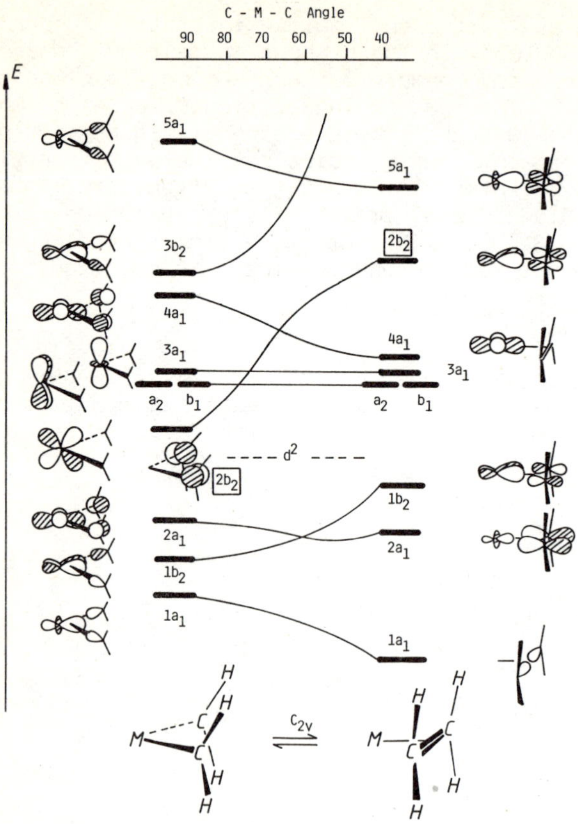

Fig.8. Orbital correlation diagram for a C_{2v} least motion coupling/decoupling process of two methylenes, suitably attached (as shown) to an unligated metal center /17/. The d^2 electron count limit is indicated.

The point relevant to our discussion is that a d^6 electron count makes the least motion process symmetry forbidden. The question then arises if this symmetry restriction still persists if the metal carries four good acceptor ligands like CO. If these four ligands, arranged as in **19** to complete an octahedron, are brought in on the bismethylene side of fig. 8, one of the three metal d levels, $3a_1$, must obviously be pushed up, the other two, a_2 and b_1, will be stabilized by back bonding to the acceptor ligands and may well end up below the critical methylene based level $2b_2$. If the same effects of the four carbonyls operate on the olefin side of fig.8 as well, then the interesting conclusion would be that a least motion coupling/decoupling process is now symmetry allowed and, depending upon the thermodynamic situation, should proceed smoothly to the preferred side. It is apparent, that these purely qualitative considerations on the basis of the "naked metal" case need to be checked by calculations. The computed electronic structure of $(CO)_4Cr(CH_2)_2$,

19, indeed turns out to agree with the qualitative expectations for the d^6 octahedral complex, a_2 and b_1 metal levels appear below $2b_2$, $3a_1$ is pushed way up by the ligands. Nevertheless the least motion transformation between **19** and **20** still turns out to be symmetry forbidden because, as shown in fig.9 below, carbonyl substitution at the metal also changes the valence levels of the olefin complex side. This happens in such a way as to put $1b_2$ of fig.8 above three occupied metal orbitals. (Note that levels $4a_1$ and $5a_1$ of fig.8 correspond to $3a_1$ and $4a_1$ of fig.9, because the $3a_1$ orbitals of fig.8 are pushed out of the picture.)

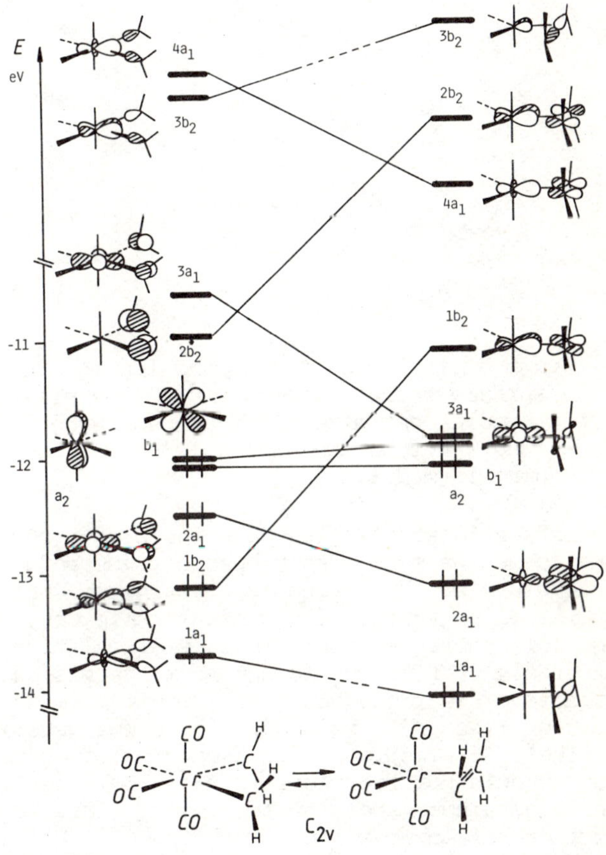

Fig.9. Orbital correlation diagram between relvant calculated valence levels of $(CO)_4Cr(CH_2)_2$ and $(CO)_4Cr(C_2H_4)$ for the least motion coupling/decoupling process.

The forbidden nature of the least motion process then implies that less symmetric pathways of carbene coupling or olefin cleavage would, although not strictly forbidden, still encounter an electronic barrier. The problem with the qualitative analysis up to this point is, that a correlation diagram as in fig.9 does not say anything about the relative thermodynamic stability of both sides, nor can the approximate height of the symmetry inposed barrier be deduced. Of course Extended Hückel calculations cannot give reliable numbers here anyway, but when we want to compare related systems in this context, trends and differences may emerge which are useful. That is why we calculated a potential energy surface for the least motion coupling/decoupling process for the pair $(CO)_4Cr(CH_2)_2/(CO)_4Cr(C_2H_4)$. For a C_{2v} pathway at least two angular variables seem necessary to properly allow for the **19** to **20** transit, and these angles, Θ and Φ, are shown in **21**.

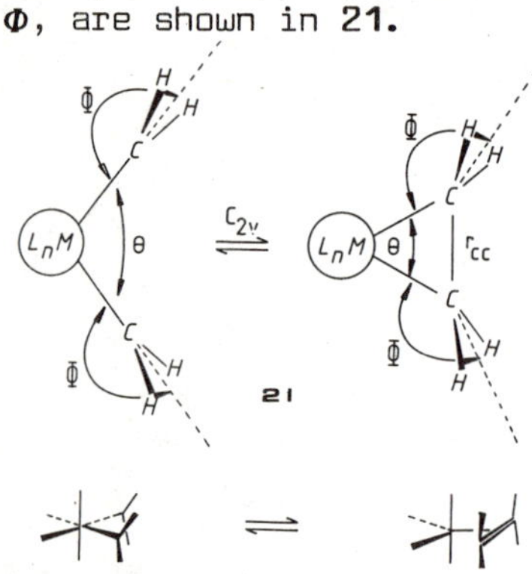

21

For a fixed $Cr(CO)_4$ fragment the variation of the two angular parameters results in the energy surface represented in fig.10.Two nice minima appear, and the olefin complex optimizes at a carbon-carbon distance of around 1.3 Å, somewhat short but not unreasonably off, if we consider the very simplified geometric model.

The potential surface interestingly shows the olefin complex with 16 valence electrons to be the more stable species! The enforced trigonal bipyramidal structure of this molecule certainly is not its ground-state geometry but will relax further (in a predictable way, analogous to $Cr(CO)_5$ /18/) and will therefore be at even lower relative energy than in the energy surface of fig. 10. In contrast the bis-carbene structure should be already near its octahedral minimum (with already noticeably close carbene ligands). The conclusion seems legitimate then, that for normal olefins a 16 electron $Cr(CO)_4$ complex, if generated from a suitable precursor, should not rearrange to a bis-carbene isomer, in spite of the 18 electron valence shell of the latter. This is indeed in accord with the experimental observation of persistent $(CO)_4Cr$(olefin) species in matrix /19/, which can be formed by irradiating the corresponding pentacarbonyls. The calculated low energy barrier for the transit from the bis-carbene structure **19** to the more stable olefin complex of only about 10 kcal/mol even for the least motion path (and probably still lower for a less symmetric, fully optimized reaction channel) suggests on the other hand, that carbene-carbene coupling in systems, which are electronically comparable to **19**, should occur readily, leading to the 16 electron olefin complexes and finally to the corresponding 18 electron olefin complexes after take-up of an additional ligand.

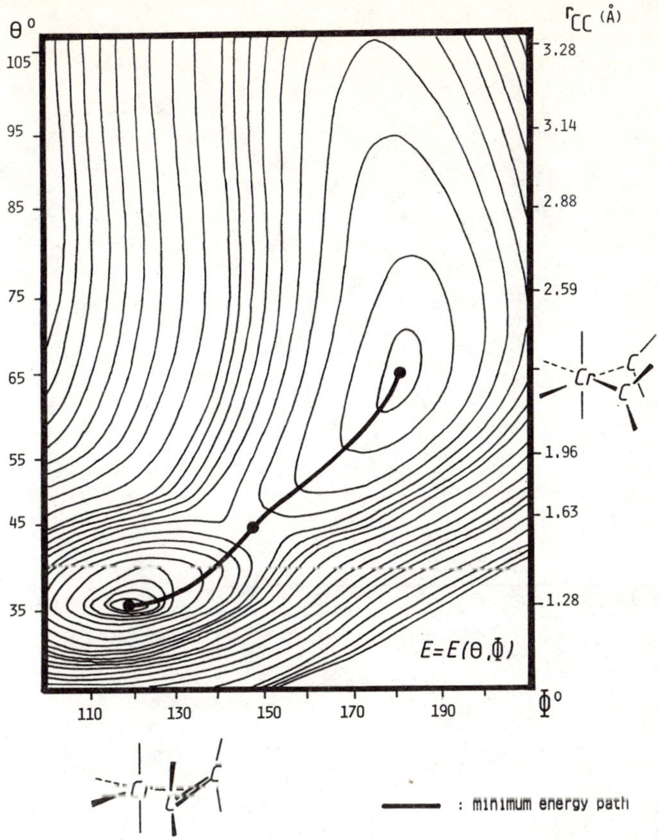

Fig.10. Potential energy surface for the least motion (C_{2v}) interconversion between $(CO)_4Cr(CH_2)_2$ and $(CO)_4Cr(C_2H_4)$ with Θ and Φ as defined in **21**. Contour lines at 0.1 eV separation.

3.3. The effect of donor substituents

Now recall,that all the above conclusions are in disagreement with the known chemistry, if the carbenes or the olefin are substituted with good π-donors. Here bis-carbene complexes of d^6 $M(CO)_4$ fragments are stable isolable compounds and are formed from the corresponding olefins. Can we say anything about this difference from model calculations? Let us look at the pair **22/23**.

A potential surface, calculated in the same manner as for the previous case, appears totally different. Only one minimum is found now, corresponding to the bis-carbene system. No 16 electron olefin complex exists, the chemical generation of such a species would immediately cause its irreversible and exothermic transformation to the bis-carbene structure; again this agrees with experimental findings.

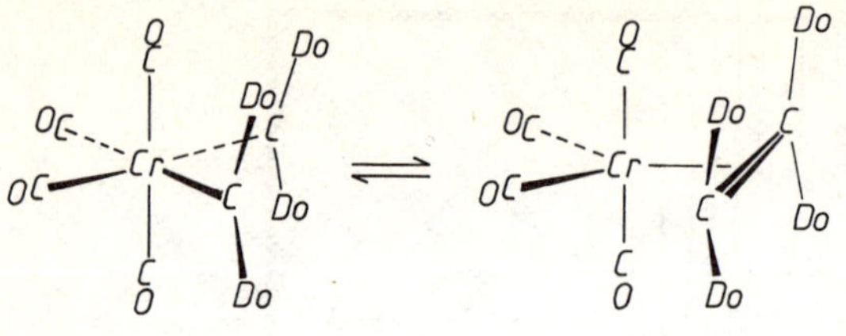

22 23

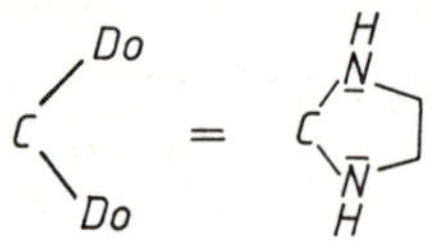

A qualitative MO correlation diagramm in fig.12 /20/ shows the **22/23** interconversion to have no symmetry restriction even in C_{2v}. The relevant effect of donor substitution is found on the olefin complex side. Essentially what happens is, that the level with predominant

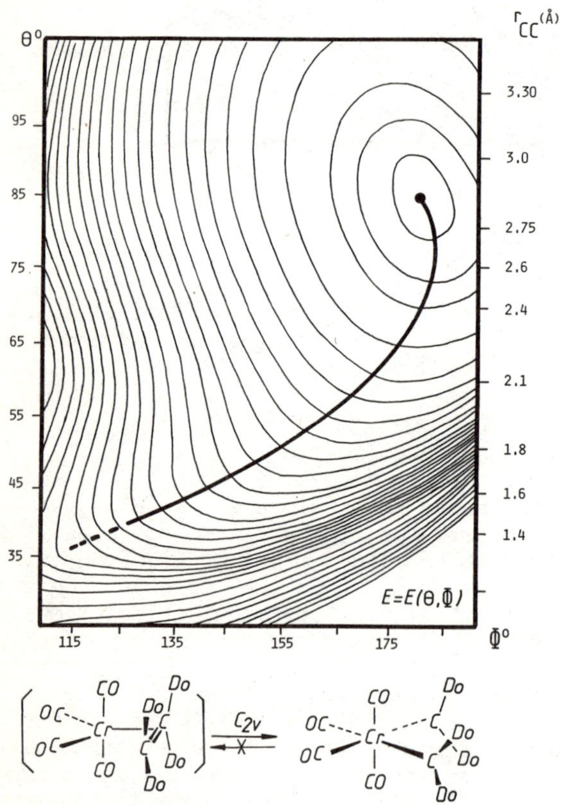

Fig.11. Potential energy surface for the interconversion of **22** and **23**, calculated as described for fig.10.

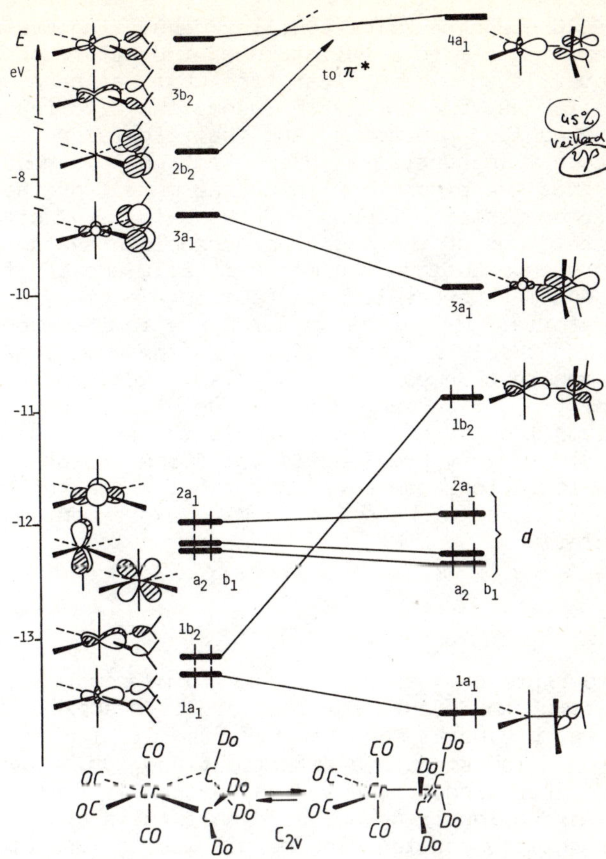

Fig.12. MO correlation diagram between calculated levels of **22** and **23**, C_{2v} symmetry, analogous to fig.9.

olefin π character ($2a_1$ in fig.9) is destabilized so much by the four donor substituents of the olefin, that it loses its electrons to a metal level ($1b_2$), which carries also some minor π* density at the olefin in the wavefunction. In other words: the "olefin complex" really has already a "broken" π bond because of electron-transfer from the electron rich olefin to the acceptor substituted metal center, and therefore does not represent a stable structure.

In comparison to fig.9, fig.12 even indicates the thermodynamic relation of both structures.

Other cases of ligand coupling/decoupling processes can and have been analyzed similarly to the example chosen here /21/ examining different ML_n fragments and other organic olefin type ligands, e.g. ketenes. More experimental work, especially mechanistically oriented, is very much needed, and ab initio studies in this field would be of great interest.

The specific example of a C_1/C_1 coupling/decoupling system discussed here only had to do with a variation of the nature of the carbene and olefin fragments. It should be clear from what has been said, however, that this is just one way to influence the electronic nature of such coupling/decoupling processes, and that others can be analyzed in an analogous manner. In general it seems from what we have seen in our studies, /21/ that the proper choice of the metal, of the number, type and steric arrangement of ligands, of the metal's electron count or its oxidation state and of the coupling/decoupling system, should allow to influence such reactions thermodynamically and kinetically in an understandable and may be predictable fashion. To give just one example: if two equatorial CO ligands in **19/20** are replaced by good donors, e.g. PH_3 in our calculations /21/, then the coupling/decoupling process becomes symmetry allowed for $(CH_2)_2/C_2H_4$. Qualitatively this can be derived from fig. 9. In many a case a detailed inspection of the specific system along the lines shown above is really necessary - the simple "naked" metal case may not lead to valid conclusions. Qualitative MO theory and EH calculations can be of great help here and hopefully can provide some guidelines along which to plan experimental work or higher level calculations.

4. ACKNOWLEDGEMENT

It is a pleasure to express my sincere thanks to my coworkers,who deserve a major part of the credit for the results described here, P.Stauffert, M.Frede, L.A.Perez Moya and H.R.Schmidt. I also thank Prof.Ulf Thewalt for X-ray structure determinations, Dr.Kazuyuki Tatsumi for valuable and helpful discussions and in particular Prof.Roald Hoffmann not only for an inspiring and enjoyable collaboration over the years but also for showing and teaching me his way to look at chemistry.

Financial support by the Deutsche Forschungsgemeinschaft and the Fonds der Chemischen Industrie is gratefully acknowledged.

5. REFERENCES

/1/ Hoffmann,R.J.Chem.Phys.**1963**,39,1397;Hoffmann,R.;Lipscomb,W.N.Ibid. **1962**,36,2179. Atomic parameters: see ref./16/ and Albright,T.A.;Hofmann, P.;Hoffmann,R.J.Am.Chem.Soc.**1977**,99,7546. A modified version of the Wolfsberg-Helmholz formula was used throughout: Ammeter,J.H.;Bürgi, H.-B.; Thibeault,J.C.;Hoffmann,R.J.Am.Chem.Soc.**1978**,100,3686.

/2/ a) Wolczanski,P.T.;Bercaw,J.E.Acc.Chem.Res.**1980**,13,121. b) Erker,G.Ibid.**1984**,17,103, and references therein. The related carbonylation chemistry of analogous actinide systems will not be discussed here.

/3/ a) Fachinetti,G.;Fochi,G.;Floriani,C.J.Chem.Soc.,Dalton Trans. **1977**,1946. b) Fachinetti,G.;Floriani,C.;Marchetti,F.;Merlino,S.J.Chem. Soc.,Chem.Commun.**1976**,522.

/4/ a) Erker,G.;Rosenfeldt,F.J.Organometal.Chem.1980,188,C1. b) Erker,G.;Rosenfeldt,F.Angew.Chem.1978,90,640.

/5/ a) Manriquez,J.M.;McAlister,D.R.; Sanner,R.D.;Bercaw,J.E.J.Am. Chem.Soc.1978,100,2716;(Zr). b) McDade,C.;Bercaw,J.E.J.Organometal. Chem.1985,279,281;(Zr and Hf). c) For Cp the reaction seems to work as well: Erker,G. private communication. See also further references in b).

/6/ Tatsumi,K.;Nakamura,A.;Hofmann,P.;Stauffert,P.;Hoffmann,R.J.Am. Chem.Soc.1985,107,4446.

/7/ Hofmann,P.;Stauffert,P.;Tatsumi,K.;Nakamura,A.;Hoffmann,R.Organometallics 1985,4,404.

/8/ Lauher,J.W.;Hoffmann,R.J.Am.Chem.Soc.1976,98,1729.

/9/ a) Hofmann P.;Stauffert,P.;Schore,N.E.Chem.Ber.1982,115,2153. b) Silver,M.E.;Eisenstein,O.;Fay,R.C.Inorg.Chem.1983,22,759.

/10/ a) Krüger,C.,Müller,G.;Erker,G.;Dorf,U.;Engel,K.Organometallics 1985,4,215 and references therein. b) Tatsumi,K.;Yasuda,H.;Nakamura,A. Isr.J.Chem.1983,23,145.

/11/ Hofmann,P.;Frede,M.;Stauffert,P.;Lasser,W.;Thewalt,U.Angew.Chem. 1985,97,693.

/12/ Hofmann,P.;Stauffert,P. unpublished; see also ref. /11/.

/13/ Tatsumi,K.;Nakamura,A.;Hofmann,P.;Hoffmann,R.;Moloy,K.G.;Marks, T.J. manuscript in preparation.

/14/ a) Casey,C.P.;Anderson,R.L.J.Chem.Soc.,Chem.Commun.1975,895. b) Herrmann,W.A.;Gimeno,J.;Weichmann,J.;Ziegler,M.L.;Balbach,B.J.Organometal.Chem.1981,213,C26. c) Mitsudo,T.;Sasaki,T.;Watanabe,Y.;Takegami,Y.; Nishigaki,A.;Nakatsu,K.J.Chem.Soc.,Chem.Commun.1978,252. d) Fischer,H. Angew.Chem.1983,95,913. e) Dorrer,B.;Fischer,E.O.Chem.Ber.1974,107,2690. f) Rüchardt,G.;Schrauzer,G.N.Chem.Ber.1960,93,1840. g) Klimes,J.;Weiss, E.Angew.Chem.1982,94,207;Angew.Chem.Suppl.1982,477. h) Miyashita,A.; Shitara,H.;Nohira,H.Organometallics 1985,4,1463. i) Miyashita,A.; Grubbs,R.H.Tetrahedron Lett.1981,1255. j) Barger,P.T.;Santarsiero,B. D.;Armantrout,J.;Bercaw,J.E.J.Am.Chem.Soc.1984,106,5178. k) Miyashita, A.;Shitara,H.;Nohira,H.J.Chem.Soc.,Chem.Commun.1985,850.

/15/ a) Birdwhistell,K.R.;Tonker,T.L.;Templeton,J.L.J.Am.Chem.Soc.1985, 107,4474. b) Churchill,M.R.;Wasserman,H.J.;Holmes,S.J.;Schrock,R.R. Organometallics 1982,1,766. c) Kreissl,F.R.;Eberl,K.;Uedelhoven,W.Chem. Ber.1977,110,3782. d) Kreissl,F.R.;Frank,A.;Schubert,U.;Friedrich, P.;Lindner,T.L.;Huttner,G.Angew.Chem.1976,88,649.

/16/ a) Öfele,K.;Heberhold,M.Angew.Chem.1970,82,775. b) Lappert,M.F.;

Pye,P.L.;McLaughlin,G.M.J.Chem.Soc.,Dalton Trans.**1977**,1272. c) Lappert,M.F.;Pye,P.L.Ibid.**1977**,1283. d) Hitchcock,P.B.;Lappert,M.F.;Pye, P.L.Ibid.**1977**,2160. A more extensive list of references is found in /17b/.

/17/ a) Hoffmann,R.;Wilker,C.N.;Eisenstein,O.J.Am.Chem.Soc.**1982**,104, 632. b) Wilker,C.N.;Hoffmann,R.;Eisenstein,O.Nouv.J.Chim.**1983**,7,535. The diagram of fig. 8 has been drawn similar to the one in these references.

/18/ Elian,M.;Hoffmann,R.Inorg.Chem.**1975**,14,1058.

/19/ Grevels,F.;Mülheim, personal communication; (olefin = trans-cyclooctene).

/20/ Note that the olefin side here does not represent an energy minimum but is calculated with a model geometry to compare it to the previous case in fig.9.

/21/ Hofmann,P.;Schmidt,H.R.;Frede,M.;Perez Moya,L.A.; unpublished results and manuscript in preparation.

CO ACTIVATION AND REACTIVITY IN ORGANOMETALLIC CHEMISTRY : THEORETICAL STUDIES

A. Dedieu * and S. Nakamura
Laboratoire de Chimie Quantique
E.R. n° 139 du CNRS
Université Louis Pasteur, 4, rue B. Pascal,
F-67000 STRASBOURG, France.

ABSTRACT. SCF and CI calculations are reported for two elementary reactions involved in CO reductive hydrogenation catalytic processes, the nucleophilic addition of a hydride to the carbonyl ligand and the carbonyl insertion into the metal hydride bond. The nucleophilic addition is found to be exothermic and with no activation barrier. The CO insertion reaction is shown to be an hydride migration toward the carbonyl ligand and is characterized by a high energy barrier. The factors which control the corresponding stereochemistries and energy profiles are analyzed in connection with experimental gas phase and solution data. The importance of correlation effects is also discussed.

1- INTRODUCTION

There is currently great interest in homogeneously catalyzed carbon monoxide hydrogenation reactions [1]. Such processes which were at first focused on the conversion of the synthesis gas (CO/H_2 mixture) into liquid fuels are now more directed toward the production of basic chemicals (such as methanol, ethanol, ethylene glycol) or even more elaborate organic products. Transition metal formyl complexes [2] have been very often postulated as key intermediates of these processes, resulting either from the nucleophilic addition of a hydride ion to the carbonyl ligand (reaction 1) or from

$$L_nM\text{-}CO + H^- \longrightarrow L_nM\text{-}C\genfrac{}{}{0pt}{}{\diagup O^-}{\diagdown H} \qquad (1)$$

the so-called CO insertion of a carbonyl ligand into the metal hydride bond (reaction 2).

$$L_n\overset{\displaystyle H}{\overset{|}{M}} - CO \rightleftharpoons L_nM - C\genfrac{}{}{0pt}{}{\diagup H}{\diagdown O} \underset{-L}{\overset{+L}{\rightleftharpoons}} L_{n+1}\,M - C\genfrac{}{}{0pt}{}{\diagup H}{\diagdown O} \qquad (2)$$

Despite an apparent similarity these two reactions turn out to be markedly different : there are many known examples of the addition of various nucleophiles either in solution [2] or in the gas phase [3]. Very

A. Veillard (ed.), Quantum Chemistry: The Challenge of Transition Metals and Coordination Chemistry, 277–299.

recent gas phase experiments indicate that the hydride affinity of various carbonyl transition metal complexes is large, ranging from 45 to 55 kcal/mol[3-5]. The hydride affinity of the bare carbon monoxide amounts to a few kcal/mol only[6] however, therefore showing the activating role of CO coordination.

On the other hand, stoichiometric examples of CO insertion into the metal hydride bond are generally lacking in organometallic chemistry[1,7], at least when leading to a monohapto formyl ligand. This is usually ascribed to the overall endothermicity of reaction (2) (an estimate of + 5 kcal/mol has been given for the $HMn(CO)_5 + CO \rightarrow Mn(CO)_4(CHO)$ reaction[8]). Nothing is known however about the kinetics of such reactions which, despite their intrinsic endothermicity, might still be kinetically feasible and operative in catalytic processes, owing to a rather moderate energy barrier.

The above conclusions refer to a monohapto carbonyl ligand 1. One may wonder whether other coordination modes such as the bridging mode 2 or the dual coordination of both the carbon and oxygen ends[9]

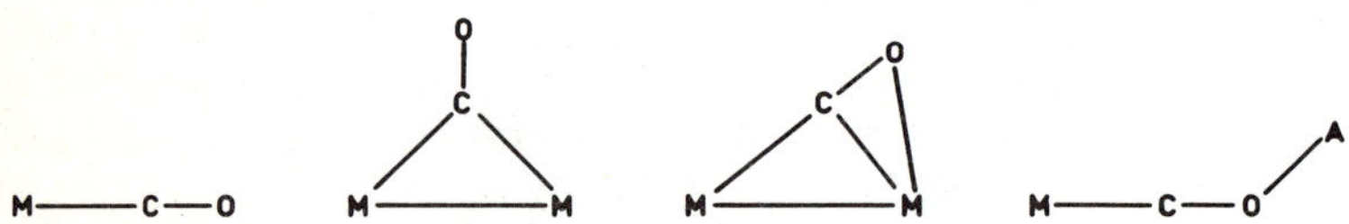

1 2 3 4

(exemplified in 3 and 4) might be more activating toward the reductive hydrogenation of CO. Experimental results are not conclusive in that respect. The presence of a Lewis acid seems to promote the homogeneously catalysed processes[10,11]. The greater efficiency of the addition of various nucleophiles on $Fe_3(CO)_{12}$ (5) compared to $Ru_3(CO)_{12}$, $Os_3(CO)_{12}$ and $Fe(CO)_5$ has been ascribed to the presence of

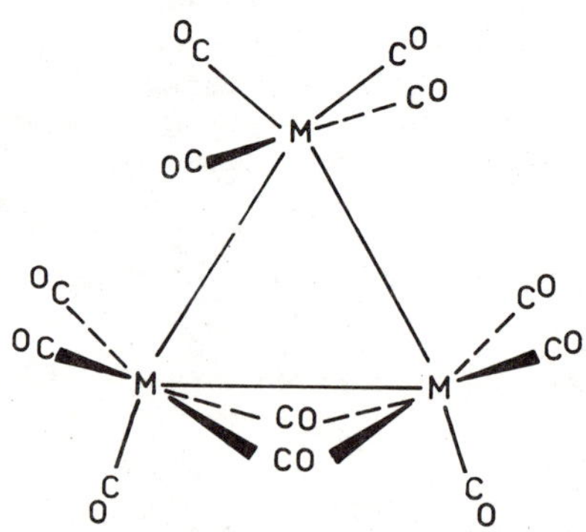

5 (M = Fe)

bridging carbonyl ligands in the former system[12]. On the other hand there is some indirect evidence that the nucleophilic addition of the hydride ion on the <u>cis</u> - $[C_5H_5Fe(CO)]_2(CO)_2$ **6**, takes place on the

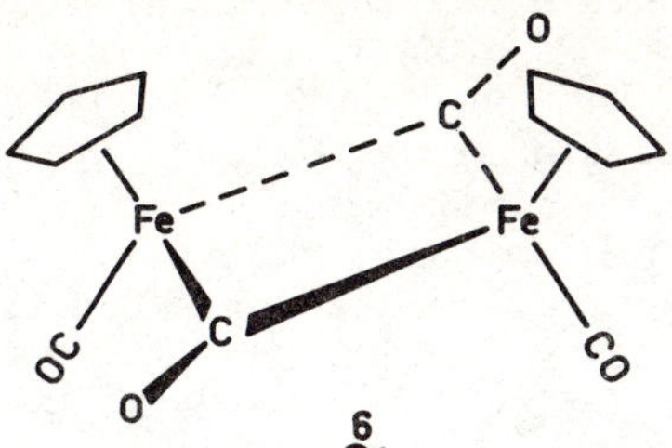

6

terminal carbonyl ligands only[13]. The $Mn_2(CO)_5(dppm)_2$ system **7** which

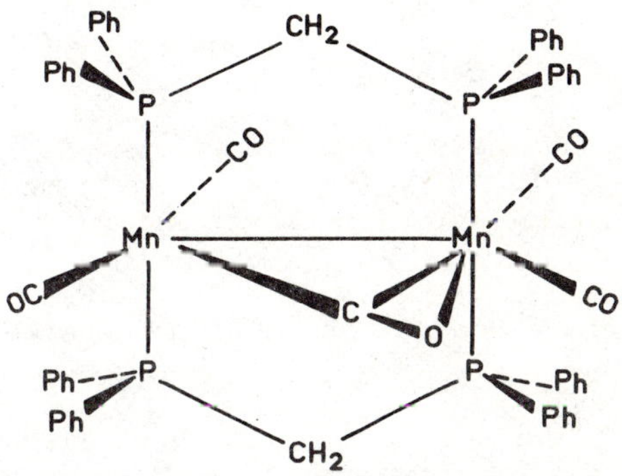

7

is often taken as a prototype of dual coordination in molecular clusters and solids[14] does not seem to be susceptible to nucleophilic attack[15].

It is noteworthy, also, that the bis (pentamethylcyclopentadienyl) thorium complex **8** undergoes an intramolecular insertion step leading

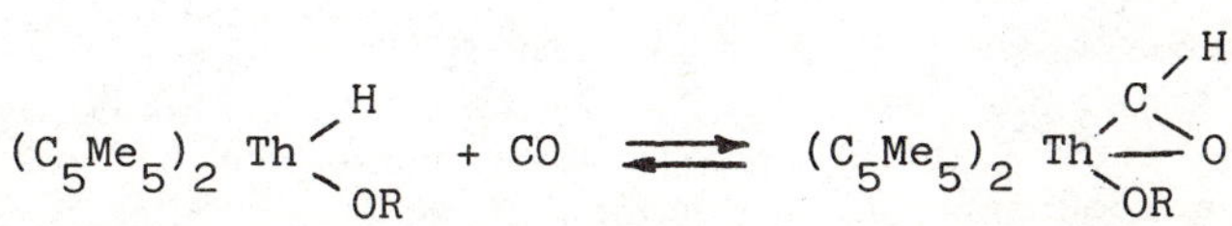

8 **9**

to the dihapto formyl system **9**[16]. The transformation **8** → **9** has been found to be slightly exothermic (-1.8 ± 1.4 kcal/mol) and with a moderate activation energy (11.7 ± 0.4 kcal/mol)[17]. Whether or not the η^2 coordination mode of the formyl moiety is essential for the ease of the insertion step remains to be firmly established.

One may therefore hope that a theoretical study of reactions such as (1) and (2) would lead to a better understanding of the factors which control CO activation and reactivity. More specifically, the points which we have adressed are the following :
(i) why is the monohapto coordination of CO to a transition metal an activating factor for the nucleophilic addition ?
(ii) would other coordination modes be more activating or less activating ?
(iii) what makes the CO insertion into the metal hydride bond difficult to observe. Is it for thermodynamic or for kinetic reasons ?
(iv) Is the CO insertion into the metal hydride bond a true insertion or is it in fact a hydride migration toward the coordinated carbon monoxide as often found, both experimentally[18-21] and theoretically[22-23] for the CO insertion into the metal alkyl bond ?
(v) what is the minimal level of theory for studying these reactions. Does one get a correct picture at the SCF level or should one include correlation effects ?

2 - GENERAL FEATURES OF THE CALCULATIONS

Two prototype reactions have been investigated, the nucleophilic addition of H^- on the $Fe(CO)_5$ system (reaction 3) and the CO insertion

$$H^- + Fe(CO)_5 \longrightarrow Fe(CO)_4(CHO)^- \qquad (3)$$

into the manganese hydride bond of the $HMn(CO)_5$ system (reaction 4).

$$HMn(CO)_5 \rightleftharpoons Mn(CO)_4(CHO) \underset{-CO}{\overset{+CO}{\rightleftharpoons}} Mn(CO)_5(CHO) \qquad (4)$$

The potential energy surfaces of interest for these reactions were determined through LCAO-MO-SCF calculations carried out with the ASTERIX system of programs[24] and using the following basis set : (13, 8, 6) contracted to [5, 3, 3] for the iron and manganese atoms[25], (9, 5) contracted to [3, 2] for the first row atoms[26] and (6) contracted to [3] for the hydrogen atom[27]. The contracted basis set is minimal for the inner shells, double zeta for the valence shells and triple zeta for the 3d of iron and manganese and for the hydrogen atom as well. The triple zeta basis set for hydrogen (BSI)[27] was chosen in order to achieve a balanced description of this atom along the reaction path, especially at the beginning of the reaction where one has either a hydride ion (for the reaction (3)) or substantial hydrido character (in $HMn(CO)_5$). Another basis set (BSII) in which the exponents were optimized for H^- was also tested for the nucleophilic addition reaction. As shown in Table I, the computed value for the hydride affinity (HA) comes in closer agreement with the experimental value. This is traced to an improvement of the description of H^- only, since the energy of CHO^- or $Fe(CO)_4(CHO)^-$ did not change appreciably on going from BSI to BSII.

Table I : Computed and experimental hydride affinities (HA) of $Fe(CO)_5$ and CO

Substrate	Basis set	HA (kcal/mol)
CO	BSI	14.9
	BSI + BSSE[a]	12.5
	BSII	8.1
	BSII + BSSE[a]	6.7
	Experimental[6]	1 - 6
$Fe(CO)_5$	BSI	73.6
	BSI + BSSE[a]	69.2
	BSII	65.3
	Experimental[3]	56.1 ± 1.0

a) The basis set superposition error (BSSE) is taken into account through the counterpoise method[28].

From these computed potential energy surfaces a reaction path and the corresponding energy profile were obtained. The influence of electron correlation was then tested, either for the model reaction $H^- + Fe(CO)_2 \rightarrow Fe(CO)(CHO)^-$ or for the reaction (4) by carrying out CI calculations, using configurations generated by single and double excitations from the reference HF wavefunction[29]. For computational reasons, the CI expansion was based on a limited set of active orbitals, the details of which will be given in the following sections, when referring to a particular calculation.

3 - THE NUCLEOPHILIC ADDITION TO THE CARBONYL LIGAND

The study began with the determination of the reaction path for the addition of the hydride ion on $Fe(CO)_5$. As shown in Figure 1 the approach of the hydride was restricted to the plane containing the two axial and one equatorial ligand. The direction of attack in the early stages of the reaction is found[30] to be almost equidistant from the axial and equatorial ligands, being controlled by forces of electrostatic nature arising from both ligands. The last stage of the reaction is controlled by molecular orbital type interactions[30] and shows a slight preference for the attack on the axial ligand. This corroborates the experimental finding of an axial formyl ligand in the phosphite substituted homolog $(CO)_3[P(ArO)_3]\ Fe(CHO)^-$[31]. The bending of the carbonyl ligand allows, as seen on the computed potential energy

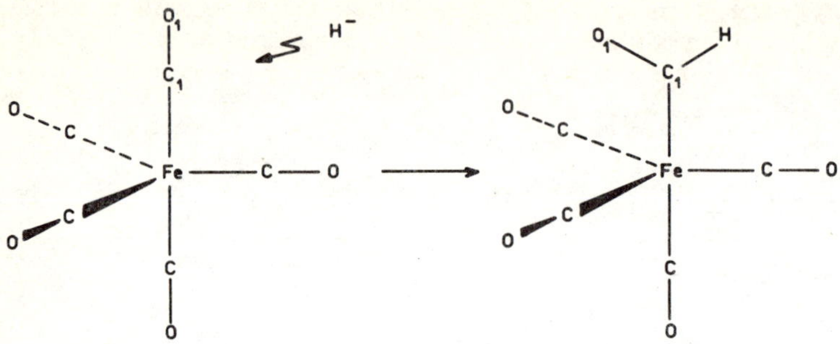

Fig. 1 : Schematic representation of the nucleophilic reaction $H^- + Fe(CO)_5 \rightarrow Fe(CO)_4(CHO)^-$

surface of Figure 2, an approach of the hydride at about 120° of the

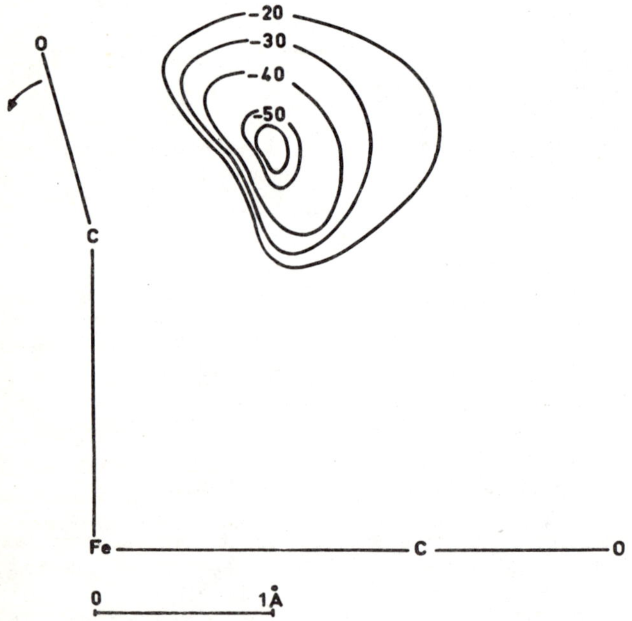

Fig. 2 : Potential energy surface for the approach of the hydride ion, the axial Fe-C-O angle being allowed to relax at each point (but keeping the Fe-C and C-O distances fixed at 1.81 Å and 1.15 Å respectively). The axes are the equatorial and axial Fe-C bonds and the metal is at the origin. The zero of energy is for the separated reactants and relative energies are in kcal/mol. The inner contour corresponds to a relative energy of - 52 kcal/mol.

Fe-C bond. The geometry of the formyl moiety - shown in **10** - is in good agreement with experimental data [31-35] (the C-O bond distance of 1.22 Å was not optimized but taken from the experimental structure of $(CO)_3$ $[P(ArO)_3]$ $Fe(CHO)^-$ [31].

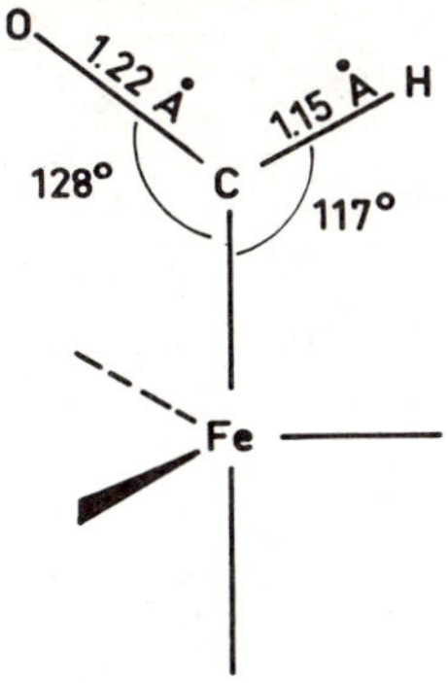

10

A glance at the Figure 2 indicates that the reaction is highly exothermic and that it proceeds without any energy barrier. A similar conclusion has been reached later on from the experimental gas phase study [3]. The much greater exothermicity of the $Fe(CO)_5$ reaction compared to the CO reaction (see Table I) also points to the activation of CO when coordinated to a transition metal. Among the various factors, which would account for this activation, the charge of the carbon atom is not found to be - at least in this case - a dominant one : its positive charge is practically unchanged on going from CO (+0.30) to $Fe(CO)_5$ (+0.32). We rather put forth [30,36,54] some orbital effects : the involvment of the $5\sigma_{CO}$ in the bonding interaction with the metal, which no longer experiences a repulsive interaction with the S_H- orbital, and the presence of a low lying metal orbital (here d_{z^2}) pointing toward the attacked carbonyl. As shown in the schematic correlation diagram of Fig. 3, the mixing of the d_{z^2} orbital with π^*_{CO} and S_H- gives rise to an allylic type three orbital mixing pattern, in which the bonding combination $\pi^*_{CO} + S_H-$ is stabilized further by the empty d_{z^2} orbital. The increased mixing of the d_{z^2} orbital, when one proceeds along the reaction path of the reaction, is best accounted for (see Table II) by the increase of the d_{z^2} orbital population which parallels the decrease of the S_H- orbital population and the increase of the $(\pi+\pi^*)_{CO}$ orbital population.

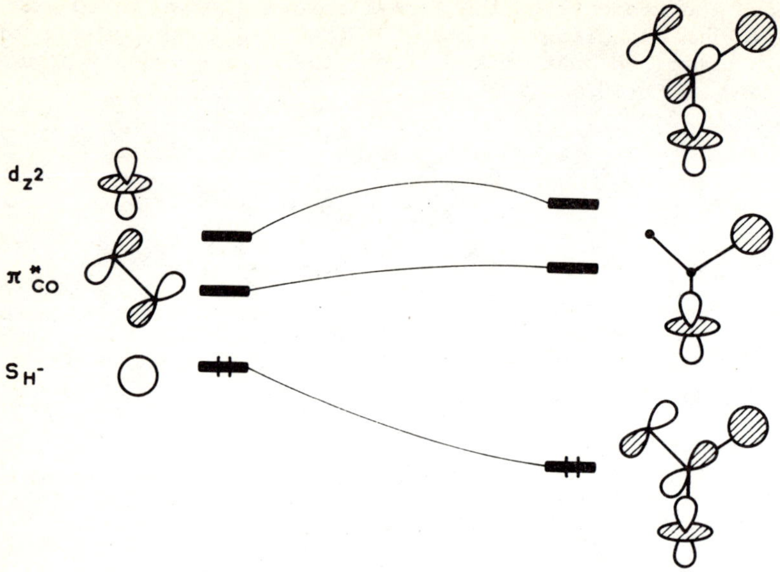

Fig. 3. Schematic correlation diagram for the nucleophilic addition of H^- on $Fe(CO)_5$

Table II : Variation of the population of the d_{z^2}, S_{H^-} and in plane $(\pi+\pi^*)_{CO}$ orbitals along the reaction path for the nucleophilic addition of H^- on $Fe(CO)_5$.

R_{C-H} (Å)	5	2	1.5	1.3	1.15
q_{z^2}	.45	.49	.54	.57	.61
q_s	1.99	1.54	1.20	1.10	1.01
$q_{(\pi+\pi^*)CO}$	2.11	2.48	2.77	2.86	2.95

The above orbital factors can explain many structural and reactivity features of several transition metal formyl or acyl complexes[36]. One may wonder whether they can also explain the reactivity pattern of other CO coordination modes.

From the wave function[37] of the trans-$[C_5H_5Fe(CO)]_2(CO)_2$ system **11** (which should not be very different from the wavefunction of the

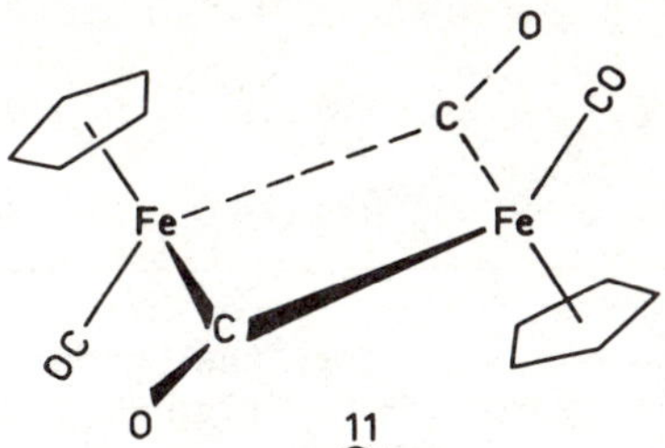

11

cis isomer **6**), it appears that the lowest π^*_{CO} orbitals - the ones which would accomodate the incoming hydride - all belong to the terminal carbonyl ligands. The lowest unoccupied molecular orbital **12**

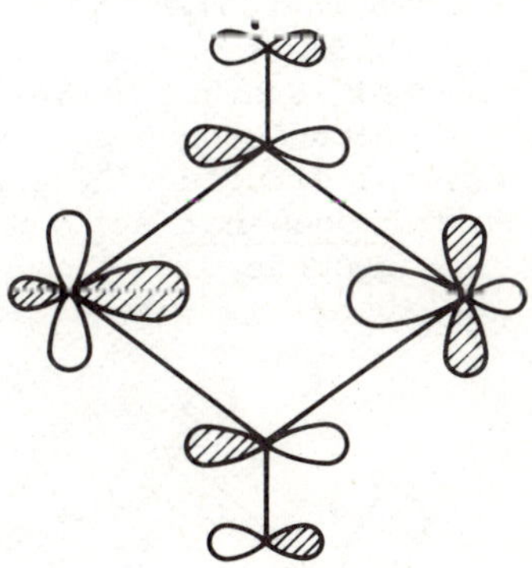

12

is mainly of metal character. Moreover there is no vacant d metal orbital of low energy pointing toward the bridging carbonyl ligands, which might stabilize the S_H- + **12** bonding combination. Finally the positive atomic charge of the carbon atom is found (from a Mulliken population analysis) to be much lower for the bridging carbonyl ligand (see table III) than for the terminal carbonyl ligand. Hence both

Table III : Atomic and overlap population of the terminal and bridging carbonyl ligands in the $[CpFe(CO)]_2(CO)_2$, $Mn_2(CO)_5(PH_3)_4$ and $Mn(CO)_5.BH_3$ systems [38].

System	Terminal CO			Bridging CO		
	q_C	q_O	C-O	q_C	q_O	C-O
$[CpFe(CO)]_2(CO)_2$	+0.49	-0.49	+1.04	+0.23	-0.59	+0.90
$Mn_2(CO)_5(PH_3)_4$	+0.27	-0.24	+1.14	+0.06	-0.36	+0.62
$Mn(CO)_5.BH_3$	+0.28	-0.28	+1.14	+0.46	-0.36	+1.03

orbital and charge effects account for the greater easiness of the nucleophilic attack on the terminal CO ligands[13]. The same type of analysis also led us[36] to propose that the isolobal[39] $Fe_3(CO)_{12}$ system **5** may react preferentially through the CO ligands of the $Fe(CO)_4$ unit and not through the bridging ones as experimentally thought[12].

The dihapto coordination mode **3** in $Mn_2(CO)_5(dppm)_2$ **7** does not activate CO toward the nucleophilic addition either. Here again orbital and charge effects are at work. The lowest π^* carbonyl orbitals which would accomodate the S_H- of the incoming hydride do not belong to the semi-bridging carbonyl ligand and an important charge transfer (in the HOMO) from the metal to the ligand results in a much lower charge of the bridging carbon atom (see Table III)[40]. In fact calculations carried out for the model system $Mn_2(CO)_5(PH_3)_4$ have shown[40] that the semi-bridging ligand, although being often considered as a prototype for dual coordination, does not exhibit any specific interaction between the oxygen end and the second manganese atom (as one would have anticipated from the X-ray crystal structure[14a]). More specifically the corresponding occupied π orbital does not act as a donor orbital for the second metallic atom. The geometry of the system merely results from the maximization of the bonding interaction in the HOMO **13**[40].

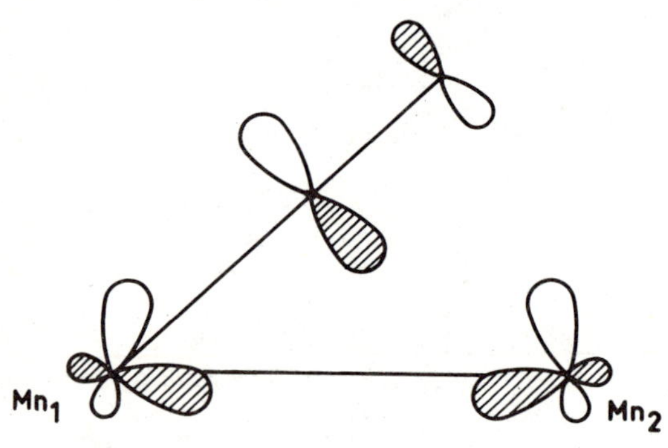

13

It is noteworthy that this bonding pattern, which has now been put forth for the semi-bridging CO ligands in $[(C_5H_5)M(CO)_2]_2$ (M = Cr, Mo, W) by Hall et al[41], has also been corroborated by the recent experimental studies of Deeming et al[15b].

The important charge transfer from the filled d metal orbitals to the empty π^* orbital of the bridging or semi-bridging CO ligand in **11** and **7** respectively, results in a weakening of the CO bond, as exemplified by the corresponding overlap populations[38] (see Table III). These two systems are therefore activated toward CO dissociation but not toward the nucleophilic addition. On the other hand, in $Mn(CO)_5.BH_3$ which one will take as a prototype of **4**, the charge transfer from the bound carbonyl toward the Lewis acid increases the charge separation on CO, making the carbon atom more positive (+0.46 vs +0.28, see Table III). The π^*_{CO} is also lowered in energy[40]. All these factors account for an easier nucleophilic attack, as already pointed out by Shriver[11].

A final question with this theoretical study of the nucleophilic addition to the carbonyl ligand pertains to the validity of the SCF approximation. As shown in the Table I, there is some remaining discrepancy (of the order of a few kcal/mol) between the SCF value and the gas phase experimental value of the exothermicity of the reaction. To check whether this could be traced to the neglect of electron correlation, CI calculations were carried out for the $H^- + Fe(CO)_2 \rightarrow Fe(CO)(CHO)^-$ reaction[42]. In these calculations the active space was built from all valence orbitals (with the exception of the 3σ orbitals of both carbonyl ligand and the 1π orbital of the non reacting CO ligand) and from the complete virtual set of the SCF closed shell wavefunction (freezing again the virtual orbitals corresponding to the 3σ of both CO's). All single and double excitations within this active space and relative to the $^1A_1$ d^8 closed shell reference configuration, i.e. the ground state configuration of $Fe(CO)_5$ and $Fe(CO)_4(CHO)^-$, were considered. Remember that our purpose was not to assess the ground state configuration of $Fe(CO)_2$ and $Fe(CO)(CHO)^-$ but rather to model the $H^- + Fe(CO)_5$ reaction.

As one might have somehow anticipated (since this type of reaction involves interactions between closed shell subspecies only) the inclusion of electron correlation does not seem to change significantly - at least with the basis set used here - the SCF results. The computed exothermicity is 68.8 kcal/mol at the SCF level and 67.5 kcal/mol at the SD-CI level (It increases up to 73.8 kcal/mol when taking into account the contribution of the unlinked quadruple excitations through the Langhoff-Davidson estimation[44]). It remains to see whether a better basis set, e.g. of double zeta quality and with polarization functions, would give better agreement with the experiment. Such calculations are now in progress.

4 - CO INSERTION INTO THE METAL HYDRIDE BOND

As said in the introduction, our main goal was to find out the stereochemistry and the energy profile of the insertion step of reaction (4), namely the rearrangement.

$$HMn(CO)_5 \rightleftarrows Mn(CO)_4(CHO) \qquad (5)$$

In order to assess whether (5) would correspond to either a migration of the hydride or to a migration of the carbonyl ligand, a two-dimensional potential energy surface function of the two critical angles α and β (see **14**) was determined[45]. For this determination the

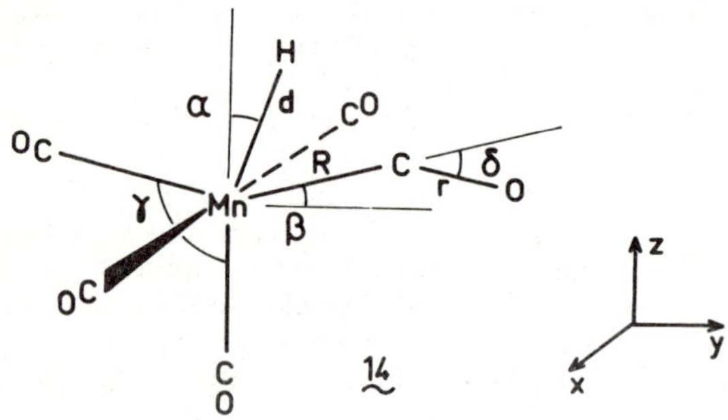

14

two out-of-plane carbonyl ligands were kept frozen (perpendicular to the reaction plane) as were also kept fixed the Mn-C and C-O bond lengths of the non reacting in-plane CO ligands (at their experimental value[46]). For each point of the surface the following parameters : the Mn-H bond length (d); the γ and δ angles and the C-O bond length were successively optimized. This surface is shown in Fig.4. The Mn-C bond length (R) was then optimized for representative points of the minimum energy path of this surface. The successive geometries (A to F) of the $HMn(CO)_5$ system along the resulting reaction path (see Fig.5) clearly show that the process (5) is best described as a hydride migration. This is not surprising since the same result has been obtained experimentally[18,19] and theoretically[47] for the related CO insertion into the manganese-alkyl Mn-R bond of the $RMn(CO)_5$ system. The recent calculations of Morokuma et al[48] for the $Pt(CH_3)(H)(CO)(PH_3)$ complex also point to a CH_3 migration. In the $HMn(CO)_5$ system (and in $RMn(CO)_5$ probably as well) the stereochemical course of (5) is governed by the orbital interactions occuring in the HOMO of the system. This HOMO, is primarily the $S_H - + \pi^*_{CO} - 5\sigma_{CO}$, i.e. the bonding combination of the s orbital of the hydride and the π^* of the carbonyl ligand somewhat destabilized by the $5\sigma_{CO}$. It interacts mainly with the empty $d_{z^2-y^2}$ metal orbital, as shown next page in **15**, when H migrates. This interaction is therefore stabilizing. On the other hand, when CO

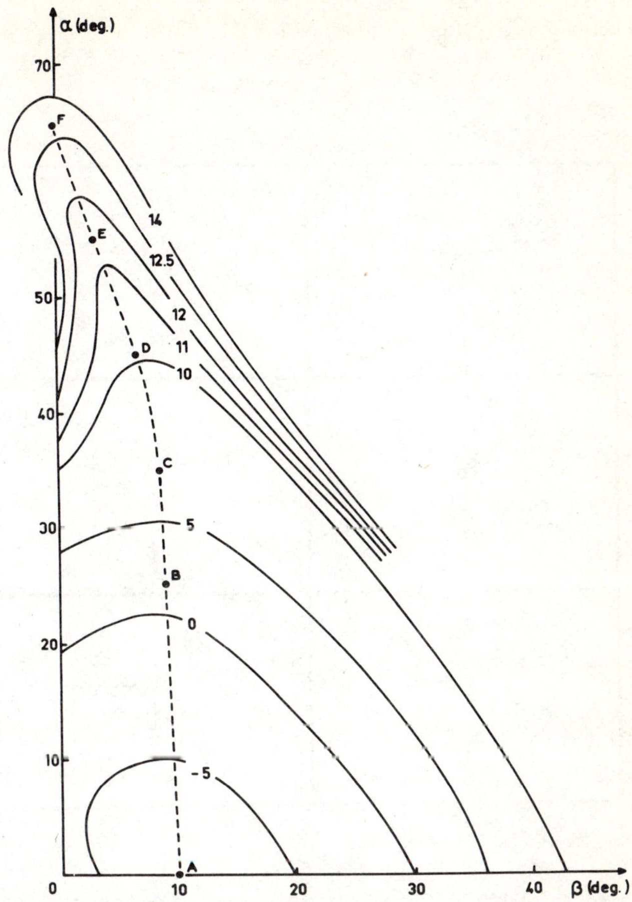

Fig. 4 : SCF potential energy surface as a function of the angles α and β, the Mn-C bond length being kept constant at 1.885 Å. The zero of energy is for α = β = 0 and the relative energies are in kcal/mol.

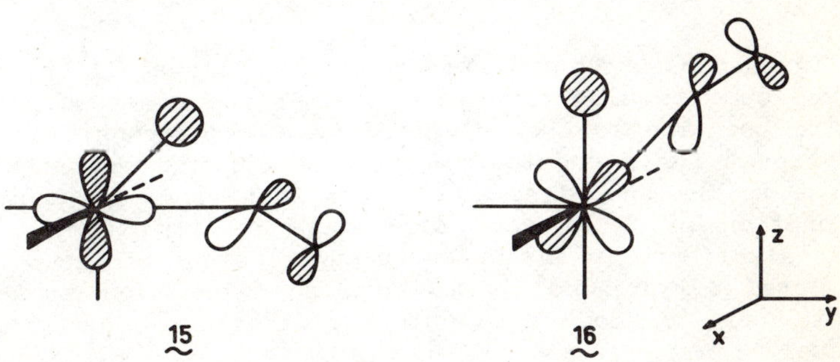

15

16

migrates, see **16**, the main interaction of the HOMO is with the occupied d_{yz} metal orbital and thus a destabilizing one.

One would have also noticed from the Figure 5 that the attacked

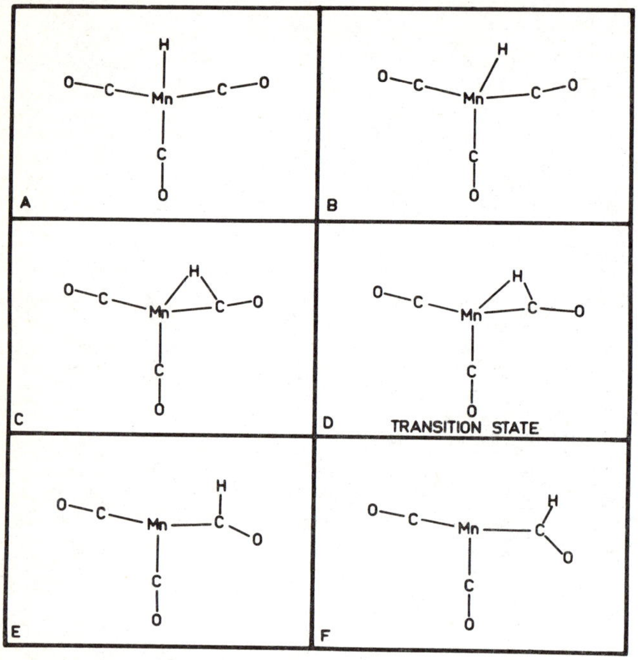

Fig. 5 : Schematic representation of the in-plane geometry changes along the reaction path of the hydride migration in the $HMn(CO)_5$ system. For the sake of clarity the two CO perpendicular to the reaction plane have been omitted.

carbonyl ligand does not move appreciably from its equilibrium position and that the same is also true for the in-plane non-reacting carbonyl ligands. The reaction is characterized by a late transition state (as one would expect for a endothermic process[49]), the geometry of which (D) is close to the one of the five-coordinate formyl intermediate F. The wavefunction of D is also indicative of its formyl character. Finally, as also found in the Extended Hückel study of the CO insertion into the metal alkyl bond[47], neither the transition state nor the intermediate seem to adopt a dihapto coordination mode, at least at the SCF level.

The SCF energy profile (see Figure 6, next page) would indicate

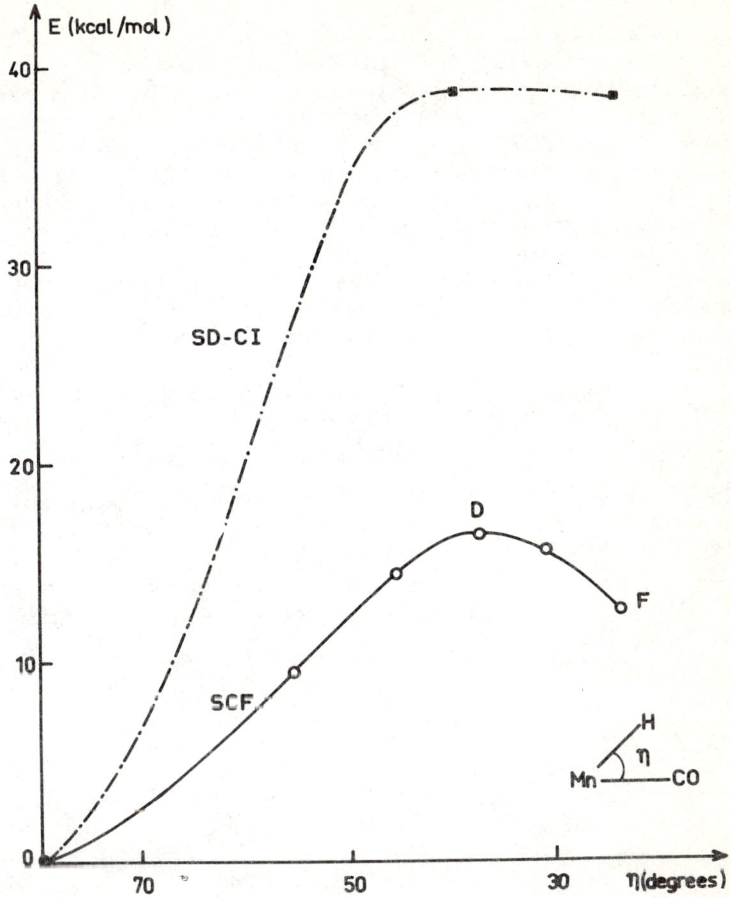

Fig 6. SCF (-o-) and CI (-·-·-) energy profiles for the hydride migration process. The reaction coordinate is the H-Mn-C angle. The zero of energy is for point A (see Fig. 5)

that the reaction is feasible, the energy barrier to overcome amounting to about 15 kcal/mol. The formyl intermediate is also close in energy to the transition state (the barrier between F and D being of the order of 4 kcal/mol). This would therefore account for the difficulty to observe formyl intermediates in such reactions.

One may worry in this case, however, about the neglect of the electron correlation since bond breaking (the Mn-H bond) and bond forming (the C-H bond) do not necessarily involve interactions between closed shell moieties. To assess this point CI calculations were carried out for the three key structures A, D and F of the SCF reaction path. The configurations of the CI expansion were generated by single and double excitations from the HF wavefunction. The CI expansion was based on a set of 41 active orbitals belonging to the atoms of the reaction plane : the 3d, 4d, 4p and 4s of Mn ; the three orbitals associated with the hydrogen atom ; the 1π, 2π, 3π, 4σ, 5σ and 6σ orbitals of the carbonyl ligands (with the exception of the two non reacting carbonyl ligands for which the 4σ and 5σ orbitals were not considered due to computational limitations). Of these 41 orbitals, 12 are doubly occupied in the closed shell configurations : the 3d (t_{2g}) of Mn ; the 1s of the hydride and the 4σ, 5σ and 1π orbitals of the reacting carbonyl ligand and the 1π of the non reacting carbonyl ligands.

A striking result, displayed on Fig.6 and in table IV is the rather important increase in the energy barrier for the forward

Table IV : Relative SCF, CI and MP2 energies (in kcal/mol) of the $HMn(CO)_5$ and HMnCO systems along the SCF reaction path of the hydride migration in $HMn(CO)_5$.

		ΔE_{SCF}	ΔE_{CI}	ΔE_{MP2}
$HMn(CO)_5$	A	0	0	-
	D	14.4	38.6	-
	F	10.5	38.4	-
HMn(CO)	A	0	0	0
	D	11.4	24.7	35.8
	F	2.3	23.3	35.5

reaction, on going from SCF to CI. One would have rather expected a decrease such as the one found for the reverse reaction (the barrier between F and D has almost disappeared under the influence of the CI). In order to check whether these features are not due to the truncation of the active space and/or to an improper choice of the active orbitals, SD-CI were also carried out for the bare HMnCO system with an imposed closed-shell $^1A'$ state arising from the same electronic configuration as in $HMn(CO)_5$, and using the corresponding geometries of points A, D and F. In this case, however, the active space was built from all the valence orbitals and from the complete virtual set of the SCF closed-shell wavefunctions (an analysis of the wavefunctions of HMnCO and $HMn(CO)_5$ indicated that the orbitals of both systems are rather similar). Here again the same trend is followed, that is an increase in the energy barrier on going from HMnCO (point A) to point D. Furthermore, it is noteworthy that MP2 calculations [50] do not differ in that respect from the SD-CI calculations. It would be of course interesting to have the results of MP2 calculations for the $HMn(CO)_5$ system. Calculations are now in progress on this point.

The present finding does not seem to be specific to a migrating hydride group : preliminary pilot calculations carried out for the $(CH_3)MnCO$ and $Mn(COCH_3)$ systems also indicate that the energy difference between these two systems is significantly increased on going from the SCF to the MP2 level[50].

We tentatively ascribe these results to an unbalanced description at the SCF level of the Mn-CO bond and Mn-CHO bond, as shown schematically in Figure 7. The Mn-C bond in $HMn(CO)_5$ and HMnCO involves a

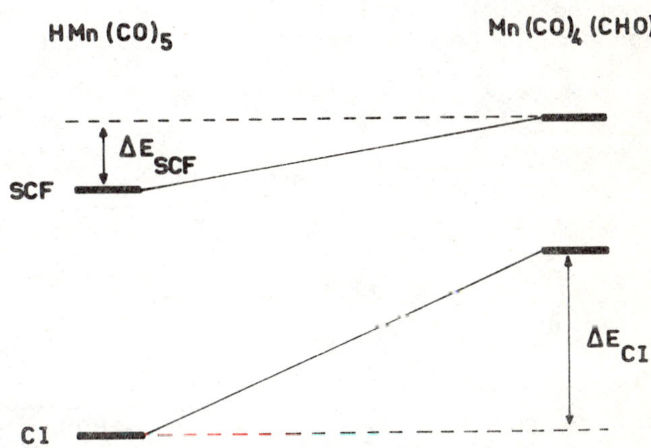

Fig. 7 : Schematic representation of the energy changes on going from SCF to CI for the $HMn(CO)_5 \rightarrow Mn(CO)_4(CHO)$ reaction

rather important degree of π back-bonding, in-plane and out of plane[51]. In the Mn-CHO bond this π back-bonding is out of plane only and is <u>smaller</u> than in the Mn-CO bond [51] (the less covalent character of the Mn-CHO bond induces a greater positive charge on the metal atom in $Mn(CO)_4(CHO)$ which in turn hampers the π back-donation). Since the π back donation is poorly described at the SCF level, the energy lowering brought up by the CI calculations is greater for the $HMn(CO)_5$ than for $Mn(CO)_4(CHO)$ or for the transition state (which, as we have already mentioned, is formyl-like).

There are several features which are consistent with this

suggestion. We first note that the Cr-C bond strength in $Cr(CO)_6$ (which is also a d^6 system) has been computed to be 17.1 kcal/mol at the SCF level and 33 kcal/mol at the CI level [52] (with a basis set and a CI expansion similar to ours). The difference between the SCF and the CI values (16 kcal/mol) provides an upper bound for the error associated with the unbalanced description of the metal-CO and of the metal-CHO bonds.

We also note that a good agreement is obtained between the SCF value [45] (3.7 kcal/mol) and the experimental estimate [8] (5 kcal/mol) for the endothermicity of the whole process $HMn(CO)_5 + CO \rightarrow Mn(CO)_5(CHO)$. This not surprising since both $HMn(CO)_5$ and $Mn(CO)_5(CHO)$ have the same number of carbonyl ligands. Another comparison with other experimental thermodynamic data is also interesting : the CO dissociation energy for iron tetracarbonyl acyl anions $Fe(CO)_4(COR)^-$ has been estimated to be about 34 kcal/mol in the gas phase [4]. Assuming the same value for CO dissociation in the neutral $Mn(CO)_5(CHO)$ system and adding it to the experimental estimate for the endothermicity of the $HMn(CO)_5 + CO \rightarrow Mn(CO)_5(CHO)$ reaction would lead (see the Scheme below) to an energy difference of about 39 kcal/mol between $HMn(CO)_5$ and $Mn(CO)_4(CHO)$. The

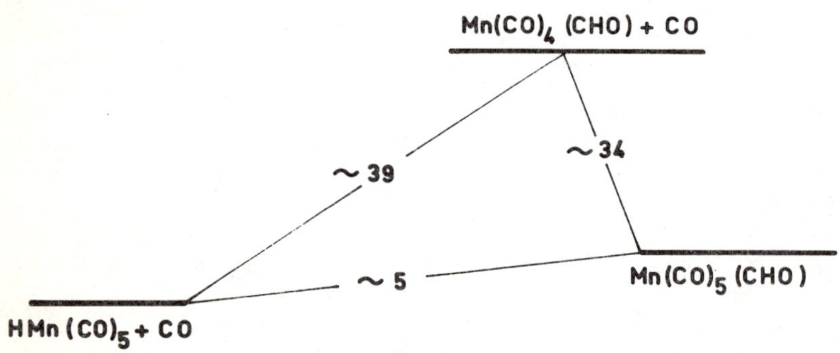

agreement with the computed value, although being somewhat fortituous, is probably meaningful.

Finally, the analysis of the CI expansions for the points A, D and F of the reaction path (see Table V) indicates that the most important configurations (i.e. with a CI coefficient greater than 0.05) involve π back donation to all carbonyl ligands (including the one which undergoes the insertion process) in $HMn(CO)_5$ but only the <u>non reacting</u> carbonyl ligands in the transition state D and in the formyl intermediate F.

There is one argument against our proposal however : one would also expect that a similar unbalanced description of the metal-CO and metal-CHO bonds would decrease the exothermicity of the nucleophilic addition of the hydride to the carbonyl ligand when going from a SCF calculation to a CI calculation. But this turns out not to be the case, at least for the $Fe(CO)_2 + H^- \rightarrow Fe(CO)(CHO)^-$ reaction which we have

TABLE V : Most important configurations [a,b,c] for the three points A, D and F of the reaction path of the $HMn(CO)_5 \rightarrow Mn(CO)_4(CHO)$ reaction.

A		D		F	
Excitation	Coefficient	Excitation	Coefficient	Excitation	Coefficient
-	.917	-	.920	-	.922
$(d_{xy})^2 \rightarrow (\pi_a^* - \pi_t^*)^2$	.065	$1s + \pi^*_a - 5\sigma_a \rightarrow d_{z^2}$	.075	$1s + \pi^*_a - 5\sigma_a \rightarrow 6\sigma_a + 6\sigma_c$	.066
$d_{yz}, d_{xz} \rightarrow \pi_c^*, \pi_c^*$	.061[d]	$d_{xy}, d_{xz} \rightarrow \pi_t^*, \pi_c^*$	.059	$d_{xy}, d_{xz} \rightarrow \pi_t^*, \pi_c^*$	.059
$(d_{yz})^2 \rightarrow (\pi_c^*)^2$	.060	$(\pi_a)^2 \rightarrow (\pi_a^* + \pi_t^*)(\pi_a^* - \pi_t^*)$	.056	$d_{xz} \rightarrow \pi_c^*$	.054
$(d_{xz})^2 \rightarrow \pi_c^*, (\pi_a^* - \pi_t^*)$	.060	$d_{yz} \rightarrow \pi_t^*$	.056	$(\pi_a)^2 \rightarrow (\pi_a^* + \pi_t^*)(\pi_a^* - \pi_t^*)$	.054
$d_{xy}, d_{xz} \rightarrow \pi_c^*, (\pi_a^* - \pi_t^*)$	.059	$d_{xz} \rightarrow \pi_c^*$	.055		
$d_{xy}, d_{yz} \rightarrow \pi_c^*, (\pi_a^* - \pi_t^*)$	.058	$1s + \pi^*_a \rightarrow 6\sigma_a + 6\sigma_c$	.053		
$1s + d_{z^2} \rightarrow 4d_{z^2}$	.056	$(d_{xy})^2 \rightarrow (\pi_t^*)^2$	.052		
$d_{xy} \rightarrow \pi_a^* - \pi_t^*$	.056	$d_{xy}, d_{yz} \rightarrow \pi_t^*, \pi_t^*$	.052[d]		
$(d_{xz})^2 \rightarrow (\pi_c^*)^2$	.052				
$1s + d_{z^2} \rightarrow 2s$	.051				
$d_{yz} \rightarrow \pi_c^*$	.051				

a) Only configurations the coefficients of which are greater than 0.05 are reported ; b) The reference configuration is the SCF ground state configuration ; c) In this table subscripts a, c and t refer to the attacked, cis to the attacked and trans to the attacked CO ligands respectively. s stands for the s orbital of hydrogen. d) both in-plane and out-of-plane π^* orbitals of the same CO ligand are involved.

reported. One might argue that π back donation to the carbonyl should decrease on going to the left of the periodic table. On the other hand, it should be greater for a metal being formally in a zero oxidation state (Fe in $Fe(CO)_2$) than for a metal with the +1 oxidation state (Mn in $HMn(CO)_5$). Additional studies are therefore necessary to clarify this problem, especially to check the relative importance of the non-dynamical[53] correlation effects mentioned above.

At the end of this section and going back to the chemistry of such processes, one may therefore conclude that the failure to observe CO insertion into a metal hydride bond leading to a monohapto formyl ligand is well accounted for by a high energy barrier to overcome and by the close proximity of the transition state and the five coordinate intermediate. It remains to see whether a dihapto coordination mode of the formyl moiety both in the transition state and in the intermediate (for systems such as 9) would lower the energy barrier and the overall endothermicity of the process.

5 - CONCLUSION

There are several conclusions which may be drawn from this study : SCF calculations appear to provide a correct picture - at least qualitatively - of many geometric and energetic features of processes involving carbon monoxide activation and reactivity. The factors which account for the stereochemistry of the nucleophilic addition of the hydride ion to the carbonyl ligand and the insertion of the carbonyl ligand into the metal hydride bond have been unravelled. The last reaction has been found to be an hydride migration toward the carbonyl ligand rather than a true CO insertion.

The nucleophilic addition reaction is a strongly exothermic reaction without any energy barrier. On the other hand the intramolecular hydride migration is an endothermic process which has to overcome a rather important energy barrier. This explains the scarcity of this process - at least when leading to a monohapto formyl ligand - in organometallic chemistry.

A rather interesting point is that recent gas phase experimental data, which have been obtained for the same reactions, allow a comparison between theory and experiment. The influence of electron correlation appears to be especially important in order to get a reliable estimate of the activation barrier for the CO insertion reaction into the Mn-H bond of the $HMn(CO)_5$ system. Additional calculations are currently under way to check whether this result - which differ from the one obtained in Pt(II) square planar complexes[48] - may be ascribed to different non dynamical correlation effects (e.g. those involving the π back donation to the CO ligand, which might be more crucial for Mn(I) than for Pt(II) or Pd(II)[56]).

The influence of the basis set, which may be checked through the comparison with the experimental data, should not be overlooked, especially when the ultimate goal of theoretical investigations is to get quantitative accuracy and a theoretical design of homogeneous catalysts.

ACKNOWLEDGMENTS

Calculations have been carried out partly on the UNIVAC 1110 of the Centre de Calcul du CNRS at Strasbourg Cronenbourg and partly on the CRAY 1 of the CCVR. We thank the staff of both computer centers for their cooperation. We also thank Profs K. Morukuma, R.R. Squires and Dr. H. Berke for communicating us results prior to publication. We are particularly grateful to Drs M. Bénard and S. Sakaki for helpful discussions. This work has been supported by the GRECO CO of the CNRS.

REFERENCES AND NOTES

1 - For a review see for instance :
a) C. Masters, Adv. Organomet. Chem., **17**, 61 (1979)
b) E.L. Mutterties and J. Stein, Chem. Rev. **79**, 479 (1979)
c) C.K. Rofer-De Poorter, Chem. Rev. **81**, 447 (1981)
d) W.A. Herrmann, Angew. Chem. Int. Ed. Engl. **21**, 117 (1982)
e) J.R. Blackborow, R.J. Daroda and G. Wilkinson, Coord. Chem, Rev., **43**, 17 (1982)
See also
f) R. Eisenberg and D.E. Hendriksen, Adv. Catalysis, **28**, 79 (1979)
g) D.R. Fahey, J. Am. Chem. Soc., **103**, 136 (1981)
2 - J.A. Gladysz, Adv. Organomet. Chem., **20**, 1 (1982)
3 - K.R. Lane, L. Sallans and R.R. Squires, J. Am. Chem. Soc., **107**, 5369 (1985)
4 - K.R. Lane, L. Sallans and R.R. Squires, submitted for publication
5 - R.R. Squires, private communication
6 - J. Chandrasekhar, J.G. Andrade and P.v.R. Schleyer, J. Am. Chem. Soc., **103**, 5612 (1981)
7 - D.L. Grimett, J.A. Labinger, J.N. Bonfiglio, S.T. Masuo, E. Shearin and J.S. Miller, Organomet. **2**, 1325 (1983)
8 - J.A. Connor, M.T. Zafarani-Moattar, J. Bickerton, N.I. El-Saïed, S. Suradi, R. Carson, G. Al Takhin and H.A. Skinner, Organomet. **1**, 1166 (1982)
9 - For a review, see C.P. Horwitz and D.F. Shriver, Adv. Organomet. Chem. **23**, 219 (1984)
10 - G.C. Demitras and E.L. Mutterties, J. Am. Chem. Soc., **99**, 2796 (1977)
11 - D.F. Shriver, ACS Symp. Ser. **152**, 1 (1981)
12 - a) D.C. Cross and P.C. Ford, Inorg. Chem., **21**, 1702 (1982)
b) D.C. Cross and P.C. Ford, J. Am. Chem. Soc., **107**, 585 (1985)
13 - M.M. Harris, J.D. Atwood, M.E. Wright and G.O. Nelson, Inorg. Chem., **21**, 2117 (1982)
14 - a) C.J. Commons and B.F. Hoskins, Aust. J. Chem., **28**, 1663 (1975)
b) R. Colton and C.J. Commons, Aust. J. Chem., **28**, 1673 (1975)
c) K.G. Caulton and P. Adair, J. Organomet. Chem., **114**, C11 (1976)
d) J.A. Marsella and K.G. Caulton, Organomet. **1**, 274 (1982)
15 - a) H.C. Aspinall and A.J. Deeming, J. Chem. Soc., Chem. Commun, 724 (1981)
b) A.J. Deeming and S. Donovan-Mtunzi, Organomet., **4**, 693 (1985)

16 - P.J. Fagan, K.G. Moloy and T.J. Marks, J. Am. Chem. Soc., **103**, 6959 (1981)
17 - K.G. Moloy and T.J. Marks, J. Am. Chem. Soc. **106**, 7051 (1984)
18 - a) K. Noack and F. Calderazzo, J. Organomet. Chem. **10**, 101 (1967)
b) K. Noack, M. Ruch and F. Calderazzo, Inorg. Chem. **7**, 345 (1968)
19 - T.C. Flood, J.E. Jensen and J.A. Statler, J. Am. Chem. Soc., **103**, 4410 (1981)
20 - F. Ozawa and A. Yamamoto, Chem. Lett., 289 (1981)
21 - For other reaction pathways see
a) R.W. Glyde and R.J. Mawby, Inorg. Chim. Acta, **5**, 317 (1971)
b) T.C. Flood and K.D. Campbell, J. Am. Chem. Soc., **106**, 2853 (1984)
22 - H. Berke and R. Hoffmann, J. Am. Chem. Soc., **100**, 7224 (1978)
23 - S. Sakaki, K. Kitaura, K. Morokuma and K. Ohkubo, J. Am. Chem. Soc., **105**, 2280 (1983)
24 - M. Bénard, A. Dedieu, J. Demuynck, M.M. Rohmer, A. Strich, A. Veillard and R. Wiest, ASTERIX, a system of programs, unpublished work
25 - I. Hyla-Kryspin, J. Demuynck, A. Strich and M. Bénard, J. Chem. Phys., **75**, 3954 (1981)
26 - S. Huzinaga, Approximate Atomic Functions, Technical Report, University of Alberta (1971)
27 - S. Huzinaga, J. Chem. Phys., **42**, 1293 (1965)
28 - a) S.F. Boys and F. Bernardi, Mol. Phys., **19**, 553 (1970)
b) W. Kolos, Theoret. Chim. Acta **51**, 219 (1979)
29 - The CI program developped originally by Brooks and Schaefer and adapted for the IBM computer by F. Brown and I. Shavitt was interfaced for use in conjunction with the ASTERIX system of programs for the Univac 1110 by J. Demuynck and R. Wiest and for the CRAY-1 computer by C. Daniel and R. Wiest.
30 - S. Nakamura and A. Dedieu, Theoret. Chim. Acta **61**, 587 (1982)
31 - C.P. Casey, S.M. Neumann, M.A. Andrews and D.R. McAlister, Pure Appl. Chem. **52**, 625 (1980)
32 - W.K. Wong, W. Tam, C.E. Strouse and J.A. Gladysz. J. Chem. Soc, Chem. Commun, 531 (1979)
33 - B.B. Wayland, B.A. Woods and R. Pierce, J. Amer. Chem. Soc., **104**, 302 (1982)
34 - G. Smith, D.J. Cole-Hamilton, M. Thornton-Pett and M.B. Hurthouse, J. Chem. Soc., Dalton Trans. 2501 (1983)
35 - H. Berke, G. Huttner, O. Scheidsteger and G. Weiler, Angew. Chem. Int. Ed. Engl. **23**, 735 (1984)
36 - A. Dedieu and S. Nakamura, Nouv. J. Chimie **8**, 317 (1984)
37 - M. Bénard, Inorg. Chem. **18**, 2782 (1979)
38 - One should not compare the results of the Mulliken population analysis for the $[Cp\ Fe(CO)]_2\ (CO)_2$ system with those of the $Mn_2(CO)_5(PH_3)_4$ and $Mn(CO)_5.BH$ since they have been computed with different basis sets : (11, 7,5/8, 4/ 4) [4, 3, 2/3, 2/2] for the diiron sytem and (13, 8, 6/9, 5/6) [5, 3, 3/3, 2/3] for the manganese systems.
39 - R. Hoffmann, Angew. Chem. Int. Ed. Engl. **21**, 711 (1982)

40 - M. Bénard, A. Dedieu and S. Nakamura, Nouv. J. Chimie, **8**, 149 (1984)

41 - B.J. Morris-Sherwood, C.B. Powell and M.B. Hall, J. Am. Chem. Soc., **106**, 5079 (1984)

42 - In order to account for the size inconsistency error[43], calculations were carried out with an hydride located at 50 Å of the attacked carbon atom for the reactant side $Fe(CO)_2 + H^-$

43 - J. A. Pople, J.S. Binkley and R. Seeger, Int. J. Quant. Chem. Symp, **10**, 1 (1976)

44 - S.R. Langhoff, S.T. Elbert and E.R. Davidson, Intern. J. Quantum Chemistry, **7**, 999 (1973)

45 - S. Nakamura and A. Dedieu, Chem. Phys. Letters, **111**, 243 (1984)

46 - E.A. McNeill and F.R. Scholer, J. Am. Chem. Soc., **99**, 6243 (1977)

47 - H. Berke and R. Hoffmann, J. Am. Chem. Soc., **100**, 7224 (1978)

48 - a) K. Morokuma, private communication
b) N. Koga and K. Morokuma, this volume

49 - G.S. Hammond, J. Am. Chem. Soc., **77**, 334 (1955)

50 - S. Sakaki and A. Dedieu, to be published

51 - S. Nakamura, Thèse de Doctorat d'Etat, Strasbourg 1984

52 - D. Moncrieff, P.C. Ford, I.H. Hillier and V. Saunders, J. Chem. Soc, Chem. Commun, 1108 (1983)

53 - See for instance I. Shavitt in 'Advanced theories and computational approaches to the electronic structure of molecules' NATO ASI, Vol 133, page 185, D. Reidel, Dordrecht (1985)

54 - NOTE ADDED IN PROOF : the activating role, in the nucleophilic addition, of an empty d metal orbital pointing toward the attacked carbonyl ligand is best exemplified by the results of a comparative calculation carried out for the model reactions $H^- + M(CO)_4 \rightarrow M(CO)_3(CHO)^-$, M being either iron or nickel[55]. For the $M(CO)_4$ system a C_{3v} geometry having the z axis as principal axis was chosen in order to single out the d_{z^2} orbital which is therefore doubly occupied in $Ni(CO)_4$ and empty in $Fe(CO)_4$. Same bond lengths and bond angles were used in both reactions. The computed SCF exothermicity is 56.2 kcal/mol for the $H^- + Ni(CO)_4$ reaction and 61.0 kcal/mol for the $H^- + Fe(CO)_4$ reaction. The difference, 4.8 kcal/mol, therefore reflects the extra stabilization brought up by the in-phase mixing of the empty d_{z^2} orbital into the $\pi^*_{CO} + s_H$ - bonding combination

55 - The calculation for the $Ni(CO)_4$ reaction was suggested to us by Prof. M. B. Hall.

56 - There is some experimental evidence that Mn(I) complexes involve a rather important degree of π back-bonding[57] whereas Pt(II) and especially Pd(II) are characterized by a weak tendency to π back-donate[58].

57 - R. Gross and W. Kaim, Angew. Chem. Int. Ed. Engl., **23**, 614 (1984)

58 - F. Calderazzo and D.B. Dell'Amico, Inorg. Chem., **20**, 1310 (1981).

STRUCTURAL DISTORTIONS AND ACTIVATION OF A C-H BOND IN d^0 ELECTRON DEFICIENT ALKYL TRANSITION METAL COMPLEXES

A.Demolliens, Y. Jean, O. Eisenstein
Laboratoire de Chimie Théorique UA 506
Bâtiment 490
Centre de Paris-Sud
91405 Orsay
France

ABSTRACT. The main structural features of d^0 hexacoordinated alkyl complexes are analyzed by means of Extended Hückel calculations. The geometrical distortions from the octahedral field around the metal as well as the distortions of the alkyl ligand (agostic group) are discussed. It is shown that the octahedral ligand field around the metal is not a minimum on the potential energy surface but that the actual minimum is very sensitive to the nature of the ligands. It is also shown that electron deficiency is not sufficient by itself to favor an agostic group. We thus analyze tetrahedral complexes and show why they do not have agostic groups.

Current interests in aliphatic C-H bond activation by transition metal arise from the need for homogeneous catalytic systems capable of selectively functionalizing aliphatic hydrocarbons. Although C-H bonds of a hydrocarbon are considered as chemically inert, numerous reactions involve the insertion of a transition metal into a C-H bond. Intramolecular reactions such as α and β eliminations commonly occur and are responsible for the instability of many alkyl-metal complexes[1]. Despite the great number of experimental facts, the factors that favor an α or β elimination are still difficult to establish. A great step forward has been achieved by the recent discovery of electron-deficient transition metal complexes that exhibit in their ground state an interaction between the metal center and a nearby C-H bond[2]. In such complexes the C-H bond has been considered as an electron donor group toward the electron-deficient metal center. The group that brings a C-H bond within bonding distance to the metal has been called 'agostic' by Brookart and Green[2].

Complexes having an agostic group can be divided in two classes. In one class, the M..H..C interaction can be geometrically achieved without any noticeable distortion of the C-H containing ligand geometry (1 is the first of such complexes characterized by neutron diffraction[3]). No strain is built in the ligand and the stabilization energy is expected to result from the partial electron donation from the C-H bond to the

A. Veillard (ed.), Quantum Chemistry: The Challenge of Transition Metals and Coordination Chemistry, 301–317

metal centre. Some of these complexes present interesting fluxional behavior[4]. The second class of compounds presents some fascinating structural aspects. The M..H..C interaction is made at the expense of a strong geometrical distortion of the organic ligand. In this category, one finds the 12-electron complexes 2 and 3.

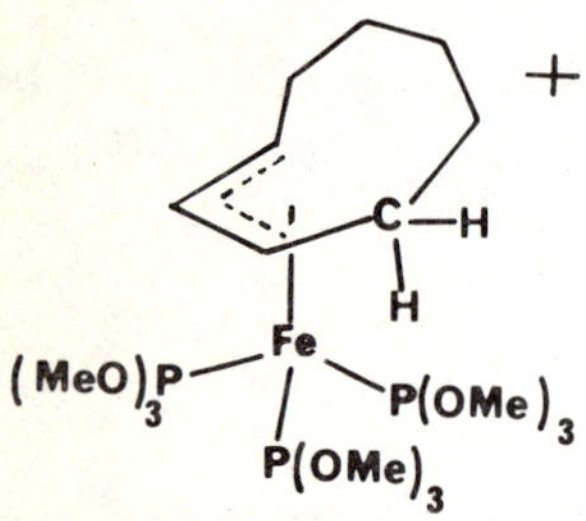

1

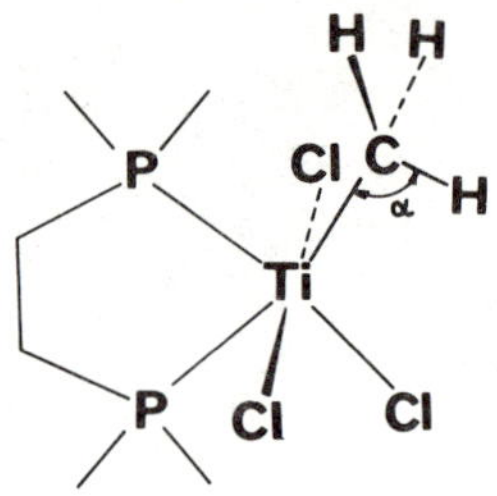

2

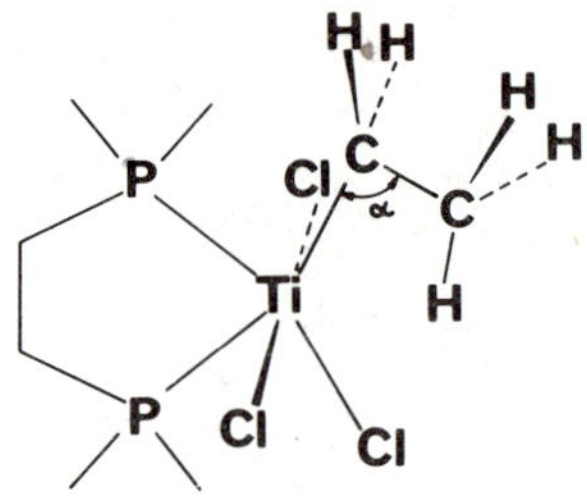

3

The Ti-C-H angle in 2 is 93.7° from neutron diffraction studies and the Ti-C-C angle in **3** is 86°[5]. Furthermore the alkyl distortion is not the only structural characteristic of these complexes. One can also notice a large deviation from the ideal octahedral field around the metal centre, both in 2 and 3. While two Ti-Cl bonds remain almost exactly perpendicular to the alkyl distortion plane, ligand angles in that plane are far from the ideal 90° value. In particular, the C-Ti-Cl angle is larger (114° in **2** and 128.4° in **3**) while the opposite phosphines angle is only 75°. It may be asked whether or not there is a relationship between the agostic nature of the alkyl group and the distortion around the metal away from the octahedral field. As a matter of fact, the first idea that comes to mind is that steric requirements impose such distortion in order to accomodate the agostic group. A very different point of view can also be taken. The octahedral structure is the optimum structure for a d^6 hexacoordinated complex. Removal of electrons from the metal can lead to a different structure.

There is another interesting property of these strongly electron - deficient complexes. Electron deficiency is a necessity to observe the agostic group (all systems that have agostic groups do not fulfill the 18-electron rule). However it is important to realize that electron deficiency at the metal is not sufficient by itself to give rise to an agostic group. Consider Cl_3TiCH_3, 4. Although it is only an 8-electron complex, there is no evidence for a Ti-C-H interaction. The methyl group is slightly flattened with all Ti-C-H angles equal to 104°[2]. Similarly the high spin 13-electron Manganese complex 5 does not present any Mn-C-H interaction[6].

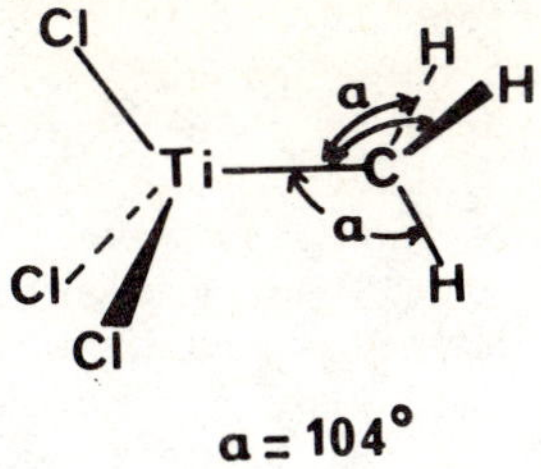

4

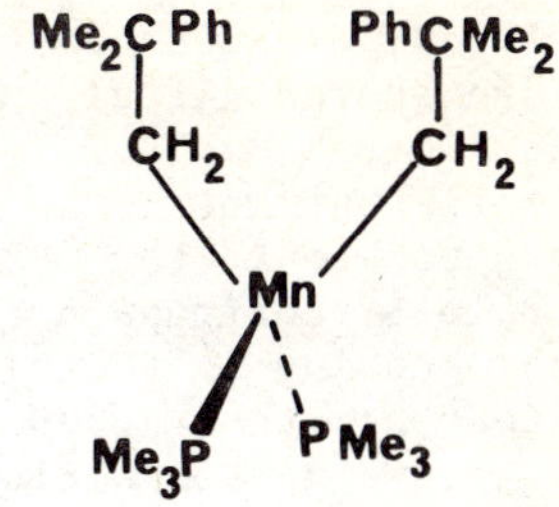

5

The relation between the presence of an agostic group and the potential activation of a C-H bond is nicely evidenced by the following set of observations obtained on Cobalt complexes. In complex 6 structural assignements shows the presence of a β hydrogen getting close to the metal (1.46 Å) and a Co-C_α-C_β angle equal to 74°[7]. Replacement of PPh_3 by an ethylene ligand leads to a fluxional system for which spectroscopic studies are consistent with the equilibrium shown in 7[8]. Finally replacement of PPh_3 by $P(OPh)_3$ leads to a system able to polymerize ethylene[9].

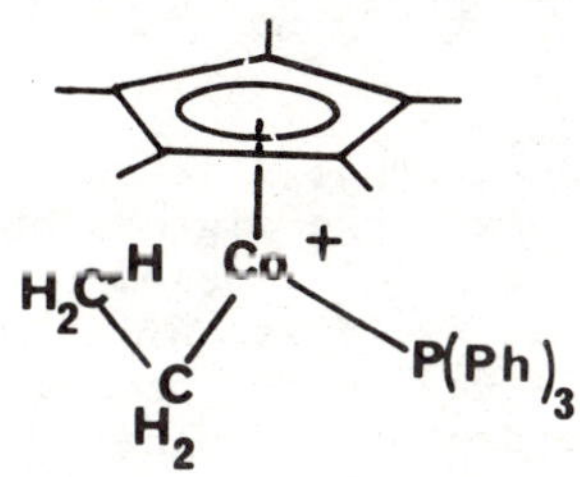

6

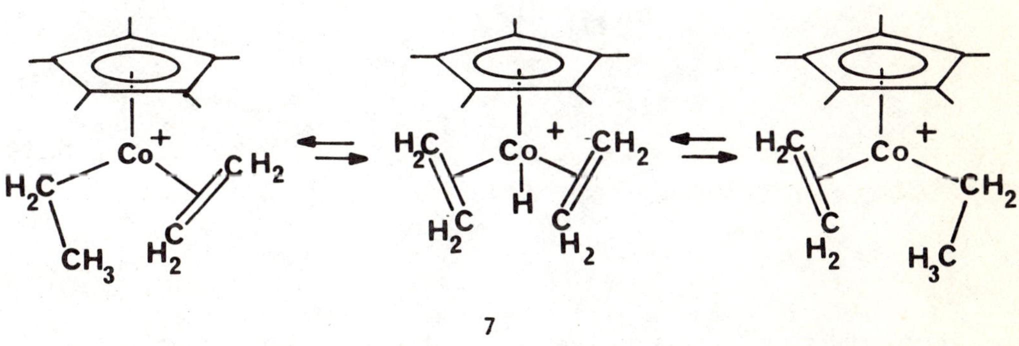

7

Theoretical studies of structural distortions in electron deficient complexes are scarce. Ittel and co-workers have shown by EHT calculations that a M..H..C interaction is favored in a 16-electron η^3 alkenyl complex but not in an 18-electron one[10]. Recent studies on the same systems show that the metal..H..C bond does not play a very important role[11]. Hoffmann and co-workers have analyzed similar distortions occuring in tantalum alkylidenes complexes and have predicted that a methyl group could in principle distort[12]. Morokuma and co-workers have done ab-initio pseudopotential calculations on models of the titanium complexes **2**[13] and **3**[14] and have obtained optimized structural parameters in good agreement with experiment. The purpose of our work is to search by means of Extended Hückel calculations the factors that lead to strong structural distortions around the metal and within the organic ligands. We will show that these distortions are very sensitive to the number of coordination around the metal as well as to the nature and precise position of the ligands. We will also show that agostic alkyl groups do not involve a large amount of C-H to metal donation. However the amount of electron donation depends on the alkyl group. It is found to be minute in the case of methyl group and significantly larger in the case of an ethyl group. A more detailed and slightly different presentation of the material presented here can be found in reference 15.

DEVIATION FROM THE IDEAL OCTAHEDRAL FIELD OF d^0 HEXACOORDINATED COMPLEXES

In a first set of calculations, we studied the d^0 TiH_6^{2-} complex **8**, a model for **2** and **3** in which all the ligands have bee replaced by hydride ions. This simplified model is used to check whether the deviation from the octahedral field is an intrinsic property of d^0 hexacoordinated system or is due to the presence of an agostic group and/or to the particular nature of the other ligands in **2** and **3**.

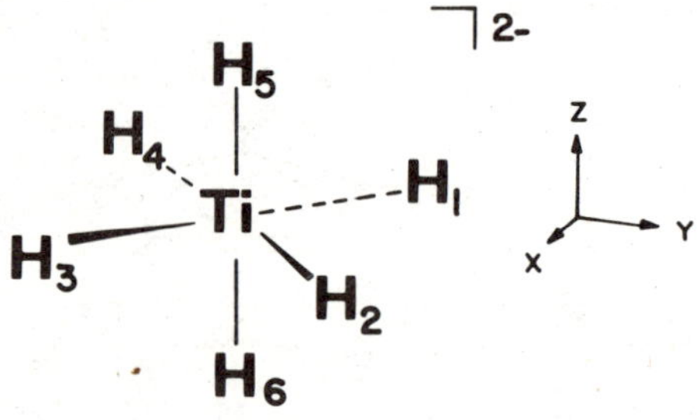

8

Let us first discuss a deviation from the octahedral field involving the four ligands in the xy (equatorial) plane, the two remaining Ti-H bonds (not shown in the transformation **9** → **10**) being kept perpendicular to that plane. **10** mimics the actual structures of complexes **2** and **3**. It is described by the angles ϕ_1 (larger than 90°) and ϕ_2 (smaller than 90°). EH calculations lead to a minimum for ϕ_1 = 150° and ϕ_2 = 70°, the stabilization energy with respect to **9** being 1.49 eV. There is clearly a strong preference for a non octahedral field in d^0 unsubstituted complex. Since such result would not have been predicted easily with the VSEPR rules because the electronic distribution is spherical, it is worth finding the origin of deformation.

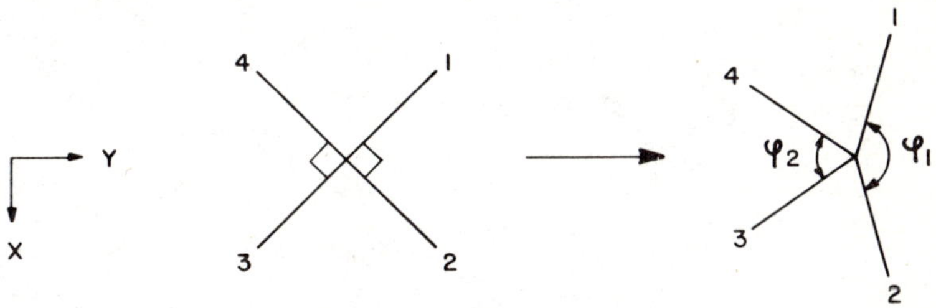

9 **10**

Changes in the shapes and energies of the molecular orbitals (MO) for the transformation **9** → **10** is reported in Figure 1. MOs $1a_1$, $3a_1$ and $2b_1$ favor **10** while $1b_1$ is in favor of **9**. As it is often the case, we will concentrate our attention on the higher set of orbitals $1b_1$, $3a_1$, $2b_1$. $1b_1$ is destabilized in **10** because the 1s hydrogen orbitals do not point anymore along the lobes of xy. On the other hand, $2b_1$ is stabilized because the reduced symmetry in **10** allows participation of xy. On the whole the energetic evolution of b_1 orbitals disfavors **10** in front of **9**. The energy lowering is thus due to the a_1 orbitals. In particular, $3a_1$ is strongly stabilized because the empty x^2-y^2, which could not mix in the symmetrical structure **9** is now allowed to come into play. The participation of x^2-y^2 into the bonding $3a_1$ causes the strong destabilization of $4a_1$. The consequence of the participation of x^2-y^2 in the bonding scheme is the reinforcement of the Ti-H bonds lying in the equatorial plane. Another way to view the problem is to consider that 2 d orbitals (xy and x^2-y^2) are involved in the bonding with the equatorial hydrogens in structure **10** instead of 1 d (xy) in structure **9**. We are confronted here to a typical second-order Jahn-Teller situation which causes the lowering of the symmetry of the molecule.

Can the system get to an even lower symmetry? In **10**, there are still two non-bonding d orbitals (xz and yz) which could well participate in the metal ligands bonds if a suitable structural distortion is allowed. To attain such situation, it is sufficient to move one of the two hydrogens which were kept on the axis perpendicular to the equatorial plane (z axis) away from that axis. Although the search for the full optimized structure is out of the scope of that study, the test

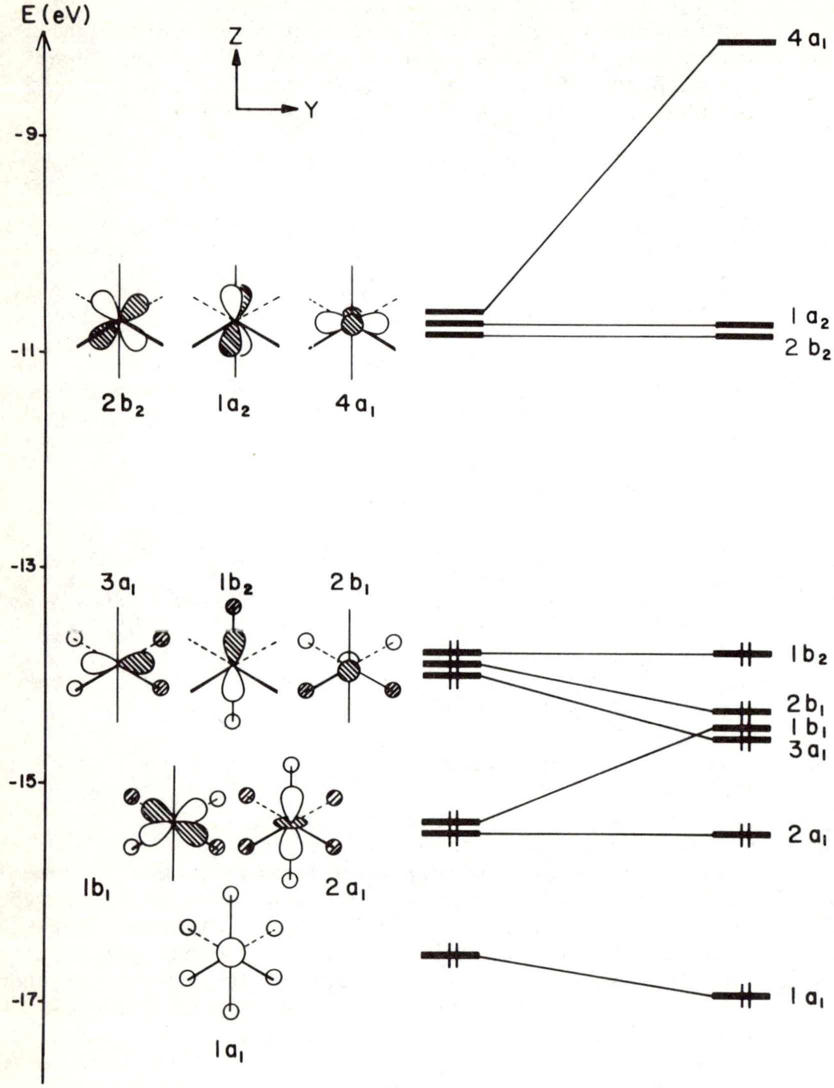

Figure 1 Molecular orbitals of TiH_6^{2-} in the ideal octahedral field ,9 (left hand side) and their evolution upon **9** → **10** transformation.

calculations have shown that such distortion causes additional stabilization.

As in every second-order Jahn-Teller situation, the occurence of the distortion strongly depends on the energy gap between the occupied ($3a_1$) and empty ($4a_1$) orbitals that are involved in the distortion. Lifting the empty orbital disfavors the distortion. For instance, consider that chlorine atoms are set on the z_2 axis as in **11**. xz and yz are destabilized by the Cl lone pairs while x^2-y^2 is left unchanged. In that case structural distortions preferably occur in the equatorial (xy) plane while the axial ligands remain perpendicular to that plane. This is the case of the actual compounds **2** and **3**. Along the same line of reasoning, a chlorine atom in the xy plane diminishes the drift away from octahedral field.

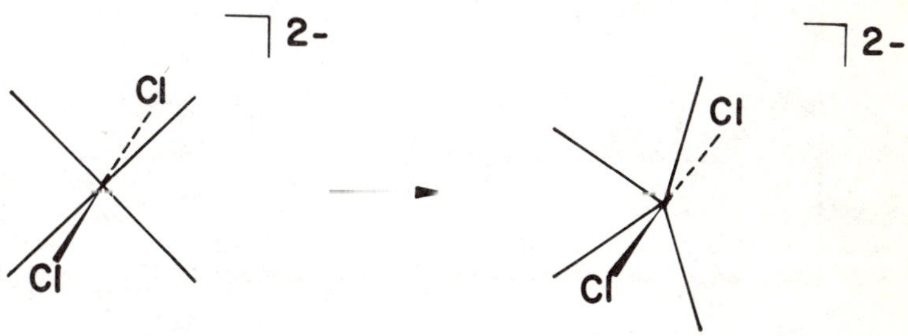

11

These calculations, done on TiH_6^{2-}, were repeated for $H_5TiCH_3^{2-}$ and $H_5TiC_2H_5^{2-}$ (keeping a non agostic alkyl group). Results were identical. Deviation from octahedral field is thus not due to the presence of an agostic group. From this analysis, it can thus be expected that d^0 systems may show interesting structural or vibrational properties which can be very sensitive to the nature of the ligands.

ALKYL DISTORTION, THE AGOSTIC GROUP IN THE HEXACOORDINATED COMPLEX

We have just seen that the drift away from the octahedral field produces a reinforcement of the metal-ligand bonds. We will show that the same phenomenon occurs upon pivoting the CH_3 group.

The total energy variation and the Walsh diagram for the pivoting of CH_3 is presented in Figure 2. The motion of CH_3 is described by a single paramater α (Ti-C-H_1 angle). When α is varied, all H-C-H angles are kept equal to 109.47°. As it appears on Figure 2, α = 109.47° represents the most unfavorable structure and any distortion away from that position, either towards smaller values of α or toward larger values leads to a neat stabilization. We will focus here on the region where α is smaller than 109.47°. The behavior of the HOMO (7a') is especially interesting. It nicely parallels the variation of the total energy. In contrast to that, one of the low-lying empty orbital behaves

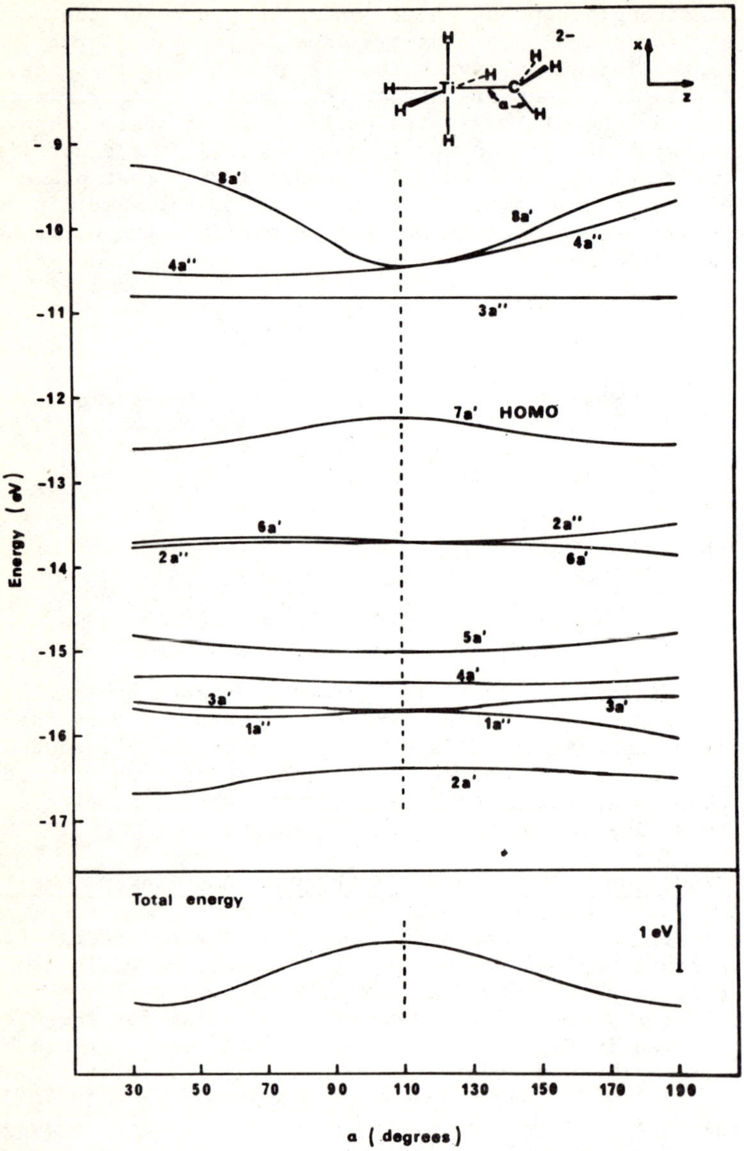

Figure 2 Walsh diagram and total energy variation as a function of α in $H_5TiCH_3^{2-}$.

in exactly the opposit fashion and is strongly destabilized by the structural change.

In order to understand the above energy scheme, we consider the molecule as made of TiH_5^- and CH_3^- fragments. The interaction diagram is shown in Figure 3. The HOMO (7a') of the non-distorted complex is the bonding combination of the high energy empty metal orbital σ_{Ti} and the HOMO n_{CH3} of the CH_3^- group. 7a' charaterizes the Ti-C σ bond. When α is moved away from 109.47°, the HOMO is affected in two manners. First, the overlap between σ_{Ti} and n_{CH3} decreases because the two fragment orbitals do not point toward each other anymore, (**12**). This is an unfavorable factor which should have resulted in a destabilization of the HOMO and of the total energy. Secondly, at α = 109.47° n_{CH3} cannot overlap with xz by symmetry. When α takes a different value, the symmetry is lowered and n_{CH3} overlaps with xz, (**13**). Since xz is a vacant orbital in the d^0 complex, this is clearly a stabilizing interaction which causes in turn the destabization of xz (8a').

12

13

These two opposite effects add when the methyl group is tilted. Calculations show that the stabilizing effect dominates in the octahedron but one could gather that it is not necessary the case for every systems (see the tetrahedron case below) and that it does depends on the relative energy of the 3 fragments orbitals that are involved (the 2 empty ones on the metal and n_{CH3} on the methyl group). In the octahedron σ_{Ti} is considerably higher than xz. Thus the pivoting of CH_3 allows n_{CH3} to switch from the high lying σ_{Ti} to the lower lying empty xz. This results in a much better Ti-C interaction.

If one compares the empty metal orbital that was responsible for the deviation away from octahedral field to the one responsible for the pivoting of the methyl group, one easily notices that it is the same d orbital (non bonding in the equatorial plane). Deviation away from octahedral field and agostic methyl group are two consequences of the presence of a low-lying empty orbital of adequate symmetry on the metal. Because of that coupling effect, it is important to optimize simultaneously these two parameters in $H_5TiCH_3^{2-}$. The results reported in Figure 4a are in good agreement with experiment (compare with 2).

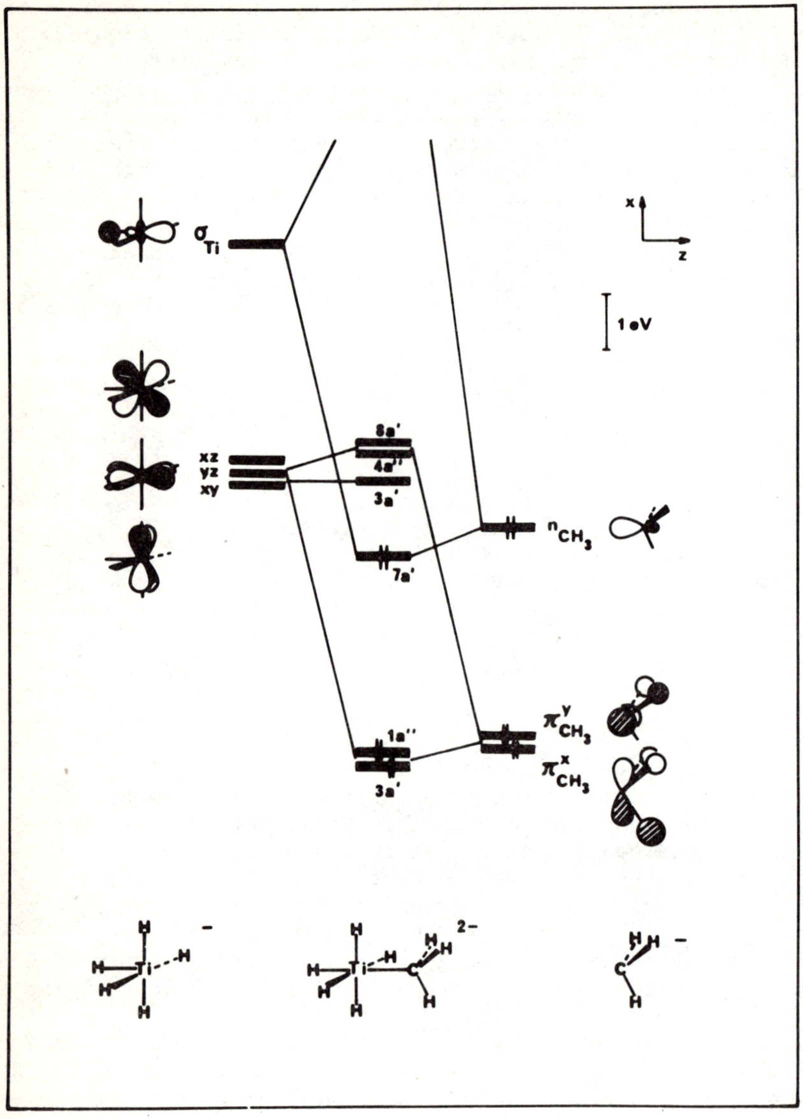

Figure 3 Interaction diagram between the H_5Ti^- and CH_3^- fragments

Figure 4 Optimal structures for a) $H_5TiCH_3^{2-}$ and b) $H_5TiC_2H_5^{2-}$.

π donor ligands which are preventing the distortion of the ligand field also diminishes the ability of the methyl group to become agostic. Thus if Cl atoms are positionned in the distortion plane of the methyl group, their lone pairs can destabilize the non-bonding d orbital and prevent the pivoting of the methyl group. One thus understand why, in the experimental structures 2 and 3, the distortion of the methyl group and the deviation away from octahedral field occurs in the plane containing only one chlorine atom.

The analysis done for the agostic methyl group can be reported in its main line to the case of the agostic ethyl group. The optimized structure is shown in Figure 4b. Note that in the case of the ethyl group we found that the reinforcing of the C-C bond contributes importantly to the stabilization of the system. The system is on its way to form partially a C-C double bond.

Up to that moment we have been able to understand all the experimental facts without explicit intervention of the C-H to metal electron donation. Calculations show that this donation is small in the case of the methyl group but is larger in the case of ethyl group. It should be noticed that the orbitals characteristic of the donating capability of the C-H bonds (mostly the π_{CH3} orbitals) are rather deep in energy and are not very powerful electron donors. The fact that some electron transfer takes place in the ethyl group is due to a better overlap between metal and π_{CH3} orbitals. This is in good agreement with the experimental spectroscopic observations. For the methyl compound 2, according to the neutron diffraction study, the C-H bond involved with the metal is not found any longer than a normal C-H bond. In addition, no change in the NMR coupling constant J(C-H) is found[3] as it has been noticed for all other agostic structures[2]. In contrast, in the ethyl case, a large change in the C-H NMR coupling constant has been interpreted in term of a significative C-H_β bond lenghtening[3].

TETRAHEDRON CASE

Calculations were performed on d^0 H_3TiCH_3 complex in a staggered conformation, **14**. As in the octaheral case, the structural change is described by the Ti-C-H (α) angle. Molecular orbital energies and total energy variations are reported in Figure 5. In contrast with the octahedron case, any departure of the α angle from 109.47° is destabilizing. Results are similar for the eclipsed conformation of the complex. This finding agrees nicely with the experimental results on Cl_3TiCH_3, **4**.

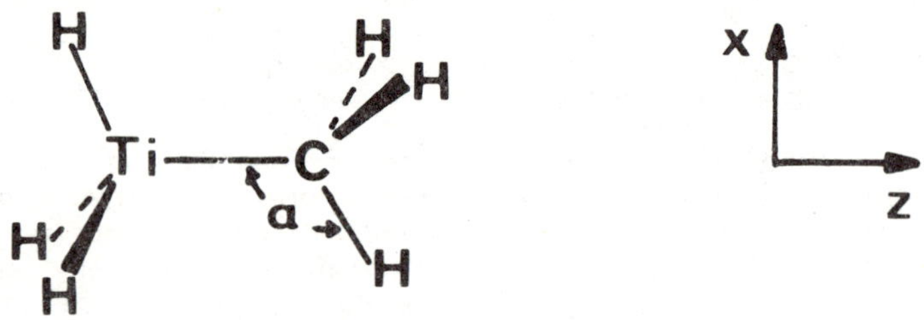

14

Figure 5 shows that the HOMO evolution closely follows that of the total energy. Why do the HOMOs of the octahedron and tetrahedron behave so differently? The reasons appear when one looks at the way H_3TiCH_3 is made from the H_3Ti^+ and CH_3^- fragments (Figure 6). The HOMO of the non distorted complex is made of the bonding combination of $1a_1$ and n_{CH3}. Upon tilting CH_3, the overlap between $1a_1$ and n_{CH3} diminishes while that between n_{CH3} and the empty $1e_s$ takes a non zero value. Qualitatively the interaction diagram looks very similar to that occuring in the octahedron. Quantitatively the situation is very different. $1a_1$ and $1e_s$ are both non bonding orbitals and are thus very close in energy. So that what is lost by diminishing the interaction between $1a_1$ and n_{CH3} upon tilting CH_3 is hardly recovered by building an interaction between $1e_s$ and n_{CH3}. Explicit consideration of the overlaps even leads to an overall loss of interaction between the metal and the CH_3 group. So the situation is very different to that of the octahedron. σ_{Ti} was a high lying antibonding (within H_5Ti^-) orbital while xz was non bonding. They were separated by a large energy gap, so that partial loss of σ_{Ti}-n_{CH3} interaction was largely compensated by the build in of the new xz-n_{CH3} interaction.

CONCLUSION

In this work, we have shown that the structure of electron deficient d^0 alkyl transition metal complexes is dominated by the attempt of the metal to involve the maximum number of d orbitals to bond

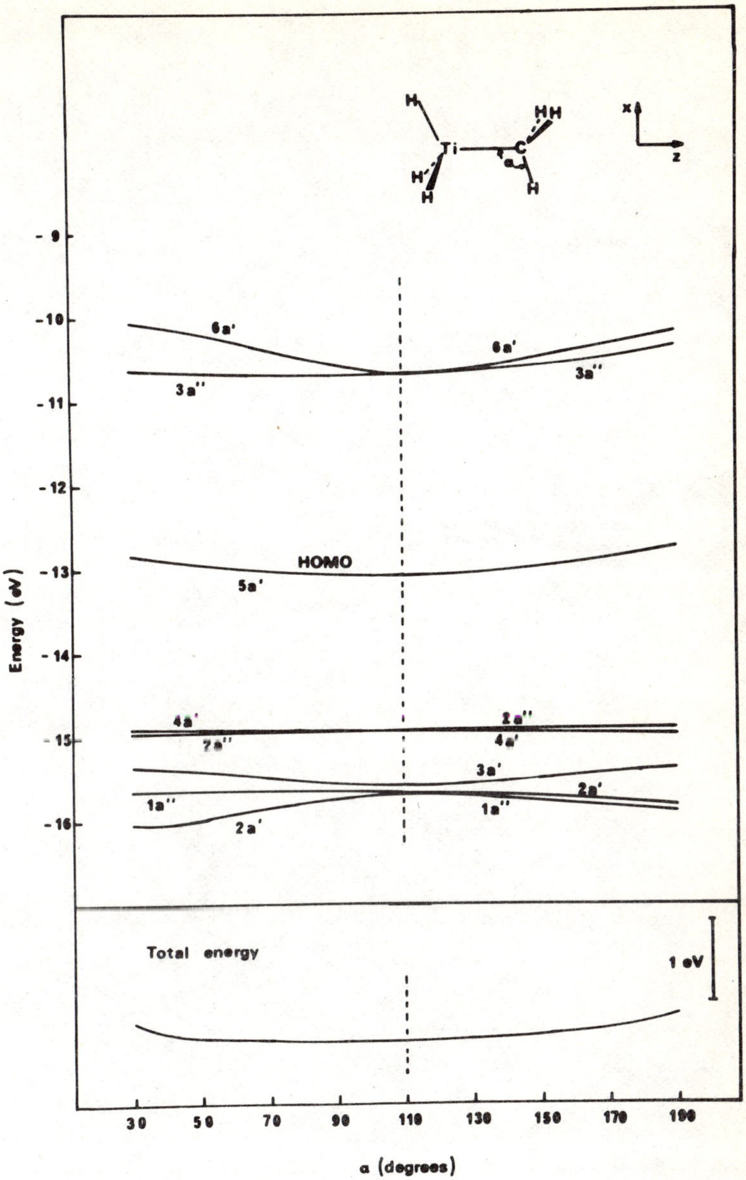

Figure 5 Walsh diagram and total energy variation as a function of α in H_3TiCH_3.

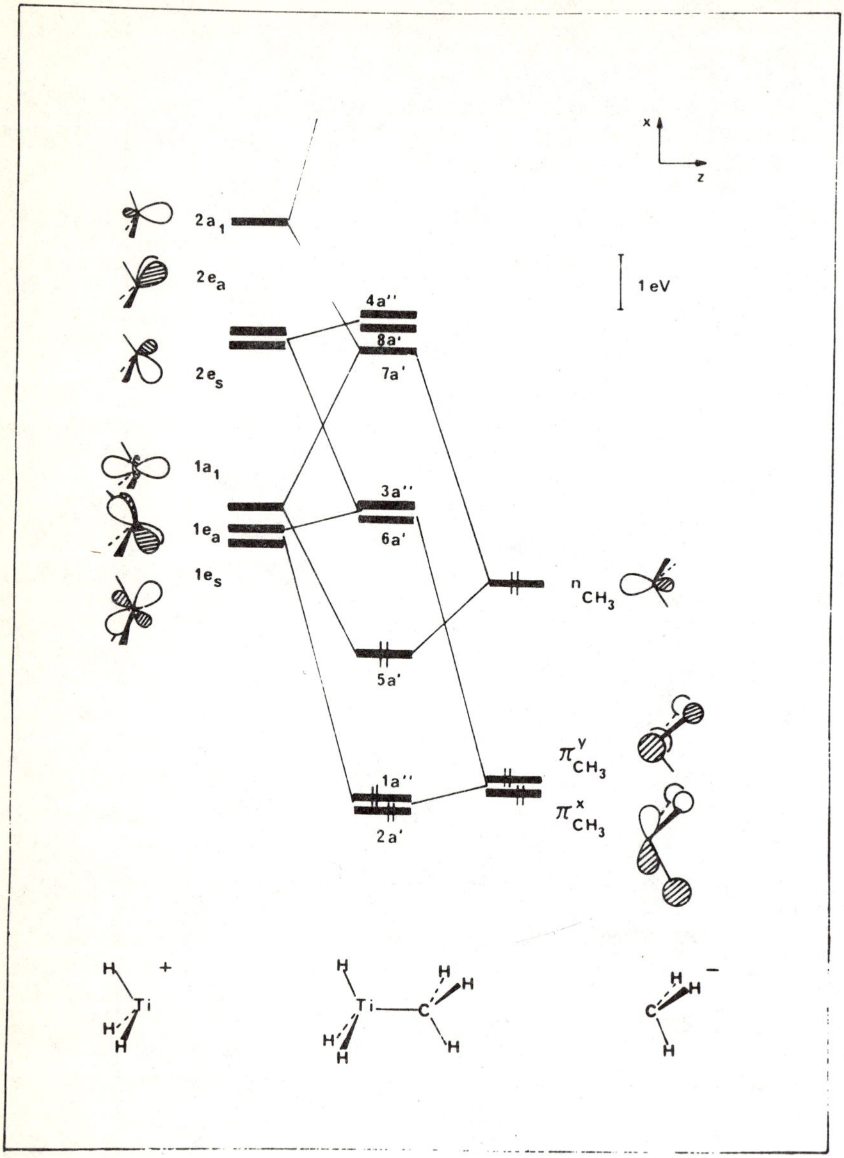

Figure 6 Interaction diagram between the H_3Ti^+ and CH_3^- fragments

with the ligands. This can result in a lowering of the symmetry of the complex and in strong distortion of the organic ligands. Whether or not such things occur depends mostly on two factors: number and arrangement of the ligands around the metal. Such systems are fascinating for that respect.

REFERENCES

1 - a) For a general discussion of these reactions, see : Collman, J.P.; Hegedus, L.S. " Principles and Applications of Organotransition Metal Chemistry " ; University Science Books, Mill Valley, CA 1980 ; p.73.
b) Cooper, N.J.; Green, M.L.H. J. Chem. Soc., Chem. Commun. 1974, 761.
c) Schrock, R.R. J. Am. Chem. Soc. 1975, 97, 6577.
d) Schrock, R.R.; Parshall, G.W. Chem. Rev. 1976, 76, 243.

2 - Brookhart, M.; Green, M.L.H. J. Organometal. Chem. 1983, 250, 395 and references therein. Girolami, G.S.; Wilkinson, G.;Galas A.M.R.; Thornton-Pett M.; Hursthouse, M.B. J. Chem. Soc. Dalton Trans. 1985, 1339. For distorted alkylidenes see for instance Schultz, A.J.; Williams, J.M.; Schrock, R.R.; Rupprecht, G.; Fellmann, J.D. J. Am. Chem. Soc. 1979, 101, 1593. Schultz, A.J.; Brown, R.K.; Williams, J.M.; Schrock, R.R. ibid, 1981, 103, 169. Wengrovius, J.H.; Schrock, R.R.; Churchill, M.R.; Wasserman, H.J. ibid, 1982, 104, 1739. For other structures with hydrogen bonded in a non classical fashion to two centers ,see for instance: Green, M.; Orpen, A.G.; Salter, I.D.; Stone, F.G.A J. Chem. Soc. Chem. Comm. 1982, 813. Jeffery, J.C.; Orpen, A.G.; Robinson, W.T.; Stone, F.G.A; Went, M.J. ibid, 1984, 396. Schubert, U.; Ackermann, K.; Kraft, G.; Wörle B. Z. Naturforsch. 1983, 38b, 1488.

3 - Brown, R.K.; Williams, J.M.; Schultz, A.J.; Stucky, G.D.; Ittel, S.D.; Harlow, R.L. J. Am. Chem. Soc. 1980, 102, 981. Williams, J.M.; Brown, R.K.; Schultz, A.J.; Stucky, G.D.; Ittel, S.D. ibid, 1978, 100, 7407.

4 - a) Ittel, S.D.; Van-Catledge, F.A.; Jesson, J.P. J. Am. Chem. Soc. 1979, 101, 6905.
b) Brookhart, M.; Lukacs, A. Organometallics, 1983, 2, 649. Brookhart, M.; Lamanna, W.; Pinhas, A.R. ibid, 638. Brookhart, M.; Lamanna, W.; Beth Humphrey, M. J. Am. Chem. Soc., 1982, 104, 2117.

5 - Dawoodi, Z.; Green, M.L.H.; Mtetwa, V.S.B.; Prout, K. J. Chem. Soc. Chem. Commun. 1982, 802, 1410. Dawoodi, Z.; Green, M.L.H.; Mtetwa, V.S.B.; Prout,K. ;Schultz, A.J.; Williams, J.M.; Koetzle, T.F. J. Chem. Soc. Dalton Trans. in press.

6 - Howard, C.G.; Girolami, G.S.; Wilkinson, G.; Thornton-Pett, M.; Hursthouse, M.B. J. Chem. Soc. Dalton Trans, 1983, 2631.

7 - Cracknell, R.B.; Orpen, A.G.; Spencer, J.L. J. Chem. Soc. Chem. Comm. 1984, 326.

8 - Brookhart, M.; Green, M.L.H.; Pardy, R.B.A. J.Chem.Soc. Chem.Comm. 1983, 691.

9 - Schmidt, G.F.; Brookhart,M. J.Am.Chem.Soc. 1985,107,1443.

10 - Harlow, R.L.; McKinney, R.J.; Ittel, S.D. J. Am. Chem. Soc. 1979, 101, 7496.

11 - Fitzpatrick, N.J.; McGinn, M.A. J.Chem.Soc. Dalton Trans. 1985, 1637.
12 - Goddard, R.J.; Hoffmann, R.; Jemmis, E.D. J. Am. Chem. Soc. 1980 102, 7667.
13 - Obara, S.; Koga, N.; Morokuma, K. J. Organomet. Chem. 1984, 270, C33.
14 - Koga, N.; Obara, S.; Morokuma, K. J. Am. Chem. Soc. 1984, 106, 4625.
15 - Eisenstein, O.; Jean, Y. J.Am Chem. Soc. 1985, 107, 1177.
Demolliens, A.; Jean, Y.; Eisenstein, O. submitted for publication.

Stereochemistry and Metal-ligand Interaction of Group VIII Low-valent Transition Metal Complexes. An ab-initio MO and Energy Decomposition Analysis Study

S. Sakaki
Department of Synthetic Chemistry, Faculty of Engineering, Kumamoto University, Kumamoto 860 Japan and Institut Le Bel, Université Louis Pasteur, rue Blaise Pascal Strasbourg, France

ABSTRACT. Coordinate bond nature and stereochemistry of several group VIII transition metal complexes have been successfully investigated with ab-initio MO method and energy decomposition analysis (EDA). $Ni(PH_3)_2(CO)_2$ takes pseudo-tetrahedral (p-Td) structure, to avoid large exchange repulsion between the Ni d and the CO lone-pair orbitals. On the other hand, $Ni(PH_3)_3(H_2CO)$ has planar (Pl) structure, because of strong back-donative interaction in this structure. In $RhCl(PH_3)_2(N_2)$, the η^1-end on mode receives larger stabilization from the $N_2 \rightarrow Rh$ donative interaction and the electrostatic interaction between N_2 and Rh than the η^2-side on mode does, and as a result, the η^1-end on mode is more stable than the η^2-side on mode. Several coordination modes are also examined in CO_2 complexes. The η^2-side on mode is stable in $Ni(PH_3)_2(CO_2)$, because large back-donative ($M\ d\pi \rightarrow CO_2\ \pi^*$) interaction is possible. The strong ($M\ d\pi \rightarrow CO_2\ \pi^*$) back-donative interaction favors the η^1-C mode in $[Co(alcn)_2(CO_2)]^-$ (alcn = NHCH-CH-CH-O^-) The electrostatic interaction is important in the η^1-O end on mode, and this mode is most stable in $[Cu(PH_3)_2(CO_2)]^+$.

1. Introduction

Transition metal complexes have received much attention in the last decade, because they present many interesting features. For example, transition metal can form new compounds including complexes of inert gases such as N_2 and CO_2, with variety of different stereochemistries, and they have novel reactivity and are useful as catalysts in many cases. These features of transition metal complexes depend largely on the nature of the central metal, such as positions in the periodic table, oxidation state, d-electron numbers and configuration, and so on. In this regard, it is worthy of investigating theoretically the metal-ligand interaction and what kind of natures of metal influences the above-mentioned features of transition metal complexes.

Ever since Dewar, Chatt, and Duncanson introduced the notion,

A. Veillard (ed.), Quantum Chemistry: The Challenge of Transition Metals and Coordination Chemistry, 319–331.

the coordinate bonds of non-Werner type transition metal complexes have been discussed in terms of donative and back-donative interactions. The concepts of donation and back-donation are still valid and are used in many discussions to explain experimental results. In such discussions, the quantitative estimation of those interactions is important. Unfortunately, the quantitative estimation is not easy but rather difficult in both experimental and theoretical fields.

Further, it is not easy to find valid information about metal-ligand interaction in ab-initio MO calculation unlike semi-empirical MO method, because many contracted Gaussians are generally used as basic functions and LCAO coefficients spread over the whole molecular orbital. By using energy decomposition analysis (EDA) [1], however, we can easily extract some valid informations about the metal-ligand interaction from results of ab-initio MO calculation, and also we can get some information about donative and back-donative interactions.

In the last few years, we have attempted to apply the EDA to investigation of coordinate bond and stereochemistry of transition metal complexes [2], and we were successful in explaining some of those issues. In this article, ab-initio MO and EDA studies of several group VIII transition metal complexes are presented. At first, it is examined how the coordinate bond is described in terms of energy decomposition analysis. Then, the stereochemistry of $Ni(PH_3)_2L$ ($L=H_2CO$ or $(CO)_2$) is investigated. Finally, serveral coordination modes of N_2 and CO_2 complexes are discussed in $Ni(PH_3)_2$-(CO_2), $[Co(alcn)_2(CO_2)]^-$ (alcn=HNCH–CH–CH–O$^-$) and $RhCl(PH_3)_2(N_2)$.

2. Computational details

Ab initio MO calculations have been carried out by using IMSPAC [3a] and Asterix programs [3b]. In these calculations, closed shell singlet state has been calculated for following reasons [4c];(1) $Ni(PR_3)_2(R'_2CO)$ has been reported to show no ESR signal and considered to be a singlet [5], (2) $RhCl(PH_3)_2L$ ($L=N_2$ or C_2H_4) has square-planar structure with d^8 electron configuration [6], in which singlet closed-shell electron configuration is well known to be stable, (3) In $[Co(alcn)_2(CO_2)]^-$, a large energy gap between the Co dz^2 and Co dx^2-y^2 orbitals was calculated, where the z-axis is taken to be perpendicular to the $Co(alcn)_2$ plane and the alcn ligands are placed on the x and y axis. Further, the doubly occupied dz^2 orbital is important for the η^1-CO_2 coordination [7].

In our old calculations of $Ni(PH_3)_2L$ ($L=CO_2$, H_2CO, or $(CO)_2$), rather small basis sets were used; for Ni, the [4s 3p 2d] set, contracted from (12s 7p 5d) primitives was employed [8], and for ligands, the usual 4-31 G basis sets were used [9]. In $RhCl(PH_3)_2L'$ ($L'=N_2$ or C_2H_4), the Rh basis set was [6s 5p 4d] contracted from (14s 9p 8d) primitives [10] and the basis sets for ligand atoms were the usual 4-31 G sets. In $[Co(alcn)_2(CO_2)]^-$, two kinds of basis sets were employed. In smaller basis set, the [5s 3p 3d], contracted from

(12s 7p 5d) primitives [8], and the usual 4-31 G sets were used for Co and ligand atoms [9], respectively. In larger basis set, the Co basis set was composed of [6s 4p 3d] contracted from (14s 9p 6d) [11] and the basis set of ligand atoms was described by [3s 2p] contracted from (9s 5p) primitives [12].

The geometries of complexes examined here were taken from experimental structure, while some important geometrical parameters, such as the coordinating distance and the geometry of the noticed ligand were optimized [2]; for example, the Rh-N and N-N distances were optimized in the case of $RhCl(PH_3)_2(N_2)$, but the geometry of $RhCl(PH_3)_2$ part was taken from the experimental structure of similar compound [6a].

The EDA method, reported elsewhere [1, 2a], is briefly described here. We consider the total complex $ML_n.L^o$ is separated into two parts, ML_n and L^o, where we notice the L^o ligand. The binding energy (BE) is defined as a stabilization of $ML_n.L^o$ relative to ML_n and L^o taking the respective equilibrium structures;

$$BE = E_t(ML_n.L^o) - E_t(ML_n) - E_t(L^o)_{eq} = DEF + INT,$$

$$DEF = E_t(L^o)_{dis} - E_t(L^o)_{eq},$$

$$INT = E_t(ML_n.L^o) - E_t(ML_n) - E_t(L^o)_{dis},$$

$$= ES + EX + FCTPLX + BCTPLX + R,$$

where the suffixes "eq" and "dis" are the abbreviation of equilibrium and distorted structures, respectively. The deformation energy (DEF) corresponds to the destabilization energy required to distort L^o from its equilibrium to the deformed structure [13], and the interaction energy (INT) represents the stabilization resulting from the interaction between ML_n and L^o taking the deformed structures as in the complex. INT is further divided into various chemically meaningful interactions; ES means the electrostatic (i.e., coulombic) interaction between ML_n and L^o, and EX is the exchange repulsion resulting from Pauli exclusion principle. FCTPLX includes the charge-transfer from L^o to ML_n, the polarization of ML_n, and their coupling term. BCTPLX also includes the charge-transfer from ML_n to L^o, the polarization of L^o, and their coupling term. These interactions are schematically shown in Scheme I. Negative values of these terms mean that stabilization results from the interaction upon the coordination of the L^o ligand.

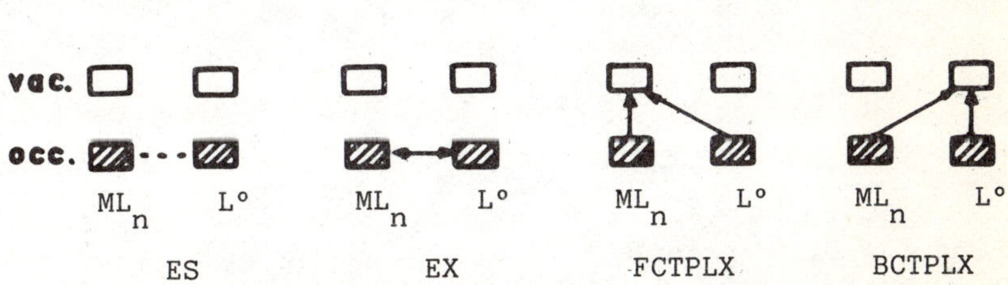

...;MO's are not mixed between fragments.——; MO's are mixed between fragments.

Scheme 1

3. Results and Discussion

3.1. Nature of each interaction term [2d].

At first, let us examine the nature of interaction terms. Difference density maps obtained by each interaction are shown in Figure 1, for $RhCl(PH_3)_2(N_2)$ as an example. The total electron density decresases in the region near to Rh but increases in the regions near to the

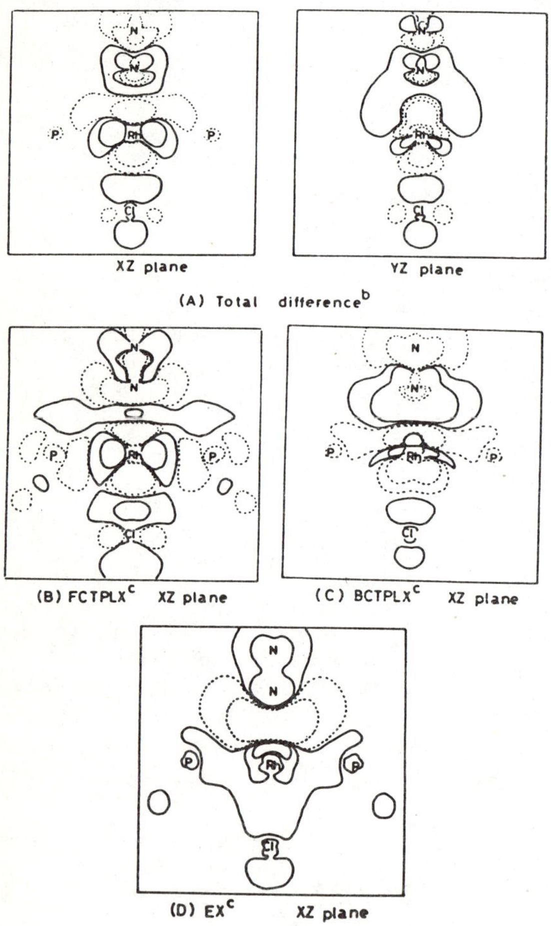

Figure 1. Difference density maps [a)] of $RhCl(PH_3)_2(\eta^1-N_2)$

a) $RhCl(PH_3)_2(\eta^1-N_2)$ – $RhCl(PH_3)_2$ – (η^1-N_2)

—— ; increased density, ... ; decreased density

b) total difference density ; ± 0,005, ± 0,0005

c) difference density by each interaction ; ± 0,001, ± 0,0001.

coordinating N atom and the Cl ligand, suggesting the charge-transfer from $RhCl(PH_3)_2$ to N_2 and polarizations of the $RhCl(PH_3)_2$ and N_2 parts. Through the EX interaction, electron density decreases in the coordinate bonding region between $RhCl(PH_3)_2$ and N_2, suggesting that this interaction is repulsive. Also, through this interaction, both electron clouds of $RhCl(PH_3)_2$ and N_2 are pushed away from each other, which seems to correspond to Pauli exclusion principle. The FCTPLX interaction decreases electron density in the N_2 part, but accumulates it in the region between $RhCl(PH_3)_2$ and N_2. By detailed inspection of this difference density map, we notice that electron density decreases in the region near the Rh atom along the Rh-N line but increases in the regions near the Rh atom along the Rh-P line and near the Cl ligand. Also, the electron density slightly increases near the terminal N atom. This movement of electron cloud seems to reduce the exchange repulsion between $RhCl(PH_3)_2$ and the N_2 lone pair orbital. That is, FCTPLX includes the charge-transfer from N_2 to $RhCl(PH_3)_2$, accompanied with exchange-polarization type electron movement. The BCTPLX interaction decreases electron density in the $RhCl(PH_3)_2$ part, in particular, near the Rh atom, but increases electron density in the N_2 ligand part. Simultaneously, electron density is decreased near the terminal N atom, probably due to the electron withdrawing effect of the positively charged Rh atom. Also, the electron cloud of the $RhCl(PH_3)_2$ part is somewhat pushed away from the N_2 ligand. These results suggest that the BCTPLX interaction includes the charge transfer from the metal part to ligand, accompanied with some polarization type electron movement in the metal and ligand parts. Thus, FCTPLX and BCTPLX correspond to donative and back-donative interactions, respectively, while both are supplemented by somo polarization type electron movement.

3.2. Stereochemistry of $Ni(PH_3)_2L$ ($L=H_2CO$ or $(CO)_2$) : planar (Pl) vs. pseud-tetrahedral (p-Td) structure [2c].

Generally, the metal complex with d^{10} electron configuration tends to have tetrahedral (Td) or p-Td structure. In fact, $Ni(PH_3)_2(CO)_2$ and $Ni(CO)_4$ have p-Td and Td structures, respectively [14]. On the other hand, $Ni(PH_3)_2L$ ($L=\eta^2$-ligand such as C_2H_4, C_2H_2, CO_2, H_2CO) has the Pl structure, in spite of large steric repulsion in the Pl structure [14]. Results of EDA are given in Table I, where values of $Ni(PH_3)_2(CO)_2$ are given for one pair of $Ni(PH_3)_2$-CO interaction. The p-Td $Ni(PH_3)_2(CO)_2$ is substantially more stable than the Pl structure, while the Pl $Ni(PH_3)_2(H_2CO)$ is more stable than the p-Td structure. As clearly shown by energy decomposition analysis, the Pl $Ni(PH_3)_2(CO)_2$ is much more destabilized by EX than its p-Td structure. The Pl structure have the optimized CNiC and PNiP angles of about 90°, perhaps trying to minimize the steric repulsion between ligands. In this structure, the occupied Ni dσ orbital overlaps well with the PH_3 lone pair orbitals, to cause anti-bonding interaction. Into this anti-bonding overlap, Ni 4pσ orbitals mix in a bonding manner, yielding the hybridization of the Ni dσ and 4pσ

orbitals [15]. Through such hybridization, the occupied Ni d_σ orbital, extends very much towards the CO ligand, unfortunately yielding a substantially large EX repulsion with it (see Figure 2). Hence, this structure is unstable. In the p-Td structure, on the other hand, the CNiC angle can become larger (114°) than 90° without large steric repulsion between ligands. Thus, two CO ligands can take up the position, in which they avoid the expanding d-electron clouds, as shown in Figure 2. As a result, EX repulsion is small in the p-Td structure. In $Ni(PH_3)_2(H_2CO)$, on the contrary, the Pl structure is more stable than the p-Td structure, due to a larger BCTPLX stabilization (see Table I). It is wellknown that the HOMO of $Ni(PH_3)_2$ is mainly composed of the Ni dxz orbital [16]. This HOMO can overlap well with the π^* orbital of H_2CO, to form strong back-donative interaction, as shown in Figure 2. In the p-Td structure, the Ni dyz orbital, lying lower in energy than the Ni dxz orbital by ca. 1 eV, interacts with the π^* orbital of H_2CO, as shown in Figure 2. Consequently, BCTPLX is larger in the Pl than in the p-Td structure. In $Ni(PH_3)_2(H_2CO)$, the EX repulsion of the Pl structure is similar to that of the p-Td structure, unlike $Ni(PH_3)_2(CO)_2$, which can be easily understood by considering the different extent of the hybridizing mixing between Ni 3d and Ni 4p in $Ni(PH_3)_2(CO)_2$ and $Ni(PH_3)_2(H_2CO)$. In $Ni(PH_3)_2(H_2CO)$, the positions of H_2CO and PH_3 ligands relative to Ni are almost the same in both structures, yielding similar degree of the hybridization, and as a result similar degree of overlap between the occupied Ni d orbital and H_2CO electron cloud. In $Ni(PH_3)_2(CO)_2$, however, degree of the hybridization of Ni 3d with Ni 4p is different very much in the Pl and p-Td structures, because the positions of CO and PH_3 ligands are very different in two structures. Thus, the EX repulsion of the Pl $Ni(PH_3)_2(CO)_2$ is much different from that of the p-Td structure, while the EX value is not different very much in the Pl and p-Td $Ni(PH_3)_2(H_2CO)$.

Table I. Energy decomposition analysis of $Ni(PH_3)_2$-L ($L = H_2CO$ or $(CO)_2$) (kcal/mol)

	$Ni(PH_3)_2(CO)_2$ a)		$Ni(PH_3)_2(H_2CO)$	
	p-Td b)	Pl c)	p-Td b)	Pl c)
BE	−23	17	−14	−42
DEF	3	14	20	20
INT	−26	3	−34	−62
ES	−102	−115	−96	−102
EX	136	175	155	157
BCTPLX	−42	−41	−55	−74
FCTPLX	−18	−16	−13	−17
R			−25	−25

a) Each value is given for one pair of Ni-CO interaction b) pseudo-tetrahedral c) Planar.

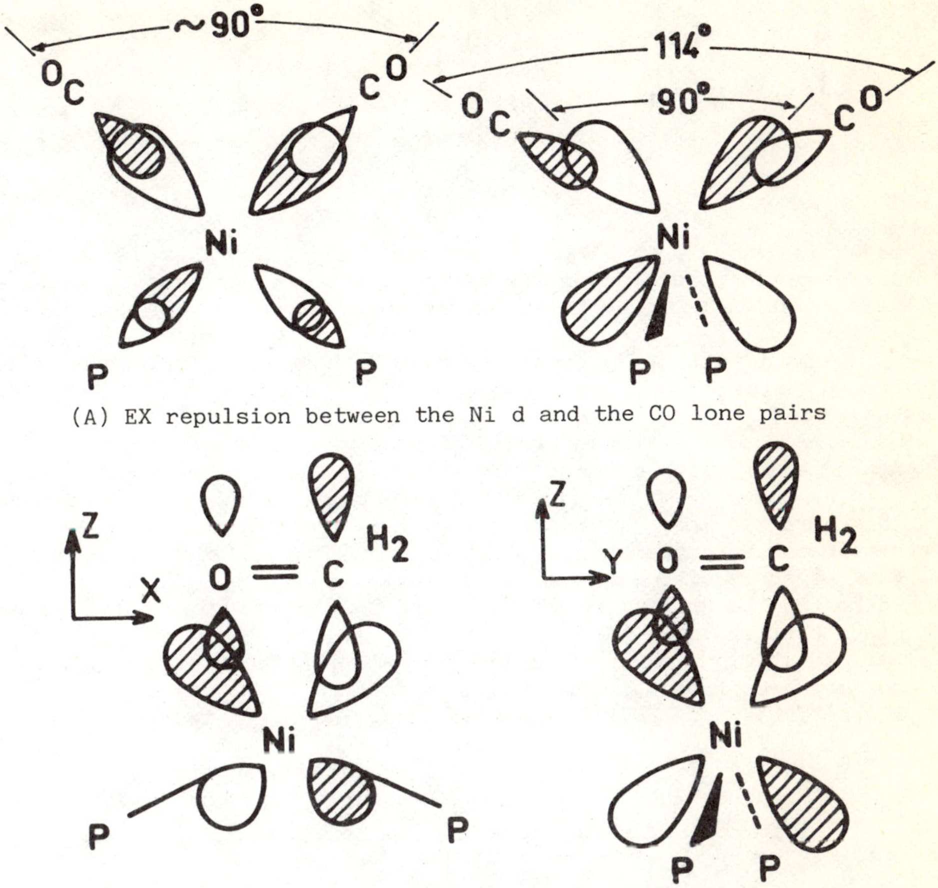

(A) EX repulsion between the Ni d and the CO lone pairs

(B) Back-donation from the Ni d_π to H_2CO π^* orbital

Figure 2. Schematical pictures of the exchange repulsion between $Ni(PH_3)_2$ and CO and of the back-donation from $Ni(PH_3)_2$ to H_2CO.

In conclusion, the stereochemistries of $Ni(PH_3)_2(CO)_2$ and $Ni(PH_3)_2(H_2CO)$ can be easily understood by considering the EX and BCTPLX interactions.

3.3. Coordination modes of the N_2 complex [2d] :

Recently, transition metal complexes with N_2 and CO_2 have received much attention from viewpoint of N_2 and CO_2 fixation [17]. One of the important features of such N_2 and CO_2 complexes is the presence of many possible coordination modes, such as η^1-end on and η^2-side on modes in the N_2 complexes, η^2-side on, η^1-C, and η^1-O end on

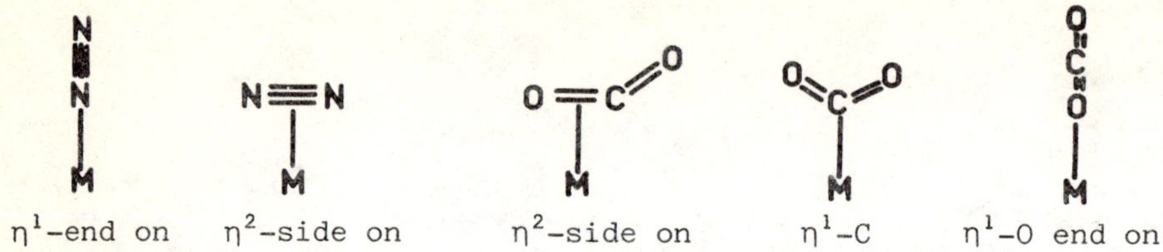

Scheme II

modes in the CO_2 complexes (see Scheme II). Thus, it is worthy of investigating what factor decides the type of coordination mode, and what relation is found between electronic structure and coordination mode.

Many η^1-end on N_2 complexes have been known [17b], whereas only one η^2-side on N_2 complex has been isolated for $\{Ni_2(N_2)(PhLi)_6(Et_2O)_2\}_2$ [18]. In order to investigate why the η^1-end on mode is common in many transition metal-N_2 complexes, both the η^1-end and η^2-side on modes of $RhCl(PH_3)_2(N_2)$ were examined.

As shown in Table II, the η^1-end on N_2 complex is more stable than the η^2-side on mode by ca. 16 kcal/mol, and this energy difference is comparable with the activation enthalpy estimated for the intramolecular linkage isomerization of N_2 in $[Ru(NH_3)_5(N_2)]^{2+}$, where the isomerization is considered to proceed via the η^2-side on structure [19].

In the optimized structure (R=2.70 A), the η^2-side on complex suffers less from EX repulsion and receives small stabilization from ES, FCTPLX, and BCTPLX interactions. The optimized η^1-end on mode

Table II. BE, DEF, INT, and energy components between $RhCl(PH_3)_2$ and L (L=N_2 or C_2H_4) (kcal/mol)

L	C_2H_4	η^2-N_2			η^1-N_2 [b]
		R= 2.446 [a]	2.70 [b]	2.50 [a]	
BE	-24.0	1.9	-5.7	0.0	-21.5
DEF	3.2	0.1	0.1	0.1	0
INT	-27.2	1.8	-5.8	-0.1	21.5
ES	-57.6	-26.7	-1.2	-21.4	-39.2
EX	66.0	66 0	7.1	54.1	54.1
FCTPLX	-16.2	-11.1	-4.1	-9.8	-13.0
BCTPLX	-17.6	-20.1	-4.6	-17.3	-17.7
R	-6.8	-6.3	-3.1	-5.7	-5.7

a) The Rh-N_2 distance is shortened keeping the geometries of the other part fixed, to give the same EX value as of that the Rh-C_2H_4 or Rh-η^1-N_2 complex. b) The optimized structure.

can receive large stabilization from ES, FCTPLX, and BCTPLX interactions, but suffers from large EX repulsion. This different features of energetics between two modes comes from the fact that in the η^2-side on mode the Rh-N_2 distance is much longer than in the η^1-end on mode. To find intrinsic difference between these two modes, we had better compare energy components at the same Rh-N_2 distance. However, it is not easy to define the same Rh-N_2 distance for these different coordination modes. The EX repulsion value can be considered as a measure of interfragment distance between ML_n and L°, because the EX value depends on the contact of electron clouds between ML_n and L°, and the same value of EX means the similar contact of these two parts [2b, c, d]. Thus, we compare each energy component at the interfragment distance giving the same EX value. The EX value of the η^1-end on N_2 mode is taken as a standard, and the Rh-N_2 distance of the η^2-side on N_2 mode is shortened to give this EX value. Although both coordination modes receive similar stabilization from BCTPLX, as compared in Table II, the η^1-end on mode has much larger ES and slightly larger FCTPLX stabilizations than the η^2-side on mode does. It is interesting to compare the η^2-side on N_2 complex with the Rh-C_2H_4 complex, because C_2H_4, unlike N_2, can ligate to Rh(I) by a way of η^2-side on coordination to form a stable Rh-C_2H_4 complex [6a]. Again, the comparison was carried out at the interfragment distance giving the same EX value. As shown in Table II, the Rh-C_2H_4 complex is more stable than the η^2-side on N_2 complex because of larger ES and FCTPLX interactions. The ES potential of N_2 and C_2H_4 was examined, to investigate the ES interaction. In the case of N_2, the positive ES potential (destabilization of positive charge) expands perpendicularly to the N-N bond but the negative region (stabilization of positive charge) expands towards the outside of N-N bond along the bond axis [2d]. This feature of ES potential, corresponding to the negative quadrupole moment of N_2 [2o], means that the ES stabilization would be large when a positive metal ion such as Rh(I) approaches N_2 along the N-N bond axis, i.e., by way of η^1-end on coordination, but that the ES stabilization would be small when a positive chemical species approaches N_2 perpendicularly to the N-N bond, i.e., by way of η^2-side on coordination. The ES potential of C_2H_4 is very different from that of N_2, in which the negative region of ES potential expands perpendicularly to the C-C bond. This features of ES potential corresponds to the positive quadrupole moment of C_2H_4 and results in the large ES stabilization when the positive Rh(I) approaches C_2H_4 perpendicularly to the C=C bond. The FCTPLX interaction depends on the energy level and expanse of donor orbital. In the η^1-end on N_2 coordination, the σ-lone pair orbital of N_2 plays a key role of a donor orbital, and in the η^2-side on N_2 and C_2H_4 coordinations, the π-orbital is important as donor orbital. Both the σ-lone pair and π-orbitals of N_2 lie similarly in energy (in the deformed structure, σ-lone pair =-17.0 eV, π-orbital =-17.1 eV), but the σ-lone pair orbital extends toward Rh(I) more than the π-orbital does, due to hybridization. The π-orbital of C_2H_4 (-10.1 e.V) lies higher in energy than that of N_2.

In conclusion, the η^2-side on N_2 complex is less stable than the η^1-end on N_2 and C_2H_4 complexes, due to smaller ES and FCTPLX interactions.

3.4. Coordination modes of the CO_2 complexes [2b, 7] ;

As shown in Scheme II there are three possible coordination modes in the CO_2 complexes [17a], whereas the η^1-O end on mode has not been isolated. Here, three coordination modes are examined in $M(PH_3)_2(CO_2)$ (M=Ni(O) or Cu(I)) and $[Co(alcn)_2(CO_2)]^-$ [2b,7]. As given in Table III, the η^2-side on mode is most stable in $Ni(PH_3)_2(CO_2)$, but the η^1-O end on mode is most stable in $[Cu(PH_3)_2(CO_2)]^+$. In both complexes, the η^1-C mode is less stable. First, the η^2-side on mode will be compared with the η^1-O end on mode. Again, a comparison is carried out at the interfragment distance giving the same EX repulsion value. We choose, somewhat arbitrarily, the EX value of 131 kcal/mol, as standard, which is obtained in the η^2-side on Ni complex. Both the η^2-side on and η^1-O end on complexes receive almost the same ES and FCTPLX stabilization. The principale difference between two's comes from BCTPLX supplemented in part by R. A similar comparison is made for the Cu complex at EX=20 kcal/mol. While BCTPLX and FCTPLX + R is almost the same in the two modes, the ES stabilization strongly favors the η^1-O end on mode. Thus, the η^2-side on mode is stable,

Table III. BE, DEF, INT, and energy components for the interaction between $M(PH_3)_2$ and CO_2 (M=Ni(O) or Cu(I)) (kcal/mol)

	$Ni(PH_3)_2(CO_2)$				$[Cu(PH_3)_2(CO_2)]^+$			
		end-on			side-on			
	side-on a)	b) 2.1A	c) 1.7A	C-mode d)	e	f	end-on a)	C-mode g)
BE	-27	1	17	5	57	38	-14	28
DE	34	4	4	1	34	34	4	1
INT	-61	-3	13	4	23	4	-18	27
ES	-76	-14	-75	-	-31	6	-22	-
EX	131	24	131	-	105	20	20	-
FCTPLX	-16	-4	-13	-	-13	-12	-4	-
BCTPLX	-70	-7	-27	-	-35	-10	-8	-
R	-30	-3	-3	-	-3		-3	-

a) The optimized structure. b) R(Ni-O) distance. R=2.0A gives almost the same BE value. c) R(Ni-O) is shortened to give the same EX value as the side-on mode. d) Only the CO_2 part is optimized at R(Ni-C)=2.0 A. e) The same structure as the side-on $Ni(PH_3)_2(CO_2)$. f) The Cu-CO_2 distance is lengthened to give the same EX value as the end-on mode, keeping the CO_2 geometry fixed. g) The same structure as the η^1-C mode of $Ni(PH_3)_2(CO_2)$.

when the back-donating interaction gives a large stabilization. HOMO of $Ni(PH_3)_2$ is mainly composed of the Ni dxz orbital, as described above, and lies high in energy (-6.9 eV). This strong dπ-donor ability results in strong back-donative interaction, yielding the stable η^2-side on complex. In $Cu(PH_3)_2^+$, the Cu 3dxz orbital lies at -15.8 eV (dxz-PH_3 anti-bonding) and at -20.2 eV (dxz-PH_3 bonding). Thus, the η^2-side on complex can not be stabilized by back-donative interaction. In this case, the Cu is formally in +1 oxidation state, and the O atom of CO_2 is negatively charged, leading to relatively strong ES interaction. Thus, the η^1-O end on mode is most stable in the Cu(I) complex. Finally, we will discuss the η^1-C coordination mode which has been isolated in d^8 complexes, $[Co(salen)(CO_2)]^-$ and $RhCl(diars)_2(CO_2)$ (diars = O-phenylenebis(dimethylarsine)) [22,23]. Here, the model compound, $[Co(alcn)_2(CO_2)]^-$, was investigated [7]. The BE value was calculated to be -6.8 kcal/mol by using the larger basis set, in which the geometry of Co-CO_2 part was optimized with the smaller basis set. The other two coordination modes were also examined by using the smaller basis set. However, no-binding was obtained for them. Only the η^1-C mode is possible in $[Co(alcn)_2(CO_2)]^-$, which agrees with the experimental result. From Mulliken population analysis, the Co dz^2 orbital population is decreased very much by the CO_2 coordination; dz^2 = 1.852 in $[Co(alcn)_2]^-$ and 1.226 in $[Co(alcn)_2(CO_2)]^-$. This result suggests the Co $dz^2 \rightarrow CO\ \pi^*$ charge-transfer is very important in $[Co(alcn)_2(CO_2)]^-$. Based on the results of EDA about $Ni(PH_3)_2(CO_2)$ and $[Cu(PH_3)_2(CO_2)]^+$, it is easily understood why the η^1-C mode is possible in $[Co(alcn)_2(CO_2)]^-$ but less stable in $Ni(PH_3)_2(CO_2)$ and $[Cu(PH_3)_2(CO_2)]^+$. In $Ni(PH_3)_2(CO_2)$, the M $d\pi \rightarrow CO_2\ \pi^*$ back donative interaction is strong, enough to stabilize the η^2-side on mode. The Ni $d\sigma \rightarrow CO_2\ \pi^*$ back donative interaction, which is important for the η^1-C mode, is less stronger than the Ni $d\pi \rightarrow CO_2\ \pi^*$ interaction, because the Ni 3dxz orbital is included in the HOMO but the Ni $3dz^2$ orbital is more stable in energy than the Ni 3dxz orbital. Thus, the η^1-C mode is less stable in this case. In $[Cu(PH_3)_2(CO_2)]^+$, only ES interaction is effective and neither Cu $d\sigma \rightarrow CO_2\ \pi^*$ nor Cu $d\pi \rightarrow CO_2\ \pi^*$ back donative interaction is effective. In the η^1-C mode, the positively charged C atom is positioned near to the positively charged Cu atom, which disfavors the η^1-C mode. In $[Co(alcn)_2(CO_2)]^-$, the positively charged C atom is placed near to the positively charged Co atom in the η^1-C mode. In this case, however, this unfavorable situation of the ES interaction is compensated by the presence of negatively charged N and O atoms of ligand. Further, the HOMO of $[Co(alcn)_2]^-$ is mainly composed of the Co dz^2 orbital, yielding strong Co $dz^2 \rightarrow CO_2\ \pi^*$ back donative interaction. As a result, only the η^1-C mode is stable in $[Co(alcn)_2(CO_2)]^-$, and this mode is less stable in $Ni(PH_3)_2(CO_2)$ and $[Cu(PH_3)_2(CO_2)]^+$.

4. Conclusion

In this article, we present the application of energy **decomposition** analysis to the investigation of several group VIII transition metal complexes. As described above, FCTPLX and BCTPLX correspond to donative and back-donative interactions of Dewar-Chatt-Duncanson scheme, while they include polarization interaction, to some extent. The stereochemistry of d^{10} Ni(0) complexes and coordination modes of N_2 and CO_2 complexes are explained successfully, on the basis of energy decomposition analysis. Besides these applications, the stereochemistries of transition metal-SO_2 complexes and non-metal SO_2 complexes have been investigated successfully by using ab-initio MO and energy decomposition analysis [24]. Thus, the energy decomposition analysis is effective method to investigate the metal-ligand interaction and stereochemistry of transition metal complexes.

Acknowledgement

The author would like to thank Dr. K. Kitaura, Prof. K. Morokuma and Prof. K. Ohkulo for their collaborations in studies of Ni(0) complexes and Rh(I) complexes. Also, he is grateful to Drs. A. Dedieu and M. Bénard for their collaborations in part of the study of CO_2 complex. The numerical calculations were carried out at the IMS Computer Center (Okazaki) and the Computer Center of CNRS (Strasbourg). This work was partially supported by Grant-in-Aid of Ministry of Education (JAPAN).

REFERENCES

1 K. Morokuma, J. Chem. Phys., 55, 1236 (1971); K. Kitaura and K. Morokuma, Int. J. Quant. Chem., 10, 325 (1976).
2 (a) K. Kitaura, S. Sakaki, and K. Morokuma, Inorg. Chem., 18, 1558 (1979). (b) S. Sakaki, K. Kitaura, and K. Morokuma, ibid., 21, 760 (1982). (c) S. Sakaki, K. Kitaura, K. Morokuma and K. Ohkubo, ibid., 22, 104 (1983). (d) S. Sakaki, K. Morokuma, and K. Ohkubo, J. Am. Chem., 107, 2686 (1985).
3 (a) K. Morokuma, S. Kato, K. Kitaura, I. Ohmine, S. Sakai, and S. Obara, IMS Computer Program Library, 1980, Program 0372. (b) M. Bénard, A. Dedieu, J. Demuynck, M.-M. Rohmer, A. Strich, and A. Veillard, unpublished.
4 References are cited in [2 b, c].
5 T. T. Tsou, J. C. Huffman, and J. K. Kochi, Inorg. Chem., 18, 2311 (1979).
6 (a) C. Busetto, A. D'Alfonso, F. Maspero, G. Perego, A. Zazzeta, J. Chem. Soc., Dalton Trans., 1977, 1828. (b) D. L. Thorn, T. H. Tulip, and J. A. Ibers, ibid., 1979, 2022.
7 S. Sakaki, A. Dedieu, and M. Bénard, to be published.
8 B. Roos, A. Veillard, and G. Vinot, Theoret. Chim. Acta, 20, 1 (1971). Some modifications were added (See ref. 2a).
9 R. Ditchfield, W. J. Hehre, and J. A. Pople, J. Chem. Phys., 54, 724 (1971).

10 (a) L. Hyla-Kryspin, J. Demuynck, A. Strich, and M. Bénard, J. Chem. Phys., 75, 3954 (1981).
(b) The (14s 9p 8d) primitives were obtained from the original (14s 8p 7d) after some modifications (see ref. 2d).
11 The (14s 9p 6d) primitives were obtained from the original (13s 7p 5d) [10a] after some modifications.
12 S. Huzinaga, "Approximate Atomic Functions", Technical Report, University Alberta, Alberta (1971).
13 In many cases, the DEF of metal part was not considered, because the coordinate bond between various ligands and the same metal fragment were compared with each other. In $Ni(PH_3)_2(CO)_2$, DEF of $Ni(PH_3)_2$ was taken into consideration, because the planar and pseud-tetrahedral structures were compared with each other and the PNiP angle is different very much in these two structures.
14 References are cited in [2b, c].
15 This type of orbital mixing is easily obtained. See the following papers; S. Inagaki and K. Fukui, Chem. Lett., 1974, 509 ; S. Inagaki, H. Fujimoto, and K. Fukui, J. Am. Chem. Soc., 98, 4054 (1976).
16 T. A. Albright, R. Hoffmann, J. C. Thibeault, and D. L. Thorn, J. Am. Chem. Soc., 101, 3801 (1979).
17 For example, (a) D. J. Darensbourg and R. A. Kudaroski, Ad. Organometall. Chem., 22, 129 (1983) ; D. A. Palmer and R. van Eldik, Chem. Rev., 83, 651 (1983). (b) J. Chatt, G. L. M. da Camara Pina, and R. L. Richards, "New Trends in the Chemistry of Nitrogen Fixation", Academic Press, London (1980).
18 C. Krüger, Y.-H. Tsay, Angew. Chem., Int. Ed. Engl., 12, 998 (1973).
19 J. A. Armor and H. Taube, J. Am. Chem. Soc., 92, 2560 (1980).
20 A. D. Buckingham, Adv. Chem. Phys., 12, 107 (1967).
21 These orbital energies were calculated for the deformed structure of ligands as in the Rh(I) complexes.
22 S. Gambarotta, F. Arena, C. Floriani, and P.F. Zanazzi, J. Am. Chem. Soc., 104, 5082 (1982).
23 C. Calabrese, T. Herskovitz, and J.B. Kinney, J. Am. Chem. Soc., 105, 5914 (1983).
24 S. Sakaki, H. Sato, Y. Imai, K. Morokuma, and K. Ohkubo, Inorg. Chem., in Press.

STABILIZATION OF PHOSPHINIDENE AND PHOSPHIRENE BY COMPLEXATION ON PHOSPHORUS. THEORETICAL AND PE STUDIES.

G. Pfister-Guillouzo,* D. Gonbeau
Laboratoire de Physico-Chimie Moléculaire UA 474
Institut Universitaire de Recherche Scientifique
Avenue de l'Université
64000 Pau
France

ABSTRACT. Electronic structures of phosphinidene and phosphirene complexes are described, using EHT, ab initio calculations and data from UV photoelectron spectroscopy in conjunction. The nature of the metal-phosphorus double bond in these compounds was investigated. This stabilization of phosphinidene and phosphirene by complexation on phosphorus was studied on the basis of the diagram of orbital correlation with the fragment orbitals. These conclusions agree with analysis of photoelectron spectra of stable phosphirene complexes.

1. INTRODUCTION

Stabilization and characterization of four π electrons systems is still a problem and numerous recent results deal with the three-membered heterocycles (1). Recently a phosphirene at stable normal pressure and temperature conditions has been isolated (2). Previously the phosphirene pentacarbonyl chromium had been characterized (3).

We have focused our work on these antiaromatic systems. The main stabilizing factors were analyzed on the basis of quantum calculations and of data from UV photoelectron spectroscopy.

Previous to the study of complexed phosphirene we were led to examine the influence of complexation by pentacarbonyl chromium on the electronic structure of phosphinidene. The nature of the P-Cr bond was first analyzed on the basis of a diagram of orbital interactions obtained from Extend Hückel calculations. This method does not lead to an exact determination of spin states ; for this reason we carried out a more sophisticated PS-HONDO ab initio calculation. These results were used as the basis for studying the interactions between this system and acetylene, and estimating the role played by the metal atom in the stabilization of phosphirene cycle.

2. COMPUTATIONAL AND EXPERIMENTAL DETAILS

EHT calculations were performed with the parameters previously utilized for Cr (4). All SCF calculations were carried out using a version of the

A. Veillard (ed.), Quantum Chemistry: The Challenge of Transition Metals and Coordination Chemistry, 333–340.

HONDO program PSHONDO (5)

For carbon, oxygen and phosphorus the pseudo potentials and basis sets were those previously determined (6). The four gaussian functions were contracted to the double ξ level (3+1) for each atom except for carbon and oxygen in $Cr(CO)_5$. For phosphorus a 3d gaussian function was added ($\alpha_d = 0.57$) as a polarization function. We have determined the non-empirical atomic pseudo potential for chromium from the atomic Hartree Fock calculations of Clementi and Roetti (7). A valence basis set (4s 4p 3d) was optimized for this atom : for 4s and 3d orbitals these gaussian functions were contracted to the double ξ level by means of a 2+1 procedure for 4s and 4+1 for 3d. The SCF valence energies for the open shell triplet states were obtained by calculation of the mean value of the H operator between the wave functions determined from a Nesbet type operator (8).

The geometric parameters were those obtained with X ray diffraction for phosphirene pentacarbonyl tungsten (3).

Spectra were recorded on a 0078 spectrometer PES Laboratories equiped with a HeI-HeII source.

All spectra were calibrated with the $^2p_{1/2}$ and $^2p_{3/2}$ lines of xenon and argon.

3. RESULTS

3.1. Preferential structures of phosphinidene complexe RP-$Cr(CO)_5$ with the E.H.T.method

We first estimated the preferential structure of the RP-$Cr(CO)_5$ entity the existence of which was demonstrated by chemical trapping, by analyzing the nature of the interactions occurring between a singlet phosphinidene and a metal fragment (fig. 1).

A RP singlet phosphinidene is characterized by two degenerated orbitals with p symmetry one being formally occupied and the other vacant. By comparison with its carbene homologue this arrangment enables us to "a priori" predict a difference in the interactions.

The $Cr(CO)_5$ fragment contains three occupied d orbitals, two with π symmetry and one with dδ symmetry. The first virtual orbital has considerable localization on the dσ orbital.

The principal two-electron-stabilizing interactions involve :

- the p(P) HOMO of the ligand and the LUMO of $Cr(CO)_5$: σ(ligand $\rightarrow$ metal transfer)
- the p*(P) LUMO of the ligand and the $d\pi_1$ of the metal : π (metal $\rightarrow$ ligand) transfer.

These transfers give a double nature to the P-Cr bond.

The most favorable arrangment should "a priori" be a θ angle of 90° between the RP and PCr bonds since this would lead to maximal p(P) dσ(Cr) overlap. A variation of the main geometric parameters (d P-Cr and θ) has enabled us to observe that the energetically preferred conformation included d P-Cr and θ values of 1,9 Å and 125° (stabilization of about 92 $kJ.mol^{-1}$).

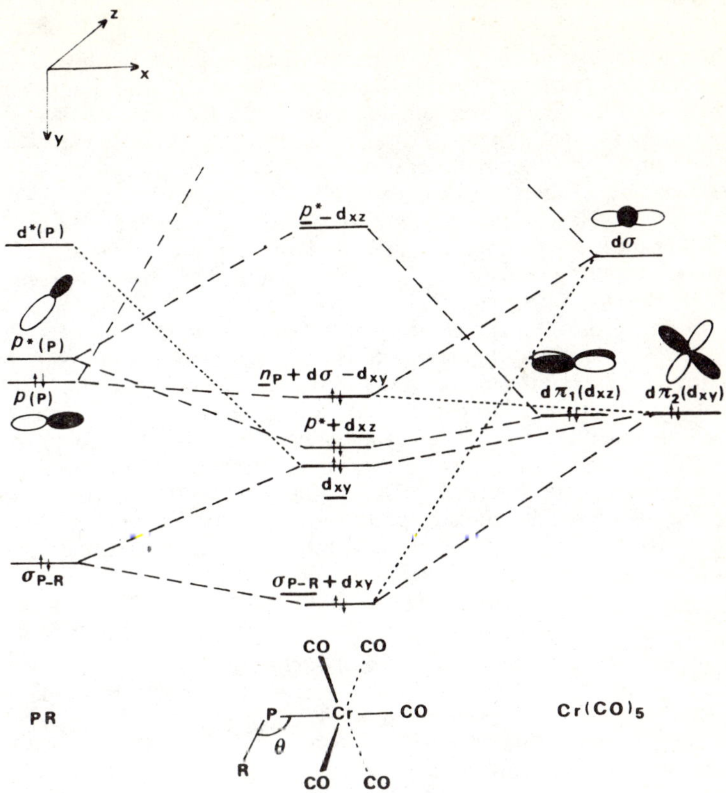

Figure 1 - Interaction diagram (EHT) for $RP{=}Cr(CO)_5$
– – – – main interactions at θ = 90° and θ = 125°
-------- secondary stabilizing interactions at θ = 125°.

The origin of the observed stabilization stems from two factors :
1) participation of secondary stabilizing interactions : σ P-R ↔ d σ (Cr) and more particularly that involving the second hybrid virtual orbital with p-d character of the RP fragment (noted d*(P)) and the occupied orbital $d\pi_2$ of $Cr(CO)_5$ (slightly greater metal ligand transfer).
2) decrease of destabilizing interactions σ PR ↔ $d\pi_2$ (Cr). Actually this effect reflects steric factors.

We also observed that the staggered RP Cr-CO position was favored by about 3.3 $kJ.mol^{-1}$.

These results correspond to the structural data obtained by X ray diffraction for phosphirene complexed with pentacarbonyl tungsten (3).

3.2. Electronic structure of the phosphinidene chromium pentacarbonyl complex

3.2.1. We extended this approach by carrying out ab initio calculations on complexed phosphinidene and corresponding fragments by considering the singlet and triplet states.

In the case of phosphinidene the ground state is a triplet one. The singlet-triplet separation is of about 111 $kJ.mol^{-1}$. On the other hand complexation of the HP entity generates considerable stabilization of the singlet state. It therefore appears to be energetically favored by 52.2 $kJ.mol^{-1}$. The ground state of complexed phosphinidene therefore corresponds to a singlet state as predicted during studies of the reactivity of this entity (9).

On the contrary the $Cr(CO)_5$ fragment presents a singlet ground state but the energy difference between this state and the triplet state of $Cr(CO)_5$ with π symmetry is much greater than that calculated for phosphinidene ($\sim$ 570 $kJ.mol^{-1}$). It seems more reasonable to consider the formation of the phosphinidene complex from singlet state of the fragments.

The orbitals of complexed phosphinidene are characterized by a HOMO with a predominant localization on the $n_p(P)$ pair of phosphorus, the following three occupied orbitals are localized primarily on the d orbitals of chromium. The LUMO involves the p* orbital on the phosphorus atom. Energetically, by comparison with phosphinidene we observe a lift of the degeneracy of the phosphorus orbitals. The present results suggest that if the reactivity on singlet phosphinidene and metal phosphinidene complexes is frontier orbitals-controlled then the behaviour of these species towards alkenes must be different.

3.2.2. Charge distributions. We examined on the basis of a Mulliken population analysis the extent of σ transfer and π back transfer.

The electronic charge on the $d\pi$ orbital of the chromium atom decreases by approximatively 0,49e when passing from $Cr(CO)_5$ to $HP{=}Cr(CO)_5$. At the same time the ligand metal σ transfer has been estimated at about 0.95 e^-. It thus appears that there is much more σ transfer than π transfer. So, the effect of complexing imparts a notable net positive charge of +0.41 to phosphorus. In comparison to carbene complexes, phosphinidene complexes are much more electropositive, a hypothesis formulated previously to interpret the behaviour of this entity in trapping reactions.

4. STABILIZATION OF PHOSPHIRENE : PART OF COMPLEXATION

4.1. Quantum study

We continued the present investigation by examining the phosphirene and the phosphirene chromium pentacarbonyl complexe. In order to understand the stabilization of phosphirene cycle by complexation more fully, we comparatively analyzed the interactions between the orbitals of acetylene HP and $HP\text{-}Cr(CO)_5$ fragments.

These diagrams are not correlation diagrams since in the complexed phosphirene case this reaction is not allowed by the symmetry (least motion electronic principle).

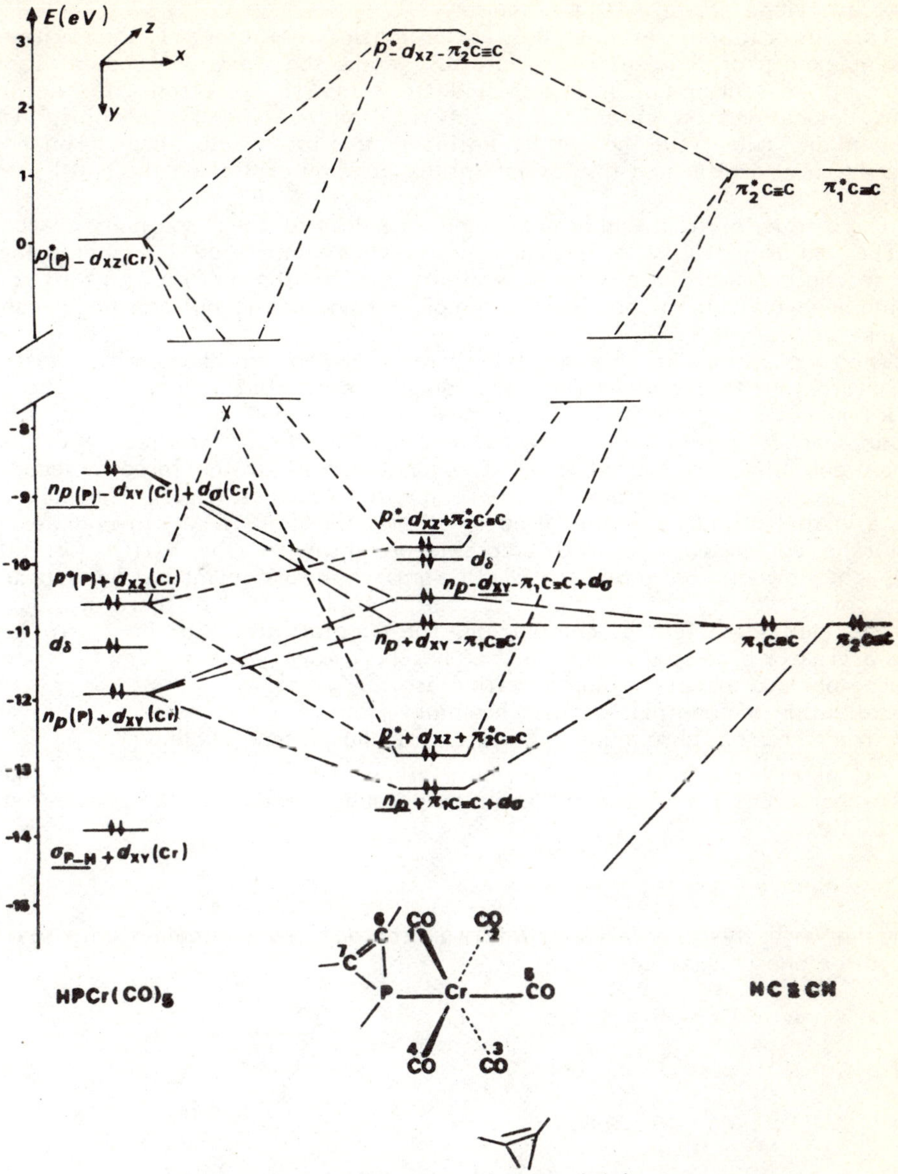

Figure 2 - Interaction diagram for HP=Cr(CO)$_5$ fragments HP=Cr(CO)$_5$ and HC≡CH. Underlined values correspond to the highest localizations.
— — — interactions with n_p(P)
- - - - - - interactions with p*(P)

Two principal interactions can be distinguished :

- the first one is a destabilizing interaction between the n_p(P) pair of phosphorus and the acetylene orbital with π symmetry.

This interaction reflects the antiaromatic character of the system. For complexed phosphinidene the n_P orbital and the dxy (occupied) and dσ (virtual) of chromium mix significantly. This six electrons interaction is thus delocalized on chromium, phosphorus and acetylene. In keeping with orbital mixing rules (10) the totally antibonding orbital is localized primarily on the dxy (Cr) orbital and the lowest totally bonding orbital on the n_P(P) pair (fig.2).

The σ transfer to the chromium induces a loss of the "lone pair" character. The antiaromaticity of phosphirene is thus diminished by complexing.
- The second characteristic of these systems involves the p (P) orbital of phosphinidene (vacant in the case of complexed phosphinidene) and the vacant acetylene orbital π^*_2 C≡C.

The occupation of this orbital corresponds to an electronic transfer from heteroatom to acetylene. The importance of σ delocalization is one of the factors governing the stability of these systems. In the case of complexed phosphinidene it should be noted that the p*(P) orbital interacts with the occupied chromium dπ orbital (dxz). The localizations of the three molecular orbitals can be deduced from the above mentioned rules (fig.2). The highest one is a virtual primarily localized on π^*_2C≡C, the next HOMO of complexed phosphirene corresponds to a considerable weighting of the dxz(Cr) orbital. Finally the lowest characteristic of PC antisymmetric combination appears with a much lower energy than for the phosphirene. We therefore have a direct stabilizing effect by complexing. The π metal phosphorus transfer causes a greater σ delocalization on the phosphorus-acetylene system.

The obvious effects of complexation are :
- delocalization of lone pair on the chromium atom
- back transfer from chromium to σ system of phosphirene cycle.

We observe on the whole a more important net positive charge on the phosphorus Q_P = 0.57 e in the complex phosphirene than in the phosphirene Q_P = 0.03 e.

4.2. Photoelectron spectra

In parallel with this theoretical study we recorded the photoelectronic spectra of three phosphirene complexes :

1 M = W R' = Et R = CH_3

2 M = W R' = Et R = phenyl

3 M = Cr R' = Et R = phenyl

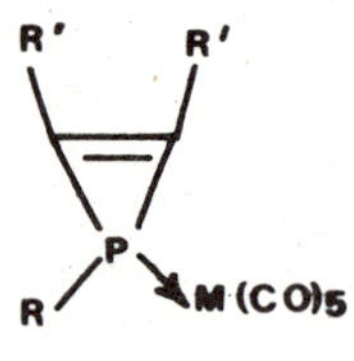

This technique is well adapted for the estimation of metal ligand interactions and leads also to verification of the validity of our conclusions.

In the conditions of temperature (more than 100°C) and pressure (5.10^{-2} mbars) required for recording the spectra, these complexes decompose with formation of the corresponding acetylenic derivatives. The values of ionization potentials are reported in table I. A comparative analysis of band intensities with HeI (21.21 eV) and HeII (40.41 eV) excitations leads to a reliable assignment of the spectra in agreement with the theoretical descriptions of different molecular orbitals.

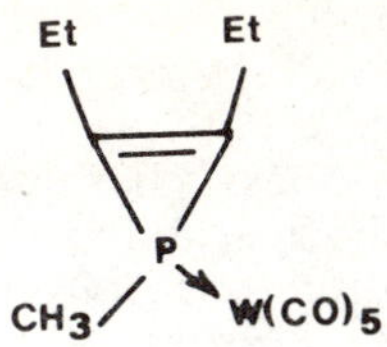	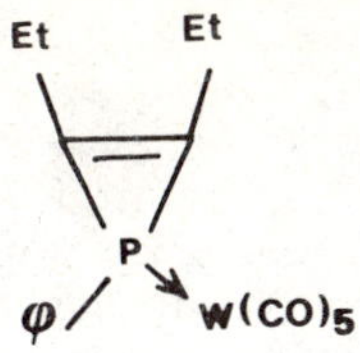	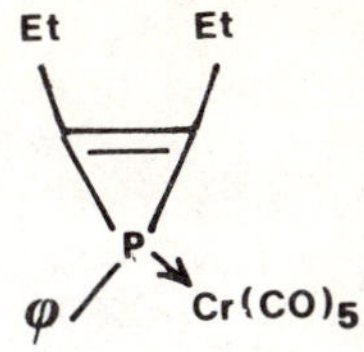	Attribution
7.35 7.70 ↗	7.30 7.50 ↗ 7.80	7.20 7.50 ↗	d_{Cr}
	9.	8.9 - 9.3	
9.87 ↘	10.30 10.75 ↘	10.20 10.65 ↘	$n_p - \pi_{C=C}$ σ^P_{P-C} (A.S)
10.60 ↘	11.35 ↘	11.35 ↘	$n_p + \pi_{C=C}$

Table I - Experimental vertical ionization potentials (eV) of phosphirene complexes.

The intensity of the band decreases ↘ (increases ↗) when going from the HeI to HeII spectrum.

For compound **1** the intensity of the first band increases considerably in going from HeI to HeII excitation. It may thus be unambiguously assigned to the ionization of the d orbitals.

The expansion obviously shows the appearance of three separate bands. The band at 9.87 eV is attributed to the ionization of the molecular orbital corresponding to the $n_p(P) \leftrightarrow \pi\ C=C$ interaction with, as seen above, significant weighting of d orbitals and a weak localization on phosphorus. This attribution is consistent with the fact that we observe weak intensity variation.

Two ionizations are attributed to the 10.60 eV band :

- that of the characteristic molecular orbital of the P-C bonds
- that corresponding to the phosphorus-pair ethylenic system interaction. The localization calculated for phosphorus is important. The considerable intensity decrease observed for this band corresponds to this attribution

For compounds **2** and **3** in addition to the intense band at about 9 eV due to the ionizations of the phenyl ring, three distinct bands are observed between 10.20 and 11.35 eV. The presence of the phenyl ring induces a stabilization of the two ionizations associated with the $n_p(P) \leftrightarrow \pi C=C$ interaction observed at 10.30 - 11.35 eV for 2 and 10.20 - 11.35 eV for 3.

The ionization potentials observed at 10.75 and 10.65 eV correspond to the molecular orbital with σ symmetry (PC antisymmetric combination).

This attribution is based on two experimental observations :

the band at 11.35 eV in these compounds is the one whose intensity decreases the most (localization on phosphorus). Also when passing from tungstene to chromium, we note that the first band, and the two following bands previously attributed to the ionization of molecular orbitals with a certain degree of the d orbitals of the metal, move towards lower energies.

Based on these experimental data we observe a destabilizing "lone pair n_p π C≡C system" interaction of about 1 eV, which does not seem to be considerable. The ionization potential associated with the most characteristic orbital of these entities, since it partially reflects their stabilization is at highest energies : 10.60 - 10.75 eV. In non substituted phosphirane, the ionization potential associated with the orbital partially presenting the same nature appears at 10.21 eV (11) reflecting a considerable overall stabilization notably taking alkyl group effects into account.

Acknowledgments - The authors thank Dr F. Mathey for gift of samples and for helpful discussion. Allowance of computer time through an A.T.P.-Cray is gratefully acknowledged.

References

(1) a) M. Torres, E.M. Lown, H.E. Gunning and O.P. Strausz, Pure Appl. Chem., **52**, 1623 (1980) ; b) A. Krantz and J. Laureni, J. Am. Chem. Soc., **103**, 486 (1981) ; c) M. Torres, J.L. Bourdelande, A. Clement and O.P. Strausz, J. Am. Chem. Soc., **105**, 1698 (1983) ; d) H. Bock and B. Solouki, Angew. Chem., **20**, 427 (1981) ; e) G. Capozzi, O. de Lucchi, V. Lucchini and G. Modena, Tetrahedron Lett., 2603 (1975) ; f) R. Destro, T.Pilati and M. Simonetta, Nouv. J. Chim., **3**, 533 (1979)

(2) A. Marinetti, F. Mathey, J. Fischer and A. Mitschler, J. Chem. Soc. Chem. Comm., 45 (1984)

(3) A. Marinetti, F. Mathey, J. Fischer and A. Mitschler, J. Am. Chem. Soc., **104**, 4484 (1982)

(4) R.J. Goddard, R. Hoffmann and E.D. Jemmis, J. Am. Chem. Soc., **102**, 7667 (1980)

(5) a) M. Dupuis, J. Rys and H.F. KIng, J. Chem. Phys., **65**, 11 (1976) ; b) P. Durand, J.C. Barthelat, Theor. Chim. Acta., 38228 (1975)

(6) Laboratoire de Physique Quantique, Toulouse, Ateliers, Octobre 1981

(7) E. Clementi and C. Roetti, At. Data and Nucl. Data Tables, **14**, 177 (1974)

(8) R.K. Nesbet, Rev. Mod. Phys., **35**, 552 (1963)

(9) A. Marinetti and F. Mathey, Organometallics, **3**, 456 (1984)

(10) S. Inagaki, H. Fujimoto and K. Fukui, J.Am.Chem.Soc., **98**, 4054 (1976)

(11) D.E. Aue, H.M. Webb, W.R. Davidson, M. Vidal, M.T. Bowers, H. Goldwhite, L.E. Vertal, J.E. Douglas, P.A. Kollman and G.L. Kenyon, J. Am. Chem. Soc., **102**, 5151 (1980)

MOLECULAR ORBITAL CALCULATIONS ON METALLOCENES WITH UNUSUAL GEOMETRIES

Thomas A. Albright, Jesus P. Lopez, and Jerome Silvestre
Department of Chemistry
University of Houston
Houston, Texas 77004
USA

ABSTRACT. An alternative structure for ferrocene is proposed on the basis of the isolobal analogy. Its bonding and stability relative to the classical structure of ferrocene has been examined by extended Hückel and ab initio calculations. Extension to other metallocenes and a strategy to stabilize this nonclassical type of structure has been carried out.

The synthesis and structural determination of ferrocene in the early fifties (1) was the catalyst of modern organometallic chemistry. It revolutionized the way organic and inorganic chemists thought about compounds (2). Today a vast number of metallocenes, wherein a transition metal is coordinated to one or more polyene ligands, have been prepared and structually determined (3). Another discovery that has helped to bridge the worlds of organic and inorganic chemistry has been the isolobal analogy (4). Two molecules or fragments within molecules are said to be isolobal if the number, symmetry properties, occupation by electrons, and approximate energy of their valence orbitals are similar. The example which we will draw from most heavily in this contribution pairs CH^{3+} with a $FeCp^{+}$ or $CoCp^{2+}$ fragment. These fragments each contain three empty frontier orbitals, 1. As we shall soon

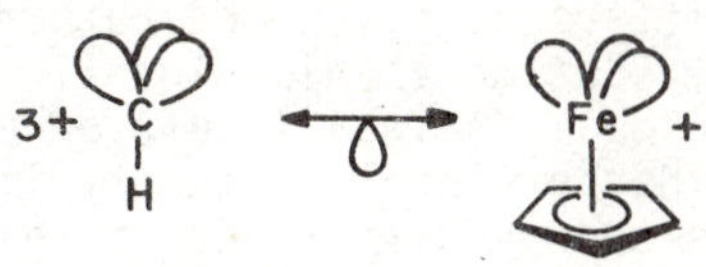

1

see the empty orbitals in each case will be of a_1+e symmetry. Part of the novelty of the isolobal analogy is that one can replace fragments in molecules with isolobal partners to generate new molecules. This is illustrated for one example in Scheme I. The cobaltacyclopentadiene complex 2 has been known for sometime (5). Replacement of a "basal" CoCp unit (within the colbaltacyclopentadienyl ring) with an isolobal CH^{+} unit generates the cobaltocenium cation, 3, which is

A. Veillard (ed.), Quantum Chemistry: The Challenge of Transition Metals and Coordination Chemistry, 341–350.

SCHEME I

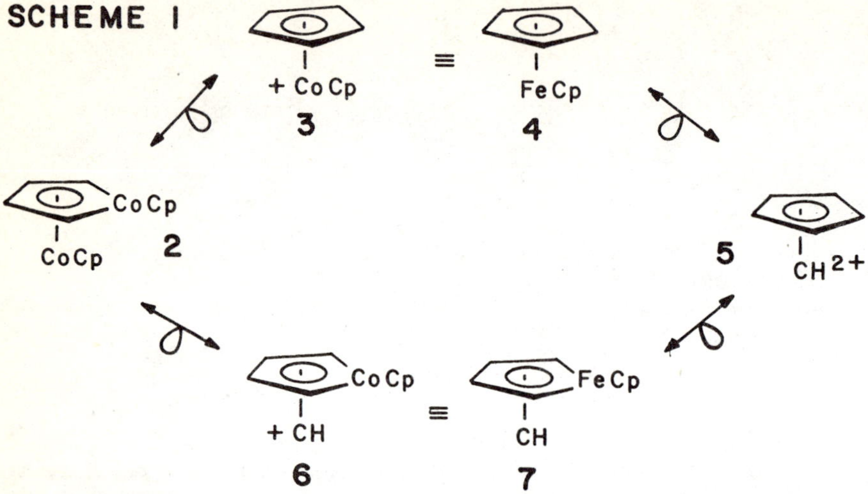

isoelectronic to ferrocene, 4. Substitution of FeCp with an isolobal CH^{2+} fragment generates the Hogeveen-Kwant dication, 5, (6). All of these compounds are well-known. The basic electronic structures of 2 (7), 4 (8), and 5 (9) have been studied extensively. But why single out the "basal" CoCp in 2 for replacement? An equivalent tranformation would be to replace the "apical" CoCp unit in 2 with CH^+. This generates 6 which is isoelectronic to 7, a structural isomer of ferrocene. We shall refer to 7 as the nonclassical isomer of ferrocene and 4, where the two cyclopentadienyl rings lie parallel to each other, as the classical isomer. There is absolutely no precedent for the nonclassical ferrocene structure in the literature. Over 50 structures of ferrocene and simple derivatives have been reported (10). None give even the slightest hint of a distortion towards the nonclassical geometry. Thus, the isolobal analogy has led us to a prediction of a unique class of compounds.

When we carried out extended Hückel calculations on the two isomers of ferrocene we found that the classical structure was 6.0 eV more stable than the nonclassical one. Admittedly, not much faith should be placed on this energy difference given the rather crude nature of the extended Hückel method. But does this gigantic energy difference signal a failure of the isolobal analogy? Before we present the results of calculations at a higher level of theory it is instructive to trace the origin of this energy difference at this stage.

Figure 1 shows an orbital interaction diagram for interacting the filled π orbitals of cyclopentadienyl anion with the valence orbitals of FeCp cation (4b,11). The lowest π level, a_2'', is stabilized by empty $2a_1$ in FeCp cation. Likewise, the e_1'' set of cyclopentadienyl anion interacts mainly with the e_1 set and is stabilized. The $1a_1$ and e_2 orbitals of FeCp cation are left nonbonding. They correspond to the "t_{2g}" set in this octahedral splitting pattern. Thus, a total of three stabilized, bonding and three nonbonding levels are produced in this union.

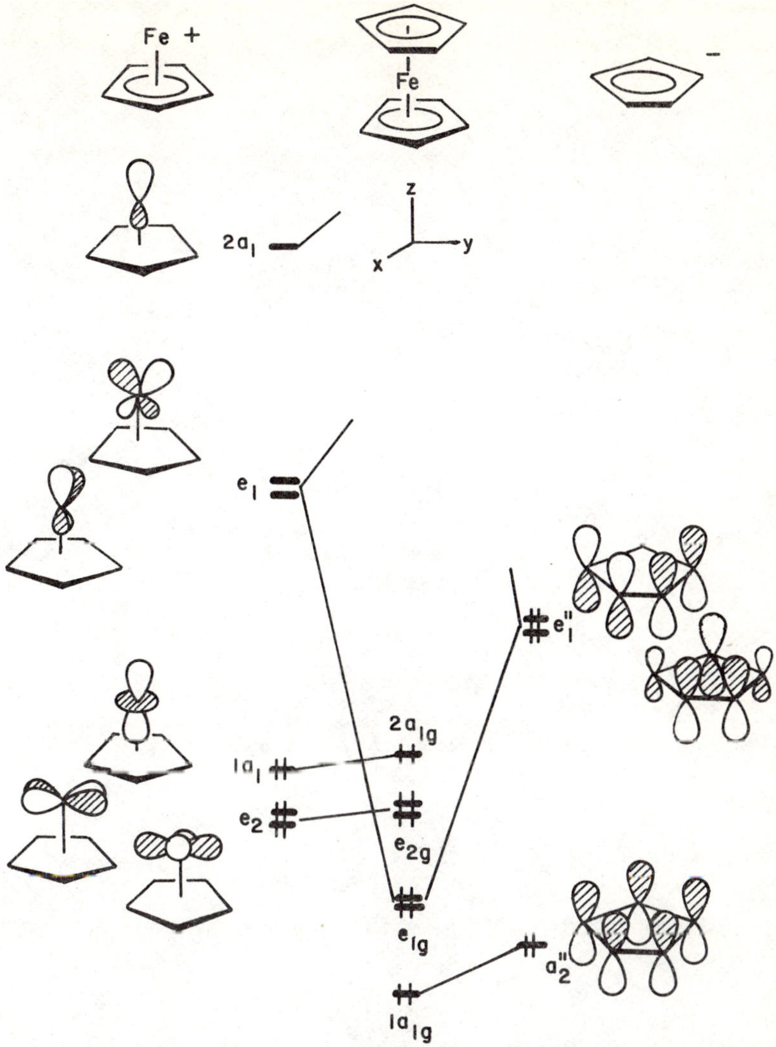

Figure 1. Building the valence orbitals of ferrocene in D_{5d} symmetry.

We can partition the nonclassical isomer of ferrocene in an analogous fashion. There are three empty valence orbitals of CH^{3+}, as shown on the left side of Figure 2. It is clear that the hybrid function, hy, corresponds to the $2a_1$ fragment orbital of FeCp cation in Figure 1. Likewise, the e set is analogous to the e_1 set of FeCp cation. The two fragments are isolobal. On the right side of Figure 2 are the valence orbitals of a ferrole^{3-} fragment (7,12). The π_1 and π_3 levels in the ferrole fragment are topologically equivalent to a_2'' and one component of e_1'' in the cyclopentadienyl anion (see Figure 1). Unfortunately one d orbital on the iron, xz using the coordinate

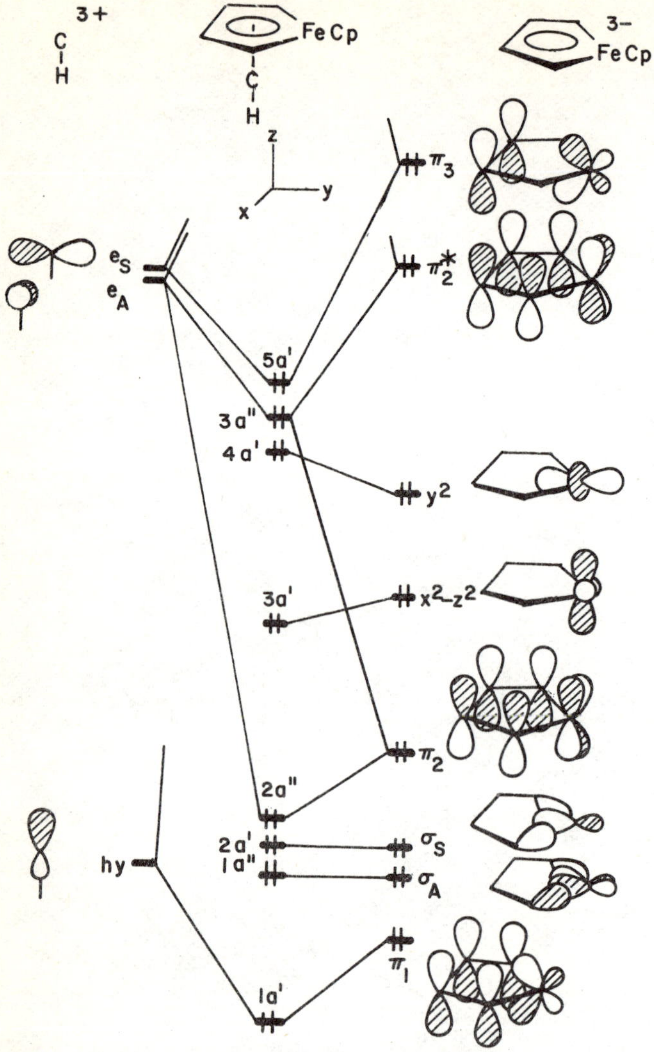

Figure 2. An orbital interaction diagram for the nonclassical isomer of ferrocene.

system at the top of Figure 2, lies nearly degenerate with a π orbital identical to the other component of e_1''. The two orbitals mix to produce π_2 and π_2^* on the right side of Figure 2. We shall see that this combination will deconvolute itself again when the CH^{3+} fragment interacts; so neglecting this technicality, it is clear that ferrole^{3-} is isolobal to cyclopentadienyl anion. The orbital interaction pattern for the classical and nonclassical isomers of ferrocene are also very similar. In Figure 2 the π_1 and π_3 orbitals of ferrole are stabilized by hy and e_s, respectively, to produce the 1a' and 5a' molecular orbitals. The e_A fragment orbital overlaps with π_2 and π_2* to produce a bonding molecular orbital, 2a", and a "nonbonding" one, 3a". The 3a" MO has almost totally metal xz character; this is a result that can be easily derived from perturbation theory considerations. Finally, the y^2 and x^2-z^2 metal functions stay primarily nonbonding. Notice that we have constructed three nonbonding, metal centered MOs (3a', 4a', and 3a") which are identical to the $2a_{1g}$ and e_{2g} set in the classical isomer of ferrocene. There are also three bonding orbitals (1a', 2a", and 5a') which are analogous to $1a_{1g}$ and e_{1g}. The strength of these three bonding interactions is quite comparable for the two structures; a result that can be determined in a number of ways using perturbation theory (13). Thus, the isolobal analogy appears to hold quite well for this example. But, where then is the energy difference between the two structures? Notice in Figure 2 that there are two other high-lying, filled orbitals, σ_S and σ_A. They are nothing more than symmetry adapted linear combinations of the two Fe-C σ bonds in the ferrole. The analogous C-C σ orbitals for ferrocene lie at a much lower energy. This is where the large energy difference between the classical and nonclassical structures comes from. The Fe-C σ orbitals lie at higher energy than their C-C counterparts since Fe is more electropositive than C and its d orbitals are considerably more diffuse than the C 2s/2p set which in turn leads to a smaller overlap in forming the σ bond.

The conversion of nonclassical to classical ferrocene which maintains a mirror plane of symmetry is symmetry allowed. Therefore, with an energy difference of 6.0 eV at the extended Hückel level, it is not likely that a nonclassical structure will be a minimum on the potential surface. We have carried out *ab initio* calculations on both isomers of ferrocene using an STO-3G basis set (14) for the metal and ligand atoms. Complete geometrical optimization was undertaken within C_s symmetry for the nonclassical and within D_{5h} symmetry for the classical isomer. The nonclassical structure lies in a twixtyl region (15), a plateau where no true minimum exists. It decays without activation to a classical structure. The orbital structure of the nonclassical structure at the *ab initio* level is precisely that given in Figure 2. In particular there are two high-lying Fe-C σ levels, analogous to 1a" and 2a' at this geometry. It is amusing that the STO-3G calculations do not predict a D_{5h} (or D_{5d}) geometry to be the minimum. A structure, shown in 8, lies 37 kcal/mol below that optimized within a D_{5h} symmetry constraint! The iron atom is coordinated in an η^5 position for one cyclopentadienyl ring and at η^3 in the other. The two noncoordinated carbon atoms are tipped by an angle of 16° away from the plane defined

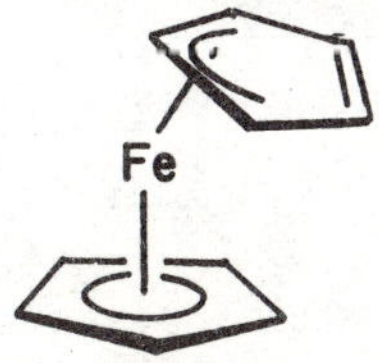

8

by the three coordinated carbons. Furthermore, the η^3 plane does not lie parallel to the η^5 ring. They make an angle of 40°. The C-C bond lengths in the η^5 ring were very close to being equivalent, with an average value of 1.42Å. In the other cyclopentadienyl ring, however, the two coordinated C-C bond lengths were 1.46Å, the other two were 1.44Å, and the unique one was 1.35Å. There has been much worry in the literature about the failure of ab initio methods at the Hartree-Fock level to produce the correct Fe-C bond lengths in ferrocene (16). All of these optimizations were carried out assuming a D_{5h} or D_{5d} symmetry. Our computations suggest that there should be even more concern. We have also used a more extensive basis set, at the 4-31G level for the ligand atoms and a locally developed (12s, 9p, 6d) set contracted to (4s, 3p, 2d) at iron. The relative energy difference between the two structures (optimized at the STO-3G level) was reduced to 5 kcal/mol. We have not yet computed the energy gradient for either geometry, but it is reasonable to assume that it still will not be at D_{5h}.

Returning back to the major theme of this article, there are two basic ways to stabilize a nonclassical metallocene of arbitrary size and type relative to the classical structure. One could lower the energy of the two metal-carbon σ bonds for the nonclassical isomer by increasing the electronegativity of the metal, etc. We doubt that this will substantially offset the huge energy difference. The alternative is to build two high-lying σ orbitals into the classical structure so that they more evenly match those at the nonclassical geometry. We have used two strategies to probe this idea. First of all, we decreased and increased the electronegativity of one carbon atom by replacing it with boron and nitrogen. The results at the extended Hückel level are shown in 9. The energy difference is cut almost

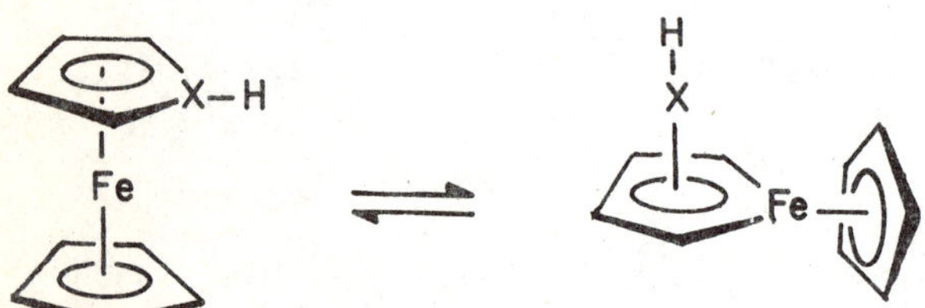

XH	ΔE (eV)
a) BH^-	3.6
b) CH	6.0
c) NH^+	7.7

9

in half by replacing carbon with the less electronegative boron atom. There are two B-C σ levels in the classical isomer that lie at a much higher energy than their C-C σ counterparts. Notice that replacement with the more electronegative nitrogen atom creates an even larger energy difference which is consistent with our ideas. Our preliminary ab initio calculations on the borole complex, 9a, also show this trend. However, the energy difference for 9a is still probably too large to hope for its isolation. Indeed, all borole and heterocyclic metallocene complexes reported to date have a classical geometry (17). The second strategy is to make the cyclic polyene ring smaller in the classical isomer. Constraining the C-C-C bond angles to be small will cause C-C σ orbitals to rise in energy. On the other hand, creating a small C-metal-C angle will not cause metal-carbon σ orbitals to rise in energy. The component on the metal is quite diffuse and contains appreciable d character. We investigated the isomers shown in 10 to check this hypothesis. At the extended Hückel level, the nonclassical

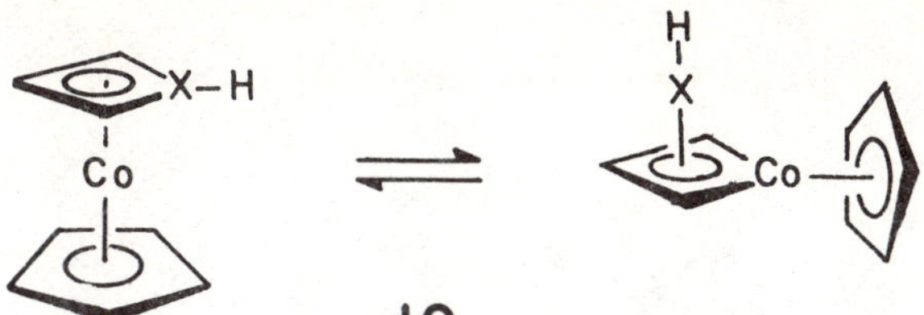

	XH	ΔE (eV)
a)	CH	1.9
b)	BH^-	1.1

10

isomer for 10a lies only 1.9 eV above the classical one. All reported compounds analogous to 10a are of the classical type (18), but one wonders if the nonclassical geometry can serve as a intermediate in pyrolytic or photolytic reactions? In fact at the ab initio level (using the STO-3G or more extended basis set) the two isomers of 10a are at an equivalent energy. This is probably not an accurate accessment of the relative energies, we shall examine correlation effects in the near future. Most importantly, however, the ab initio calculations find a true minimum for the nonclassical structure. At the STO-3G level the cobalt atom lies 0.67Å out of the plane defined by the three carbons in the metallacycle. The apical CH unit lies roughly in an η^4 position with one C-C distance of 1.44Å and two others at 1.55Å. The apical C-Co distance was predicted to be somewhat long - 2.33Å. Compound 10b or any derivative is, to our knowledge, unknown. The extended Hückel calculations predict an even smaller difference of 1.1 eV. While we have not completed geometry optimizations for 10b at the ab initio level, our preliminary results put the nonclassical geometry at a substantially lower energy.

Are there any examples of compounds which have been shown to undergo a transformation analogous to that in 9 or 10? Indeed there are many. A number of dinuclear complexes including 2 (5) and the isolobal analog where both CoCp units have been replaced by $Fe(CO)_3$ groups (19) have been shown to undergo this rearrangement. A set of trinuclear clusters, isolobal to 10, have recently been shown to undergo this isomerization (20) and a binuclear complex is actually caught at the transition state in the solid state (21). A compound with the empirical formula $BC_4H_5Fe(CO)_3$ has been reported on the basis of NMR data (22) to have a structure like 7 with an $Fe(CO)_3$ in the basal plane and a CH group at the apex! We doubt that this is the case for the reasons given earlier concerning the nonexistence of 7. There are, however, several well-documented cases of the isomerization reaction in metalloboranes (23). For example, $B_5H_{10}FeCp$ and B_4H_8CoCp undergo the rearrangements shown in 11 and 12, respectively. Here the solid circles stand for BH units and there are respectively five and four bridging hydrogens in the ring. The compounds in 11 are isolobal to the nonclassical and classical isomers of ferrocene while those in 12 are isolobal to cyclobutadiene-CoCp.

We still are not in a position to give an accurate assessment of

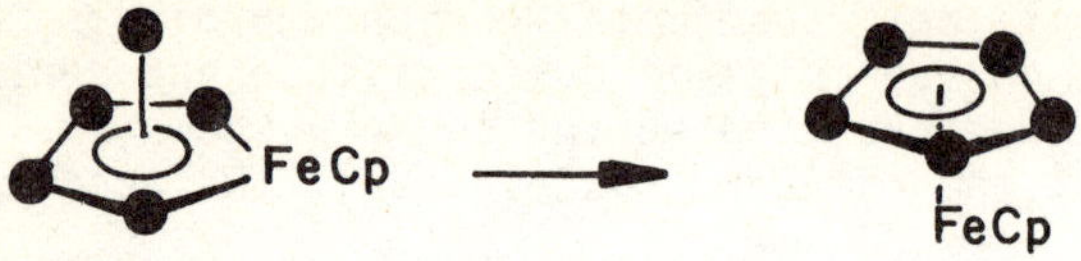

11

12

the relative energy differences for these metallocenes. We do think, however, that nonclassical structures can exist, particularily for 10b, and we encourage experimental work in this direction.

ACKNOWLEDGEMENTS. We thank the Robert A. Welch, Alfred P. Sloan, and Petroleum Research Foundations for support of this work. The idea for the work reported here originated while one of the authors (TAA) was a postdoctoral student with Professor Roald Hoffmann. We thank him for his insight and encouragement.

REFERENCES.

1. a) T.J. Kealy and P.L. Pauson, Nature,168,1039(1951).
   b) G. Wilkinson, M. Rosenblum, M.C. Whiting and R.B. Woodward, J. Am. Chem. Soc.,74,2125(1952).
   c) R.B. Woodward, M. Rosenblum, and M.C. Whiting, ibid,74,3458 (1952).
2. A perhaps typical reaction to this discovery may be found in A.S. Smith, Chem. Ind.,353(1955).
3. a) C.M. Lukehart, "Fundamental Transition Metal Organometallic Chemistry" Books/Cole, Belmont, CA (1985).
   b) J.P. Collman and L.S. Hegedus, "Principles and Applications of Organotransition Metal Chemistry", University Science Books, Mill Valley, CA (1980).
   c) M. Tsutsui, M.N. Levy, A. Nakamura, M. Ichikawa, and K. Mori, "Introduction to Metal π-Complex Chemistry", Plenum Press, New York, NY (1970).
   d) G. Deganello, "Transition Metal Complexes of Cyclic Polyolefins", Academic Press, New York, NY (1979).
4. a) R. Hoffmann, Angew. Chem.,94,725(1982); Angew. Chem., Int. Ed. Engl.,21,711(1982).
   b) T.A. Albright, J.K. Burdett and M.-H. Whangbo, "Orbital Interactions in Chemistry", John Wiley, New York, NY (1985).
   c) T.A. Albright, Tetrahedron,38,1339(1982).
5. a) M. Rosenblum and B. North, J. Am. Chem. Soc.,90,1060(1968).

b) M. Rosenblum, B. North, D. Wells, and W.P. Giering, ibid,94, 1239(1972).
c) M. Rosenblum, W.P. Giering, B. North, and D. Wells, J. Organomet. Chem.,28,C17(1971).
d) W.S. Lee and H.H. Brintzinger, ibid,127,93(1977).

6. H. Hogeveen and P.W. Kwant, Accts. Chem. Res.,8,413(1975); H. Carnadi, C. Giordano, R.F. Heldeweg, H. Hogeveen, and E.M.G.A. van Kruchten, Isr. J. Chem.,21,229(1981); H. Hogeveen and E.M.G.A. van Kruchten, J. Org. Chem.,46,1350(1981).
7. D.W. Thorn and R. Hoffmann, Inorg. Chem.,17,126(1978).
8. E.M. Shustorovich and M.E. Dyatkina, Doklady Akad. Nauk. SSSR, 128,1234(1959); O.Ya. Lopatko, N.M. Klimenko, and M.E. Dyatkina, ibid,192,1083(1970); M.-M. Rohmer and A. Veillard, Chem. Phys.,11, 349(1975); P.S. Bogus, U.E. Walgren, and J. Almlof, J. Chem. Phys., 64,2324(1976).
9. H.T. Jonkman and W.C. Nieuwpoort, Tetrahedron Lett.,1671(1971); E.D. Jemmis and P.V.R. Schleyer, J. Am. Chem. Soc.,104,4781(1982); K. Lammertsma and P.V.R. Schleyer, ibid,105,1049(1983); V.I. Minkin and R.M. Minyaev, Usp. Khim.,51,586(1982); M.J.S. Dewar and M.K. Holloway, J. Am. Chem. Soc.,106,6619(1984).
10. C. Kruger, B.L. Barnet, and D. Brauer in "The Organic Chemistry of Iron" Vol. 1, E.A. Koerner von Gustorf, F.-W. Grevels, and I. Fischler, eds., Academic Press, New York, NY (1978), pp. 1-112.
11. M. Elian, M.M.L. Chen, D.M.P. Mingos, and R. Hoffmann, Inorg. Chem., 15,1148(1976).
12. D.L. Thorn and R. Hoffmann, Nouv. J. Chim.,3,39(1979).
13. J. Silvestre, PhD Dissertation University of Houston (1983).
14. W.J. Pietro and W.J. Hehre, J. Comput. Chem.,4,241(1983).
15. R. Hoffmann, S. Swaminathan, B.G. Odell, and R. Gleiter, J. Am. Chem. Soc.,92,7091(1970).
16. H.P. Luthi, J.H. Ammeter, J. Almlof, and K. Korsell, Chem. Phys. Letts.,69,540(1980); H.P. Luthi, J.H. Ammeter, J. Almlof, and K. Faegri, J. Chem. Phys.,77,2002(1982); J. Almlof, K. Faegri, B.E.R. Schilling, and H.P. Luthi, Chem. Phys. Letts.,106,266(1984); T.E. Taylor and M.B. Hall, ibid,114,338(1985).
17. For reviews see, G.E. Herberich in "Comprehensive Organometallic Chemistry", G. Wilkinson, F.G.A. Stone, and E.W. Abel, eds., Vol. 1, Pergamon Press, Oxford (1982), pp. 381-410; K.H. Pannell, B.L. Kalsotra, and C. Parkanyi, J. Heterocyclic Chem.,15,1057(1978).
18. A. Efraty, Chem. Rev.,77,691(1977); G.M. Reissner, I. Bernal, M.D. Rausch, S.A. Gardner, and A. Clearfield, J. Organomet. Chem.,184, 237(1980); P.E. Riley and R.E. Davis, Organometallics,2,286(1983) and references therein.
19. R. Case, E.R.H. Jones, N.V. Schwartz, and M.C. Whiting, Proc. Chem. Soc.,256(1962); see also M.E.E. Veldman, H.R. vanderWal, S.J. Veenstra, and H.J. DeLiefde, J. Organomet. Chem.,197,59(1980) for a related example.
20. M. Mlekuz, P. Bougeard, B.G. Sayer, S. Peng, M.J. McGlinchey, A. Marinetti, J.-Y. Saillard, J.B. Naceur, B. Mentzen, and G. Jaouen, Organometallics,4,1123(1985).
21. H. Hoberg, R. Krause-Going, C. Kruger, and J.C. Sekutowski,

Angew. Chem.,89,179(1977); Angew. Chem. Int. Ed. Engl.,16,183(1977).
22. T.P. Fehlner, J. Am. Chem. Soc.,100,3250(1978).
23. V.R. Miller and R.N. Grimes, ibid,95,5079(1973); 97,4213(1975); D.C. Beer, V.R. Miller, L.G. Sneddon, R.N. Grimes, M. Mathew, and G.J. Palenik, ibid,95,3046(1973); R.Weiss and R.N. Grimes, ibid, 99,8087(1977); Inorg. Chem.,18,3291(1979); W.T. Robinson and R.N. Grimes, ibid,14,3056(1975); R.N. Grimes, E. Sinn, and R.B. Maynard, ibid,19,2384(1980); L.G. Sneddon and D. Voet, J. Chem. Soc., Chem. Commun.,118(1976); V.R. Miller, R. Weiss, and R.N. Grimes, J. Am. Chem. Soc.,99,5646(1977).

TRANSITION STATE FOR CARBONYL AND OLEFIN INSERTION REACTIONS

Nobuaki Koga and Keiji Morokuma
Institute for Molecular Science
Myodaiji, Okazaki, 444
Japan

ABSTRACT. The optimized structures will be reported for the transition states as well as the products and the reactants for two important elementary organometallic reactions, carbonyl insertion reaction and olefin insertion/β-elimination reaction . The model reactions studied are $M(CH_3)(H)(CO)(PH_3) \rightarrow M(COCH_3)(H)(PH_3)$ (M=Pd,Pt) and $M(H)_2(CH_2CX_2)(PH_3) \rightleftarrows M(CH_2CHX_2)(H)(PH_3)$ (M=Ni,Pd,Pt for X=H and M=Pd for X=F). The migrating group has been found to be the methyl group in the former and the hydride in the latter. In each reaction, the activation energy and the transition state structure depend strongly on the transition element as well as ligand. Factors that control such changes will be discussed.

INTRODUCTION

Insertion reaction in which unsaturated ligands insert into metal-hydrogen or metal-carbon bonds is one of the most important reactions in the organometallic chemistry, since it creates often-needed new C-H or C-C bonds [1].

Many catalytic cycles are considered to include the insertion reaction as a key process. Therefore, many experimental studies have appeared in order to clarify the mechanism of insertion reactions [2,3]. One of the most interesting and controversial points is whether R migrates to the unsaturated ligand or L migrates to insert into the metal-R bond. Of special interest are the effects of substituents and transition metals on the ease of reaction. Here we report the transition state structures for the carbonyl insertion reaction and the olefin insertion/β-hydrogen elimination reaction of group 10 transition metal complexes.

A. Veillard (ed.), Quantum Chemistry: The Challenge of Transition Metals and Coordination Chemistry, 351–361.

The ab initio restricted Hartree-Fock (RHF) method with energy gradient technique was used for geometry optimization [4]. The fully optimized geometries of stable complexes (reactant and product) and the transition state for each model reaction system have been obtained. Energies are calculated, in some cases, with the frozen-core Møller-Plesset second-order perturbation theory (MP2) at the RHF optimized geometries.

CARBONYL INSERTION REACTION [5]

The model reactions studied are

M=Pd	**1**	**2**	(1)
M=Pt	**3**	**4**	(2)

The optimized structures of **1**, **2**, **3** and **4** and the transition states are shown in Figure 1 and their relative energies in Table 1. The basis sets used for geometry optimizations are valence double-zeta basis functions of Hay and Wadt [6] and Noell and Hay [7] for Pd and Pt, respectively, with the relativistic effective core potential, the 3-21G basis set [8] for C, H and F and the STO-2G [9] for PH_3. For MP2 calculation the Huzinaga-Dunning basis set was used for all the ligand atoms [10].

During the course of reaction, the structure of the H_3PMH fragments remains nearly unchanged, the angle PMH being about 90°. The methyl group, with its pseudo C_3 axis kept nearly horizontal, moves up halfway toward the carbonyl group at the transition state and proceeds further to form a CC bond with the carbonyl group in the product. The carbonyl group moves slightly toward the methyl group as if to facilitate its migration. The situation is essentially the same for both M=Pd and Pt. These results indicate that the reaction coordinate is methyl migration through a three-center transition state. The methyl migration is what has been found experimentally for $Pd(CH_3)_2(CO)(PR_3)$ [11]. Although it has been proposed that an agostic interaction might assist carbonyl insertion reaction [12], the CH^1 bond lengths of the product and the transition state are 1.089Å and the 1.090Å, respectively, for the Pd reaction and thus the agostic interaction which would have made the CH bond longer is absent.

In order to clarify the reason why the carbonyl group does not migrate, we assume the transition state for the carbonyl group migration in which the angles (CO)Pd(H) and $(CH_3)Pd(PH_3)$ of the true

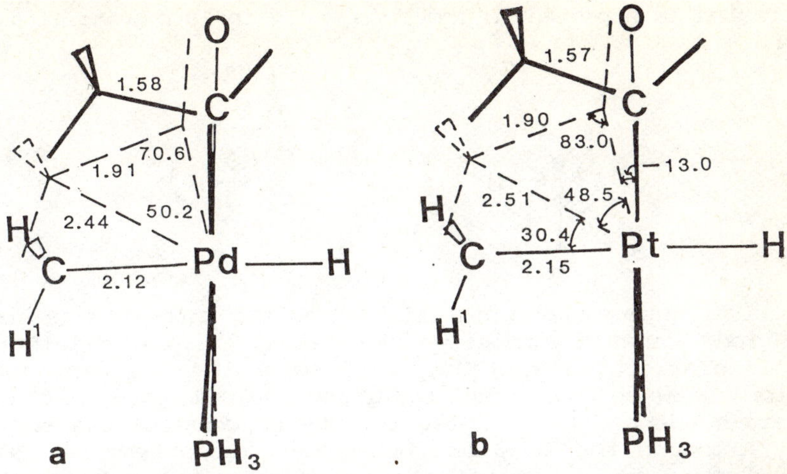

Figure 1. Optimized geometries (in Å and degrees) of $M(CH_3)(H)(CO)(PH_3)$, the transition state and $M(COCH_3)(H)(PH_3)$ for (a) M=Pd and for (b) M=Pt. The position of the metal atom and the direction of the MH bond are fixed in the illustration. The illustrated positions of atoms are those in $M(CH_3)(H)(CO)(PH_3)$. The positions shown by the broken lines and the thick solid lines are those at the transition state and in $M(COCH_3)(H)(PH_3)$, respectively. For clarity, only essential geometrical parameters are shown.

Table 1. Relative Energies (in kcal/mol) for the process $M(CH_3)(H)(CO)(PH_3) \rightarrow M(COCH_3)(H)(PH_3)$

M	method	reactant	transition state	product
Pd	RHF	0.0	25.8	19.1
	MP2 a)	0.0	13.5	8.8
Pt	RHF	0.0	31.3	23.0
	MP2 a)	0.0	21.8	17.6

a) The geometries optimized at RHF are used.

transition state are exchanged. The assumed transition state is less stable by about 7 kcal/mol than the true transition state. This instability is ascribed to the repulsion between two carbonyl lone pair electrons and Pd d_{xy} electrons, as shown in **5**. The repulsion in the true transition state, as shown in **6**, is weaker, because the carbonyl lone pair is no longer directed toward the Pd d_{xy} electrons and because the long distance and the small electron density (~1) keep the increase of the methyl sp^3 - Pd d_{xy} repulsion to a minimum. Therefore, the alkyl group migration is expected to occur commonly in late transition metal complexes.

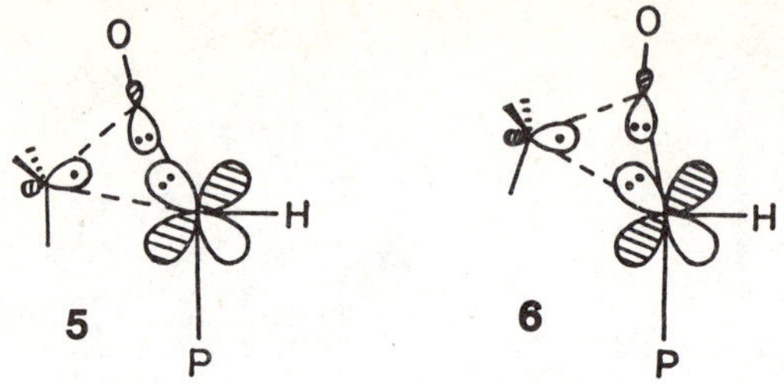

5 6

The orbital interaction which stabilizes the transition state has been analyzed. For this purpose we calculated the paired interactive hybrid molecular orbitals(IHMOs), into which the dominant orbital interaction between the methyl group and the remaining part of the complex are condensed [13]. Those for the Pd complex are shown in Figure 2. Although the IHMO of the methyl group remains almost unchanged during the course of reaction, that of the $Pd(H)(CO)(PH_3)$ fragment changes its shape smoothly. While the pair of IHMOs for the reactant describes CH_3-Pd bond and that for the product describes CH_3-CO bond, at the transition state the pair of IHMOs has a character between the two and shows that the three-center transition state is stabilized by the interaction between the methyl sp^3-like orbital and CO π^* and Pd (d + p) orbitals. The IHMOs calculated for the Pt complexes have similar characters to those for the Pd complexes.

The structural behaviors of the reactions of the Pd and the Pt complexes are almost the same. However, energetics is quite different. While the activation energy of the Pd complex is 13 kcal/mol at the MP2 level, that of the Pt complex is higher, consistent with the experimental facts [14]. The reaction of the Pt complex is also more endothermic. These results can be explained by the differences of the bond strengths of M-CH_3 and M-CO between the Pd and the Pt complexes. These bonds of the Pt complex are stronger by 7 and 11 kcal/mol, respectively, than those of the Pd complexes, while Pt-$COCH_3$ is stronger by 9 kcal/mol than Pd-$COCH_3$. Therefore, the Pt reaction is more endothermic by 9 (=7+11-9) kcal/mol and hence has a higher barrier.

Figure 2. Interactive Hybrid Molecular Orbitals of CH_3 and $Pd(H)(CO)(PH_3)$ for (left) the reactant, (middle) the transition state and (right) the product.

Table 2. Relative Energies (in kcal/mol) for the process $Pd(R)(H)(CO)(PH_3) \rightarrow Pd(COR)(H)(PH_3)$ [a)]

R	reactant	transition state	product
CHF_2	0.0	40.9	31.2
C_2H_5	0.0	23.1	14.6

a) The geometries optimized for $R=CH_3$ are used.

The effect of the bond strength on the reaction energetics can be assessed in the cases where the methyl group is replaced by difluoromethyl and ethyl (Table 2). While the electron-withdrawing fluorine makes the CHF_2-Pd bond stronger, the Pd-$COCHF_2$ bond is not much different from Pd-$COCH_3$, and therefore, the reaction becomes more endothermic with a higher barrier. In addition, the fluorine substitution makes electron delocalization $CHF_2 \rightarrow M$ and CO π^* at the transition state more difficult, contributing to a higher barrier. On the other hand, the electon-donative CH_3 substitution produces the reverse effect.

OLEFIN INSERTION/β-ELIMINATION REACTION [15]

The model reactions for insertion/β-elimination are

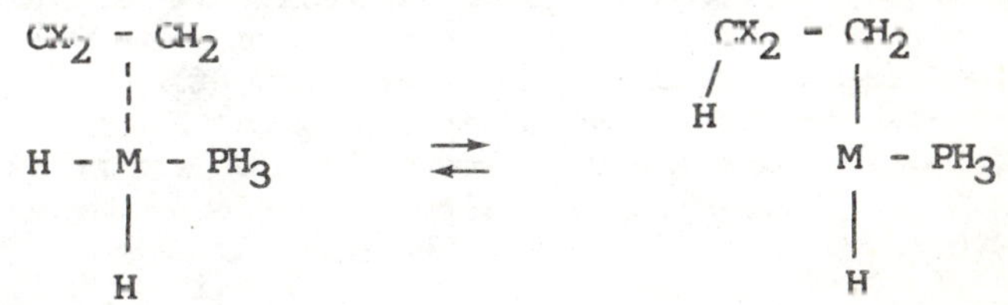

M=Ni, X=H	7	8	(3)
M=Pd, X=H	9	10	(4)
M=Pt, X=H	11	12	(5)
M=Pd, X=F	13	14	(6)

We show the optimized structures of the transition states for reactions (3), (4) and (5) as well as those of the ethyl and ethylene complexes in Figure 3 and their relative energies in Table 3. The basis sets used are valence double-zeta basis functions of Hay [16] and Hay and Wadt [6] for Pd and Ni, respectively, together with the core basis functions of MINI-1 [17], that of Noell and Hay [7] for Pt with the relativistic effective core potential, the 3-21G basis set [8] for C, H and F and the STO-2G [9] for PH_3.

The structure of the transition state and its approximate force constant matrices indicate that in insertion/β-elimination reaction the hydrogen atom transfers between the β-carbon atom and the transition metal via a four-center transition state. At the transition state of the Pd complex, the distance between the Pd atom and the migrating hydrogen is not much longer than the usual Pd-H bond distance and the CC bond length is closer to that of the corresponding ethylene complex. These structural features indicate that the transition state of insertion reaction is located 'early'. On the other hand, the PdC$^{\alpha}$ length is not so long as that of the corresponding ethylene complex, and the distance between the Pd atom and the β-carbon is short, showing that the transition state is 'tight' as well. On the other hand, the structure of the transition state of the Ni complex is very different from that of Pd complex, located extremely early and very close to 7. The transition state for the Pt reaction is located in a later stage than that for the Pd reaction.

These structural features are reflected in the energy profiles of reactions (Table 3). The energy barriers for the insertion and β-elimination reactions connecting 9 and 10 are 8.0 and 11.0 kcal/mol, respectively, with the HF method, and become even lower when the electron correlation is taken into account. The low energy barriers for both directions point to the reversibility of insertion/β-elimination reactions [1].

The situation for the Ni complex is different from that of the Pd complex. The insertion reaction is very exothermic and its activation energy is very low. In a better calculation the ethylene complex may not even be a stable species. The instability of the Ni ethylene complex is due to the weak ethylene-Ni bond. The ethylene-Pt bond is stronger than the ethylene-Pd bond. Therefore, the Pt ethylene complex is more stable than the ethyl complex. Overall, the stability of the ethyl complex increases in the order Pt<Pd<Ni, and the barrier height for the insertion reaction decreases in the order Pt>Pd>Ni.

Although the metal-ethylene bond strength varies with the metals, the CC bond distances of all the ethylene complexes are nearly as short as that of a free ethylene and the distances between ethylene and the central metal are long, indicating that the bonds between

Table 3. Relative Energies (in kcal/mol) of the process $M(H)_2(CH_2CX_2)(PH_3) \rightleftarrows M(CH_2CHX_2)(H)(PH_3)$

M	X	ethylene complex	transition state	ethyl complex
Pd	H	0.0	8.0	-3.0
		0.0 a)	5.1 a)	3.0 a)
Pd	F	0.0	7.3	-11.1
Ni	H	0.0	0.6	-31.5
Pt	H	0.0	12.5	4.2

a) calculated by frozen core second-order Møller-Plesset perturbation calculation by using the geometries obtained at the RHF level.

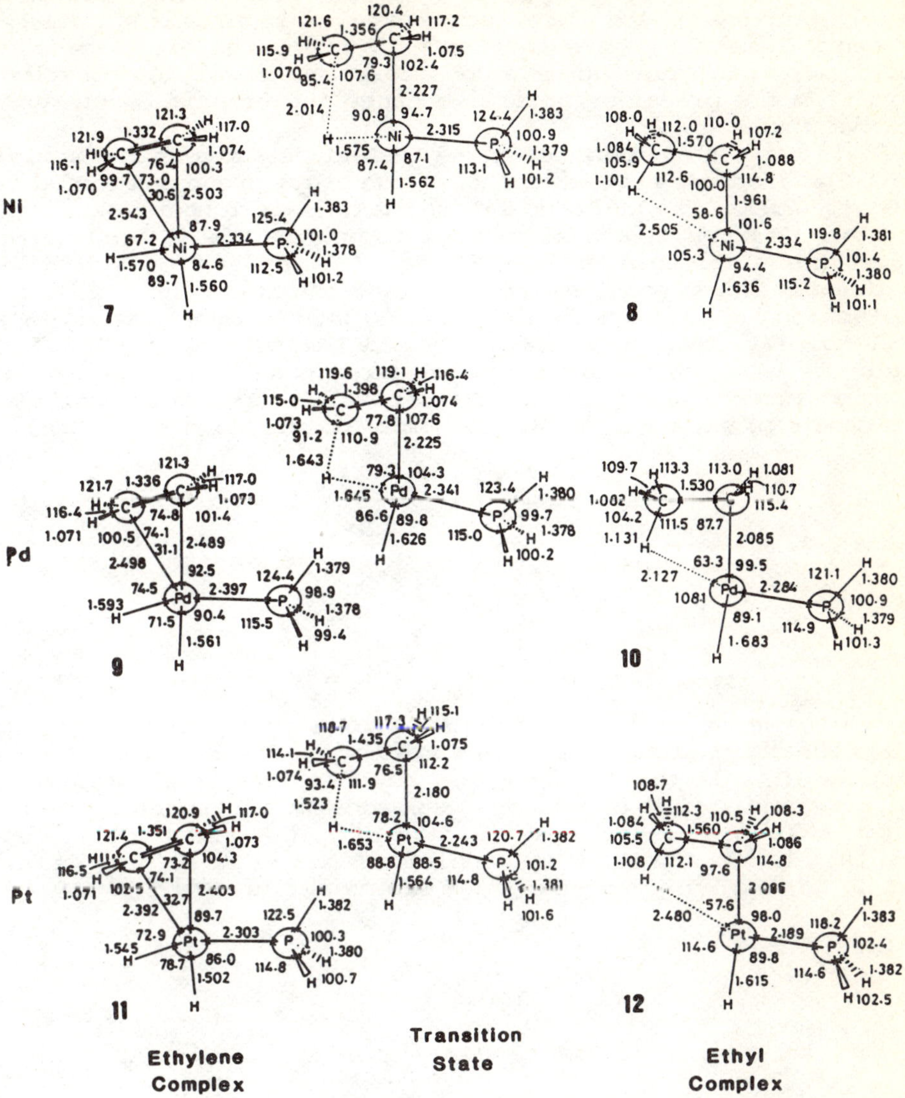

Figure 3. Optimized structures of (left) ethylene complexes, (right) ethyl complexes and (middle) the transition states for (top) the Ni, (middle) the Pd and (bottom) the Pt complexes. The bond lengths and angles are in Å and degrees.

metal and ethylene are weak. This may be in part due to a strong trans influence of the hydride ligand. In general, as was shown previously [16,18], ethylene of the planar ethylene complex binds only weakly to the central metal, where the upright structures are more stable [16,18]. In this study we have considered the insertion reaction to start from the planar structure, because the transition states are coplanar four-centered.

It is noted that there are three unusual features in the structure of **10**. The distance between the H^{β} of the ethyl group and the Pd atom is short, 2.13Å. The CH^{β} bond length of 1.13Å is longer by 0.05Å than the other CH bonds in the same ethyl group. The Pd-C-C bond angle is 88°, much smaller than that expected in the sp^3 hybridization(109°). All these features point to the existence of direct CH^{β}...Pd agostic interaction [19] and are very similar to what have been found in a Ti ethyl complex both experimentally and theoretically [20,21]. The origin of an agostic interaction has been found to be the electron-donating interaction from the $CH^{\beta}\sigma$ orbital to the empty orbital (in this case d+p) extending around the empty coordination site (**15**).

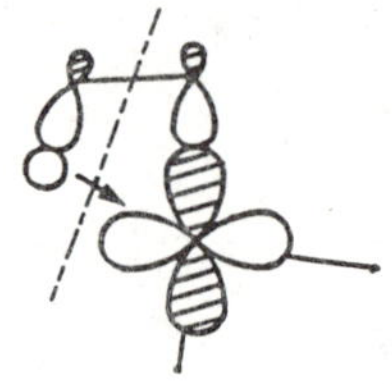

15

On the other hands, the difluoroethyl group in **14** shown in Figure 4 and the ethyl group in **8** and **12** are not as distorted as the ethyl group in **10**. All the M-C-C angles, however, are still smaller than 109°, suggesting that they have a weak agostic interaction. One finds that the CH^{β} bond lengths in **8** and **12** of 1.10Å, are slightly longer than that in **14**, and the M-C-C angle in **8** and **12** is slightly smaller than in **14**, implying that the CH...M interaction in **8** and **12** is

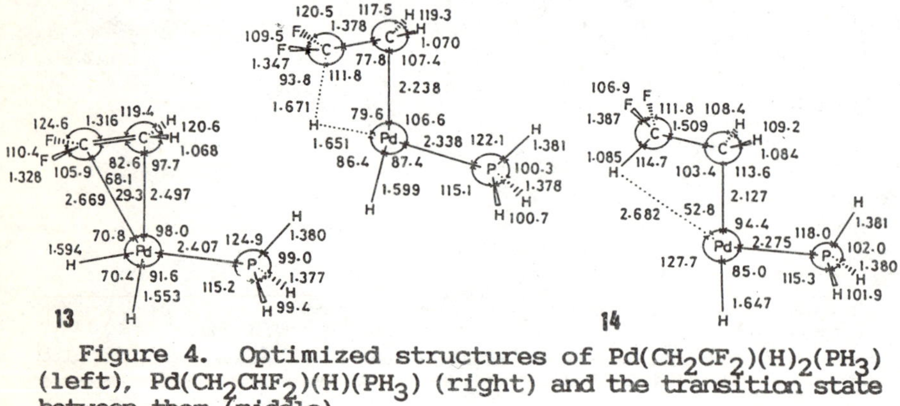

Figure 4. Optimized structures of $Pd(CH_2CF_2)(H)_2(PH_3)$ (left), $Pd(CH_2CHF_2)(H)(PH_3)$ (right) and the transition state between them (middle).

slightly stronger in **14.** In **14** the electron-withdrawing fluorines on the β-carbon atom make the electron-donating ability of the CH bond weaker. The energy of vacant d orbital in the Ni complex (0.1069 hartree) is higher than that (0.0505 hartree) in **10** and the Ni d orbital is tighter and has a smaller overlap with CH than the Pd d orbital. Consequently, the donative interaction in **8** is weaker in the Ni complex. In the Pt complex, the diffuse d_{xy} elctrons prevents the CH^{β} from approaching toward the Pt atom, so that an agostic interaction cannot be take place.

While the energy barrier for the insertion reaction **13**→**14** is comparable to **9**→**10**, the barrier for β-elimination reaction **14**→**13** is higher by 7 kcal/mol than that of **10**→**9**. The β-elimination barrier for **10** that has an agostic interaction is lower than for **14** that has little agostic interaction, suggesting that the barrier and the interaction are closely related. It looks as though the agostic interaction has incipiently activated the CH^{β} bond in the reactant, and therefore, the reactant **10** requires less energy to break the bond.

From the above discussion and the detailed analysis of the wavefunction, it is concluded that for β-elimination reaction the donative interaction from $CH^{\beta}\sigma$ to the metal vacant orbital activates the CH^{β}. bond. But this is not sufficient. In addition, the back donative interaction from the metal-C^{α} bond and metal d lone pair orbital to $CH^{\beta}\sigma^{*}$ orbital is also necessary to lower the barrier and stabilize the product. In the Pt reaction the back donative interaction is probably responsible to the low activation energy for the β-elimination reaction, while in the Ni complex the difficulty of backdonation gives rise to a high activation energy. In Table 4 the extents of the agostic interaction and the back donation in the β-elimination reaction of each complex are summarized.

Table 4. Agostic Interaction and Backdonative Interaction in β-Elimination Reaction $M(CH_2CHX_2)(H)(PH_3) \rightarrow M(H)_2(CH_2CX_2)(PH_3)$

M	X	activation energy	agostic interaction	back donation
Pd	H	11.0	excellent	good
	F	18.4	fair	good
Ni	H	42.1	good	poor
Pt	H	8.3	good	excellent

CONCLUSION

In this paper we reported the optimized structures of the transition states for the insertion/β-hydrogen elimination reaction and carbonyl insertion reaction and discussed factors controlling the ease of reaction by varying transition metals and substituents. In the last few years, using the energy gradient method, we have studied various elementary organometallic reactions including oxidative addition/reductive elimination reactions [22] and those presented in this paper. With a combination of these reactions, we will soon be able to study theoretically a complete cycle of homogeneous catalytic reactions and predict how to change barrier heights and heats of reaction. A theoretical design of homogeneous catalysts is our goal.

ACKNOWLEDGMENT

The authors are grateful to Drs. S. Obara and K. Kitaura for collaborations in the part of the olefin insertion/β-elimination reaction. The numerical calculations were carried out at the IMS Computer Center.

REFERENCES

[1] F.A.Cotton and G. Wilkinson, Advanced Inorganic Chemistry, Wiley, New York, 1980.
[2] Ref. 1 and references cited therein.
[3] J. J. Alexander, in The Chemistry of the metal-carbon bond, Vol.2, F.R. Hartley and S. Patai Eds., Wiley, New York, 1985, p 339. and R. J. Cross, p 55
[4] We used the GAUSSIAN80 program implemented with the L. R. Kahn's ECP program. J. S. Binkley, R. A. Whiteside, R. Krishnan, R. Seeger, D. J. DeFrees, H. B. Schlegel, S. Topiol, L. R. Kahn and J. A. Pople, QCPE 11 406 (1981).
[5] N. Koga and K. Morokuma J. Am. Chem. Soc. in press. N. Koga and K. Morokuma to be published.
[6] P. J. Hay and W. R. Wadt, J. Chem. Phys. 82 270 (1985).
[7] J. O. Noell and P. J. Hay, Inorg. Chem. 21 14 (1982).
[8] J. S. Binkley, J. A. Pople and W. J. Hehre, J. Am. Chem. Soc. 102 939 (1980).
[9] W.J.Hehre, R. F. Stewart and J. A. Pople, J. Chem. Phys. 51 2657 (1969).
[10] T.H. Dunning and P. J. Hay, In Methods of Electronic Structure Theory, H. Schafer, Ed., Plenum, New York, 1977, p 1.
[11] F. Ozawa and A. Yamamoto, Chem. Lett. 289 (1981).
[12] E.Carmona, L. Sanchez, J. M. Marin, M. L. Poveda, J. L. Atwood, R. D. Priester and R. D. Rogers, J. Am. Chem. Soc. 106 3214 (106).
[13] H.Fujimoto, N.Koga and K.Fukui, J. Am. Chem. Soc. 103 7452 (1981).
[14] P. E. Garrou and R. F. Heck, J. Am. Chem. Soc. 98 4115 (1976).
[15] N.Koga, S. Obara, K. Kitaura and K. Morokuma, J. Am. Chem. Soc. in press.
[16] P. J. Hay, J. Am. Chem. Soc. 103 1390 (1981).
[17] Y. Sakai, H. Tatewaki and S. Huzinaga, J. Comp. Chem. 3 6 (1982); H. Tatewaki and S. Huzinaga, J. Chem. Phys. 72 4339 (1979).

[18] J.-E. Bäckvall, E. E. Björkman, L. Pettersson and P. Siegbahn, *J. Am. Chem. Soc.* **106** 4369 (1984).; T. A. Albright, R. Hoffmann, J. C. Thibeault and D. L. Thorn, *J. Am. Chem. Soc.* **101** 3801 (1979)

[19] M. Brookhart and M. L. H. Green, *J. Organomet. Chem.* **250** 295 (1983) and references cited therein.

[20] Z. Dawoodi, M. L. H. Green, V. S. B. Mtetwa and K. J. Prout, *J. Chem. Soc. Chem. Commun.* 802 (1982).

[21] N. Koga, S. Obara and K. Morokuma *J. Am. Chem. Soc.* **106** 4625 (1984).

[22] K. Kitaura, S. Obara and K. Morokuma, *J. Am. Chem. Soc.* **103** 2891 (1981).; S. Obara, K. Kitaura and K. Morokuma, *J. Am. Chem. Soc.* **106** 7482 (1984).

THEORETICAL ASPECTS OF THE PHOTOCHEMISTRY OF ORGANOMETALLICS

C. Daniel and A. Veillard
E.R. no 139 du CNRS
Institut Le Bel, Université Louis Pasteur
4, rue Blaise Pascal, 67000 Strasbourg, France

ABSTRACT. Three photochemical reactions of organometallic molecules

$$Fe(CO)_5 \xrightarrow{h\nu} Fe(CO)_4 + CO \qquad (1)$$

$$HCo(CO)_4 \xrightarrow{h\nu} Co(CO)_4 + H \qquad (2)$$

$$Mo(CO)_5L + L' \xrightarrow{h\nu} Mo(CO)_4LL' + CO \quad L=NH_3, PMe_3, C(OMe)H, C_2H_4 \qquad (3)$$

have been studied theoretically through ab initio configuration interaction (CI) calculations of the state correlation diagrams and of the potential energy curves connecting the reactant and products. These potential energy curves are cross-sections of the many-dimensional potential energy surfaces, corresponding for instance to C_{2v} or C_{4v} symmetry constraints for reactions (1) and (2) respectively. The CI calculations are of the single- and double-excitations type. A common mechanism has been proposed for reactions (1) and (2), where the excitation to a spin-allowed singlet state is followed by population of a triplet state through intersystem crossing and dissociation to the products of the reaction along this triplet potential energy surface. The mechanism advanced for the photosubstitution reaction (3) accounts for its cis-stereospecificity through excitation of $M(CO)_5L$ into the ^1E LF state followed by elimination of a carbonyl ligand, subsequent Berry pseudorotation (in the case of axial elimination) and internal conversion to the ground state of the square pyramid $M(CO)_4L$ with L basal, which can then react with an incident nucleophile.

1. INTRODUCTION

There is an extensive photochemistry of organometallic compounds[1]. However, "one unfortunate characteristic of much of organometallic photochemistry is a relative absence of detailed mechanistic information"[2]. Current theoretical interpretation is based on an analysis in terms of bonding and antibonding orbitals, with the corollary that exciting an electron from a bonding to an antibonding orbital should result in a dissociative photochemical reaction[1]. The

A. Veillard (ed.), Quantum Chemistry: The Challenge of Transition Metals and Coordination Chemistry. 363–375.

limitations of this approach are obvious : for instance it does not distinguish, for a given electronic excitation, between the excited states of different spin multiplicities. A proper understanding of photochemical reactions should be based on a knowledge of the potential energy surfaces which connect the reactants to the primary photochemical products, or their qualitative counterparts, the state correlation diagrams [3]. We have been engaged, for the last few years, in the derivation of state correlation diagrams and the calculation of potential energy curves for selected photochemical reactions of organometallics [4-9]. We present here a survey of the methodology used and of the results achieved for three photochemical reactions :

i) the loss of a carbonyl ligand from $Fe(CO)_5$ [7]

$$Fe(CO)_5 \xrightarrow{h\nu} Fe(CO)_4 + CO \qquad (1)$$

ii) the photolysis of the Co-H bond in $HCo(CO)_4$ [8]

$$HCo(CO)_4 \xrightarrow{h\nu} Co(CO)_4 + H \qquad (2)$$

iii) the photosubstitution of d^6 metal carbonyls [9]

$$M(CO)_5L + L' \xrightarrow{h\nu} M(CO)_4LL' + CO \qquad (3)$$

which is known to proceed through a primary photodissociation

$$M(CO)_5L \xrightarrow{h\nu} M(CO)_4L + CO \qquad (4)$$

with the coordinatively unsaturated species $M(CO)_4L$ undergoing a subsequent thermal reaction

$$M(CO)_4L + L' \longrightarrow M(CO)_4LL' \qquad (5)$$

For reactions (1) and (2) we have been able to compute the potential energy curves representing cross sections of the potential energy surfaces, with a number of simplifying assumptions. For reaction (3) the calculation of the potential energy curves for a series of ligands L ($L=NH_3$, PMe_3, C(OMe)H and C_2H_4 for M=Mo or L=Cl for M=Tc) was considered to exceed our present computational facilities. For this reason we have restricted the calculations to a number of remarkable points on the potential energy surfaces and from this we have generated the corresponding state correlation diagrams.

2. THEORETICAL APPROACH

2.1 Defining the cross-sections of the potential energy surfaces

The potential energy curves calculated for reactions (1) and (2) are cross sections of the many-dimensional potential energy surfaces, corresponding to a number of restrictive assumptions :

- the bond lengths, except for the bond under dissociation,

are kept equal to a fixed value ;

- the geometry may be slightly idealized (for instance we have used the same metal-carbon bond length for the axial and equatorial carbonyl ligands in $Fe(CO)_5$ and $HCo(CO)_4$) ;

- the highest possible symmetry is maintained along the reaction path. For the photodissociation of $Fe(CO)_5$, this is achieved by considering the dissociation of an equatorial ligand under C_{2V} symmetry (Scheme 1) since the product $Fe(CO)_4$ is of C_{2V} symmetry (we

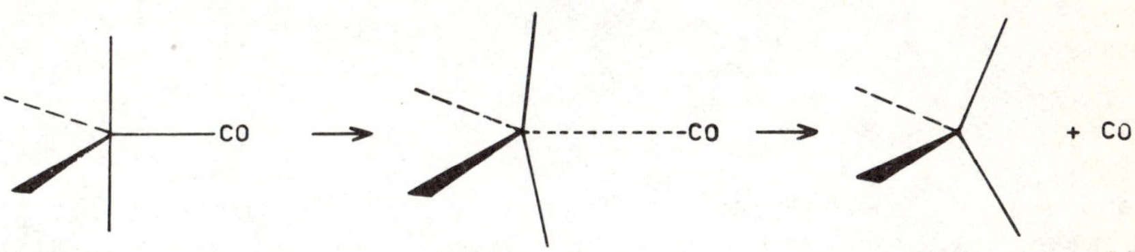

Scheme 1

were able to show, through state correlation diagrams, that qualitatively similar conclusions would be reached by considering either the dissociation of an equatorial ligand under a lower (C_s) symmetry or the dissociation of an axial ligand [7]). For the photolysis of $HCo(CO)_4$, both the reactant and the product $Co(CO)_4$ belong to the C_{3V} point group in their ground state, and we have assumed that this symmetry is retained along the reaction path, with the H atom departing along the C_3 axis of $HCo(CO)_4$ (Scheme 2). For the five-coordinate

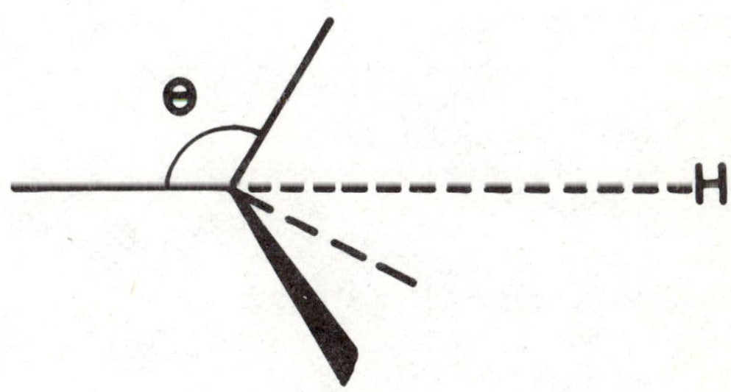

Scheme 2

systems $Tc(CO)_4Cl$ and $Mo(CO)_4L$ (L=NH_3, PMe_3, C(OMe)H and C_2H_4), we considered the regular square pyramid structures 1 and 2 and the regular trigonal-bipyramid structure 3 (with the angles in the equatorial plane equal to 120°).

- additional assumptions are sometimes needed. When the carbonyl ligand dissociates from $Fe(CO)_5$, the fragment $Fe(CO)_4$ rearranges. We have assumed that this rearrangement, which depends on the distance of the dissociating ligand, is the same for all the electronic

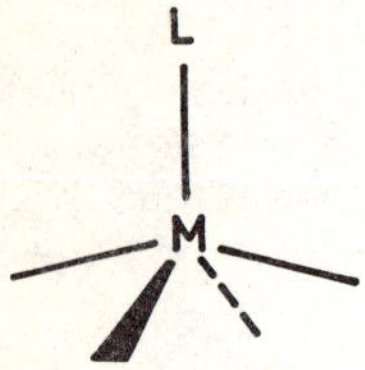

1

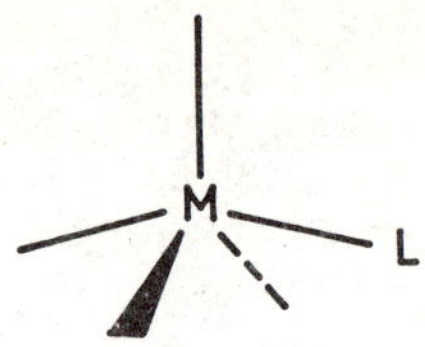

2

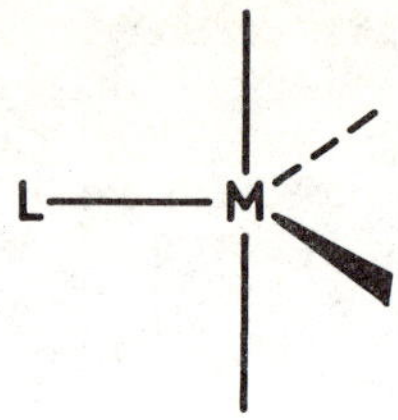

3

states and we have calculated it at the SCF level for the potential energy surface $^1A_1$ corresponding to the ground state $^1A'_1$ of $Fe(CO)_5$, i.e. for thermal elimination (some justification for this assumption may be found in the fact that the states $^3B_2$ and $^1A_1$ of $Fe(CO)_4$ have rather similar geometries). For the photodissociation of $HCo(CO)_4$, this rearrangement is defined by the angle θ. Since the value of θ is close to 100° for both the reactant $HCo(CO)_4$ and the product $Co(CO)_4$, we have retained this value for all the points along the reaction path.

2.2 The calculations

The SCF calculations were carried out with the system of programs ASTERIX[10] using the Gaussian basis sets of Table I (those used for

Table I. Gaussian basis sets (metal atom/1st row atoms/hydrogen)

Gaussian basis set	Contracted basis set	
(13,8,6/9,5/4)	[5,3,3/3,2/2]	BSI (Potential energy curves for $Fe(CO)_5$ and $HCo(CO)_4$)
(15,11,6/10,6/6)	[9,6,3/4,2/3]	BSII (Excited states of $HCo(CO)_4$)
(16,11,9/9,5/4)	[7,5,5/3,2/2]	$Mo(CO)_4L$, $Tc(CO)_4Cl$

the calculation of the potential energy curves are of split-zeta quality for the valence shells). A proper description of the homolytic dissociation of a covalent bond like the Co-H bond requires at least a two-configurations (DC) wavefunction and for this reason the molecular orbitals for the system $HCo(CO)_4$ were obtained from DC-SCF calculations[11]. The potential energy curves for the photoreactions of $Fe(CO)_5$ and $HCo(CO)_4$ were generated from C.I. calculations, using the m.o'.s from the SCF or DC-SCF calculations. All single and double excitations relatively to either one or a few reference states were considered within a given set of active orbitals (between 15 and 48). These calculations were carried out with the CI program developed originally by Brooks and Schaefer[12] using the graphical unitary group approach[13]. The CI calculations for the molecule $Mo(CO)_4NH_3$ were based on m.o'.s from three-configurations SCF (TC-SCF) calculations (in order to accurately represent the ligand-field (LF) excited states).

3. The results

Fig. 1 represents the potential energy curves for the photochemical dissociation of an equatorial ligand of $Fe(CO)_5$, under C_{2v} constraint, from CI calculations with a set of 15 active orbitals. Fig. 2 represents the potential energy curves for a limited number of electronic states, but now with a set of 47 active orbitals. CI potential energy curves for the dissociation of the Co-H bond in $HCo(CO)_4$ are shown in Fig. 3. The state correlation diagrams for the photoelimination of a carbonyl ligand from $M(CO)_5L$ (based on the calculations for $Mo(CO)_4NH_3$) are represented in Fig. 4.

4. Discussion

4.1. Some general criteria

In order to ensure that the potential energy surfaces obtained from the above calculations are qualitatively correct, one should check the following points :

i) does the calculation reproduce correctly the sequence and energetics of the excited states of the reactant ?

ii) is the assignment of the electronic ground state of the products correct ?

iii) are the relative stabilities of the reactant and product correctly described by the calculations ?

We shall successively scrutinize these three points.

The obvious way to answer the first question, regarding the accuracy of the calculation of the excited states for the reactant , is to compare the calculated excitation energies with their experimental values. Unfortunately, the electronic spectra of organometallics are poorly resolved in general. The absorption spectrum of $Fe(CO)_5$ in solution exhibits two shoulders at 35500 and 41500 cm-1[14],in reasonable agreement with the excitation energies of 35800, 37600, 40800 and 41000 cm-1 calculated for the lowest singlet states. The first electronic transition observed for $HCo(CO)_4$ is at 44050 cm-1 [15],to be compared with the value of 47100 cm-1 which we have obtained for the lowest allowed excitation $^1A_1 \rightarrow {}^1E$. $Cr(CO)_6$ is one organometallic molecule with a well resolved electronic spectrum and we have compared in Table II the calculated excitation energies to the corresponding experimental values. The sequence of excited states is reproduced correctly by the calculation, but the theoretical excitation energies are too high by

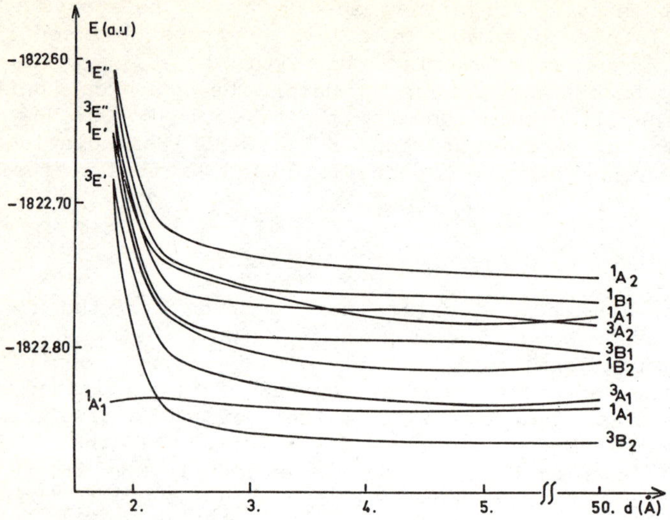

Figure 1. Potential energy curves for the dissociation of an equatorial ligand of $Fe(CO)_5$ under C_{2v} constraint (CI calculations with 15 active orbitals).

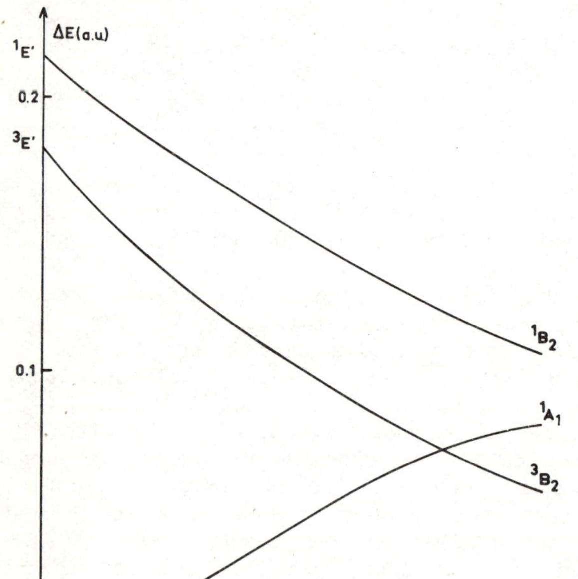

Figure 2. Potential energy curves for the dissociation of an equatorial ligand of $Fe(CO)_5$ under C_{2v} constraint (CI calculations with 47 active orbitals).

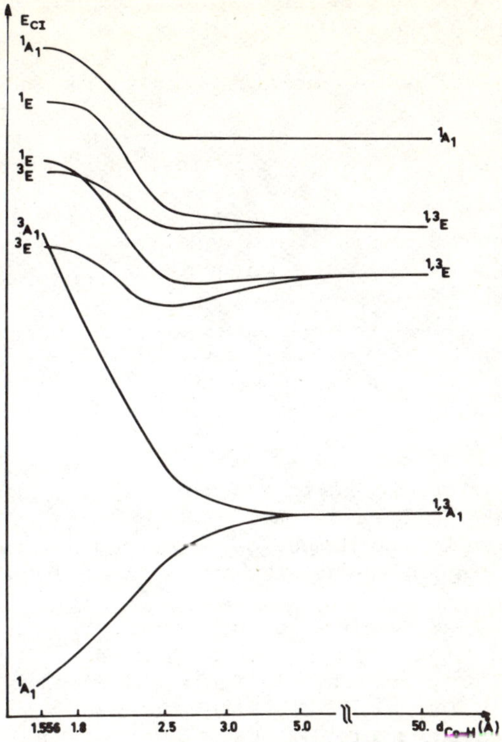

Figure 3. CI potential energy curves for the dissociation of the Co-H bond in $HCo(CO)_4$ under C_{3v} constraint.

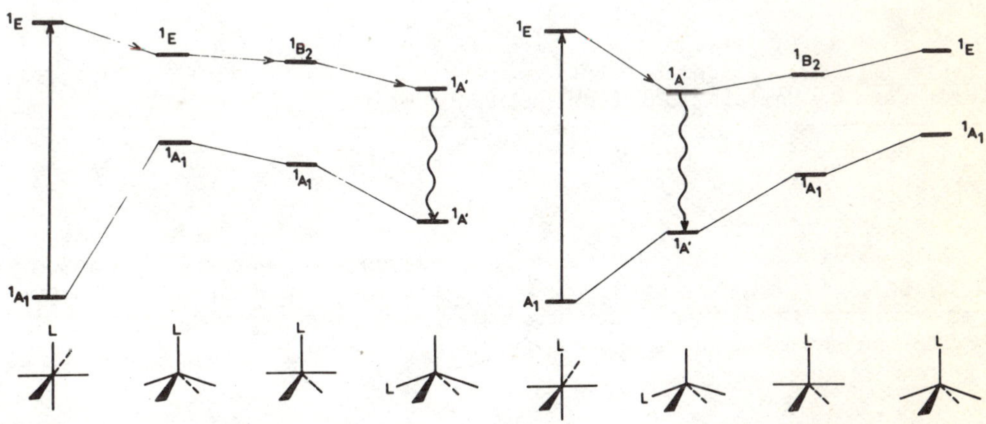

Figure 4. State correlation diagrams for the photoelimination of a carbonyl ligand from $M(CO)_5L$ (left, axial ligand; right, equatorial ligand) (this ligand does not appear in the diagram for the sake of clarity).

3000 to 7000 cm-1. This lack of quantitative agreement should be traced to :

Table II. Experimental and theoretical excitation energies (in cm-1) for $Cr(CO)_6$

Electronic excitation		Theoretical[a]	Experimental
$^1A_{1g} \rightarrow {}^1T_{1g}$	LF	37900	29500-31500
$^1A_{1g} \rightarrow {}^1T_{1u}$	MLCT	41650	35700
$^1A_{1g} \rightarrow {}^1T_{2g}$	LF	44930	38850
$^1A_{1g} \rightarrow {}^1T_{2u}$	MLCT	47000	4360C

[a]CI calculation with 66 active orbitals (only the 3d electrons being correlated) and 7 or 12 reference states for the states T_g and T_u respectively.

- the use of virtual orbitals from the SCF calculations rather than MC-SCF or CAS-SCF orbitals ;
- the limitation on the number of reference states ;
- the lack of higher spherical harmonics of f- and g-type in the gaussian basis set.

Another way to assess the accuracy of the calculation of the excited states is to carry out a series of CI calculations of increasing size. We have reported in Table III the results of a series of CI calculations for $HCo(CO)_4$, with the following conclusions : i) the lowest excited state of $HCo(CO)_4$ must be a ^3E state (corresponding to a $d_\sigma \rightarrow \sigma^*$ excitation) ; ii) there is a relatively low-lying $^3A_1$ state (corresponding to a $\sigma_{Co-H} \rightarrow \sigma^*_{Co-H}$ excitation), probably below 45000 cm-1.

Table III. Theoretical excitation energies (in cm-1) for $HCo(CO)_4$, from different CI calculations (one reference state).

Basis set	nb. of active orbitals	$^1A_1 \rightarrow {}^3E$	$^1A_1 \rightarrow {}^3A_1$	$^1A_1 \rightarrow {}^3E$
BSI	18	39600	41000	46100
BSII[a]	31	43100	49300	49300
BSII[a]	48	34100	44600	40400

[a]With one set of p functions on the hydrogen atom.

In order to characterize the products of the reaction, one needs to know both their geometry and electronic state. For $Fe(CO)_4$ and $Co(CO)_4$, this was investigated mostly through SCF calculations. The ground state of $Fe(CO)_4$ at the SCF level is a $^3B_2$ of C_{2v} symmetry.

Since the SCF calculation may be biased in favor of the triplet state (relatively to a singlet state), CI calculations were also carried out for the two states $^3B_2$ (C_{2v} symmetry) and $^1A_1$ (C_{3v} symmetry) but did not change the above conclusion. Experimentally, $Fe(CO)_4$ has been shown to be paramagnetic[16] and the analysis of its infrared spectrum indicates a C_{2v} structure very close to the one determined theoretically[17].

Finally, one should check that the relative stabilities of the reactant and the products are correctly calculated, since this will insure that the potential energy curves have the right slope. The dissociation of a carbonyl ligand from $Fe(CO)_5$ is known to be endothermic by a value which is probably close to 40 kcal/mole (the experimental value of 41.5 kcal/mole corresponds to the dissociation to the $^1A_1$ state of $Fe(CO)_4$[18], which is probably close to the ground state $^3B_2$ as seen from Fig. 1 and 2). This condition is not fullfilled in the first series of calculations (Fig. 1) but is satisfied in the second ones (Fig. 2) (the reaction is calculated to be endothermic by 42.8 kcal/mole). For the dissociation of the Co-H bond in $HCo(CO)_4$, we obtained a dissociation energy of 45 kcal/mole, to be compared to an estimated value of 58 kcal/mole[19].

The state correlation diagrams of Fig. 4 are based on the CI calculations for the system $Mo(CO)_4NH_3$. Calculations for the other d^6 systems $M(CO)_4L$ were restricted to the SCF level. The corresponding results may be used with some confidence, since the relative stabilities of the three regular structures (trigonal bipyramid and square pyramids) of $Mo(CO)_4NH_3$ are practically unchanged when going from the SCF to the CI level. For $Mo(CO)_4NH_3$ (square pyramid with NH_3 apical), the excited states calculated at 9300 cm-1 correspond to the state ^1E of $Cr(CO)_5$ which was found by Hay at 12000 cm-1[20].

4.2 The mechanisms of these photoreactions.

Reactions (1) and (2) share one common feature, namely the existence of a <u>triplet</u> potential energy surface which connects, without any barrier, an excited state of the reactant ($^3E'$ for $Fe(CO)_5$ and $^3A_1$ for $HCo(CO)_4$) to the ground state of the products ($^3B_2$ for $Fe(CO)_4$ + CO and $^3A_1$ for H + $Co(CO)_4$) (Fig. 2 and 3). Thus the molecule can evolve from this excited state to the ground state of the primary products along an adiabatic surface. One difficulty is that the transition from the singlet ground state of the reactant to this triplet excited state is spin-forbidden. Although this forbiddenness may be removed in principle through spin-orbit coupling, it is usually considered that, in the 3d metal complexes and organometallics, population of spin-forbidden states by direct light absorption can be seldom achieved[21]. We propose that these reactions take place through the mechanism depicted in Fig. 5 :

i) the reactant, initially in its ground state, is excited to a spin-allowed singlet state (for $Fe(CO)_5$, this could be the LF excited state $^1E'$) ;

ii) intersystem crossing from this singlet excited state populates the triplet state ;

iii) from there the molecule can dissociate to the products of the

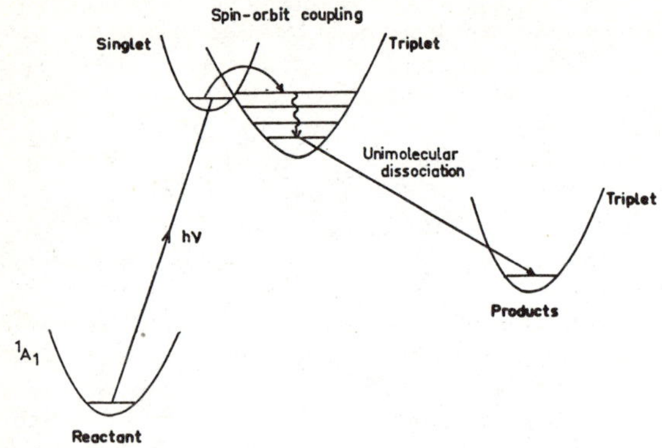

Fig 5. The mechanism proposed for the photochemical dissociations of $Fe(CO)_5$ and $HCo(CO)_4$.

reaction along the triplet potential energy surface (without any barrier).

This mechanism can be extended to many other photochemical reactions :

i) the photochemical homolysis of a σ bond, like the photodissociation of hydrogen from other metal hydrides including $HMn(CO)_5$[22], $HReCp_2$[23], $HW(CO)_3Cp$[24] or the photochemical cleavage of the metal-carbon bond in $RMn(CO)_5$ ($R=CH_3$, C_6H_5)[1] and of the metal-silicon bond in $R_3SiCo(CO)_4$ (R=Et, Ph)[25] and the photoinduced metal-metal bond cleavage in dinuclear complexes like $Mn_2(CO)_{10}$[1]. For all these photoreactions the triplet corresponding to the $\sigma \rightarrow \sigma^*$ excitation must correlate with the triplet component of the ground state of the products[5] (either directly or through an avoided crossing if the reactant has another excited state of the same symmetry but lower in energy).

ii) the photodissociation of a carbonyl ligand from other metal carbonyls like $HCo(CO)_4$[8] .

The occurence of intersystem crossing has been postulated by several authors to interpret the photochemistry of transition metal complexes [21,26–28].

The mechanism which we propose for the photosubstitution reactions of d^6 metal carbonyls is based on the state correlation diagrams of Fig. 4. Our criterion for regarding this mechanism as satisfactory will be its ability to account for the high stereospecificity of this reaction, namely the fact that the product $M(CO)_4LL'$ of reaction (3) is usually a cis disubstituted derivative (for a review of the experimental data, see for instance Ref. [7]).

We have restricted our analysis to the state correlation diagrams for the singlet states since :

i) the reactant may be considered as being initially in a singlet excited state (at least when spin-orbit coupling is not too important) ;

ii) the ground state of the products $M(CO)_4LL'$, a d^6 hexacoordinate system, should be a singlet.

We shall consider successively both axial and equatorial elimination of a carbonyl ligand (relatively to the heteroligand L) from $M(CO)_5L$.

From the diagram of Fig. 4a, we propose the following mechanism for the photosubstitution of the axial carbonyl in $M(CO)_5L$:

i) excitation of $M(CO)_5L$ into the ^1E ligand field state is followed by elimination of the carbonyl ligand with the species $M(CO)_4L$ formed as a square pyramid (SP) with L apical in the ^1E state (assuming approximative C_{4v} symmetry) ;

ii) from there $M(CO)_4L$ evolves along a Berry pseudorotation path first to a trigonal bipyramid (TBP) in the $^1B_2$ state, then to a SP with L basal in the $^1A'$ excited state ;

iii) the molecule $M(CO)_4L$, being trapped in the potential well corresponding to this excited state $^1A'$ of the SP with L basal, can evolve through internal conversion to the ground state $^1A'$;

iv) $M(CO)_4L$ as a SP (with L basal) in its ground state will react with an incident nucleophile L' to give $M(CO)_4LL'$ with the cis structure (Scheme 3).

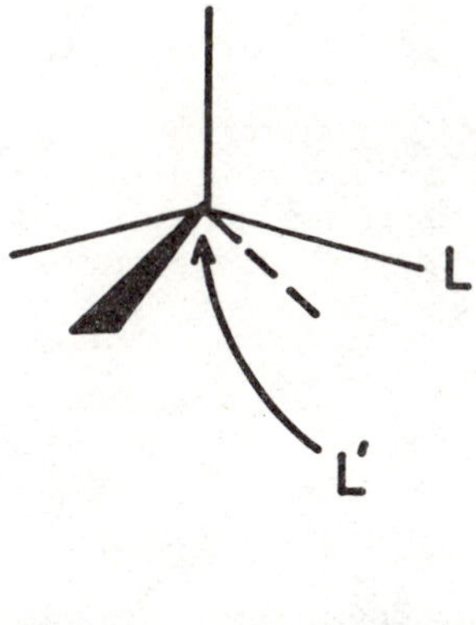

Scheme 3

If we consider now the photoelimination of an equatorial carbonyl ligand, the mechanism, based on the state correlation diagram of Fig. 4b, is similar to the one for the axial elimination but does not require a rearrangement of the $M(CO)_4L$ species :

i) excitation of $M(CO)_5L$ into the ^1E ligand field excited state followed by elimination of an equatorial carbonyl ligand produces the species $M(CO)_4L$ as a SP with L basal in the excited state $^1A'$;

ii) internal conversion brings the SP with L basal into its ground state $^1A'$;

iii) the molecule reacts with an incident nucleophile L' (Scheme 3) to give $M(CO)_4LL'$ with the cis structure.

This mechanism is consistent with our knowledge of these reactions, namely the fact that the ligand field excitation is responsible of the photoreaction. Although simple, it accounts for the *cis*-stereospecificity of the reaction with a variety of ligands. It can be extended probably to other photochemical reactions, for instance those of $M(CO)_4L$ where the ligand L occupies two coordination sites in *cis* position [9].

5. CONCLUSION

From the above examples, it seems clear that theoretical studies can contribute to the elucidation of the mechanism of the photochemical reactions of organometallics. Further work should aim at more accurate calculations of the excitation energies (this will help understanding the wavelength dependence) altogether with the calculation of other potential energy curves (for instance for the photolytic cleavage of the metal-metal bond). Mapping of potential energy surfaces will certainly contribute to a better understanding of concurrent photochemical reactions (*cf.* the competition between the photolytic cleavage of the metal-hydrogen or metal-metal bond and the photochemical loss of a carbonyl ligand in $HCo(CO)_4$ and $Mn_2(CO)_{10}$ [2,15]). Further experimental studies would be useful to assess the proposed mechanisms (regarding the role of singlet and triplet states).

ACKNOWLEDGEMENTS

Most of the calculations have been carried out on the CRAY-1 computer of the CCVR (Palaiseau), through a grant of computer time from the Conseil Scientifique du Centre de Calcul Vectoriel pour la Recherche.

REFERENCES

1. G.L Geoffroy and M.S. Wrighton, *Organometallic photochemistry*, Academic Press, New-York, 1979.
2. T.J. Meyer and J.V. Caspar, *Chem. Rev.*, *85*, 187 (1985).
3. See for instance N.J. Turro, *Modern molecular photochemistry*, Benjamin, Menlo Park, California, 1978.
4. A. Veillard, *Nouv. J. Chim.*, *5*, 599 (1981).
5. A. Veillard and A. Dedieu, *Nouv. J. Chim.*, *7*, 683 (1983).
6. A. Veillard and A. Dedieu, *Theoret. Chim. Act.*, *63*, 339 (1983).

7. C. Daniel, M. Benard, A. Dedieu, R. Wiest and A. Veillard, J. Phys. Chem., 88, 4805 (1984).
8. C. Daniel, I. Hyla-Kryspin, J. Demuynck and A. Veillard, Nouv. J. Chim., 9, 581 (1985).
9. C. Daniel and A. Veillard, Nouv. J. Chim., in press.
10. M. Benard, A. Dedieu, J. Demuynck, M.-M. Rohmer, A. Strich, A. Veillard and R. Wiest, unpublished work.
11. M.H. Wood and A. Veillard, Mol. Phys., 26, 595 (1973).
12. (a) B.R. Brooks and H.F. Schaefer, J. Chem. Phys., 70, 5092 (1979) ; (b) This program was adapted for the IBM computer by F. Brown and I. Shavitt.
13. (a) J. Paldus, Theor. Chem. Adv. Perspect., 2, 131 (1976) ; (b) I. Shavitt, Int. J. Quant. Chem. Symp., 11, 131 (1977).
14. M. Dartiguenave, Y. Dartiguenave and H.B. Gray, Bull. Soc. Chim. Fr., 12, 4223 (1969).
15. R.L. Sweany, Inorg. Chem., 21, 752 (1982).
16. T.J. Barton, R. Grinter, A.J. Thomson, B. Davies and M. Poliakoff, J.C.S. Chem. Commun., 481 (1977).
17. M. Poliakoff and J.J. Turner, J. Chem. Soc. Dalt. Trans., 2276 (1974).
18. K.E. Lewis, D.M. Golden and G.P. Smith, J. Am. Chem. Soc., 106, 3905 (1984).
19. F. Ungvary, J. Organomet. Chem., 36, 363 (1972).
20. P.J. Hay, J. Am. Chem. Soc., 100, 2411 (1978).
21. E. Zinato, in Concepts of inorganic photochemistry, A.W. Adamson and P.D. Fleischauer eds., Wiley, New-York, 1975, p. 143.
22. S.P. Church, M. Poliakoff, J.A. Timney and J.J. Turner, J. Am. Chem. Soc., 103, 7515 (1981).
23. J. Chetwynd-Talbot, P. Grebenik and R.N. Perutz, J.C.S. Chem. Com., 452 (1981).
24. N.W. Hoffman and T.L. Brown, Inorg. Chem., 17, 613 (1978).
25. C.L. Reichel and M.S. Wrighton, Inorg. Chem., 19, 3858 (1980).
26. M.A. Jamieson, N. Serpone and M.Z. Hoffman, Coord. Chem. Rev., 39, 121 (1981).
27. C. Poon, T. Lau and C. Cho, Inorg. Chem., 22, 3893 (1983).
28. J.V. Caspar and T.J. Meyer, Inorg. Chem., 22, 2444 (1983).

ELECTRONIC STRUCTURE OF METALLLOPORPHYRINS. AB INITIO CI CALCULATIONS

M.-M. Rohmer
E.R. n° 139 du C.N.R.S.
Institut Le Bel, Université Louis Pasteur,
4, rue Blaise Pascal, 67000 Strasbourg, France

ABSTRACT. Ab initio configuration interaction (CI) calculations are reported for different electronic configurations of the two model iron(II) porphyrin complexes, the four-coordinate FeP molecule and its dioxygen complex $FePO_2$, in order to assign the ground state configurations. For FeP (in an intermediate spin state S=1) the calculations predict a $^3A_{2g}$ ground state configuration. For $FePO_2$ an analysis of the results suggests a singlet ground state of Fe(II)-O_2 structure having some Fe(III) (d^5)-O_2^- character. Several low-lying triplet and singlet states are also found for the $FePO_2$ system. The electron deformation density maps computed from the CI wavefunctions for the $^3A_{2g}$ and ^3E_g states of the FeP system are discussed, in connection with the ground state assignment.

1. INTRODUCTION

Among the many metalloporphyrins, the iron porphyrins have attracted much experimental and theoretical studies, since they represent a model of the heme in hemoglobin[1,2]. Despite this, there are yet many unknowns regarding their electronic structure, either in the biological molecules or in their synthetic models. For instance, the nature of the bond between iron and oxygen and the ground state assignment of oxy-iron porphyrins are still a controversial problem, with either an iron(II)-O_2 neutral dioxygen or an iron(III)-O_2^- superoxide formulation of the FeO_2 unit. The Fe(II)-O_2 formulation is in agreement with the diamagnetism of oxyhemoglobin[3] and gains support from the rather short O-O bond length (1.16-1.23 Å) in model compounds[4,5] and from practically all calculations[6,10], whereas the superoxide formalism Fe(III)-O_2^- (Weiss formalism)[11] has been assigned to these systems on the basis of the oxygen stretching frequency[12] and of the Mössbauer quadrupole splitting of oxyhemoglobin[13]. Results obtained from theoretical work have been reviewed recently [9,14]. We have reported previously LCAO-MO-SCF ab initio calculations for a number of possible electronic configurations of the oxyferroporphyrin system $FePO_2L$ (P=porphyrin, L=NH_3) which favored a closed-shell type singlet state

A. Veillard (ed.), Quantum Chemistry: The Challenge of Transition Metals and Coordination Chemistry, 377–390.

$^1A'$[configuration of the $d^6\pi^2$ type Fe(II)(S=0)-O_2(S=0)][9]. Later on, Nozawa et al suggested on the basis of the calculated Mössbauer parameters that the open-shell singlet state $^1A''$ [configuration of the $d^5\pi^3$ type Fe(III)(S=1/2)-O_2^-(S=1/2)] was a better candidate for the ground state[15]. INDO-SCF-CI calculations of Herman an Loew point to an Fe(II)-O_2 type configuration with a diamagnetic singlet ground state $^1A'$ although the calculation yields a net negative charge for the O_2 unit[8]. On the other hand Pariser-Parr-Pople (PPP) and Xα calculations suggest that the FeO_2 unit is well represented as a an equal mixture of Fe(II)(S=0)-O_2(S=0) and Fe(II)(S=1)-O_2(S=1) valence-state pairs . This description Fe(II)(S=1)-O_2(S=1) based on the ozone model was also proposed by Goddard et al on the basis of ab initio GVB calculations[6]. More recently, generalized molecular orbital calculations with configuration interaction on a model Fe$[(NH_2)_2CH]_2$ $(NH_3)(O_2)$ led to the conclusion that the FeO_2 system is best described as a singlet dioxygen σ donating to and π accepting from an Fe^{2+}[10].

The electronic configuration of the four-coordinate iron(II) porphyrins which represent the simplest iron-porphyrin systems has not been established conclusively, although the ground state has been confirmed to be a triplet with the iron in an intermediate spin state (S=1)[16]. A $^3A_{2g}$ $(xy)^2(xz, yz)^2(z^2)^2$ ground state has been suggested by Goff et al from the analysis of the proton NMR spectra of iron(II) tetraphenylporphyrin FeTPP which indicates large π contact shifts (thus requiring unpaired spins in d_{xz} and d_{yz})[17]. The same ground state has been proposed by Collman et al on the basis that it could fit both the magnetic susceptibility and Mössbauer data[16], while Boyd et al on the basis of magnetic measurements and a ligand field calculation conclude that the lowest state is $^3A_{2g}$ followed closely by the states ^3E_g and $^3B_{2g}$[18]. Yet, the Mössbauer spectra of FeTPP[19] and the proton NMR spectra of a number of four-coordinate (S=1) ferrous porphyrins[20] could as well be accounted for by assuming that in these complexes the ground-state results from a spin-orbit mixing of the two states $^3A_{2g}$ and ^3E_g. On the other hand, the resonance Raman spectra of ferrous octaethylporphyrin has been interpreted in terms of a ^3E_g ground state[21]. A similar assignment was proposed on the basis of extended Hückel calculations by Zerner et al[22]. From ab initio SCF calculations carried out by the group of Kashiwagi and by our group the $^3A_{2g}$ state was found to be the lowest SCF state, although the theoretical values of the electric field gradient and of the quadrupole splitting in the Mössbauer spectrum calculated from these SCF wavefunctions for ferrous porphyrins are more in agreement with a ^3E_g ground state[24,25]. In the Hartree-Fock approximation the wavefunction is represented by a single determinant, but there are several configurations of ^3E_g symmetry or of $^3A_{2g}$ symmetry which can mix with the above configurations in the ground-state wavefunction. Admixing of these configurations in the $^3A_{2g}$ or ^3E_g CI wavefunctions may be of significant importance. A CI calculation limited to the excitations within the d orbitals was carried out by Obara and Kashiwagi. A $^3A_{2g}$ ground state was also proposed on the basis of orbital energies obtained recently from Xα multiple scattering calculations[23].

We report here the results of SCF and large CI calculations for different electronic states of the model systems FeP (P=porphyrin ligand $N_4C_{20}H_{12}$) and $FePO_2$ (representing the heme in oxyhemoglobin).

2. CALCULATIONS

The SCF calculations were carried out with the system of programs ASTERIX[26] using the Gaussian basis sets of Table I. Those used for the

Table I. Gaussian basis sets (iron atom[27]/1st row atoms[28]/hydrogen[29])

System	Gaussian basis set	Contracted basis set
FeP	(15,11,6/9,5/4)	[9,6,3/4,2/2]
$FePO_2$	(14,9,6/9,5/4)	[6,4,3/3,2/2]

calculation on FeP are of double-zeta quality for the inner and valence shells and triple-zeta quality for the 4s and 3d shells of iron. For $FePO_2$ they are of split-zeta quality for the valence shells.

The porphyrin ligand was assumed to be planar (D_{4h} symmetry) and the bond lengths were taken from the experimental geometry of FeTPP[16]. The experimental structure of FeTPP points to a strong deviation from planarity of the porphyrin core which present an S_4 (quasi-D_{2d}) ruffling. In our previous SCF calculation we have discussed the ground state assignment of FeP for both the planar and ruffled structures and shown that such a distortion does not change appreciably the conclusions regarding the relative energies of the two states and the values of the computed properties (ionization potentials - electron distribution - electric field gradient at the iron nucleus)[25]. Therefore such a ruffled geometry has not been considered in the present CI calculations. For the FeP system the iron atom sits at the center of the molecule.

The choice of a five-coordinate system $FePO_2$ for the dioxygen complex of iron porphyrin was fully justified by two experimental evidences : i) the ^1H NMR spectroscopic detection of the base free O_2 adduct of the meso-tetrakis(α,α,α,o-pivalamidophenyl)porphyrin $TpivPPFeO_2$ [34] ; ii) the matrix isolation infrared spectra of oxy(tetraphenylporphyrinato) iron(II)[35]. The geometry of the FeO_2 unit was based on the experimental bond lengths and bond angles of the dioxygen complex of the picket-fence porphyrin[4,5], namely Fe-O 1.75 Å, O-O 1.22 Å and an FeOO angle of 129°. While for most of the hexacoordinated metalloporphyrins the metal atom sits in or near the coordination plane (plane of the four pyrrole nitrogens), the out-of-plane displacement of the metal is an ubiquitous feature of the five-coordinate porphyrins. For the oxyheme model $FePO_2$ (S=0) this displacement was optimized at the SCF level for the closed shell state $^1A'$ $d^6(\pi_g^a)^2$ and the potential curve has a minimum for an iron out-of-plane displacement of 0.15 Å. This value is close to the displacement

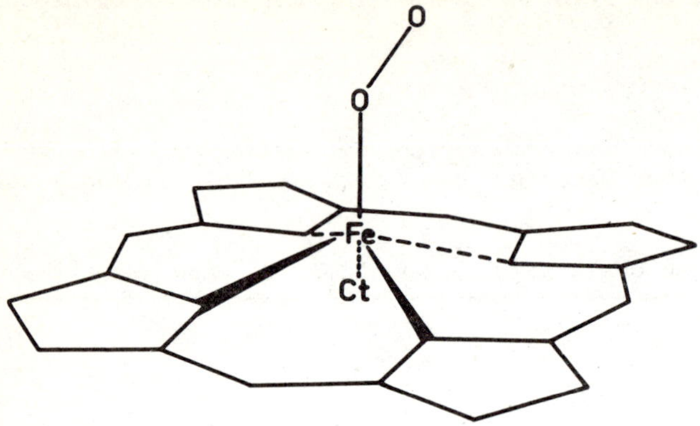

Figure 1. The $FePO_2$ model

of 0.23 Å reported from X-ray diffraction for a low-spin 5-coordinate iron(II)porphirinate Fe(OEP)(CS)[36] and has then been used in the SCF and CI calculations for all the electronic states. The porphyrin plane is the xOy plane with the nitrogen atoms of the pyrrole rings being along the axes Ox and Oy and with OX and OY the bisectors of the axes. The Fe-O axis is the z axis and the O-O axis projects along OX. The molecular point group is C_s.

For FeP SCF and CI calculations have been carried out for the two states $^3A_{2g}$ $(xy)^2(xz,yz)^2(z^2)^2$ and ^3E_g $(xy)^2(xz,yz)^3(z^2)^1$ which are thought from our previous study to be the best candidates for the ground state. For each state the CI calculation is based on the canonical m.o.'s from the corresponding SCF calculation. All single and double excitations relatively to the reference state were considered within a given set of active orbitals. Two sets of active orbitals respectively labeled CI56 (56 active orbitals) and CI69 (69 active orbitals) were considered and chosen on the basis of an energy criterion [37]. In the CI56 calculation we have correlated the 6 iron d electrons whereas in the CI69 calculation 32 electrons were correlated, namely the 6 iron d electrons and the 26 π electrons of the porphyrin ring. This resulted in 2300 and 58000 configurations for CI56 and CI69 respectively.

For $FePO_2$ we have considered states for which the leading electronic configuration is either $d^6\pi^2$ or $d^5\pi^3$. We denote π_g^a and π_g^b the two $1\pi_g$ antibonding orbitals of dioxygen in the $FePO_2$ complex which are respectively symmetrical and antisymmetrical with respect to the FeO_2 plane. The closed-shell singlet state $d^6(\pi_g^a)^2$ with the $^1A'$ symmetry corresponds to an $Fe(II)-O_2$ formal configuration. The configurations $d^5(\pi_g^a)^2(\pi_g^b)^1$ give rise to open-shell singlet and triplet of $^{1,3}A'$ and $^{1,3}A''$ symmetry corresponding to an $Fe(III)-O_2^-$ formal configuration. SCF calculations have been carried out for the closed-shell singlet state a^1A' $(xy)^2(Xz)^2(Yz)^2(\pi_g^a)^2$ and for the open-shell triplet states $^3A'$ $(xy)^2(Xz)^2(Yz)^1(\pi_g^a)^2(\pi_g^b)^1$ and $^3A''$ $(xy)^2(Xz)^1(Yz)^2$ $(\pi_g^a)^2(\pi_g^b)^1$. For the a^1A', $^3A'$ and $^3A''$ states the CI calculation is based on the canonical m.o.'s from the corresponding SCF calculations. For the CI calculation of the singlet open-shell states b^1A' $(xy)^2(Xz)^2(Yz)^1(\pi_g^a)^2(\pi_g^b)^1$ and $^1A''$ $(xy)^2(Xz)^1(Yz)^2(\pi_g^a)^2(\pi_g^b)^1$ the SCF m.o.'s of the corresponding triplet were used. The CI expansion was based on a set of 54 active obitals in which 14 electrons were correlated corresponding to the iron orbitals $3d_{xy}$, $3d_{Xz}$, $3d_{Yz}$ and to the dioxygen orbitals $3\sigma_g$, $1\pi_u^a$, $1\pi_u^b$, $1\pi_g^a$ for the $^1A'$ $d^6\pi^2$ configuration that is to the upper valence orbitals of the FeO_2 unit. The 47 unoccupied active orbitals were chosen among the lowest molecular orbitals having iron, dioxygen or π (porphyrin) character[38].

These CI calculations were carried out with the CI program developed originally by Brooks and Schaefer[39] using the graphical unitary group approach.[40] Unitary transformations were performed on the occupied space in order to maximize the weight of the iron atom and of the dioxygen ligand in the active orbitals.

3. RESULTS AND DISCUSSION

3.1. The FeP system

Table II and Figure 2 give the total SCF and CI energies and the respective weights of the main configuration for the two states $^3A_{2g}$ $(xy)^2(xz,yz)^2(z^2)^2$ and ^3E_g $(xy)^2(xz,yz)^3(z^2)^1$ of FeP. At the SCF level

Table II. Total energies (in a.u.) of the $^3A_{2g}$ and ^3E_g states of FeP.

Calculation		$^3A_{2g}$	^3E_g
SCF		-2244.2018	-2244.1903
CI 56	1 reference state	-2244.2071	-2244.2035
	2 reference states		-2244.2043
CI 69	1 reference state	-2244.4780	-2244.4679

the lowest triplet state is the $^3A_{2g}$ and the ^3E_g state is 0.31 eV higher, a value close to the one of 0.33 eV found in previous SCF calculations[25].

In the two CI calculations the $^3A_{2g}$ remains the lowest state. When only the iron d electrons are correlated (CI56) the energy lowering is larger for the ^3E_g state than for the $^3A_{2g}$ state (0.0132 versus 0.0053 a.u). The weight of the reference state configuration is 99.5 % in the $^3A_{2g}$ state wavefunction and 89 % in the ^3E_g state wavefunction. Another configuration $(xy)^1(xz,yz)^3(z^2)^2$ corresponding to the $d_{xy} \rightarrow d_{z^2}$ single excitation contributes 10 % in the ^3E_g state wavefunction. A CI calculation carried out for the ^3E_g state with two reference states, namely ^3E_gA $(xy)^2(xz,yz)^3(z^2)^1$ and ^3E_gB $(xy)^1(xz,yz)^3$ $(z^2)^2$ shows that

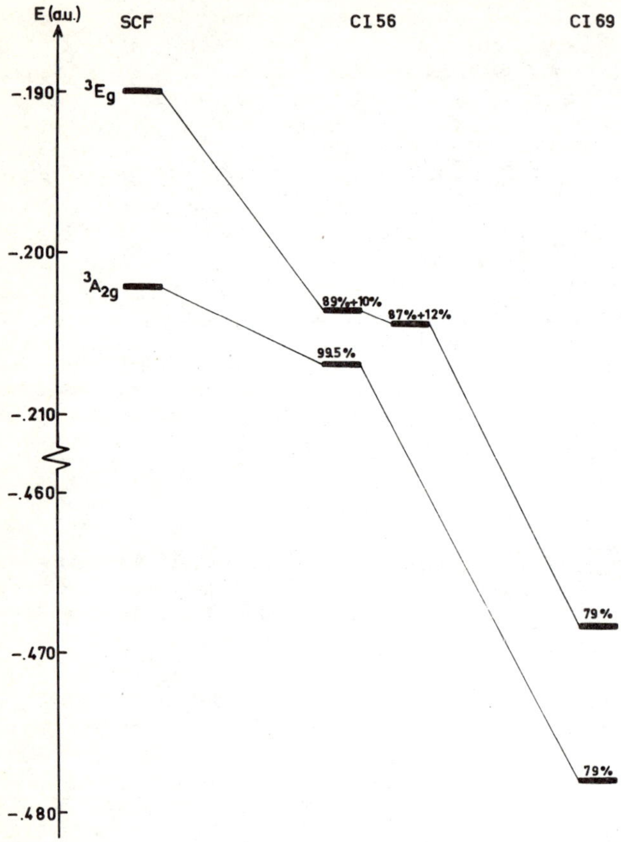

Figure 2. Relative energy levels (in a.u. and relative to -2244.) of the $^3A_{2g}$ and ^3E_g states of FeP.

the ^3E_g state is still less stable than the $^3A_{2g}$ state. The resulting gap between these two states is reduced to 0.08 eV and the configurations ^3E_gA and ^3E_gB contribute 87 % and 12 % , respectively to the ^3E_g state wavefunction. Comparable values for the respective energy lowering of the two states and for the weights of the two configurations in the ^3E_g state were also found by Obara and Kashiwagi in a CI calculation limited to excitations within the d orbitals [24].

However this result may not be conclusive, owing to the small number of electrons which have been correlated. When the CI expansion is extended to CI69 the energy lowering of both the two states $^3A_{2g}$ and ^3E_g is of the same order of magnitude, namely 7.5 eV and the $^3A_{2g}$ state remains more stable than the ^3E_g state, the gap between these two states being now 0.27 eV. For both states the primary configuration is responsible for 79 % of the total wavefunction and the largest weight of other configurations is 1 % and corresponds to a double excitation within the π system of the porphyrin $(\pi,\pi) \rightarrow (\pi^*, \pi^*)$. The configuration ^3E_gB $(xy)^1(xz,yz)^3(z^2)^2$ contributes now only 0.5 % in the ^3E_g wavefunction. These energy results show the importance of having correlated the 26 π electrons of the porphyrin ring in the CI calculations.

Our assignment of the ground state to the configuration $^3A_{2g}$ $(xy)^2(xz,yz)^2(z^2)^2$ is identical to the one made for FeTPP on the basis of NMR spectra, magnetic susceptibility measurements and some interpretation of the Mössbauer data[16,17], but differs from the ^3E_g $(xy)^2(xz,yz)^3(z^2)^1$ ground state assignment made by Obara and Kashiwagi on the basis of both their calculations and their analysis of the Mössbauer parameters[24].

Recently it has been pointed out by Benard that theoretical and experimental analysis of the electronic density deformation could contribute to the elucidation of the ground state of FeTPP[41]. The

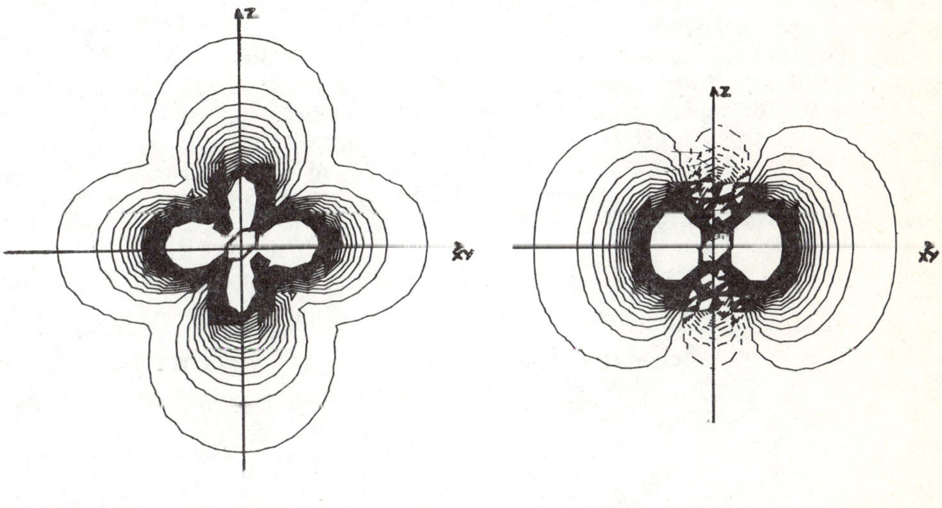

Figure 3. Computed deformation density maps from the CI69 calculation for (a) the $^3A_{2g}$ state (b) the ^3E_g state of FeP in a plane perpendicular to the porphyrin and bisecting the N-Fe-N angle - Contour interval 0.10 e$\mathring{A}^{-3}$. Negative contours dashed.

electron deformation density distribution - defined as the difference between a molecular electron density distribution and the superposition of spherically averaged atomic distributions - is frequently accessible to X-ray and neutron diffraction experiments and is closely related to the wavefunction. We have computed the deformation density maps for both the $^3A_{2g}$ and ^3E_g states from the corresponding CI69 wavefunctions. The deformation density maps along the z axis should be quite different for the two states since the population of the d_{z^2} orbital is two for the $^3A_{2g}$ state and one for the ^3E_g state. In fact the section of the computed distributions by a plane perpendicular to the porphyrin ring and bisecting two nitrogen-iron-nitrogen planes (figure 3) accounts for this difference. For the $^3A_{2g}$ state the computed map displays four accumulation regions similar in height and shape, corresponding to the filling of the xy and z^2 orbitals in this configuration. On the other hand, the computed map for the ^3E_g state shows an electron - deficient area along the z axis, but not along the xy direction. Therefore it can be predicted that any significant dissymmetry between these two directions in a forthcoming experimental map would represent a strong argument in favor of a spin-orbit mixing between the $^3A_{2g}$ and ^3E_g states.

An analysis of the electron density distribution obtained from X-ray diffraction data for the iron(II) phthalocyanine(FePc) indicates the ^3E_gA $(xy)^2(xz,yz)^3(z^2)^1$ state to be the leading contributor to the ground state of FePc[42]. However these results cannot be extended to the FeP system and further analysis must be postponed until a charge density study on FeTPP is reported.

The deformation density maps computed from either SCF or from CI56 or from CI69 wavefunctions for a given state are nearly identical and this should be related to the small variations brought in the electronic distribution by the configuration interaction. A quantitative rationalization of this evidence is directly available from the electron distributions of table III obtained from a Mulliken population analysis.

It is commonly accepted that a moderate π back-bonding from metal to porphyrin helps to relieve the charge build-up on the metal atom

Table III. d and π populations of the iron and π populations of the porphyrin for the two states $^3A_{2g}$ and ^3E_g of FeP

State	Iron						Porphyrin[a]
	x^2-y^2	xy	xz	yz	z^2	4p	
$^3A_{2g}$ (SCF)	0.18	2.02	0.98	0.98	1.94	0.05	25.98
$^3A_{2g}$ (CI69)	0.18	2.02	0.99	0.99	1.94	0.06	25.96
^3E_g (SCF)	0.18	2.03	1.47	1.47	0.99	0.06	25.99
^3E_g (CI69)	0.18	2.02	1.48	1.48	0.99	0.06	25.98

[a]Total 2pπ population of the carbon and nitrogen atoms of the porphyrin

that results from σ donation. Experimental evidence in support of or against a significant iron-to-porphyrin π backbonding has been discussed by several authors [17,43]. From the π populations of the iron and of the porphyrin ligand, the FeP system shows a small π charge transfer from the porphyrin to the iron.

The energy levels of iron porphyrins have been discussed by many authors [9,44], in relation with photoelectron spectroscopic studies [45]. We have reported previously the ionization potentials calculated for the $^3A_{2g}$ state of FeP with two geometrical structures (planar and ruffled) both on the basis of Koopmans theorem and by the ΔSCF method[9] For the two highest π orbitals of the porphyrin ligand, a_{1u} and a_{2u}, both the Koopman's and the ΔSCF values of the ionization potentials were in good agreement with experimental values. Our calculation indicated that the a_{1u} orbital has a higher energy than a_{2u} and the experimental separation of 0.04 eV between these two states of the ion was well reproduced by the previous calculation (0.04 eV). In the present calculation the energy levels of a_{1u} and a_{2u} are respectively -0.235 and -0.252 a.u. for both the two states $^3A_{2g}$ and ^3E_g of FeP. These values are close to the ones of our previous calculations (-0.234 and -0.248 a.u. for the $^3A_{2g}$ state with a ruffled structure). They are identical for the two states $^3A_{2g}$ and ^3E_g and show that the a_{1u} and a_{2u} ionization potentials are not expected to be sensitive to the nature of the ground state of FeP.

3.2 The $FePO_2$ system

According to the results of Table IV and Figure 4 the lowest SCF energy for $FePO_2$ corresponds to a triplet $^3A'$ with the formal configuration $(xy)^2(Xz)^2(Yz)^2(\pi_g^a)^2(\pi_g^b)^1$. The $^3A''$ state is calculated to be only 0.001 a.u. (0.03 eV) above the $^3A'$ state. Then at 0.052 a.u. (1.42 eV) above the lowest triplet, one finds the singlet closed-shell state a^1A' corresponding to the Pauling model with the configuration $(xy)^2(Xz)^2(Yz)^2(\pi_g^a)^2$. The calculated gap 1.42 eV between the closed-shell state a^1A' and the open-shell state $^3A'$ is of the same order of magnitude as found in our previous calculation[9].

Configuration interaction lowers all of the states significantly, but by different amounts : the a^1A' (closed-shell) state by 2.44 eV, the $^3A''$ state by 1.27 eV and the $^3A'$ state by 1.14 eV resulting in the order shown in Figure 4. These quantitative differences in the SCF and CI results indicate the importance of configuration interaction in determining the relative energies of the ground and low-lying excited states in the model dioxygen complex of iron (II) porphyrin $FePO_2$. From the results of the CI calculations, the lowest CI energy for $FePO_2$ corresponds to an open-shell singlet state b^1A' with the formal configuration $(xy)^2(Xz)^2(Yz)^1(\pi_g^a)^2(\pi_g^b)^1$. The open-shell states $^{1,3}A''$ and $^3A'$ stay just above and the closed-shell a^1A' state lies 0.36 eV above the open-shell singlet state b^1A'. Table V describes the important configurations contributing in the CI wavefunctions of the different singlet and triplet states of $FePO_2$. In the CI wavefunctions of the a^1A', $^3A'$ and $^{1,3}A''$ states the primary configuration (reference state configuration) is responsible for 90- 93 % of the wavefunction. On the

Table IV. Total energy (in a.u.) for $FePO_2$

Electronic configuration	State	Total energy SCF	CI
$(xy)^2(Xz)^2(Yz)^2(\pi_g^a)^2$	a^1A'	-2390.50480	-2390.59454
$(xy)^2(Xz)^2(Yz)^1(\pi_g^a)^2(\pi_g^b)^1$	$^3A'$	-2390.55710	-2390.59887
	b^1A'		-2390.60769
$(xy)^2(Xz)^1(Yz)^2(\pi_g^a)^2(\pi_g^b)^1$	$^3A''$	-2390.55577	-2390.60232
	$^1A''$		-2390.60119

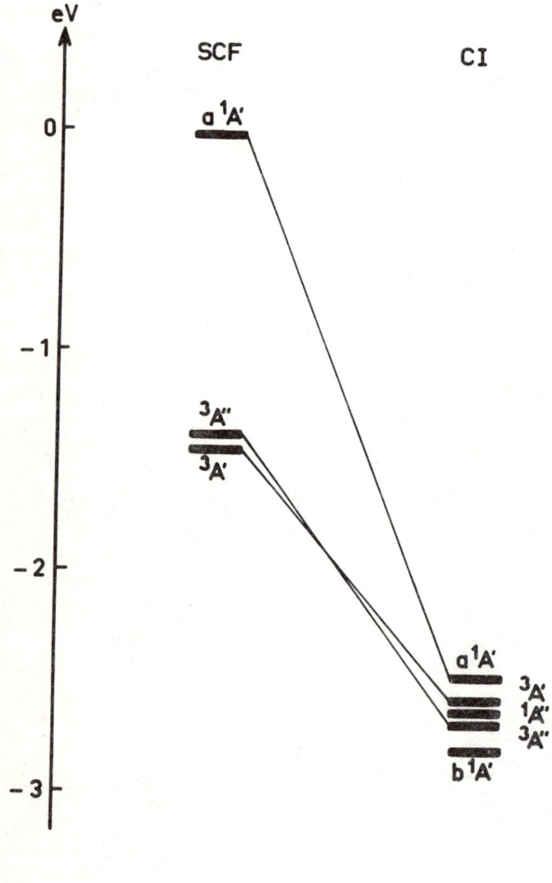

Figure 4. Relative energy levels for $FePO_2$

Table V. The CI wavefunctions for the different states of $FePO_2$

a^1A' wavefunction		
$(\pi_u^a)^2(3\sigma_g)^2(\pi_u^b)^2(\pi_g^a)^2(xy)^2(Xz)^2(Yz)^2$		92 %
$(\pi_u^b)^2(Yz)^2$	$\rightarrow (\pi_u^b)^1(Yz)^1(\pi_g^b)^2$	1 %
$(\pi_u^b)^2$	$\rightarrow (\pi_g^b)^2$	2 %
$^3A'$ wavefunction		
$(\pi_u^a)^2(3\sigma_g)^2(\pi_u^b)^2(\pi_g^a)^2(xy)^2(Xz)^2(Yz)^1(\pi_g^b)^1$		93 %
$(\pi_g^a)^2$	$\rightarrow (\pi_g^a)^1(4s)^1$	< 1 %
$(\pi_g^a)^2$	$\rightarrow (\pi_g^a)^1(z^2)^1$	2 %
b^1A' wavefunction		
$(\pi_u^a)^2(3\sigma_g)^2(\pi_u^b)^2(\pi_g^a)^2(xy)^2(Xz)^2(Yz)^1(\pi_g^b)^1$		59 %
$(Yz)^1(\pi_g^b)^1$	$\rightarrow (\pi_g^b)^2$	5 %
$(Yz)^1(\pi_g^b)^1$	$\rightarrow (Yz)^2$	30 %
$(\pi_g^a)^2$	$\rightarrow (\pi_g^a)^1(4s)^1$	< 1 %
$(\pi_g^a)^2$	$\rightarrow (\pi_g^a)^1(z^2)^1$	2 %
$^3A''$ wavefunction		
$(\pi_u^a)^2(3\sigma_g)^2(\pi_u^b)^2(\pi_g^a)^2(xy)^2(Xz)^1(Yz)^2(\pi_g^b)^1$		90 %
$(\pi_u^b)^2(\pi_g^b)^1$	$\rightarrow (\pi_u^b)^1(\pi_g^b)^2$	< 1 %
$(\pi_g^a)^2$	$\rightarrow (\pi_g^a)^1(4s)^1$	< 1 %
$(\pi_g^a)^2$	$\rightarrow (\pi_g^a)^1(z^2)^1$	2 %
$^1A''$ wavefunction		
$(\pi_u^a)^2(3\sigma_g)^2(\pi_u^b)^2(\pi_g^a)^2(xy)^2(Xz)^1(Yz)^2(\pi_g^b)^1$		91 %
$(\pi_u^b)^2(\pi_g^b)^1$	$\rightarrow (\pi_u^b)^1(\pi_g^b)^2$	< 1 %
$(\pi_g^a)^2$	$\rightarrow (\pi_g^a)^1(4s)^1$	< 1 %
$(\pi_g^a)^2$	$\rightarrow (\pi_g^a)^1(z^2)^1$	2 %

other hand, the primary configuration which places one electron in both the Yz and π_g^b orbitals is responsible for only 59 % of the wavefunction in the b^1A' state. Single excitation $(Yz)^1(\pi_g^b)^1 \rightarrow (Yz)^2$ giving the a^1A' closed-shell state makes up 30 % of the b^1A' wavefunction. The b^1A' state is the lowest state in the present study and is described as a strong mixing of an iron(II)-O_2 type configuration and an iron(III)-O_2^- type configuration. In the a^1A' closed-shell SCF wavefunction the π_g^b antibonding dioxygen orbital is a virtual molecular orbital and has then not been optimized, but in the dioxygen complex $FePO_2$ the two antibonding dioxygen orbitals π_g^a and π_g^b are nearly degenerate. Since the gap between the two $^1A'$ states is small (0.36 eV), any improvement in the description of the virtual π_g^b molecular orbital in the a^1A' SCF wavefunction may lower appreciably the CI energy of the a^1A' state. We are presently working in that direction.

4. CONCLUSION

The results of the configuration-interaction calculations on the four-coordinate iron(II)porphyrin FeP predict a $^3A_{2g}$ ground state. Further refinements in the basis set size or in the size of the CI expansion will probably not change appreciably the results since the two states $^3A_{2g}$ and ^3E_g have electronic configuration similar in nature, i.e. they differ only by the occupation of the iron d orbitals and since all the orbitals having d character have been correlated in the CI calculations, as well as the π orbitals of the porphyrin ring in the CI69 calculation. It is possible that explicit consideration of the spin-orbit coupling will be needed to identify definitely the ground state of FeP and to account for its electronic properties. The FeP molecule is a model for FeTPP and the influence of the substituents of the porphyrin ring may also have to be considered.

The quantitative differences in the SCF and CI results for the dioxygen iron(II)porphyrin complex $FePO_2$ indicate the importance of configuration interaction in determining the relative energies of the ground and low-lying excited states in these dioxygen systems. Further work is now underway to obtain a more accurate determination of these relative energies.

ACKNOWLEDGEMENTS

Calculations have been carried out on the CRAY-1 computer of the CCVR (Palaiseau), through a grant of computer time from the Conseil Scientifique du Centre de Calcul Vectoriel pour la Recherche.

REFERENCES

1. M. Gouterman, The porphyrins, Vol.3, Academic Press, New-York, 1978
2. G.H. Loew, Iron Porphyrins, Part I, A.B.P. Lever and H.B. Gray eds., Addison - Wesley, Reading, Mass., 1983

3. L. Pauling and C.D. Coryell, Proc. Natl. Acad. Sci. U.S.A., **22**, 210 (1936)
L. Pauling, Hemoglobin, Butterworth, London, 1949, p.57
4. G.B. Jameson, G.A. Rodley, W.T. Robinson, R.R. Gagne, C.A. Reed and J.P. Collman, Inorg. Chem., **17**, 850 (1978)
5. G.B. Jameson, F.S. Molinaro, J.A. Ibers, J.P. Collman, J.I. Brauman, E. Rose and K.S. Suslick, J. Amer. Chem. Soc., **102**, 3224 (1980)
6. W.A. Goddard III and B.D. Olafson, Proc. Natl. Acad. Sci. U.S.A., **72**, 2335 (1975)
B.D. Olafson and W.A. Goddard III, Proc. Natl. Acad. Sci. U.S.A., **74**, 1315 (1977)
7. D.A. Case, B.H. Huynh and M. Karplus, J. Amer. Chem. Soc., **101**, 4433 (1979)
8. Z.S. Herman and G.H. Loew, J. Amer. Chem. Soc., **102**, 1815 (1980)
9. (a) A. Dedieu, M.-M. Rohmer and A. Veillard, Adv. Quantum Chem., **16**, 43 (1982)
(b) A. Dedieu, M.-M. Rohmer and A. Veillard, Metal-Ligand Interactions in Organic Chemistry and Biochemistry, B. Pullman and N. Goldblum eds., Reidel, Dordrecht, 1977, p.101
10. J.E. Newton and M.B. Hall, Inorg. Chem., **23**, 4627 (1984)
11. J.J. Weiss, Nature, **202**, 83 (1964)
J.J. Weiss, Nature, **203**, 183 (1964)
12. J.P. Collman, J.I. Brauman, T.R. Halbert and K.S. Suslick, Proc. Natl. Acad. Sci. U.S.A., **73**, 3333 (1976)
13. K. Spartalian, G. Lang, J.P. Collman, R.R. Gagne and C.A. Reed, J. Chem. Phys., **63**, 5375 (1975)
14. M.H. Gubelmann and A.F. Williams, Structure and Bonding, **55**, 1 (1983)
15. T. Nozawa, M. Hatano, U. Nagashima, S. Obara and H. Kashiwagi, Bull. Chem. Soc. Jpn., **56**, 1721 (1983)
16. J.P. Collman, J.L. Hoard, N. Kim, G. Lang and C.A. Reed, J. Am. Chem. Soc., **97**, 2676 (1975)
17. H. Goff, G.N. La Mar and C.A. Reed, J. Amer. Chem. Soc., **99**, 3641 (1977)
18. P.D.W. Boyd, A.D. Buckingham, R.F. McMeeking and S. Mitra, Inorg. Chem., **18**, 3585 (1979)
19. G. Lang, K. Spartalian, C.A. Reed and J.P. Collman, J. Chem. Phys., **69**, 5424 (1978)
20. J. Mispelter, M. Momenteau and J.M. Lhoste, J. Chem. Phys., **72**, 1003 (1980)
21. T. Kitagawa and J. Teraoka, Chem. Phys. Letters, **63**, 443 (1979)
22. M. Zerner, M. Gouterman and H. Kobayashi, Theoret. Chim. Acta, **6**, 363 (1966)
23. S.F. Sontum, D.A. Case and M. Karplus, J. Chem. Phys., **79**, 2881 (1983)
24. S. Obara and H. Kashiwagi, J. Chem. Phys., **77**, 3155 (1982)
25. M.-M. Rohmer, A. Dedieu and A. Veillard, Chem. Phys., **77**, 449 (1983)
26. M. Benard, A. Dedieu, J. Demuynck, M.-M. Rohmer, A. Strich, A. Veillard and R. Wiest, unpublished work.

27. The (15, 11, 6) basis set is made from the (14, 9, 5) basis of Wachters by adding an additional s function with an exponent of 0.2985, two p functions with exponents of 0.2308 and 0.0899 and one d diffuse function of exponent 0.1133 . The (14, 9, 6) basis set is made from the (13, 7, 5) basis of Hyla-Kryspin et <u>al</u> by adding an additional s function with an exponent of 0.345, two p functions with exponents 0.247 and 0.08 and by replacing the last d function by two functions with exponents 0.332 and 0.106. All these exponents were chosen according to the even-tempered criterion of Raffenetti et <u>al</u> .
28. S. Huzinaga, Approximate Atomic Functions, Technical Report, University of Edmonton (1971)
29. S. Huzinaga, J. Chem. Phys., **42**, 1293 (1965)
30. A.J.A. Wachters, J. Chem. Phys., **52**, 1033 (1970)
31. P.J. Hay, J. Chem. Phys., **66**, 4377 (1977)
32. I. Hyla-Kryspin, J. Demuynck, A. Strich and M. Benard, J. Chem. Phys., **75**, 3954 (1981)
33. R.C. Raffenetti, R.D. Bardo and K. Ruedenberg in "Energy structure and reactivity, D.W. Smith and W.B. McRac eds., Wiley, New-York, 1973, p.164
34. L. Latos-Grazynski, R.J. Cheng, G.N. La Mar and A.L. Balch, J. Amer. Chem. Soc., **104**, 5992 (1982)
35. T. Watanabe, T. Ama and K. Nakamoto, J. Phys. Chem., **88**, 440 (1984)
36. W.R. Scheidt and D.K. Geiger, Inorg. Chem., **21**, 1208 (1982)
37. Of these active orbitals, 52 are unoccupied in the SCF calculations, 1 has a 3d character, 3 a 4s character, 1 a 3d, 4s character, 6 a 3d, σ^* character, 2 a 4p, σ^* character, 2 a 3d, π^* character, 33 a π^* character.
38. Of these 47 orbitals, 6 have an O_2 character (1 being the π_g^b) 30 have iron 4s, 4p and 3d character, 11 have a π^* character.
39. (a) B.R. Brooks and H.F. Schaefer, J. Chem. Phys., **70**, 5092 (1979)
(b) This program was adapted for the IBM computer by F. Brown and I. Shavitt
40. (a) J. Paldus, Theor. Chem. Adv. Perspect., **2**, 131 (1976)
(b) I. Shavitt, Int. J. Quant. Chem. Symp., **11**, 131 (1977)
41. M. Benard, Angew. Chem. Suppl., 1845 (1982)
42. P. Coppens and L. Li, J. Chem. Phys., **81**, 1983 (1984)
43. J. Mispelter, M. Momenteau and J.M. Lhoste, Chem. Phys. Letters, **57**, 405 (1978)
44. K. Ohno, Horizons of Quantum Chemistry, K. Fukui and B. Pullman eds., Reidel, Dordrecht, 1979, p.254
45. S. Kitagawa, I. Morishima, T. Yonezawa and N. Sato, Inorg. Chem., **18**, 1345 (1979)
46. M. Cerdonio, A. Congiu-Castellano, F. Mogno, B. Pispisa, G.L. Romani and S. Vitale, Proc. Natl. Acad. Sci. U.S.A., **74**, 398 (1977)
M. Cerdonio, A. Congiu-Castellano, L. Celabrese, S. Morante, B. Pispisa and S. Vitale, Proc. Natl. Acad. Sci. U.S.A., **75**, 4916 (1978)
47. L. Pauling, Proc. Natl. Acad. Sci. U.S.A., **74**, 2612 (1977).

MULTIPLE METAL-METAL AND METAL-CARBON BONDS

Michael B. Hall
Department of Chemistry
Texas A&M University
College Station, Texas 77843
U.S.A.

ABSTRACT. *Ab initio*, generalized molecular orbital, configuration interaction calculations are reported for chromium-chromium quadruple bonds, for molybdenum-molybdenum triple bonds, and for niobium-carbon and molybdenum-carbon double bonds. The results for the potential energy curves of Cr-Cr quadruple bonds suggest that the nature of the bridging ligand is a major factor in determining the Cr-Cr bond distance. The predicted dissociation energy of the model molybdenum triple bond molecule, Mo_2H_6, is 284 kJ mol^{-1}. This value is at the lower range of the possible experimental values. The difference in the chemistry and bonding between the Fischer-type carbene, $Mo(CO)_5(C(OH)H)$, and the Schrock-type carbene, $Nb(\eta^5-C_5H_5)Cl_2(CH_2)$, are shown to arise from the nature of the metal fragments. The $Mo(CO)_5$ fragment with ligand stabilized π orbitals causes the carbene to be nucleophilic, while the $NbCpCl_2$ fragment with destabilized π orbitals causes the carbene to be electrophilic. The pivoting of the methylene in the electron deficient Nb complex is examined and involves both a rotation of the Nb-C bond and a direct Nb-H interaction.

1. INTRODUCTION

Molecules that have either a multiple bond between two transition metals or between a transition metal and a main-group element continue to arouse the interest of chemists. No longer a novelty, these molecules are gaining utility as building blocks for inorganic and organometallic reactions in ways that parallel the utility of carbon-carbon multiple bonds in organic chemistry.

Although hydrated dichromium(III) tetraacetate was prepared in 1844, it was not recognized as having a quadruple bond until 1970.[1] Since that time a large number of quadruply bonded chromium dimers have been prepared with Cr-Cr bond distances ranging from 2.54Å to 1.83Å.[2] Our interest in these compounds stems from this enormous range of bond lengths for molecules with the same formal bond order.

A. Veillard (ed.), Quantum Chemistry: The Challenge of Transition Metals and Coordination Chemistry, 391–401.

Another interesting class of multiply bonded molecules are the metal-metal triple bonds typified by molybdenum compounds of the formula Mo_2X_6 with X = OR, NR_2, or CR_3.[2] The first molecule of this type was structurally characterized in 1971.[3] As in most other second and third row transition metal systems, the M-M bond distances in these compounds span a rather small range around 2.2Å.[2] In an attempt to determine the bond energy of triple-bonded metal complexes the heat of formation of $Mo_2(N(CH_3)_2)_6$ was measured by microcalorimetry. In spite of accurate measurements of $\Delta H_f°$ the assignment of a Mo≡Mo bond energy was hampered by a large range for Mo-NR_2 bond energies. The bond energy for Mo≡Mo was reported as 592 ± 196 kJ mol^{-1}.[4] This spans a range from the strength of the C≡C triple bond to that of a C-C single bond. Clearly, a more accurate value for the strength of all metal-metal multiple bonds would be helpful in understanding their chemistry.

The first molecule containing a multiple transition metal-to-carbon bond, $W(CO)_5(CPh(OMe))$, was discovered by Fischer in 1964.[5] Since that time systems with multiple metal-carbon bonds have grown in importance not only as reagents but as intermediates in both homogeneous and heterogeneous catalysis. A chemically different type of metal-carbon double bond was discovered by Schrock in 1975.[6] Unlike the Fischer-type carbenes which are nucleophilic at the carbon, these Schrock-type carbenes are electrophilic at the carbon.[5-7] While the Fischer-type obey the 18-electron rule; the Schrock-type are usually electron dificient with 14, 12, or even 10 metal ligand valence electrons. Apparently this electron deficiency often causes a structural distortion where the CHR group pivots about the carbon atom in the plane of the M-CHR unit.[7]

2. THEORETICAL APPROACH

The generalized molecular orbital method was used for all of the calculations reported in this chapter. This method consists of a limited type of multi-configuration self-consistent-field calculation followed by a configuration interaction calculation.[8] The basis sets, which are described in detail elsewhere[9,10,11], are double-zeta or better on the atoms involved in the multiple bond but only minimal on the remaining atoms.

3. RESULTS AND DISCUSSION

3.1 Chromium-chromium Quadruple Bonds

All of the tetracarboxylate derivatives of dichromium characterized to date have rather long Cr-Cr bonds (> 2.29Å) and ligands in axial positions. The first *ab initio* calculations on one of these species, $Cr_2(O_2CH)_4(H_2O)_2$, found that the lowest energy single determinant corresponded to the non-bonded configuration $\sigma^2\delta^2\delta^{*2}\sigma^{*2}$.[12] Subsequently it was shown that the correct description of these systems is not possible within the single determinant approximation.[13] Our initial work showed that at a fixed Cr-Cr distance $Cr_2(O(NH)CH)_4$ has a "stronger" quadruple bond than $Cr_2(O_2CH)_4$.[14] Others have shown that the theoretical bond distance for $Cr_2(O_2CH)_4$ and $Cr_2(O_2CH)_4(H_2O)_2$ is nearly the same.[15] Thus, the theoretical work appears to be supporting the idea that the nature of the bridging ligand dominates the length of the chromium-chromium bond.

Our results for the chromium-chromium bond distance for a series of dimers are shown in Table I. The dimer with the most electropositive

Table I. Equilibrium Bond Lengths.[a]

	Calculated	Predicted[b]	Experimental[e]
$Cr_2(O_2CH)_4$	2.41	2.27	
$Cr_2((HN)_2CH)_4$	1.93	1.79	1.84
cis - $Cr_2(O(NH)CH)_4$[c]	2.01	1.87	1.87
cis - $Cr_2(O(NH)CH)_4$[d]	1.97	1.83	
trans - $Cr_2(O(NH)CH)_4$	2.16	2.02	

[a]In angstroms.
[b]Derived from the calculated value by including the basis set error.
[c]CI performed on HF wave function.
[d]CI performed on a wave function in which the sigma and delta orbitals were optimized.
[e]Experimental bond lengths accurate to better than ± 0.01 A.

donor atoms, $Cr_2((NH)_2CH)_4$, shows a very short Cr-Cr distance close to that observed for its experimental analog $Cr_2((NMe)_2CPh)_4$. The predicted bond distance for the tetracarboxylates, however, remains long even in the absence of any axial ligands. Our predicted value for $Cr_2(O_2CH)_4$ is 2.27Å. The two systems with one ligand atom of each type, $Cr_2(O(NH)CH)_4$, must be treated with some caution, because they have lower symmetry. In the lower symmetry molecules the strongly occupied molecular orbitals will localize on one Cr or the other. Although this may actually happen to some degree, the single determinant and to a lesser degree the GMO orbitals will over emphasize this localization. In fact even the higher symmetry tetracarboxylates have single determinant symmetry-broken solutions which are lower in energy than the high symmetry solutions.[16]

Table II compares the results for several symmetry-broken solutions at both the Hartree-Fock level and configuration interaction level. The lowest energy symmetry broken solution has D_{2d} symmetry with the π orbitals localized one on each Cr and the σ and δ orbitals

Table II. Symmetry-broken Solutions for $Cr_2(O_2CH)_4$

Sym.	Cr_1 Local.	Cr_2 Local.	Delocal.	E_{HF} at 2.3A	E_{CI} at 2.3A	R_{MIN} from CI
C_{4v}	$\sigma^2\delta^2$	$\pi_x^2\pi_y^2$	--	2811.589	2811.591	2.61
C_{2v}	$\sigma^2\pi_x^2$	$\pi_y^2\delta^2$	--	2811.514	2811.525	--
C_s	σ^2	δ^2	π^4	2811.241	--	--
D_{2d}	π_x^2	π_y^2	$\sigma^2\delta^2$	2811.600	2811.610	2.61
D_{4h}	--	--	$\sigma^2\pi^4\delta^2$	2811.077	2811.775	2.36

delocalized. This solution is much lower in energy than the one with the σ and δ localized and the π delocalized. Although these broken-symmetry solutions are lower in energy than the symmetric solution for the single-determinant, configuration interaction is not as effective using localized orbitals and after CI the lowest energy solution is the delocalized one. Most important for our comparisons are the predicted Cr-Cr distances for the localized solutions from the CI results. These are all longer than the delocalized result. Thus, any bias in the $Cr_2(O(NH)CH)_4$ models should be toward a prediction of a bond distance which is too long.

Our theoretical results[9,14] and the theoretical results of others[15] all predict a long Cr-Cr bond for the tetraformate derivative without axial ligands. Our results also suggest that the driving force for the short Cr-Cr bonds is the nature of the bridging ligand and not the absence of axial ligands. However, the effect of axial ligands is not negligible. In several studies Cotton and coworker have shown that the short Cr-Cr bonds can be lengthened by the addition of axial ligands. For strong donors such as THF or pyridine the lengthening is 0.28Å and 0.42Å, respectively.[17] However, the tetracarboxylates do not seem as sensitive to the donor strength of the axial ligand. In a series of studies where the pKa of both the bridging carboxylate and the axial donor were compared the slope of the curves, Δ Cr-Cr/Δ pKa was five times larger for changes in the bridging ligands than for changes in the axial ligands.[18]

Very recently, the gas-phase electron diffraction has been reported to show that $Cr_2(O_2CCH_3)_4$ has a Cr-Cr bond distance of 1.97Å.[19] This is significantly shorter than the distance predicted by any theoretical calculation. The only other gas-phase experimental work on these compounds is the ultra-violet photoelectron (PE) spectra.[20] In all of these studies the PE spectra of the compounds with short Cr-Cr bonds distances have a characteristically sharp δ ionization well separated from the π ionization. However, the PE spectrum of the tetraacetate is very different and would be most easily interpreted as arising from a Cr-Cr dimer with a long, relatively weak Cr-Cr quadruple bond.[21] Thus, the structure of the tetracarboxylate dichromium molecules without axial interactions remains an open question. Either the electron diffraction has been incorrectly interpreted or the theoretical calculations are in error. The difference between the two, 0.3Å, is too great to be ignored.

3.2 Molybdenum-molybdenum Triple Bonds

Here, we were interested in determining a value for the Mo-Mo triple bond energy. Our hope was to reduce the error bar in the experimental value, 592 ± 198 kJ mol^{-1}, so that we could establish the correct position of the Mo≡Mo bond in a table of bond energies. Our approach was to calculate the Mo≡Mo bond energy by calculating the dissociation energy of Mo_2H_6 into MoH_3 fragments. Of course, this quantity is not exactly the same as the thermochemical bond energy, but one would expect it to be fairly close. In order to obtain a value which is more accurate than the experimental value, we calculated the bond energies of a series of triply bonded molecules: N_2, P_2, As_2 and Sb_2. We calculated the average error in these results and corrected the Mo_2H_6 dissociation energy by that value. The results are shown in Table III. The predicted value 284 kJ mol^{-1} places the Mo≡Mo triple bond energy close to that of a B-B single bond, somewhat weaker than a C-C single bond but somewhat stronger than a N-N single bond.

Table III. Dissociation Energies for Group 5A Diatomics and Mo_2H_6 Based on CI Including Differential Correlation from Other Valence and Semi-valence Electron Pairs[a]

	exptl[b]	calcd	error
N_2	944	876	68
P_2	487	355	132
As_2	381	288	93
Sb_2	301	212	89
Mo_2H_6	284[d]	188	96[c]

[a]All values in kJ mol^{-1}. [b]Reference 22.
[c]Average of above values. [d]Our prediction.

This bond energy, which is at the low end of the experimental range, is consistent with the chemistry of these compounds.[2] Here one finds some reactions in which the Mo≡Mo bond is preserved intact and other where it is cleaved. Many of the cleavage reactions do not require extreme conditions. As shown in Figure 1 our value is also consistent with the other known Mo-Mo bond energies. Recent thermochemical experiments on other Mo-Mo multiple bonded compounds also favor a value near 300 kJ mol^{-1}.[23]

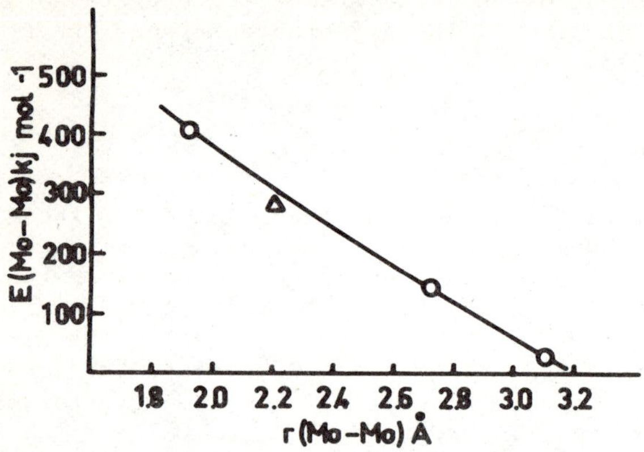

Figure 1. Bond enthalpy against bond length plot for Mo-Mo bonds. The circle at the short distance represents the value for diatomic Mo_2, while those at long distance are estimates for nearest and next nearest neighbors in solid Mo. The open triangle is the calculated value reported in this paper.

3.3 Transition Metal to Carbon Double Bonds

In a previous publication[11] we discussed the major differences between a Fischer-type and a Schrock-type carbene. In a Fischer-type carbene such as $Mo(CO)_5(C(OH)H)$, the carbene is usually stabilized by heteroatoms or phenyl groups and the metal is usually a low oxidation state metal stabilized by π-acceptor ligands. In the Schrock-type carbene such as $Nb(\pi\text{-}C_5H_5)Cl_2(CH_2)$, the carbene is usually a simple alkylidene and the metal is usually in a high oxidation state with π-donor ligands.

If we consider the fragments which make up these species we find some interesting differences. Comparing the two carbenes, we find that the heteroatom carbene, C(OH)H has a singlet ground state $\sigma^2\pi^0$ while the methylene, CH_2, has a triplet ground state $\sigma^1\pi^1$. For the metal fragments the $Mo(CO)_5$ fragment has a singlet ground state $\pi^2\sigma^0$ while the $NbCpCl_2$ fragment has a triplet ground state $\sigma^1\pi^1$. Thus, the fragments are matched. The bond between $Mo(CO)_5$ and C(OH)H is of the donor acceptor type. The σ lone pair on the carbene donates to the empty σ an the metal and the filled π on the metal donates to the empty π on the carbene. The M-C (carbene) π bond is localized toward the metal, while the empty π^* (the LUMO) is localized toward C. Hence, this carbene carbon is susceptible to attack by nucleophiles. The bond between $NbCpCl_2$ and CH_2 is substantially different. Here the two singly occupied σ orbitals form a σ bond and the two singly occupied π orbitals form a π bond. The double bond between Nb and C is much more like a typical double bond than that between Mo and C. Our calculations even suggest that the π bond is localized slightly toward C, just where electrophiles attack.

If one exchanges the carbene ligands to produce $NbCpCl_2(C(OH)H)$ and $Mo(CO)_5CH_2$ we find that the Nb-C is still electrophilic at carbon and that the Mo-C is still nucleophilic at carbon. Thus, the chemical nature of the carbene is controlled by the metal fragment to which it is attached.

Many of the electron-deficient, Schrock-type carbenes show a geometry in which the alkylidene has pivoted about C as shown at the top of Figure 2. The total energy of the system as a function of this pivot angle is shown at the bottom of Figure 2. Many of the complexes show small M-C-H angles, but these often arise in molecules which have a bulky group on carbon which provides a steric force to close this angle further. Previous theoretical work on this deformation suggested that it arises because the metal has several empty, unused δ orbitals that

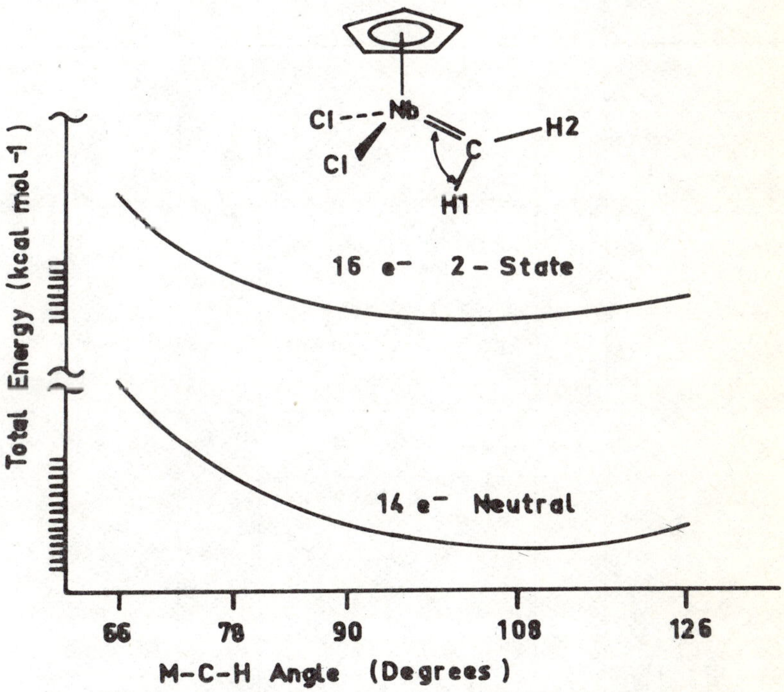

Figure 2. Pivoting potential energy curve for the $NbCpCl_2(CH_2)$ molecule and its (2-) ion. Tick marks on the left represent kcal mol^{-1}.

can be used to bond with the σ orbital of the carbene.[24] In some systems it appears as if the H actually transfers to the metal center. Thus, it has been suggested that the driving force for the bend involves the formation of a three center M-H-C bond.[25]

The orbital analysis shown in Figure 3 suggests that both effects are operative. The top three plots show the localized Nb-C σ bond at Nb-C-H angles of 126° (symmetric) 90° and 66° (180° for other Nb-C-H angle). Notice as the CH_2 pivots the Nb d orbital also pivots, and the Nb-C bond remains intact as it becomes a bent Nb-C bond. As the C-H

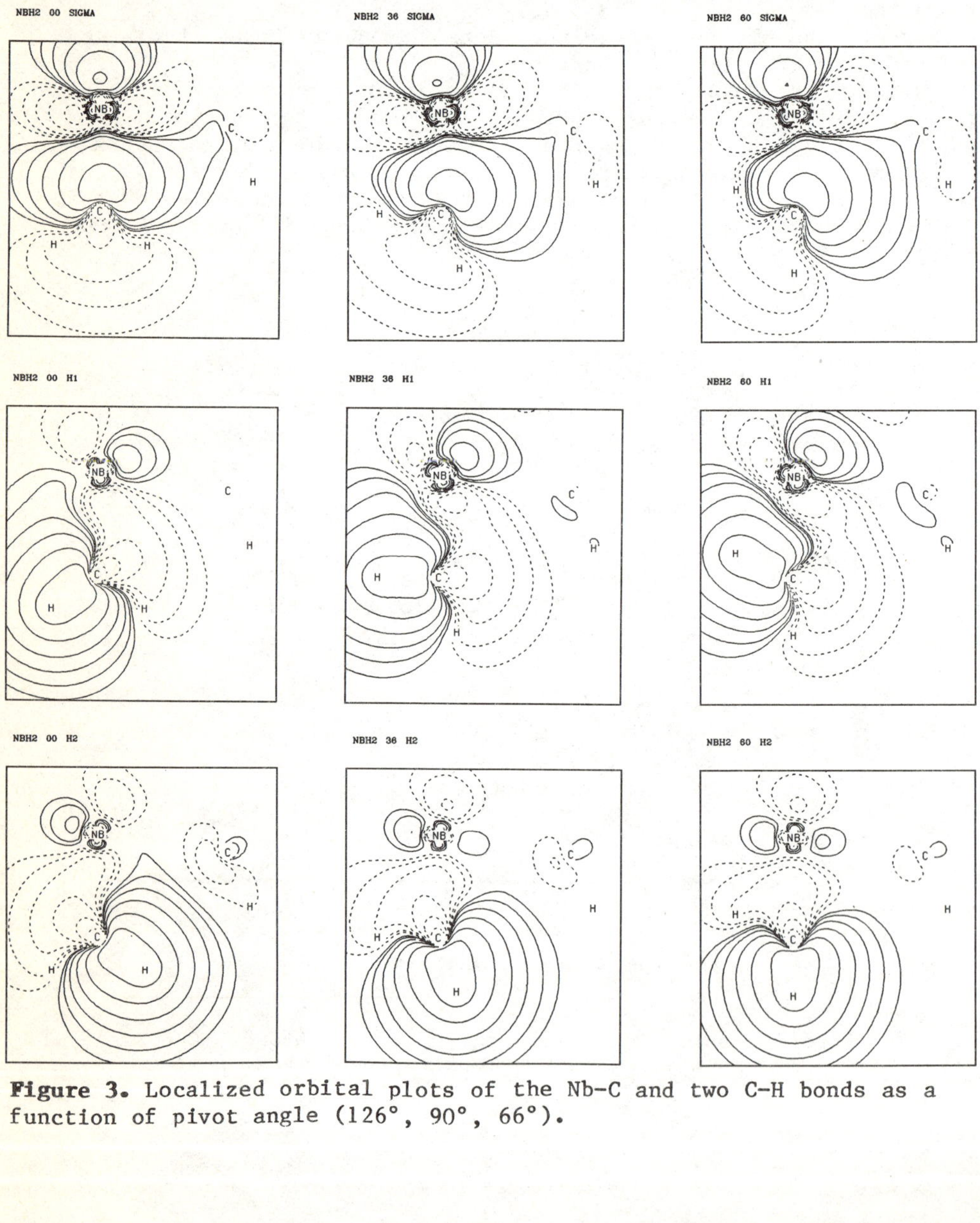

Figure 3. Localized orbital plots of the Nb-C and two C-H bonds as a function of pivot angle (126°, 90°, 66°).

bonding orbital moves toward the Nb it begins to acquire Nb character. At the smallest angle (66°) there is significant positive Nb-H overlap population. Yet, the C-H is still dominating the interaction. If one compares the two C-H bonds during the rotation, one sees weakening of the C-H bond which approaches the Nb and strengthening of the C-H bond which moves away from the Nb. This difference in C-H bonding is confirmed in structures of these complexes where differences in bond length parallels the orbital differences.

4. ACKNOWLEDGEMENT

This work was supported by the National Science Foundation (Grant No. CHE 83-09936). The calculations on the multiple metal-metal bonds were carried out by Dr. Randall Kok while those on the multiple metal-carbon bonds were carried out by Dr. Trenton Taylor.

5. REFERENCES

(1) F. A. Cotton, B. G. DeBoer, M. P. LaPrade, J. R. Pipal, and D. A. Ucko, J. Am. Chem. Soc. **93** , 2926 (1970).

(2) F. A. Cotton and R. A. Walton, Multiple Bonds Between Metal Atoms, Wiley; (New York, 1982).

(3) (a) F. Hug, W. Mowat, A. Shortland, A. C. Skapski, and G. Wilkinson, Chem. Commun. 1079 (1971).
(b) W. Mowat, A. Shortland, G. Yagupski, N. J. Hill, M. Yagupski, and G. Wilkinson, J. Chem. Soc. Dalton, 533 (1972).

(4) J. A. Connor, G. Pilcher, H. A. Skinner, M. H. Chisholm, and F. A. Cotton, J. Am. Chem. Soc. **100,** 7738 (1978).

(5) (a) E. O. Fischer and A. Massböl, Angew. Chem. Int. Ed. Engl. **3,** 580 (1964).
(b) E. O. Fischer, Adv. Organomet. Chem. **14,** 1 (1976).

(6) (a) R. R. Schrock, J. Am. Chem. Soc. **97,** 6578 (1975).
(b) R. R. Schrock, Acc. Chem. Res. **12,** 98 (1979).

(7) Transition Metal Carbene Complexes, Verlag Chem. (Weinheim, FRG, 1983).

(8) (a) M. B. Hall, Chem. Phys. Lett. **1979,** 61, 461.
(b) M. B. Hall, Int. J. Quantum Chem. **1978,** 14, 613.
(c) M. B. Hall, Int. J. Quantum Chem., Quantum Chem. Symp. **1979,** No. 13, 195.

(d) M. B. Hall, In Recent Developments and Applications of Multiconfiguration Hartree-Fock Methods, NRCC Proceedings No. 10. M. Dupuis, Ed.; National Technical Information Service: (Springfield, VA, 1981), p. 31.

(9) R. A. Kok and M. B. Hall, Inorg. Chem. **24,** 1542 (1985).

(10) R. A. Kok and M. B. Hall, Inorg. Chem. **22,** 728 (1983).

(11) T.E. Taylor and M.B. Hall, J. Am. Chem. Soc., **106,** 1576 (1984).

(12) C.D. Garner, I.H. Hillier, M.F. Guest, J.C. Green, and A.W. Coleman, Chem. Phys. Lett. **41,** 91 (1976).

(13) (a) M.F. Guest, I.H. Hillier, and C.D. Garner, Chem. Phys. Lett. **48,** 587 (1977).
(b) M. Benard and A. Veillard, Nouv. J. Chim. 1, 97 (1977).

(14) R. A. Kok and M. B. Hall, J. Am. Chem. Soc. **105,** 676 (1983).

(15) M. Benard, J. Am. Chem. Soc. **100,** 2354 (1978).

(16) (a) M. Benard, J. Chem. Phys. **71,** 2546 (1979).
(b) R. Wiest and M. Benard, Theoret. Chim. Acta, in press.

(17) F. A. Cotton, W. H. Ilsley, and W. Kaim, J. Am. Chem. Soc. **102,** 3464 (1980).

(18) F. A. Cotton and W. Wang, Nouv. J. Chim. **8,** 331 (1984).

(19) S.N. Ketkar and M. Fink, J. Am. Chem. Soc. **107,** 338 (1985).

(20) (a) A. W. Coleman, J. C. Green, A. J. Hayes, E. A. Seddon, D. R. Lloyd, and Y. Niwa, J. Chem. Soc. Dalton, 1057 (1979).
(b) C. D. Garner, I. H. Hillier, A. A. MacDowell, I. B. Walton and M. F. Guest, J. Chem. Soc. Faraday II, **75,** 485 (1979).
(c) B. E. Bursten, F. A. Cotton, A. H. Cowley, B. E. Hanson, M. Lattman, and G. G. Stanley, J. Am. Chem. Soc. **101,** 6244 (1979).
(d) C. D. Garner, I. H. Hillier, M. J. Knight, A. A. MacDowell, I. B. Walton, and M. F. Guest, J. Chem. Soc. Faraday II, **76,** 885 (1980).
(e) P. M. Atha, J. C. Campbell, C. D. Garner, I. H. Hillier, A. A. MacDowell, J. Chem. Soc. Dalton, 1085 (1983).
(f) M. Berry, C. D. Garner, I. H. Hillier, A. A. MacDowell, I. B. Walton, Chem. Phys. Lett. **70,** 350 (1980).
(g) P. M. Atha, I. H. Hillier, M. F. Guest, Mol. Phys. **46,** 437 (1982).

(21) I. H. Hillier, personal communication.

(22) (a) L. V. Gurvich, G. V. Karachevstev, V. N. Kondrat'yev, Y. A. Lebedev, V. A. Mendredev, V. K. Potapov, Y. S. Khodeev, "Bond Energies, Ionization Potentials and Electron Affinities"; Nauka: Moscow, 1974 (in Russian).
(b) A. G. Gaydon, "Dissociation Energies and Spectra of Diatomic Molecules", 3rd ed.; Chapman and Hall: London, 1968.

(23) K. J. Cavell, C. D. Garner, G. Pilcher, and S. Parkes, J. Chem. Soc. Dalton, 1714 (1979).

(24) R. J. Goddard, R. Hoffmann, E. D. Jemmis, J. Am. Chem. Soc. **102**, 7667 (1980).

(25) M. M. Francl, W. J. Pietro, R. F. Hout, Jr., W. J. Hehre, Organomet. 2, 281 (1983).

DESIGN OF FERROMAGNETICALLY COUPLED POLYMETALLIC SYSTEMS : TOWARD THE MOLECULAR FERROMAGNETS.

O. Kahn
Université Paris Sud
Laboratoire de Spectrochimie des Eléments de Transition,
U.A. 420, Bâtiment 420,
91405 ORSAY
France

ABSTRACT. One of the main challenges in the field of the molecular materials is the design of molecular ferromagnets. The first step along this line is the synthesis of ferromagnetically coupled polymetallic systems. The strategy of strict orthogonality of the magnetic orbitals allows to design such systems. Several examples of this kind are presented. The role of the overlap density maps between magnetic orbitals to estimate the magnitude of the ferromagnetic interaction is emphasized. An alternative strategy leading to molecular systems in which the ground state has an high spin multiplicity is also presented. This strategy consists to impose the parallel alignment of S = 5/2 local spins (Mn^{2+} or Fe^{3+}) through an antiferromagnetic coupling with S = 1/2 local spins (Cu^{2+}). The magnetic properties of 1 D $Mn^{2+}Cu^{2+}$ systems are discussed.

1. INTRODUCTION

The goal of this paper is to show that today it is possible to know what kind of transition ions we have to choose,what kind of bridging network we have to design, what kind of whole geometry we have to achieve in order to obtain a new polymetallic system exhibiting magnetic properties predictable in nature and order of magnitude.In a certain sense,we would like to lay the foundations of a molecular engineering of the polymetallic systems(1). Since the space is limited,we shall focuse on one aspect of the problem,namely the design of ferromagnetically coupled systems, i.e. of systems of which the ground state has the highest spin multiplicity.This aspect is important for at least two reasons:(i) less than 5% of the polymetallic systems are reported as ferromagnetically coupled.In fact, the normal situation is the antiferromagnetic interaction leading to a ground state of lowest spin multiplicity. To some extend,such an antiferromagnetic interaction corresponds to a weak bond between the magnetic centers;(ii) one of the main challenges in the field of the molecular materials is the design of molecular ferromagnets. The first step along this line is to find a strategy leading to a ferromagnetic interaction between nearest neighbor magnetic centers.

A. Veillard (ed.), Quantum Chemistry: The Challenge of Transition Metals and Coordination Chemistry, 403–412.

In the next section, we shall recall some basic concepts appropriate to describe the interaction between magnetic centers. From these concepts, we shall extract the strategy of orthogonality of the magnetic orbitals. In the following section, we shall present some experimental results along this line. Then, we shall propose an alternative strategy allowing to stabilize a state of high spin multiplicity. Finally, we shall briefly discuss the perspectives in this field.

2. BASIC CONCEPTS.

Let us consider an A-B binuclear complex where A (or B) symbolizes a metal ion with one unpaired electron surrounded by its ligands. Some of the ligands are common to A and B and constitute the bridging zone. The unpaired electrons in the monomeric fragments A and B are described by the ϕ_A and ϕ_B orbitals respectively. ϕ_A and ϕ_B are defined as the *magnetic orbitals* (M.O.). From the ground configuration $\phi_A\phi_B$, a spin singlet and a spin triplet arise, separated by J. By assuming that the metal-metal charge transfer configurations $\phi_A\bar{\phi}_A$ and $\phi_B\bar{\phi}_B$ are too high in energy to coupled significantly with the ground configuration, the singlet-triplet (S-T) energy gap J may be easily calculated as (2) :

$$J = 2j + 4tS + \text{terms in } S^2 + \ldots \qquad (1)$$

with :

$$\begin{aligned} S &= \langle\phi_A(1)|\phi_B(1)\rangle \\ \alpha_A &= \langle\phi_A(1)|h(1)|\phi_A(1)\rangle \\ \beta &= \langle\phi_A(1)|h(1)|\phi_B(1)\rangle \\ t &= \beta - \frac{\alpha_A + \alpha_B}{2} S \\ j &= \langle\phi_A(1)\phi_B(2)|\frac{1}{r_{12}}|\phi_A(2)\phi_B(1)\rangle \end{aligned} \qquad (2)$$

t and S are of opposite sign, so that 4tS is negative and favors the antiferromagnetic interaction ($J < 0$). j is positive and favors the ferromagnetic interaction ($J > 0$). It is clear from (1) that the way to obtain a ferromagnetic interaction with a triplet ground state is to minimize or better to cancel the overlap integral S. In this latter case the magnetic orbitals are orthogonal and the triplet state is stabilized by 2j with regard to the singlet state. This orthogonality can be achieved in two ways. It can be symmetry imposed when the two M.O.'s transform as different irreducible representations of the molecular symmetry group, or it can be accidental when S=0 although the M.O.'s have the same symmetry. This accidental orthogonality can only occur for very peculiar values of the structural parameters.

The factor j which governs the stabilization of the triplet state in case of orthogonality of the M.O.'s may be reexpressed as :

$$j = \iint_{space} \frac{\rho(i)\,\rho(j)}{r_{ij}} \, d\tau(i) d\tau(j) \qquad (3)$$

with

$$\rho(i) = \phi_A(i)\phi_B(i) \qquad (4)$$

where $\rho(i)$ is the overlap density between M.O.'s. j may then be defined as the self repulsion of this overlap density and may be large only when the overlap density presents zones of strong extrema. We shall see in the next section the importance of this idea.

3. SYMMETRY IMPOSED ORTHOGONALITY IN $Cu^{2+}VO^{2+}$ HETEROBINUCLEAR COMPLEXES.

The situation of strict orthogonality has been realized for the first time in the heterobinuclear complex $CuVO(fsa)_2en.CH_3OH$ (3), the structure of which is shown in Figure 1. The main feature of this structure lies in the fact that the molecular symmetry is very close to C_s, with a mirror plane containing the metal ions, perpendicular to the plane of

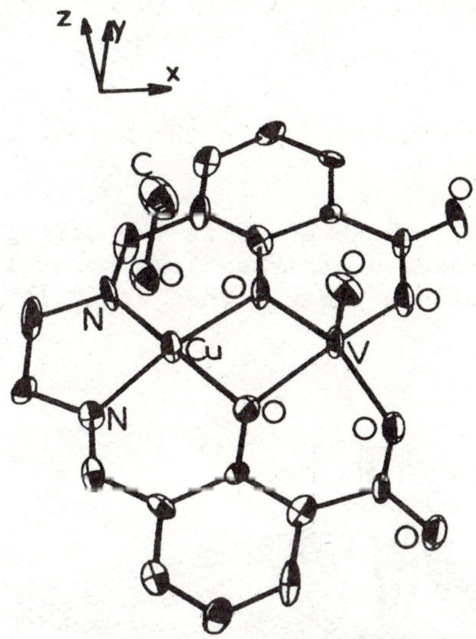

Figure 1. Molecular structure of $CuVO(fsa)_2en.CH_3OH$.

the macrocycle. The M.O. ϕ_{Cu} centered on a copper(II) ion in a 4 + 1 environment is constructed from the d_{xy} metallic orbital and is partially delocalized towards the oxygen and nitrogen atoms surrounding the copper owing to σ-type antibonding overlaps $\langle d_{xy}|p_\sigma\rangle$. This M.O. is antisymmetric with regard to the mirror plane. It transforms as a" in the C_s group. The M.O. ϕ_{VO} is constructed from the $d_{x^2-y^2}$ metallic orbital and is partially delocalized towards the oxygen atoms of the macrocycle owing to the π-type antibonding overlaps $\langle d_{x^2-y^2}|p_\pi\rangle$. This M.O. is symmetric with regard to the mirror-plane and transforms as a'. Thus, the overlap integral $\langle\phi_{Cu}|\phi_{VO}\rangle$ is identically zero. The magnetic properties, not only confirm that the triplet state is the ground state, but also indicates that the S-T separation is as large as J = 118 cm^{-1}.

The shape of the M.O.'s shown in Figure 2 allows to understand why the stabilization of the S = 1 state is so pronounced. From the Figure 2, one can obtain the overlap density map in the plane of the

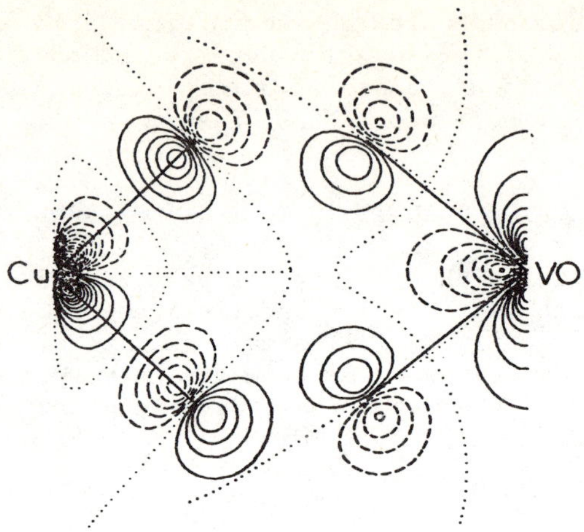

Figure 2. Magnetic orbitals $\emptyset_{Cu}$ and $\emptyset_{VO}$ in the bridging network of $CuVO(fsa)_2en.CH_3OH$.

macrocycle shown in Figure 3, which exhibits two strongly positive lobes around one of the bridging oxygen atoms and two strongly negative lobes around the other one. It follows that j, then J, is large in $CuVO(fsa)_2en.CH_3OH$.

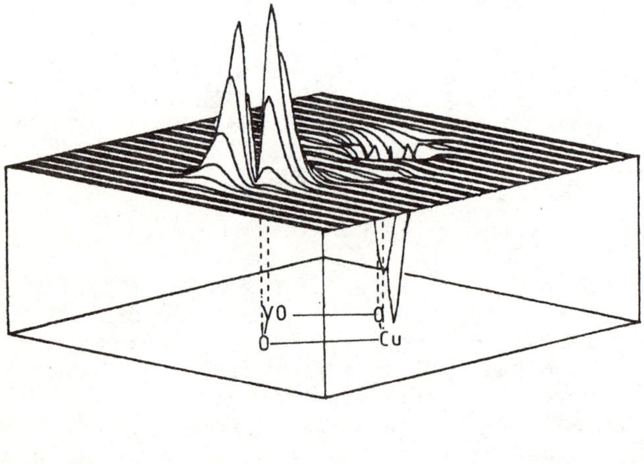

Figure 3. Overlap density map between the magnetic orbitals in $CuVO(fsa)_2en.CH_3OH$.

To understand the role of the overlap density, it is interesting to compare the complex of Figure 1 with $[tmenCu(C_2O_4)VO(C_2O_4)]$ shown in Figure 4. In this latter compound, again the $\emptyset_{Cu}$ and $\emptyset_{VO}$ M.O.'s are strictly orthogonal but, owing to the nature of the bridging network, the overlap density drawn at the same scale as in Figure 3 is completely flat so that j is expected to be very small. The magnetic properties show that the S-T splitting is actually smaller than 1 cm^{-1} (4).

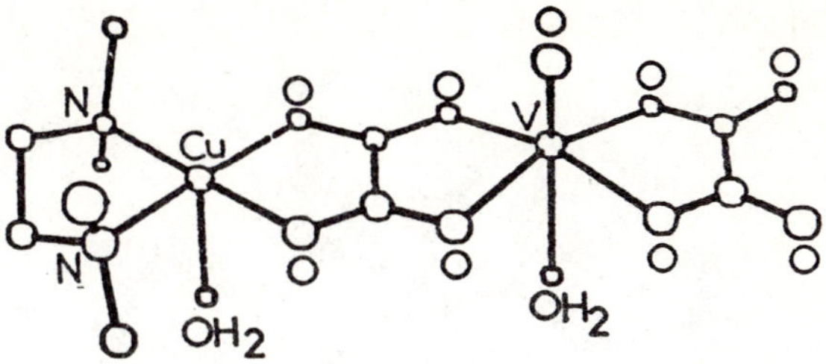

Figure 4. Molecular structure of $[tmenCu(C_2O_4)VO(C_2O_4)]$.

Finally, the best way to realize how this strategy of orthogonality of the M.O.'s is efficient is probably to compare the $CuVO(fsa)_2en.CH_3OH$ compound to $Cu_2(fsa)_2en.CH_3OH$ prepared from the same macrocycle. In this latter compound, we have again a mirror plane containing the metal ions. The outside copper(II) ion is in a square planar environment so that the two M.O.'s centered on the metal ions are antisymmetric with regard to the mirror plane. The overlap $\langle \phi_{Cu} | \phi_{Cu} \rangle$ may be important and the term 4tS in (1) becomes dominant. The interaction is antiferromagnetic with a ground singlet state stabilized by 650 cm^{-1} with regard to the triplet state.This comparison between $Cu^{2+}VO^{2+}$ and $Cu^{2+}Cu^{2+}$ is schematized in Figure 5.

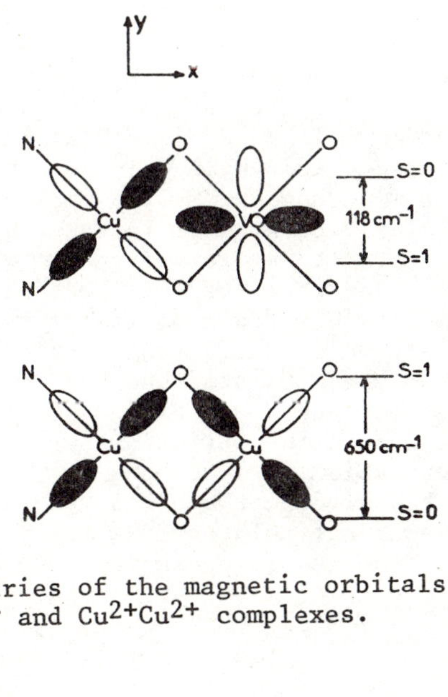

Figure 5. Symmetries of the magnetic orbitals and nature of the interaction in $Cu^{2+}VO^{2+}$ and $Cu^{2+}Cu^{2+}$ complexes.

4. COMPARISON BETWEEN $Cu^{2+}Fe^{3+}$ AND $Cu^{2+}Cr^{3+}$ HETEROBINUCLEAR COMPLEXES.

The concepts presented above can be extended to systems with more than one unpaired electron per magnetic center. For instance, let ut consider the $Cu^{2+}Fe^{3+}$ complex of which the structure is shown in Figure 6.

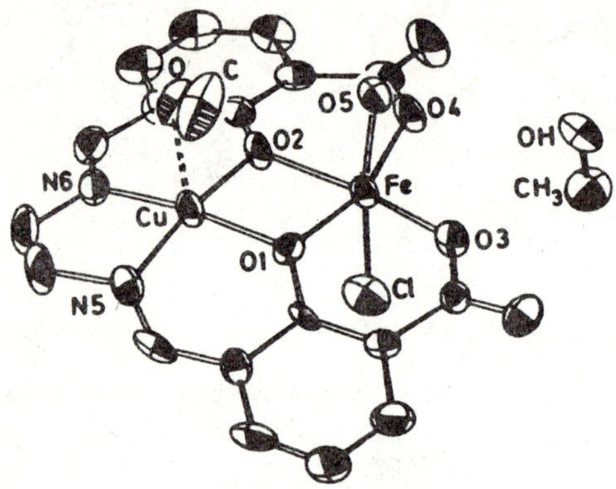

Figure 6. Molecular structure of $CuFe(fsa)_2en\ Cl(H_2O).2CH_3OH$.

Fe^{3+} is in an high spin state with five unpaired electrons occupying the five d-type M.O.'s. As for the single M.O. around copper(II), it is of the d_{xy}-type (see Figure 7) and gives a non zero overlap with the M.O. around Fe^{3+} which has the same symmetry. This overlap is sufficient to lead to an antiferromagnetic interaction. The local states are characterized by the spins $S_{Cu} = 1/2$ and $S_{Fe} = 5/2$ respectively, so that the two low lying molecular states will be $S = 2$ and $S = 3$. The $S = 2$ state is expected to be lower in energy.

If now we replace Fe^{3+} by Cr^{3+} in the outside site, the interacting centers have the spins $S_{Cu} = 1/2$ and $S_{Cr} = 3/2$ and the low lying molecular states will be characterized by $S = 1$ and $S = 2$. The three unpaired electrons around Cr^{3+} occupy the M.O.'s built from the t_{2g} metal orbitals. They are strictly orthogonal to the single M.O. around Cu^{2+}. The interaction is expected to be ferromagnetic with a $S = 2$ ground state. Therefore, one sees that both $Cu^{2+}Fe^{3+}$ and $Cu^{2+}Cr^{3+}$ complexes are expected to exhibit a $S = 2$ ground state. In the former case, this ground state will arise from an antiferromagnetic interaction due to the overlap between one of the M.O.'s around Fe^{3+} and the single M.O. around Cu^{2+}. In the latter case, the $S = 2$ ground state will arise from a ferromagnetic interaction due to the strict orthogonality of the four M.O.'s involved in the problem.

The $Cu^{2+}Cr^{3+}$ complex has also been synthesized and the magnetic properties of the two complexes entirely confirm the prediction. The low lying states as deduced from the magnetic data are as shown hereunder (5) :

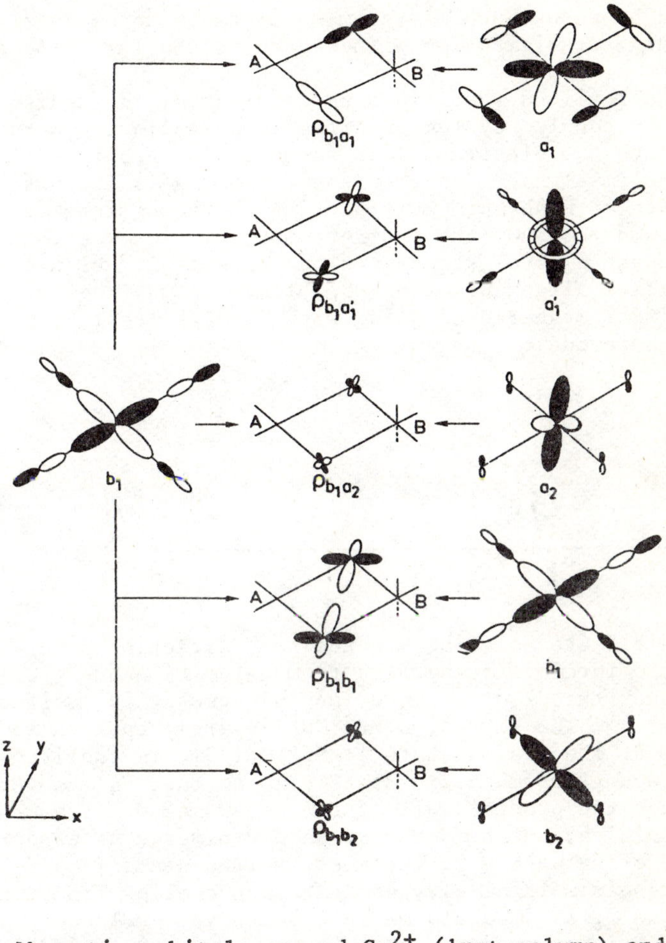

Figure 7. Magnetic orbitals around Cu^{2+} (1rst column) and around Fe^{3+} or Cr^{3+} (3rd column), and overlap densities between pairs of magnetic orbitals in $Cu^{2+}Fe^{3+}$ and $Cu^{2+}Cr^{3+}$.

5. ORDERED BIMETALLIC CHAINS.

The strategy of strict orthogonality of the magnetic orbitals has borne witness of its efficiency to stabilize the state of highest spin multiplicity. However, from the viewpoint of the synthesis, this strategy is difficult to use. For instance we have not yet been able to synthetize a $Cr^{3+}Ni^{2+}$ system in which the two metal ions would be in octahedral surroundings. In such a system, the three unpaired electrons around Cr^{3+} would occupy the t_{2g} orbitals and the two unpaired electrons around Ni^{2+} the e_g orbitals so that all the $\langle t_{2g}|e_g\rangle$ overlap integrals would be zero. Such a system would exhibit a S = 5/2 ground state with a large stabilization with regard to the low lying excited states.

It must be realized that, in a certain sense, the design of a ferromagnetically coupled system requires to go against a natural trend according to which the electrons pair in molecular orbitals of low energy. The problem at hand was then the following:is it possible to stabilize a state of high spin multiplicity by taking into account this natural trend to favor the antiferromagnetic interaction? In this section, we present a novel strategy along this line, which consists to impose the parallel alignment of S = 5/2 local spins (Mn^{2+} or Fe^{3+}) through an antiferromagnetic coupling with S = 1/2 local spins (Cu^{2+}) as schematized hereunder.

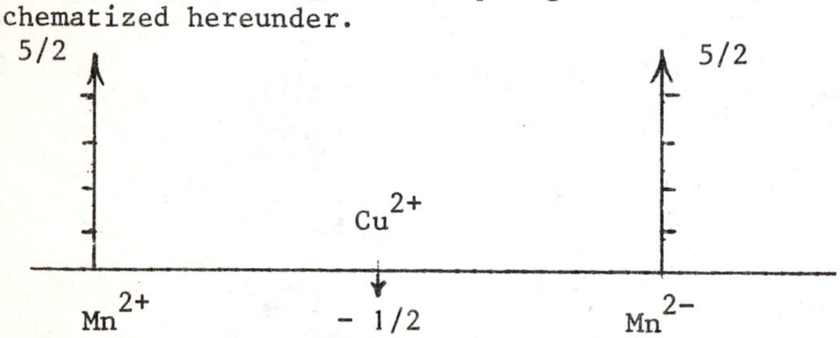

The large spins 5/2 are polarized in the same direction owing to the antiferromagnetic interaction between them and small spins 1/2. It can be recalled here that the d^5-d^1 interaction is necessarily antiferromagnetic (cf. Section 4). The first 1-D $Mn^{2+}Cu^{2+}$ system reported so far is shown in Figure 8. The ground state $|+5/2 - 1/2|_n$ is qualitatively similar to a pseudo ferromagnetic state $|2|_n$ so that in the very low temperature range the product $\chi_M \cdot T$ of the molar magnetic susceptibility (per $Mn^{2+}Cu^{2+}$ unit) by the temperature should diverge as expected for a 1-D ferromagnet of spins S = 2. The most excited state $|+5/2 + 1/2|_n$ has the highest spin multiplicity so that upon cooling from the high temperatures, one first depopulates this state and $\chi_M \cdot T$ decreases. It follows that $\chi_M \cdot T$ is expected to exhibit a minimum which is the signature of this new type of 1-D compounds. For the compound of Figure 7, the $\chi_M \cdot T$ versus T plot exhibits this minimum. Below this minimum, $\chi_M \cdot T$ increases very quickly as expected but a catastrophe occurs at 7 K due to an antiferromagnetic interaction between nearest neighbor chains (6).

In a new $Mn^{2+}Cu^{2+}$ ordered bimetallic chain synthetized in our group (7), the divergence of $\chi_M \cdot T$ upon cooling down below the temperature

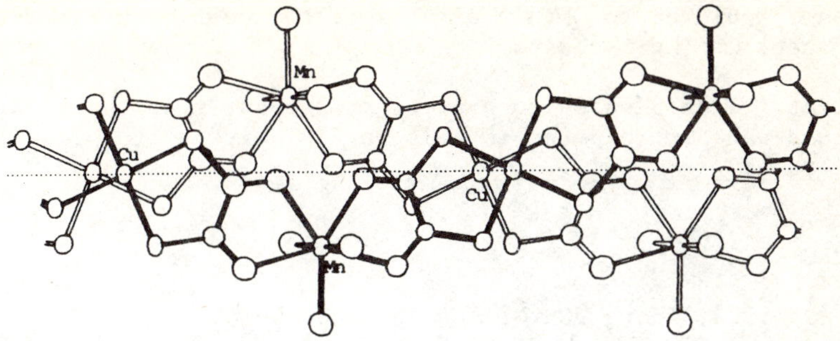

Figure 8. Crystal structure of a 1-D $Cu^{2+}Mn^{2+}$ ordered bimetallic chain.

of the minimum is more pronounced and the catastrophe only occurs at 2.1 K. Just above this transition, $\chi_M.T$ takes a very large value, unknown in the classical coordination chemistry. The system is not yet a molecular ferromagnet. There is no magnetic moment in zero field but a very large moment may be obtained with a very small field.

6. CONCLUSION.

Unlike most of the colleagues attending this NATO workshop, we are not a pure theoretical chemist. What we are trying to do is to use both our brain and our two hands, our brain to extract from the theoretical models some key concepts and our two hands to design new polymetallic systems exhibiting the properties we want to obtain. In the field of the molecular magnetism, such an approach has been really successful (8). The symmetry rules allow to predict the nature of the interaction, and the relative orientations of the involved magnetic orbitals allow to estimate the magnitude of the interaction. For the theoreticians, these simple models cannot replace in any way the ab initio calculations with large CI of the type of those which recently appeared (9,10). But, reciprocally, theses calculations cannot yet be considered as efficient weapons for the experimentalists. Indeed, even when the output is close to the experimental data, they keep a form of a black box. This is particularly true when the correct sign and order of magnitude of the interaction only appear at the term of the CI process

If we come back to the challenge of the molecular ferromagnets, we may think that the way will be still long. Indeed, with spin momenta there is no magnetic order at one or two dimensions. Consequently, if we use the strategy of orthogonality of the M.O.'s, we must achieve ferromagnetic interactions between nearest neighbor magnetic centers in the three directions, at the scale of the crystal lattice. Unfortunatly, if the molecular engineering may work, the crystal engineering is still almost a dream. Maybe, the second strategy consisting to polarize the 5/2 local spins in a ferromagnetic fashion through an antiferromagnetic coupling with 1/2 local spins could be easier to use. The problem to solve is that of the interaction between chains. The ideal (or the dream) would be to obtain a small ferromagnetic interaction between adjacent $Cu^{2+}Mn^{2+}$ chains.

Acknowledgment : I am pleased to express my deepest gratitue to all the colleagues who participated to this work. Their names appear in the references listed.

7. REFERENCES

(1) - Kahn, O. 1982, Inorg. Chim. Acta 62, pp. 3-14.
(2) - Kahn, O. and Charlot, M.F. 1980, Nouv. J. Chim. 4, pp. 567-576.
(3) - Kahn, O., Galy, J., Journaux, Y., Jaud, J. and Morgenstern-Badarau, I. 1982, J. Am. Chem. Soc. 104, pp. 2165-2176.
(4) - Julve, M., Verdaguer, M., Claude, R., Charlot, M.F. and Kahn, O. 1982, Inorg. Chimica Acta 82, pp. 5-12.
(5) - Journaux, Y., Kahn, O., Zarembowitch, J., Galy, J. and Jaud, J. 1983, J. Am. Chem. Soc. 105, pp. 7585-7591.
(6) - Gleizes, A. and Verdaguer M. 1984, J. Am. Chem. Soc. 106, pp. 3727-3737.
(7) - Yu, P., Sletten, J. and Kahn, O. to be submitted to J. Am. Chem. Soc.
(8) - Kahn, O. 1985, Angew. Chem. Int. Ed. Engl., in press.
(9) - De Loth, P., Cassoux, P., Daudey, J.P. and Malrieu, J.P. 1981, J. Am. Chem. Soc. 103, pp. 4007-4016.
(10) - Charlot, M.F., Verdaguer, M., Journaux, Y., De Loth, P. and Daudey, J.P. 1984, Inorg. Chem. 23, pp. 3802-3808.

A SPECTROSCOPIC AND MAGNETIC INVESTIGATION OF M_4X_4 CLUSTERS: LOCALIZED OR ITINERANT ELECTRONS?

Cathryn E. Davies, Jennifer C. Green and
Graham H. Stringer
Inorganic Chemistry Laboratory
South Parks Road
Oxford OX1 3QR
England

ABSTRACT. The bonding in a number of cubane, M_4X_4, clusters has been investigated. Photoelectron spectra of $[Mo_4S_4(C_5H_4Pr^i)_4]$, $[Cr_4S_4(C_5H_4Me)_4]$, $[Cr_4O_4(C_5H_5)_4]$ and $[Cr_4O_4(C_5H_4Me)_4]$, electronic absorption spectra of $[Mo_4S_4(C_5H_4Pr^i)_4]$ and $[Mo_4S_4(C_5H_4Pr^i)_4]^+BF_4^-$, $[Cr_4S_4(C_5H_4Me)_4]$ and $[Cr_4S_4(C_5H_4Me)_4]^+PF_6^-$ and magnetic measurements on $[Mo_4S_4(C_5H_4Pr^i)_4]^+BF_4^-$, $[Mo_4S_4(C_5H_4Pr^i)_4]^{2+}(I_3^-)_2$ are reported. The results are interpreted in terms of delocalized metal-metal bonding for the molybdenum sulphur and chromium sulphur cubanes, but localized electrons for the chromium oxygen cluster with only weak metal-metal interactions. The electrochemical behaviour of the clusters is examined in the light of these bonding models.

INTRODUCTION.

The cubane unit, M_4X_4, shown in Figure 1, is an important building block in many solid state structures. When M is a transition metal, the interaction between the metal d-electrons is of particular interest as they may either be itinerant, leading to metallic conductivity, or localized, leading to insulator properties for the bulk solid. In the area of metal cluster chemistry, the cubane unit plays a key role, acting as an electron reservoir and transfer reagent in such biological systems as bacterial ferrodoxins. Pioneering work by Dahl [1] on metal cubane clusters has demonstrated that the degree of interaction between the metal atoms may vary as a function of the electron number of the cluster.

We report below an experimental investigation of the bonding in Mo_4S_4, Cr_4S_4 and Cr_4O_4 clusters. Part of this work has been communicated previously [2].

A. Veillard (ed.), Quantum Chemistry: The Challenge of Transition Metals and Coordination Chemistry, 413–424.

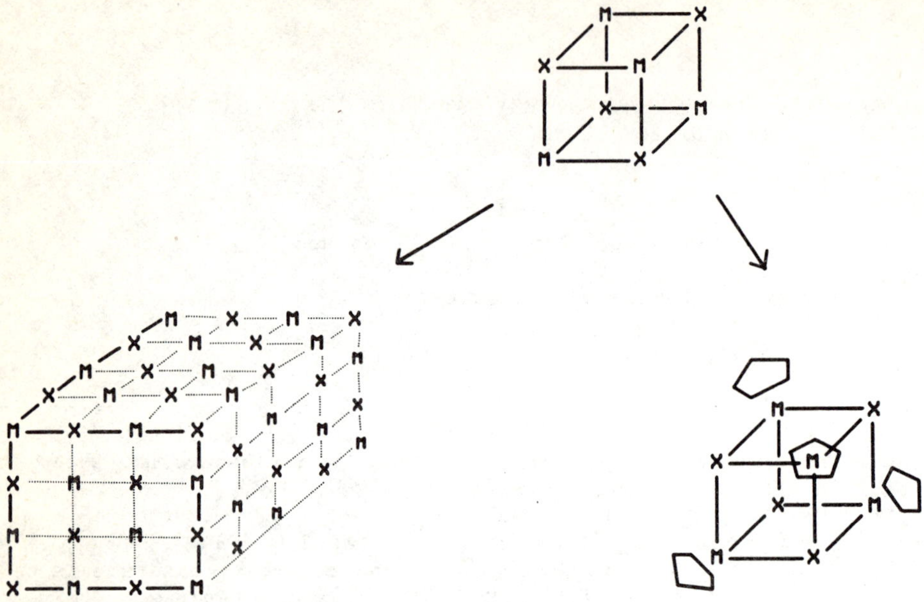

Figure 1. The cubane unit found in the rock salt structure and in molecular complexes.

MATERIALS AND METHODS.

The compounds $[Mo_4S_4(C_5H_4Pr^i)_4]$, $[Mo_4S_4(C_5H_4Pr^i)_4]^+BF_4^-$, $[Mo_4S_4(C_5H_4Pr^i)_4]^{2+}(I_3^-)_2$ [2], $[Cr_4S_4(C_5H_4Me)_4]$, $[Cr_4S_4(C_5H_4Me)_4]^+PF_6^-$ [3], $[Cr_4O_4(C_5H_5)_4]$ and $[Cr_4O_4(C_5H_4Me)_4]$ [4] were prepared by literature methods or modifications thereof.

Photoelectron (P.E.) spectra were measured on a PES Laboratories 0078 interfaced with a RML 380Z microprocessor. The magnetic measurements on $[Cr_4S_4(C_5H_4Me)_4]^+PF_6^-$ were made on an Oxford Instruments Faraday balance and those on $[Mo_4S_4(C_5H_4Pr^i)_4]^+BF_4^-$ and $[Mo_4S_4(C_5H_4Pr^i)_4]^{2+}(I_3^-)_2$ were carried out by Dr. Nelson at Belfast University. Electronic absorption spectra were measured using a Perkin-Elmer 330 spectrometer. Cyclic voltammetry measurements on the two chromium clusters were carried out in THF solution with 0.2-0.5 M tetrabutylammonium hexafluorophosphate as the supporting electrolyte.

RESULTS.

Representative photoelectron spectra are shown in Figures 2, 4 and 6 and prominent ionization energy (IE) features given in Table 1.

TABLE 1. Ionization energies (eV) of the d-bands of some cubane clusters; $[MoS(C_5H_4Pr^i)]_4$,I; $[CrS(C_5H_4Me)]_4$,II; $[CrO(C_5H_5)]_4$,III; $[CrO(C_5H_4Me)]_4$,IV.

	I	II	III	IV
Adiabatic	5.0	5.4	6.0	5.9
	5.28	5.72	6.48	6.32
	5.58	6.09	7.05	6.77
	6.47	6.97	7.39	7.29
	6.90			
Band width	2.4	1.9	1.7	1.7

The magnetic data on $[Cr_4S_4(C_5H_4Me)_4]^+PF_6^-$, shown in Figure 5, demonstrate that the compound obeys the Curie-Weiss law in the range of 15-40°K with a magnetic moment of 1.96 Bohr magnetons. $[Mo_4S_4(C_5H_4Pr^i)_4]^+BF_4^-$ has an effective magnetic moment which varied with temperature over the range 93-323 °K between 2.10 and 2.53 Bohr magnetons, see Figure 3. $[Mo_4S_4(C_5H_4Pr^i)_4]^{2+}(I_3^-)_2$ was shown to be diamagnetic at 4°K and at 293°K.

The monocation complexes $[Mo_4S_4(C_5H_4Pr^i)_4]^+BF_4^-$ and $[Cr_4S_4(C_5H_4Me)_4]^+PF_6^-$ showed absorption bands in the near infra-red at 12,900 and 7,000 cm^{-1} respectively. There were no corresponding bands in the absorption spectra of the neutral molcules.

TABLE 2. $E_{1/2}$ (volts) relative to the standard calomel electrode.

	I [2]	II	IV
First	-0.33	0.36	0.32
Second	0.32	1.15	[1.09 irreversible]

The cyclic voltammograms of $[Mo_4S_4(C_5H_4Pr^i)_4]$ and $[Cr_4S_4(C_5H_4Me)_4]$ in THF showed two reversible one electron oxidations, while that of $[Cr_4O_4(C_5H_4Me)_4]$ showed only one together with a second irreversible oxidation. The $E_{1/2}$ values, relative to the standard calomel electrode, are given in Table 2.

Reported structural data on the M_4X_4 cubane clusters investigated here are summarized in Table 3.

TABLE 3. Selected distances reported for cubane clusters.

Compound	M-M	M-X	M-ring	M-C	ref.
I	2.912(1)x2 2.892(1) 2.902(1)x2 2.905 (1) Mean 2.904(3)	2.344(2)	2.357(8) 2.355(7)		[2]
$I^{+}BF_4^{-}$	2.893(1) 2.887(1) 2.860(1) 2.900(1) 2.901(1) 2.923(1) Mean 2.894(8)	2.343(3)	2.360(12) 2.346(9) 2.339(7) 2.346(9)		[2]
$I^{2+}(I_3^{-})_2$	2.861(1) 2.902(1) 2.897(1) 2.820(1) 2.790(1) 2.879(1; Mean 2.858(18)	2.343(2)	2.335(13) 2.340(12) 2.319(18) 2.329(12)		[2]
II	2.848(2)x2 2.822(2)x2 2.822(2)x2	2.248(2)		2.240(9)	[4]
III	2.900(6) 2.897(5) 2.841(6) 2.811(6) 2.712(2) 2.702(6)	1.937(4)	1.920		[3]

DISCUSSION

Bonding

It is a relatively simple matter to treat the bonding of the framework of these clusters and assign electrons and orbitals, either on an ionic or covalent basis, to the role of binding the metals to the cyclopentadienyl rings and to the non-metal atoms of the cluster. The results from such treatments are summarized in Table 4. The separate treatment of the framework is justified in part by the insensitivity of its dimensions to oxidation in the Mo_4S_4 series and by the photoelectron spectral results.

TABLE 4. Bonding in $M_4X_4(C_5H_4R)_4$ clusters.

Basis set	Symmetry adapted combination	Ionic occupancy
S (3s)	$a_1 + t_2$	8
S (3p)	$a_1+2xt_2+t_1+e$	24
C_5H_4R (a_1 pi)	$a_1 + t_2$	8
C_5H_4R (e_1 pi)	$e+t_1+t_2$	16
M (ns)	$a_1 + t_2$	
M (np)	$a_1+2xt_2+t_1+e$	
M ((n-1)d; xz+yz)	$e+t_1+t_2$	
M ((n-1)d; z^2)	$a_1 + t_2$	12
M ((n-1)d; $xy+(x^2-y^2)$)	$e+t_1+t_2$	

The principal interest in the bonding in these cubane clusters lies in the metal d-electrons. For the molecules under examination here, where the metal is formally in the III oxidation state, and therefore d^3, there remain 12 electrons associated with the metal atoms. Each metal atom has three low lying orbitals available for accomodation of these electrons. In the local three fold symmetry of the metal site, these orbitals have a_1 and e symmetry and can be regarded as principally d_{z^2} and d_{xy} and $d_{x^2-y^2}$ in character. In the T_d symmetry of the molecule the resultant MO are of symmetry a_1, e, t_1 and $2xt_2$ (see Table 4). Maximum bonding of the M_4 tetrahedron would result from full occupancy of the a_1, e and $1t_2$ orbitals.

The diamagnetism and metal-metal distances found for $[Mo_4S_4(C_5H_4Pr^i)_4]$ and $[Cr_4S_4(C_5H_4Me)_4]$ are consistent with this bonding model. However Bottomley and Grein [4] reject such a model for

$[Cr_4O_4(C_5H_5)_4]$ on the grounds of the paramagnetic nature of the cluster, and its distortion from tetrahedral symmetry. On the basis of an Extended Hückel calculation they suggest an orbital occupancy of $e^4t_2^6t_1^2$.

$[Mo_4S_4(C_5H_4Pr^i)_4]$

The PE spectrum of $[Mo_4S_4(C_5H_4Pr^i)_4]$ may be divided into two regions. That above 7.5 eV is due to ionization of the M-ring, M-S and ring ionizations; some assignments are given in Table 1 but these bands and corresponding ones in the other spectra will not be considered further here. The relative intensity increase in the He II spectrum of the three bands below 7.5 eV indicates that they are associated with ionization of the metal electrons. Band intensity data also suggest an assignment to ion states in the order $^2T_2 < ^2E < ^2A_1$. If differential relaxation of the metal levels can be neglected an orbital ordering of $t_2 > e > a_1$ may be inferred. The assignment gains further support from the observed asymmetric splitting of the $^2T_2$ band, a shoulder on the leading edge being apparent at 5.28 eV. For a second row atom, the splitting is too great for spin orbit coupling and is probably a result of a Jahn-Teller splitting of the orbitally degenerate $^2T_2$ state to ^2B and ^2E states, the ^2B state being the ground state for the monocation. It is possible that the second band at 6.47 eV also shows a splitting, as would be expected for a ^2E state.

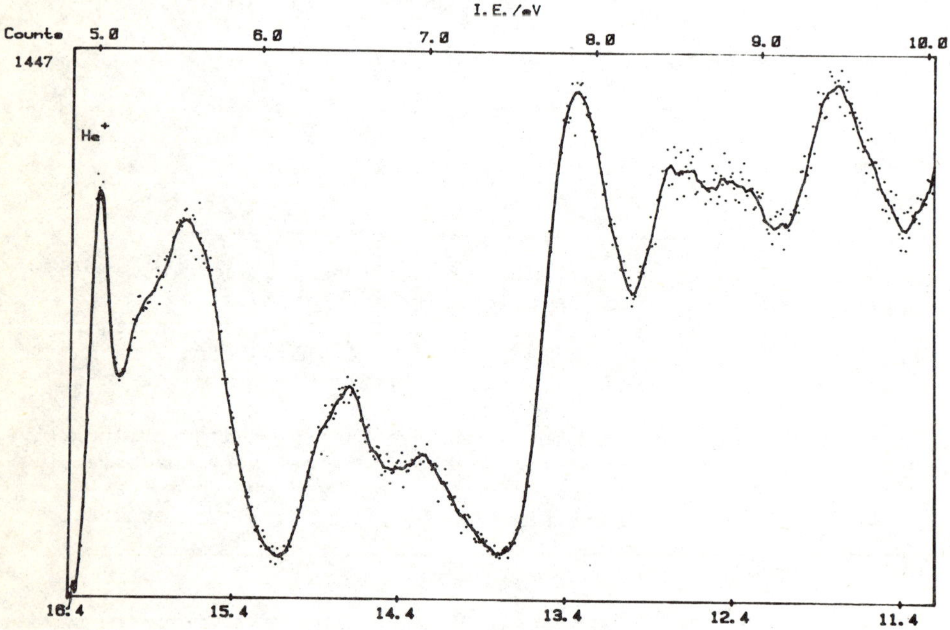

Figure 2. He I P.E. spectrum of $[Mo_4S_4(C_5H_4Pr^i)_4]$

The PE spectrum of $[Mo_4S_4(C_5H_4Pr^i)_4]$ is therefore consistent with an orbital model for the bonding of the tetrahedron and a ground state configuration $a_1^2e^4t_2^6$. The substantial separation of the $^2T_2$ state from the ^2E and $^2A_1$ states suggests that the t_2 orbitals are less stongly bonding than the e or a_1 orbitals.

The structural data on $[Mo_4S_4(C_5H_4Pr^i)_4]$, $[Mo_4S_4(C_5H_4Pr^i)_4]^+$ and $[Mo_4S_4(C_5H_4Pr^i)_4]^{2+}$ show that the Mo-S and Mo-ring distances are invariant, within experimental error, on ionization. Also the Mo-Mo distances are only slightly affected by the charge on the cluster, confirming the non-bonding role of the t_2 orbitals. Though distortion and lowering of symmetry of the Mo_4 tetrahedron dos occur in the cations, there is no strong indication of a static Jahn-Teller distortion.

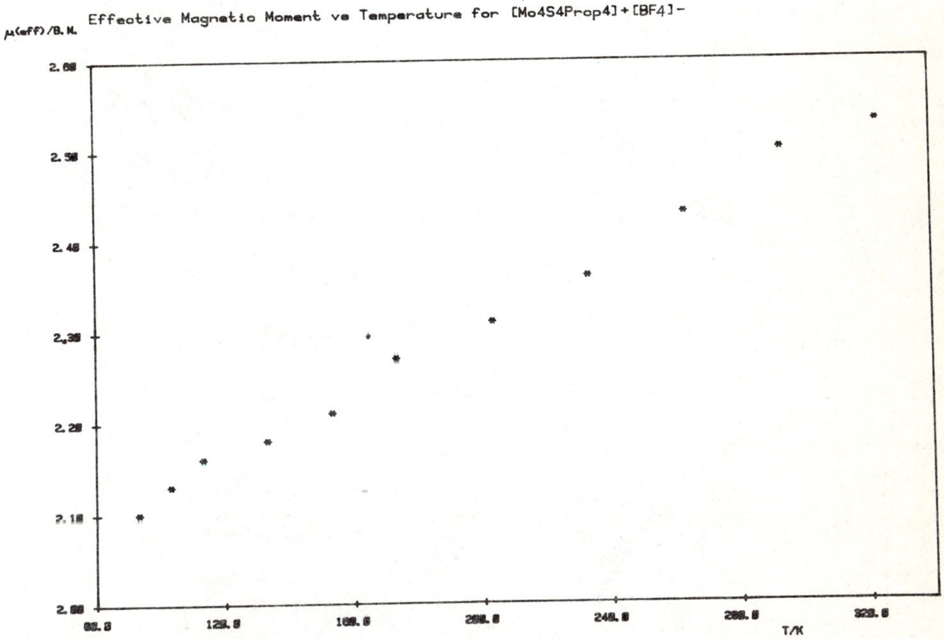

Figure 3. Plot of effective magnetic moment vs. temperature for $[Mo_4S_4(C_5H_4Pr^i)_4]^+BF_4^-$

The magnetic measurements on $[Mo_4S_4(C_5H_4Pr^i)_4]^+BF_4^-$ indicate thermal population of a state with orbital angular momentum as the temperature is raised. This is consistent with a ^2B ground state and a thermally accessible ^2E excited state. The PE spectrum of $[Mo_4S_4(C_5H_4Pr^i)_4]$ provides us with a mapping of the vibronic envelope of these two states and the relative energies are such that thermal population of the ^2E state seems reasonable. The diamagnetic nature of

$[Mo_4S_4(C_5H_4Pr^i)_4]^{2+}$ at low and at room temperature would suggest a static Jahn-Teller distortion with a $^1A_1$ ground state (e^4 configuration). In view of the structural similarity of the mono- and di-cations, it is suprising that there is no detectable thermal population of a triplet excited state at room temperature.

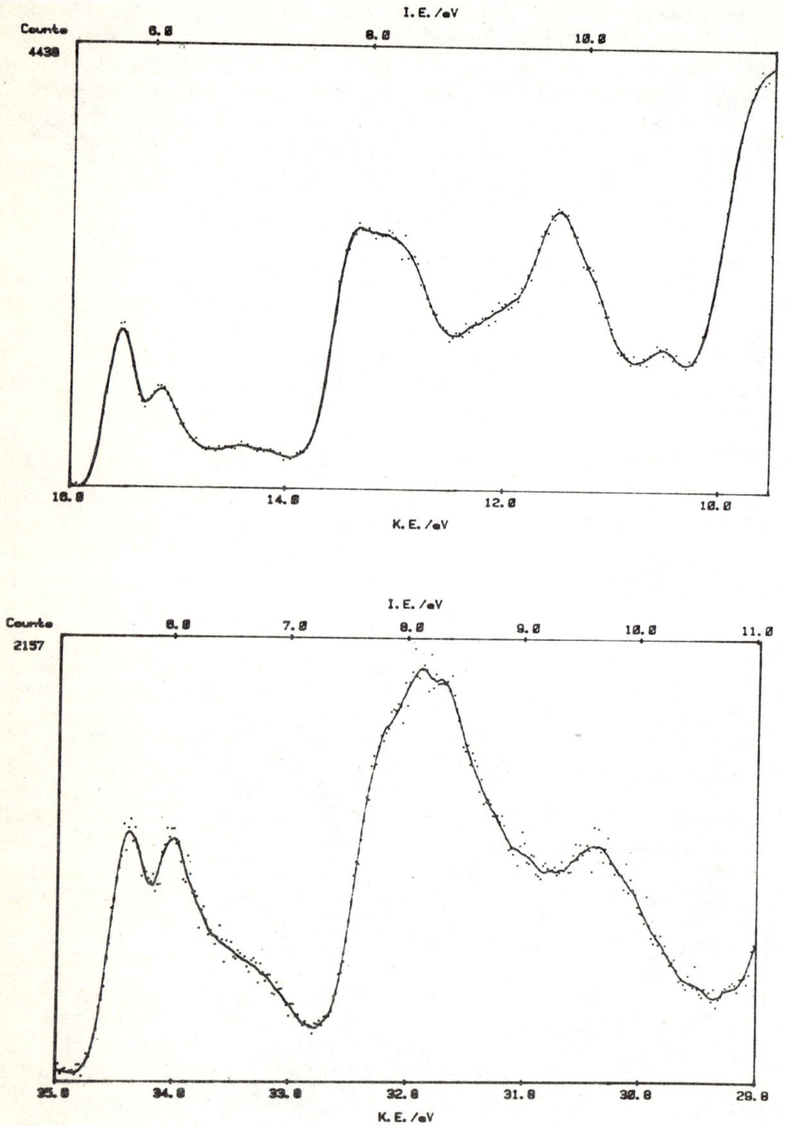

Figure 4. He I and He II spectra of $[Cr_4S_4(C_5H_4Me)_4]$.

As mentioned above, the PE spectrum of the neutral molecule

provides an energy mapping of the ground and some of the excited states of the corresponding monocation. From the observed PE spectrum of $[Mo_4S_4(C_5H_4Pr^i)_4]$ we would predict in the absorption spectrum of the monocation bands due to excitation from the $^2T_2$ state to either the ^2E or $^2A_1$ states, which should occur in the near infrared. A band occurs at 12,900 cm^{-1} (1.6 eV), which is absent in the spectrum of the neutral molecule. This matches best with the separation of the adiabatic IE of the $^2T_2$ and $^2A_1$ bands. The monocation is by definition a mixed valency compound and absorption bands in the near infrared in such compounds are often taken to be intravalence transitions, with the valencies localized in the time scale of the experiment. In this case, however, the occurence of such a band is entirely consistent with a delocalized orbital model for bonding within the cluster.

$[Cr_4S_4(C_5H_4Me)_4]$

The PE spectrum of $[Cr_4S_4(C_5H_4Me)_4]$ is similar to that of $[Mo_4S_4(C_5H_4Pr^i)_4]$ and may be assigned in an analagous fashion. Again three distinct metal ionization bands are visible. The third band is very broad and of low intensity and in the He I spectrum is difficult to distinguish from the base line; however, it is more prominent in the He II spectrum. No shoulder is detectable on the first band. The differences from the PE spectrum of $[Mo_4S_4(C_5H_4Pr^i)_4]$ are interesting.

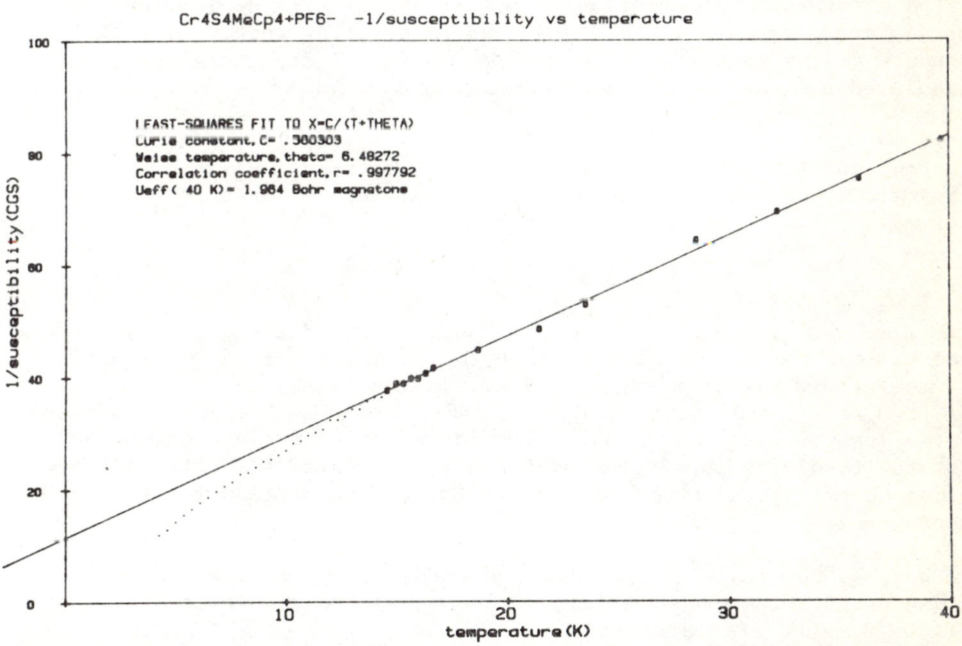

Figure 5. Plot of 1/susceptibility vs T for $[Cr_4S_4(C_5H_4Me)_4]^+PF_6^-$

The first IE is higher and the bands are closer together. The observed decrease in the spread of d-ionizations from Mo to Cr indicates a decrease in the metal-metal bonding interaction. Such a trend is common from the second to the first row, and is attributed to the lack of radial extent of the 3d orbitals compared with the 4d orbitals.

$[Cr_4S_4(C_5H_4Me)_4]^+PF_6^-$ shows an absorption band in the near infrared at 7000 cm^{-1} (0.87 eV), a lower frequency than $[Mo_4S_4(C_5H_4Pr^i)_4]^+BF_4^-$. Comparison with the PE spectrum of $[Cr_4S_4(C_5H_4Me)_4]$ again suggests assignment to the $^2T_2$ to $^2A_1$ transition.

$[Cr_4O_4(C_5H_5)_4]$ and $[Cr_4O_4(C_5H_4Me)_4]$

The PE spectra of $[Cr_4O_4(C_5H_5)_4]$ and $[Cr_4O_4(C_5H_4Me)_4]$ are very similar to one another and, in striking contrast to those of the M_4S_4 clusters, show no resolved d-band structure. The d-band does show features in its profile and is spread over ca 1.6 eV. Assignment and interpretation of these spectra are problematic. The change in the PE spectra between the M_4S_4 cubanes and the Cr_4O_4 clusters is reminiscent of that between $Mo_2(CH_3COO)_4$ and $Cr_2(CH_3COO)_4$ [5]. In the latter case, to retain an orbital picture of bonding between the metal atoms, extensive configuration interaction is invoked [6]. We propose that a similar situation holds for the Cr_4O_4 clusters. The observed d-band would then contain states of the molecular ion involving occupancy of all five cluster MO. Alternatively, one may regard the d-electrons as primarily localized on the Cr^{3+} ions and interacting weakly with their neighbours as the magnetic measurements suggest. A structural study on the isolated monocation is clearly called for here.

If we borrow from solid state theory, we may apply the Mott-Hubbard criterion, which states that for itinerant electrons the ratio of the band width, D, to the electron-electron repulsion, U, must be greater than 1

$$D/U > 1$$

If the observed band width of $[Cr_4O_4(C_5H_5)_4]$ and $[Cr_4O_4(C_5H_4Me)_4]$ does indeed reflect the complete band width, D, it seems reasonable in this case that the d electons are localized. In the cases of $[Mo_4S_4(C_5H_4Pr^i)_4]$ and $[Cr_4S_4(C_5H_4Me)_4]$ the total d band width is greater than the span of observed d ionization bands, as we are not observing ion states involving population of the higher t_1 and $2t_2$ orbitals. Even so, in both cases, the observed band width is greater than for the Cr_4O_4 clusters.

Thus, on the basis of the observed PE data, we propose that the Mo_4S_4 and Cr_4S_4 clusters have delocalized d electrons, whereas those of the Cr_4O_4 cluster are localized.

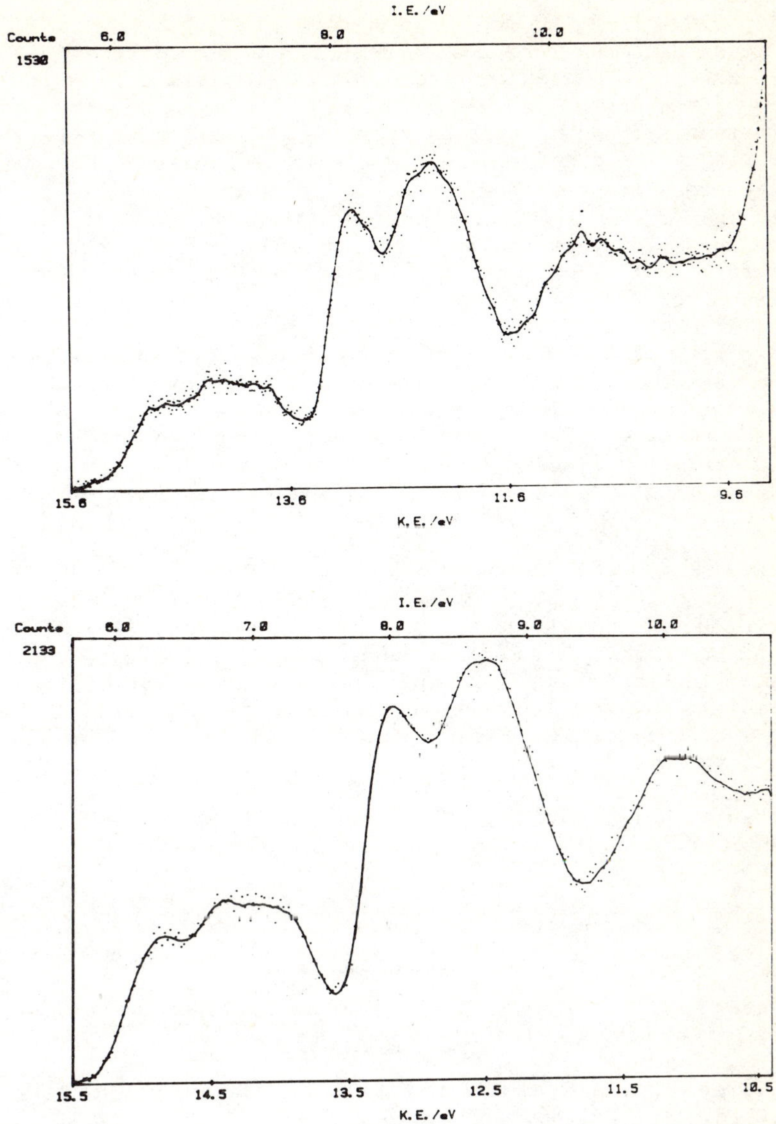

Figure 6. He I P.E. spectra of $[Cr_4O_4(C_5H_5)_4]$ and $[Cr_4O_4(C_5H_4Me)_4]$.

Redox Chemistry.

The redox data on these compounds has been summmarized in Table 2. $[Mo_4S_4(C_5H_4Pr^i)_4]$ is the easiest to oxidize, consistent with its possession of the lowest adiabatic IE. This is compatible with its being the most strongly bound cluster as there is a division of roles between the d-electrons. The e and a_1 electrons hold the molybdenum atoms together, whereas the non-bonding t_2 electrons are available for redox chemistry. The lack of structural change within the molecule on two successive oxidations, as shown by the crystallographic study, is thus accounted for. It also has the implication that electron transfer could be very fast from this system.

The ability of the M_4S_4 clusters to undergo two successive oxidations may be associated with the delocalized nature of the metal electrons, the positive charge being dispersed over the whole cluster. In contrast the Cr_4O_4 cluster can only sustain one reversible oxidation, the second oxidation leading to destruction of the unit. We attribute this to the more localized nature of the d-electrons in this case.

REFERENCES.

[1] G.L. Simon, L.F. Dahl: J. Am. Chem. Soc., **95**, 2164 (1973); G.L. Simon, L.F. Dahl: J. Am. Chem. Soc., **95**, 2175 (1973); C.H. Wei, G.R.Wilkes, P.M. Treichel, L.F. Dahl: Inorg. Chem, **5**, 900 (1966); Trinh-Toan, B.K. Teo, J.A. Ferguson, T.J. Meyer, L.F. Dahl: J. Am. Chem. Soc., **99**, 408 (1977); Trinh-Toan, W.P. Fehlhammer, L.F. Dahl: J. Am. Chem. Soc., **99**, 402 (1977).

[2] J.A. Bandy, C./E. Davies, J.C. Green, M.L.H. Green, C.K. Prout, D.P.S. Rodgers: J. Chem. Soc. Chem. Comm., 1395 (1983).

[3] O.G. Ellert, I.L. Eremenko, G.S. Gasanov, V.T. Kalinnikov, T.K. Kurbanov, S.V. Lindeman, V.M. Novotortsev, A.A. Pasynskii, T.V. Rakitin, V.E. Shklover, Y.T. Struckov: J. Organomet. Chem., **248**, 309 (1983).

[4] F. Bottomley, D.E. Paesz, P.S. White: J. Am. Chem. Soc., **103**, 5581 (1982); F. Bottomley, F. Grein: Inorg. Chem., **21**, 4170 (1982).

[5] A.W. Coleman, J.C. Green, A.J. Hayes, E.A. Seddon, D.R. Lloyd, Y. Niwa: J. Chem. Soc. Dalton, 1057 (1979).

[6] M. Benard: J. Am. Chem. Soc., **100**, 2354 (1978).

MOVING FROM DISCRETE MOLECULES TO EXTENDED STRUCTURES: A CHEMICAL AND THEORETICAL APPROACH TO THE SOLID STATE

Roald Hoffmann and Chong Zheng
Department of Chemistry
Cornell University
Ithaca, NY 14853
USA

ABSTRACT. Solid state inorganic chemistry, broadly encompassing novel conducting systems of varying dimensionality, surfaces, minerals etc. is one of the areas of chemistry in which activity will grow exponentially in the next decade. Theoretical chemists can make a contribution to problems of electronic structure in the solid state through the experience they have acquired in chemical interpretations of the wave functions they calculate. The following problem will provide an illustration of our approach to these problems. Among the many examples of the $ThCr_2Si_2$ structural type is a set of A B_2X_2 structures, B=transition metal, X=group 14 or 15 element, in which one X···X contact varies over the range of bonding between no bond and a fully formed X-X single bond. The tuning is provided by variation in the transition metal, X-X bonding promoted by metals on the right-hand side of the periodic table. We begin with chemical construction of the bonding in B_2X_2 layers. When these two-dimensional layers are stacked narrow X-X σ and σ* bands result. The σ* level is filled (no X-X bond) or empty (X-X bonding) depending on the Fermi level of the transition metal B layer.

Solid state chemistry is a growth point of our science. It also constitutes a frontier area in which interactions between chemists, physicists and engineers are remarkably easy. The development of new areas - surface science, novel conducting materials to name just two hot ones - has brought physicists and chemists closer together than they have ever been in this century.

It is essential for theoretical chemists to join in the exciting work in this area. And we can do so without any feeling of inferiority to our colleagues in theoretical solid state physics who are venturing into these fields from their side. Perhaps they have more experience with the electronic structure of solids, and indeed some of their methods (I'm thinking here of the various density functional procedures) have filtered into molecular computations. But what we have as an advantage is decades of interpretational work. We, as quantum chemists, embedded within the predominantly experimental science of chemistry, have learned very well how to talk to our fellow chemists, how to make connections between our wave functions and their experimentally based notions of

A. Veillard (ed.), Quantum Chemistry: The Challenge of Transition Metals and Coordination Chemistry, 425–443.

bonds, charge distributions, what determines reactivity, stability etc.

It is this ability to interpret wave functions, honed under the subtle pressure of our chemical colleagues which is our strength. Not the calculations of wave functions, which our physicist friends probably do better. Why not make use of that interpretational advantage? Let's learn some of their methodology, the strong dictates of translational symmetry. There are some chemical arguments for an advantage of a cluster modelling of local chemical action on a surface or in the solid. But there are also problems with finite cluster models, and I would recommend, in part for reasons of enlarging the meeting ground with physicists, a movement toward delocalized band calculations. Let us carry out then the calculations in their way, but let us use the stength of our experience in chemistry, in chemical thinking, to make a contribution to these new sciences of solids, surfaces and materials which are exploding around us.

We would like to illustrate such an approach for a specific example, one of medium complexity on the scale of modern solid state chemistry. This is a remarkable structural transformation in a group of AB_2X_2 structures in which a X-X bond is in a continuous manner made or broken in the solid state.

X-X Pair Formation in $ThCr_2Si_2$ Type AB_2X_2 Structures

More than 400 compounds of AB_2X_2 stoichiometry adopt the $ThCr_2Si_2$ type structure[1]. But you are not likely to find any mention of these in any modern textbook of general inorganic chemistry, which just tells us something about the ascendancy of molecular inorganic chemistry, especially transition metal organometallic chemistry, in the last three decades. But these compounds are there, we know their structures and they have interesting properties. A is typically a rare earth, alkaline earth, or alkali element; B is a transition metal or main group element, and X comes from group 15, 14 and occasionally 13. Since the synthesis of AB_2X_2 with A = a rare earth element, by Parthé, Rossi and their coworkers the unusual physical properties exhibited by these solids have attracted much attention. Physicists speak with enthusiasm of valence fluctuation, p-wave or heavy fermion superconductivity and of many peculiar magnetic properties of these materials. But the underlying nature of the chemical bonding in these remarkable compounds has not been explored in detail.

The $ThCr_2Si_2$ structure type for AB_2X_2 stoichiometry compounds is shown in 1. It consists of $B_2X_2^{2-}$ layers interspersed with A^{2+} layers. The bonding between A and B_2X_2 layers appears largely ionic, which is why we write the charge partitioning as A^{2+} and $B_2X_2^{2-}$. But in the $B_2X_2^{2-}$ layer there is indication not only of covalent B-X bonding, but also some metal-metal B-B bonding. Typical metal-metal distances are in the range of 2.7 - 2.9 Å.

A way to describe the B_2X_2 layers in these compounds is to imagine a perfect square-planar two-dimensional lattice of metal atoms, above and below the four-fold hollows of which lie the main group X atoms. This is shown in 2 below. The coordination environment of the metal (B) is approximately tetrahedral in the main group elements (X)

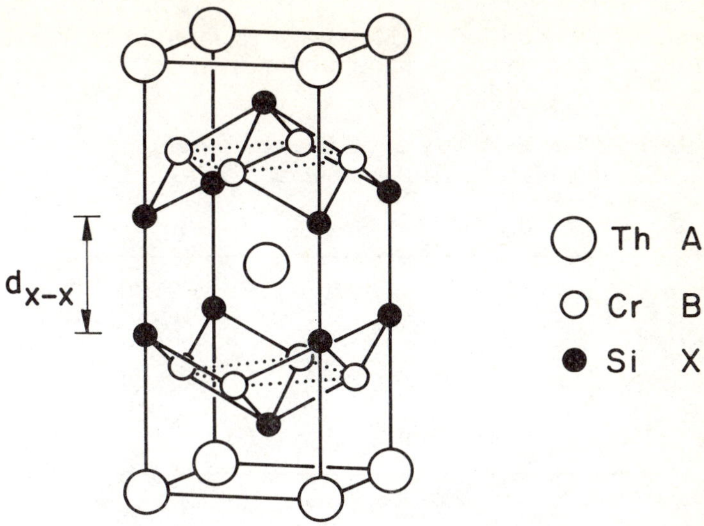

1

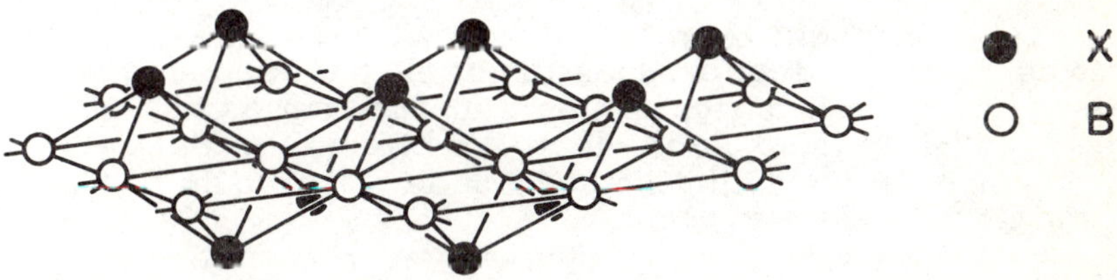

2

with four additional square planar near neighbor metals. The coordination of the X atoms is much more unusual - they reside at the apex of a square pyramid.

There is a long X···X contact within a layer, but what becomes the main focus of this paper is a remarkable tunable X···X contact between two layers, along one of the edges of the tetragonal unit cell depicted in **1**. This contact, d_{X-X}, is the primary variable geometrical factor in these molecules.

If one examines the hundreds of known $ThCr_2Si_2$ type structures a trend becomes apparent: for the same cation A and the same main group element X, d_{X-X} decreases as the transition metal B moves from the left-hand side to the right-hand side in the periodic Table. One series of these compounds, studied by Mewis[3], is singled out in Table 1. The element X in each of these is phosphorus.

Table 1. The X-X Distance in Some Phosphide Compounds of the AB_2X_2 Type.

AB_2X_2	d_{X-X}(Å)	AB_2X_2	d_{X-X}(Å)
$CaCu_{1.75}P_2$	2.25	$SrCu_{1.75}P_2$	2.30
$CaNi_2P_2$	2.30	$SrCo_2P_2$	3.42
$CaCo_2P_2$	2.45	$SrFe_2P_2$	3.43
$CaFe_2P_2$	2.71		

For reference the P-P distance in P_4 is 2.21 Å and 2.192 Å in Me_2P-PMe_2. The P-P single bond distance in many compounds is remarkably constant at 2.19-2.26 Å. The P=P double bond and P≡P triple bond lengths are around 2.03 Å and 1.87 Å respectively. It is clear that the short distances in the $ThCr_2Si_2$ type phosphides are characteristic of a full P-P single bond. The long contacts, such as 3.43 Å imply essentially no bonding at all. All the compounds known with a nonbonding X•••X separation contain metals from the left-hand side of the Periodic Table. In fact examination of all the structures reveals a trend; as one moves left to right in the transition series the P-P contact shortens. Clearly there is an electronic effect of work here -- a P•••P bond is made or broken in the solid state. We would like to understand how and why this happens.

Bonding within a B_2X_2 Layer

We approach the AB_2X_2 structure, represented by a typical $BaMn_2P_2$ compound, in stages. First we will look at a single two-dimensional $Mn_2P_2^{2-}$ layer. Then we will form a three-dimensional $Mn_2P_2^{2-}$ sublattice by bringing many such layers together in the third dimension. Finally, we may insert the Ba^{2+} counterions into the structure.

Consider a single Mn_2P_2 layer (2). The Mn-P distance is 2.455 Å, and the Mn-Mn distance in the square metal lattice is 2.855 Å. The latter is definitely in the metal-metal bonding range, so a wide band, delocalized picture is inevitable. But in some hierarchy or ranking of interactions it is clear that Mn-P bonding is stronger than Mn-Mn. So let's construct this solid conceptually, or think of it, in terms of first turning on Mn-P bonding, and then Mn-Mn interaction.

The local coordination environment at each Mn is approximately tetrahedral. If we had a discrete tetrahedral Mn complex, e.g. $Mn(PR_3)_4$, we might expect a qualitative bonding picture such as 3.

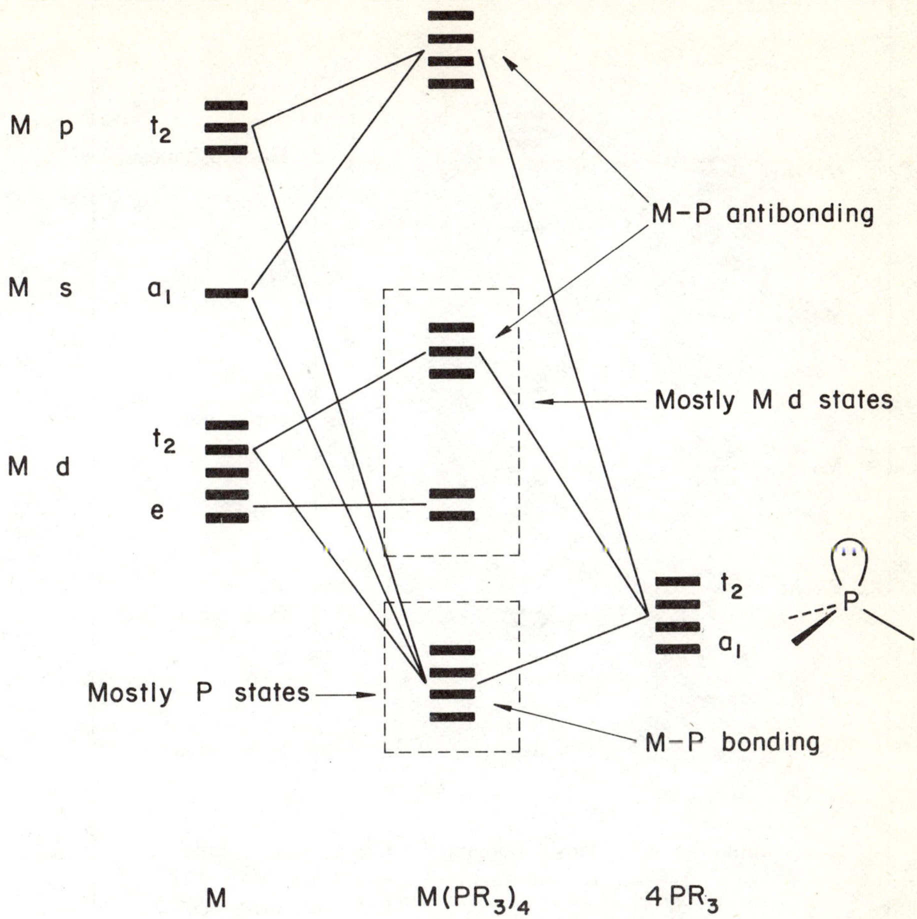

3

Four phosphine lone pairs, $a_1 + t_2$ in symmetry, interact with their symmetry match, mainly Mn 4s and 4p, but also with the t_2 component of Mn 3d set. Four orbitals, mainly on P, P-Mn σ bonding, go down. Four orbitals, mainly on Mn, P-Mn σ antibonding, go up. The Mn d block splits in the expected two below three way.

Something like this <u>must</u> happen in the solid. In addition there are Mn-Mn bonding contacts in the layer, and these will lead to dispersion in those bands which are built up from orbitals containing substantial metal character. The combined construction is shown in Figure 1.

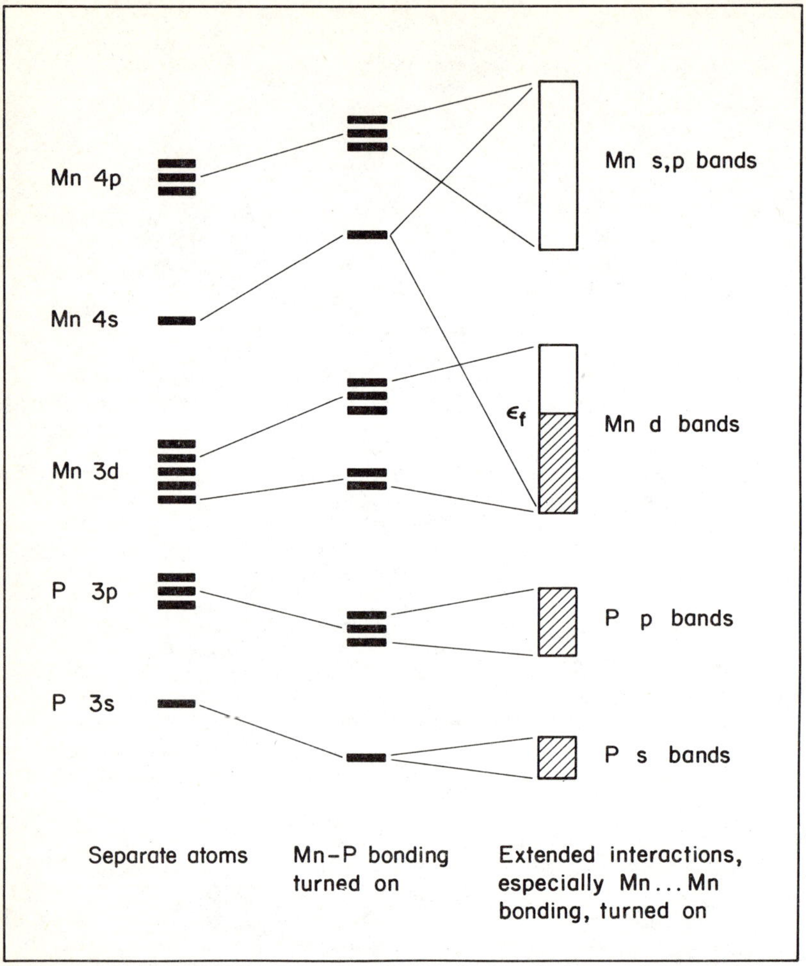

Figure 1. A schematic picture of the $Mn_2P_2^{2-}$ layer band structure as derived by first turning on local Mn–P interactions and then the two-dimensional periodicity and Mn–Mn interactions.

At this point a little backtracking is in order to explain the physics that is behind the argument used.[4] Let us review some of the basic determinants of band structure, and in the process get in hand some of the analytical tools we need. Consider a one-dimensional chain of hydrogen atoms 4, or the topologically equivalent π-system of a polyene (no bond alternation), 5. We indicate schematically the energy levels and wave functions of some oligomers and the infinite polymer. In the oligomers the number of nodes in the wave-functions rises with increas-

4 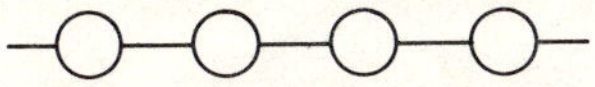

5 

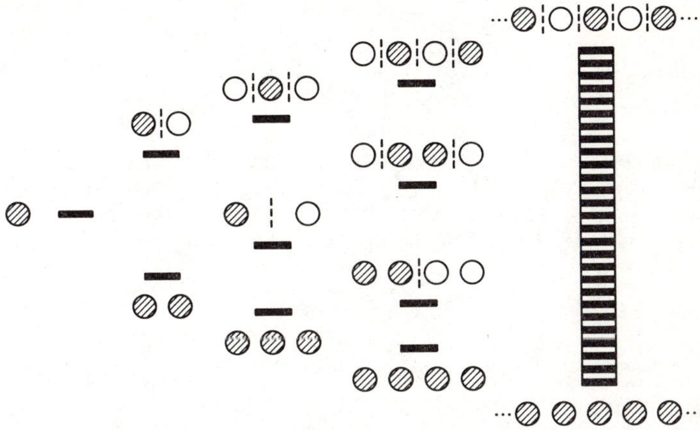

ing energy, and so it does in the polymer. What determines the band width or dispersion of the polymer, the difference in energy between the highest and lowest level in the band? The same thing that determines the level splitting in the dimer (H_2 or ethylene), namely the overlap. The greater the overlap between neighbors, the greater the band width. A chain of hydrogen atoms nailed down 10 Å apart would have a very small band dispersion.

That is for the trivial system of one orbital per unit cell, 5. What about a more complicated unit cell, three-dimensional, several atoms per unit cell? The physics is the same - one finds the appropriate wave functions in the unit cell, develops them into bands, one band per unit cell MO. The width of the resulting bands will be determined by the inter-unit-cell overlap. In the specific case of our Mn_2P_2 two-dimensional polymer each unit cell MO will be some linear combination of Mn and P AO's. The inter-unit-cell contact is through metals mainly, the P$\cdots$P distance (between neighboring apices of square pyramids) is ~4 Å. So those levels which are heavy on P will give narrow bands, while those concentrated on Mn will yield wide bands.

"Develops [the unit cell wave functions] into bands" was the language we used above. What we had in mind, of course, is the powerful apparatus of Bloch functions, the use of translational symmetry. The proper symmetry-adapted linear combinations of a chain of basis functions χ_i set on lattice points 0, 1, 2, 3$\cdots$i of an infinite one-dimensional lattice with unit cell size a are given in 6.
Note the labeling by an index k, here appearing as a group theoretical symbol. k is of course much more - a node counter, a wave vector, related to momentum.

6

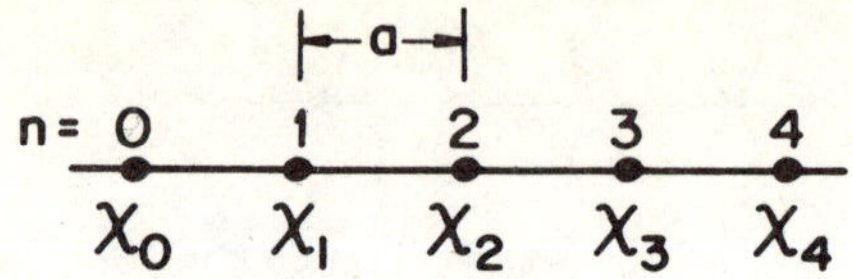

$$\psi_k = \sum_n e^{ikna} \chi_n$$

Let us see the form of the wave function for two extreme values of k, k = 0 (the zone center), k = π/a (zone edge). The reference is to the Brillouin zone, which is the range of unique k. The space of k is reciprocal space. The two wave functions are given in 7.

7

$$k = 0 \qquad \psi_0 = \sum_n e^0 \chi_n = \sum_n \chi_n$$

$$= \chi_0 + \chi_1 + \chi_2 + \chi_3 + \cdots$$

$$k = \frac{\pi}{a} \qquad \psi_{\frac{\pi}{a}} = \sum_n e^{\pi i n} \chi_n = \sum_n (-1)^n \chi_n$$

$$= \chi_0 - \chi_1 + \chi_2 - \chi_3$$

We see the top and bottom of the band here; values of k in between 0 and π/a have intermediate energies. Furthermore it can be shown that $E(-k)=E(k)$, so that the energy values can be plotted in the interval $0 \le |k| \le \pi/a$ instead of $-\pi/a \le k \le \pi/a$. The absolute sign is often dropped, and the result displayed as in 8.

8

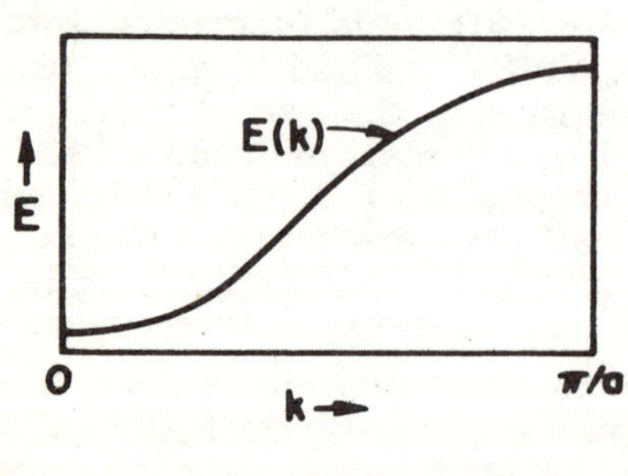

k is quantized by cyclic boundary conditions, so that the number of unique values of k is equal to the number of microscopic unit cells in the macroscopic crystal, or the number of distinct translations. The energy levels are equally spaced along the k axis.

An important point is that while the lovely form of 6 applies in general, it does not say anything about the energy of the orbitals at the center of the zone relative to those at the edge. For a chain of H atoms it is clear that $E(k=0) < E(k = \pi/a)$. But consider a chain of p functions, 9. The same combinations are given to us by the translational symmetry, but now it is clearly $k = 0$ which is high energy, the most antibonding way to put together a chain of p orbitals.

9

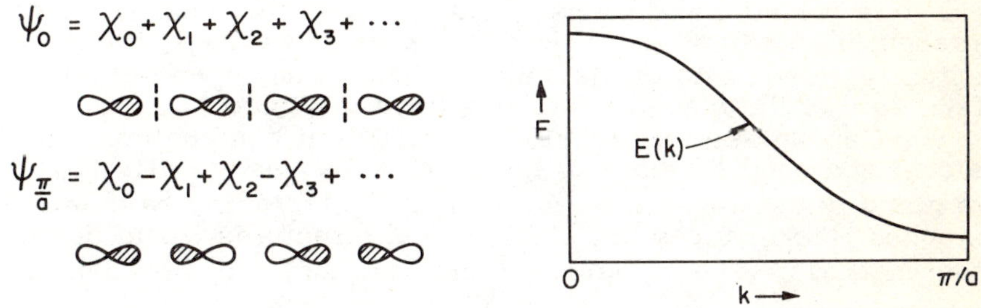

The band of s functions for the hydrogen chain "runs up", the band of p orbitals "runs down" (from zone center to zone edge). In general it is the topology of orbital interactions which determines which way bands run.

To summarize: bandwidth is set by inter-unit-cell overlap, and the way the bands run is determined by the topology of that overlap.

One more concept and we're ready to proceed. In the solid state we obviously cannot get away with talking of a single level (out of Avogadro's number) doing this or that. We must devise a language of bunches of levels, and one obvious way to group such bunches is by energy. The density of states DOS(E) is defined such that

DOS(E)dE = number of levels between E and E + dE

For the simple band of 8 the DOS curve is given in 10. The characteristic shape of the DOS curve is generated by the cosine-type curve of E(k) and the fact that levels are spaced at equal intervals of k. Obviously DOS(E) is related to the inverse of the derivative dE/dk.

Densities of states are going to be our primary interpretational tool. They represent a return from reciprocal space to real space, always welcomed by experimentalists. They will allow us to construct frontier orbital argument in the solid, as we will see. For the moment it is important to note that if one has the band structure E(k) one can sketch qualitatively DOS(E) following the simple guideline: flat bands → large DOS.

10

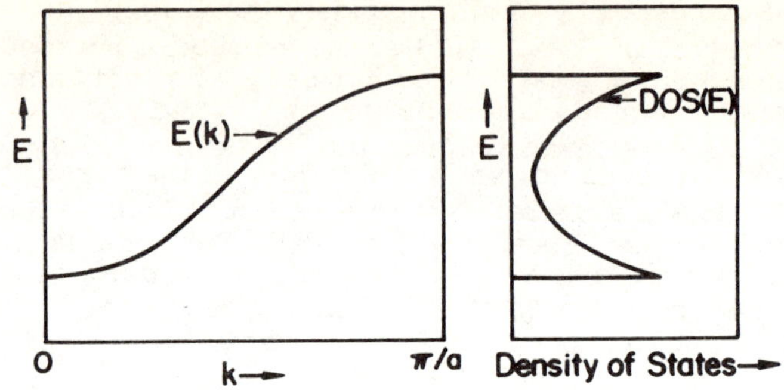

Or one can have a computer do it, more accurately.

Now we must return to our Mn_2P_2 layer. Arguing from the distances in its structure we layed out a hierarchy of interactions to follow in an interpretational scheme. First there is the local environment at Mn, tetrahedral coordination by P. Then these localized orbitals were conceptually spread into bands, and we had a simple way of guessing the dispersion of the bands. This was done in Figure 1.

Can we see this local, very chemical bonding construction in a delocalized band structure? Most certainly, as we will now show. We are going to use throughout this paper tight-binding calculations of the extended Hückel type, but similar band structures would be obtained from other methods. The calculated band structure and total density of states

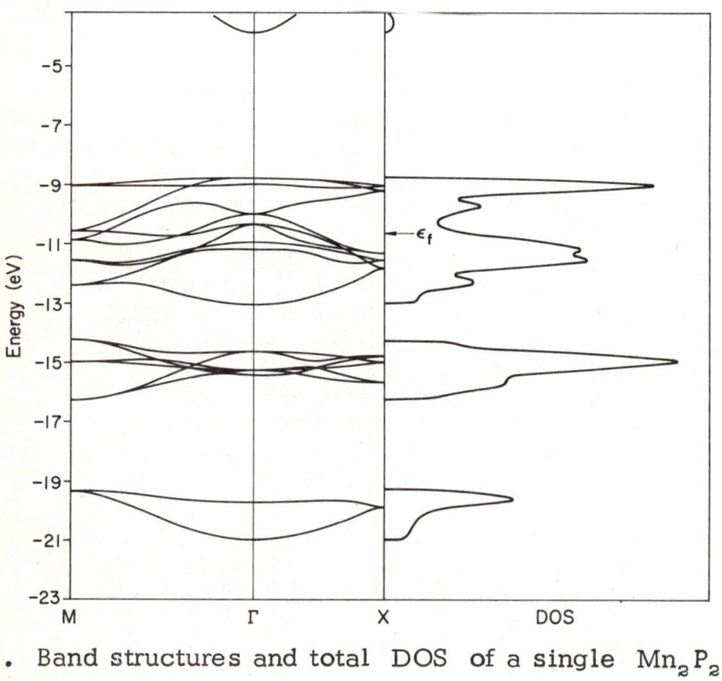

Figure 2. Band structures and total DOS of a single $Mn_2P_2^{2-}$ layer.

of a single $Mn_2P_2^{2-}$ layer is illustrated in Figure 2. The bands correspond to the final translationally delocalized picture, the right-hand side of Figure 1. They are not very informative chemically until we apply some decomposition and partitioning analyses to them.

Let us begin by constructing the Mn_2P_2 layer from separate Mn and P films, in an attempt to model the primary bonding effect, Mn-P interaction. This is done in Figure 3. At left is the P sublattice. We see P 3s (around -19 eV) and P 3p (around -14 eV) bands. Both are narrow, because the P atoms are ~4 Å apart. The Mn sublattice

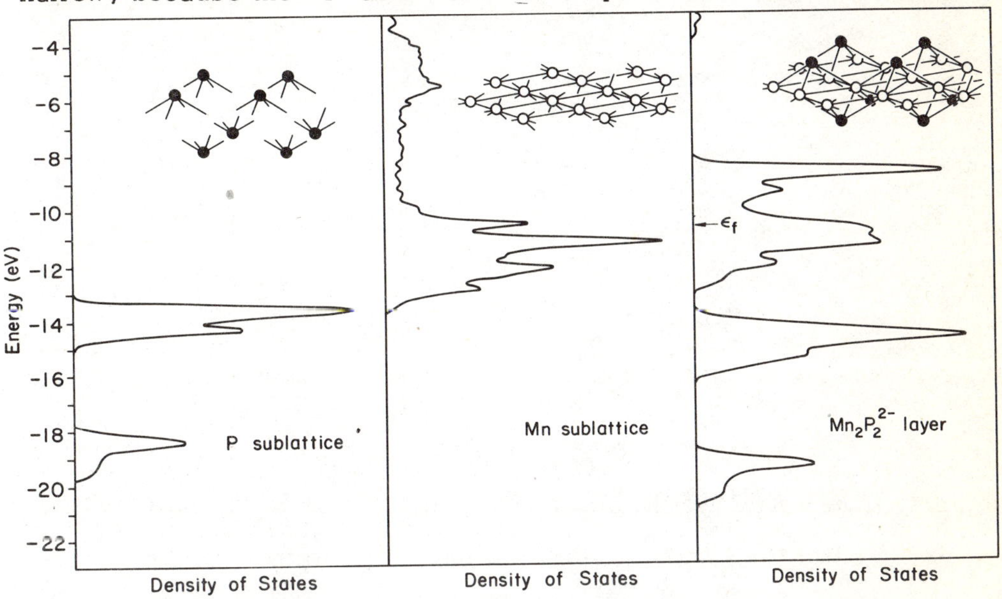

Figure 3. Total DOS of the P sublattice (left), the Mn sublattice (middle) and the composite $Mn_2P_2^{2-}$ layer lattice (right).

(middle of Figure 3) shows a nicely dispersed density of states (DOS). The Mn-Mn separation is only 2.855 Å. Thus we have a two-dimensional metal, with a familiar wide s, p plus narrow d band pattern. The bottom part of the DOS in the middle of Figure 3 is the 3d band, the top is the lower part of the 4s, 4p band. At the right in Figure **3** is the density of states of the composite Mn_2P_2 layer. Note how the individual P and Mn bunches of states repel each other on forming the composite lattice. There is no more graphic way of showing that what happens in the inorganic solid is similar to what happens in an isolated inorganic molecule.

But are the clumpings of levels in the $Mn_2P_2^{2-}$ layer really behaving the way we expect? We can do some analytical detective work on the DOS. First we can decompose it into the part that is on Mn and the part that is on P. This is done in Figure 4, which in the solid state trade is called a projection of the DOS on Mn or the local DOS on Mn. In our case this is implemented by a simple Muliken population analysis - Figure 4 is just that, a Mulliken population analysis of all (or a representative sample; there is an art here) the levels in the Brillouin zone at a given energy. It gives us exactly the same information as a population analysis on a molecule, telling us what part of the DOS is on Mn. What is not on Mn is, of course, on P. Clearly the lower bands are mainly P, the higher ones are mainly Mn.

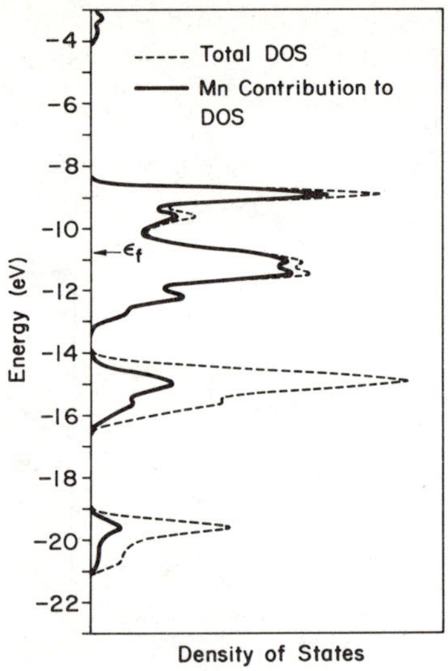

Figure 4. Total DOS of the composite $Mn_2P_2^{2-}$ layer lattice (dashed line) and the contributions of Mn orbitals to that DOS (solid line). The dotted line is the integrated Mn DOS.

What about the bonding characteristics predicted in the qualitative bonding scheme in Figure 1? We have devised a bonding index called COOP, technically the overlap population weighted density of states,[5] which allows us to investigate this. All the states in a given energy interval are interrogated as to their bonding character (as measured by an overlap population) and the resulting curve plotted as function of energy. The integral of this COOP curve, which has positive (bonding) and negative (antibonding) regions, up to the Fermi level gives the total over overlap population.

Figure 5 shows Mn-P and Mn-Mn COOP curves for the Mn_2P_2 layer. Note that the two lower bands (at -15 and -19 eV), which by the previous decomposition were seen to be mainly P, are Mn-P bonding, whereas the mainly metal bands around -12 eV are Mn-P nonbonding. The bunch of levels at ~ -9 eV is Mn-P antibonding — it corresponds to the crystal-field destabilized t_2 level in Figure 1. The bottom of the mainly metal band is Mn-Mn bonding, the top Mn-Mn antibonding.

Everything is as expected. And it could not be otherwise, for there is no new physics in the bonding in the infinite solid that is not there in the discrete molecule.

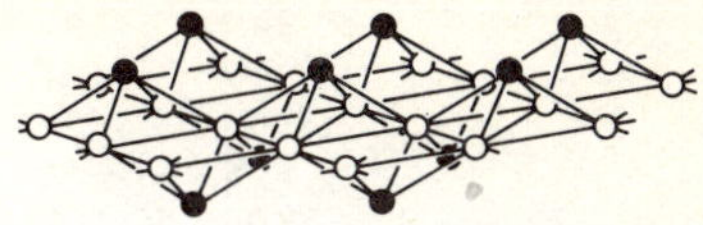

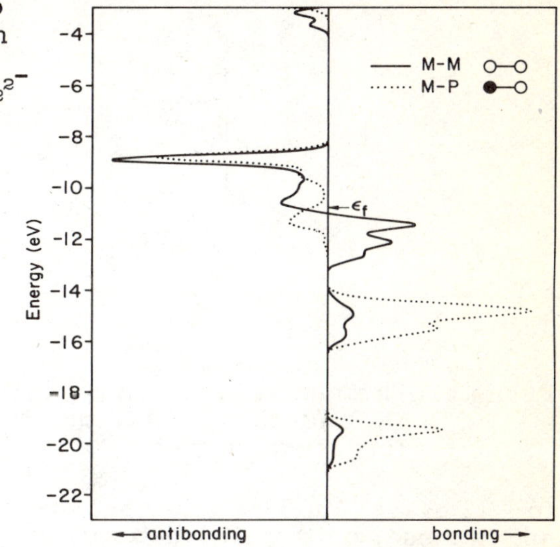

Figure 5. Crystal orbital overlap population curves for the Mn-Mn bonds (solid line) and Mn-P bonds (dotted line) in the $Mn_2P_2^{2-}$ single layer.

Three-Dimensional $Mn_2P_2^{2-}$

In preparation for putting the $Mn_2P_2^{2-}$ layers together, let us look at the bonding situation around P. Each phosphorus in a two-dimensional slab is in an unusual coordination environment, 11, at the apex of a square pyramid of four Mn atoms. Such four-coördinate pyramidal phos-

11

phido groups are as far as we know unknown, though they have been suggested theoretically.[6] One would expect something like a localized pair at P pointing away from the Mn atoms. One can look for this by asking for the contribution of P 3 p_z to the total DOS, Figure 6. The $3p_z$ orbital is indeed very well localized in a band at ~-15 eV, > 70% of the $3p_z$ being in that region.

On to the three-dimensional solid. When the two-dimensional $Mn_2P_2^{2-}$ layers are brought together to form the three-dimensional solid ($Mn_2P_2^{2-}$, still without the counterions), the P $3p_z$ orbitals or lone pairs in one layer form bonding and antibonding combinations with the corresponding orbitals in the layers above or below. Figure 7 shows the P $3p_z$ density of states at interlayer P-P=2.4 Å. The wide band at -8

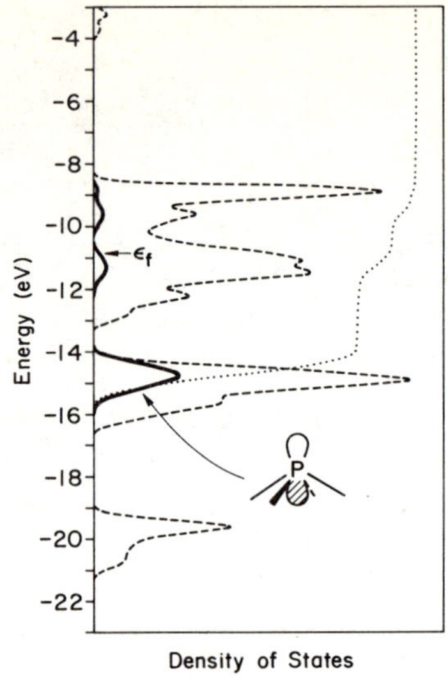

Figure 6. Phosphorus $3p_z$ orbital contribution (dark line) to the total DOS (dashed line) of the $Mn_2P_2^{2-}$ single layer. The dotted line is an integration of the dark line, on a scale of 0 to 100%.

to − 12 eV is Mn 3d. Below and above this metal band are P bands, and in these, quite well localized, are P–P σ and σ* combinations, 12.

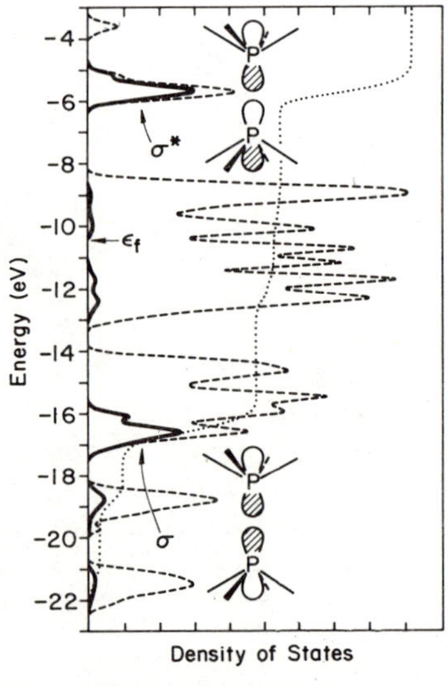

Figure 7. Phosphorus $3p_z$ orbital contribution (dark line) to the total DOS (dashed line) of the three-dimensional $Mn_2P_2^{2-}$ lattice, without the cation. The phosphorus-phosphorus bond length here is 2.4 Å. The dotted line is an integration, on a scale of 0 to 100%.

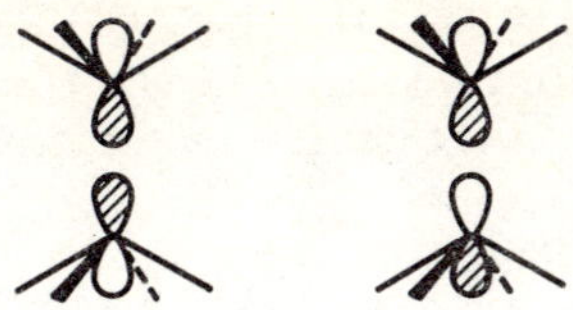

12

The bands are narrow because the lateral P-P distance is long, as 13, a representation at the zone center of one of the two bands contributing to σ^*, shows

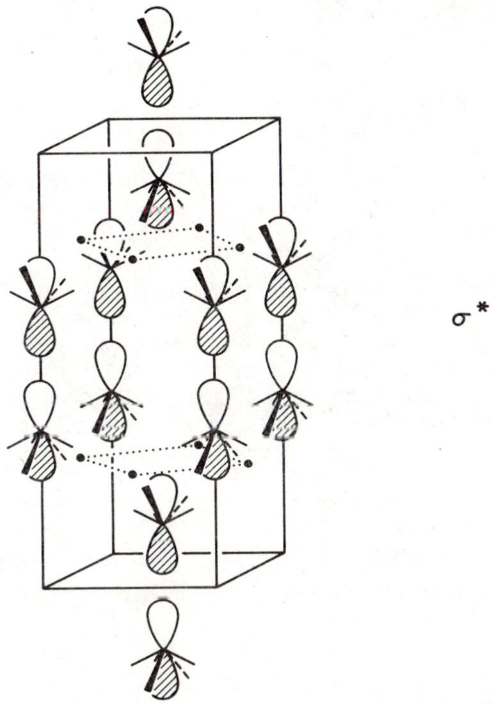

13

If the three dimensional calculation is repeated at different interslab or P···P distances all that happens is that the localized P-Pσ and σ^* bands occur at different energies. Their splitting decreases with increasing P···P separation, as one would expect from their respective bonding and antibonding nature.

We are now in a position to explain simply the effect of the transition metal on the P-P separation. What happens when the transition metal moves to the right hand side of the Periodic Table? The increased nuclear charge will be more incompletely screened and the d electrons more tightly bound. As a result, the d bands come down in energy and become narrower, as shown in 14. At the same time the band filling increases. Two factors thus compete to determine the trend in the

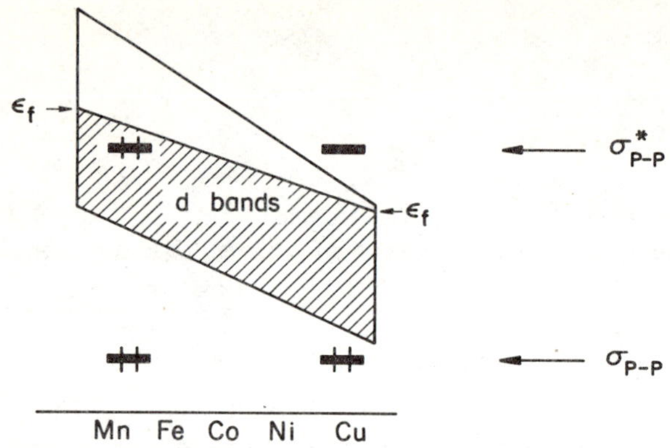

14

Fermi level, and it turns out that the first, the strength of the d-electron bonding, wins out. The Fermi level sinks as one moves to the right, the work function increases.

Now imagine superimposed on this variable energy sea of electrons the P-P σ and σ^* bands for some typical, moderately bonding P-P distance. On the left of the transition series the metal Fermi level is above the P-P σ^*. Both σ and σ^* are occupied, there is no resultant P-P bond. As P-P stretches in response the σ^* only becomes more filled. On the right side of the transition series the P-P σ^* is above the Fermi level of the metal, and so is unfilled. The filled P-P σ makes a P-P bond. Making the P-P distance shorter only improves this situation.

Another way to put this is to think in terms of the Zintl concept[7] of a full P-P bond and diatomic P_2^{4-}, 15, in compounds of the late transition metals. As one moves to the left in the transition series one reduces the P_2^{4-} molecule to P_2^{6-}, breaking the σ bond. Alternatively,

15

starting from two isolated phosphide ions in compounds of the middle transition series, the metal oxidizes two such ions to form a diatomic P_2^{4-}. Either way, in any of a number of complimentary pictures, what we have before us is the remarkable phenomenon of the tunable breaking and making of a diatomic bond in the solid state.

Let us examine the bond formation in a little more detail. Figure 8 shows the total energy per unit cell as a function of P-P separation.

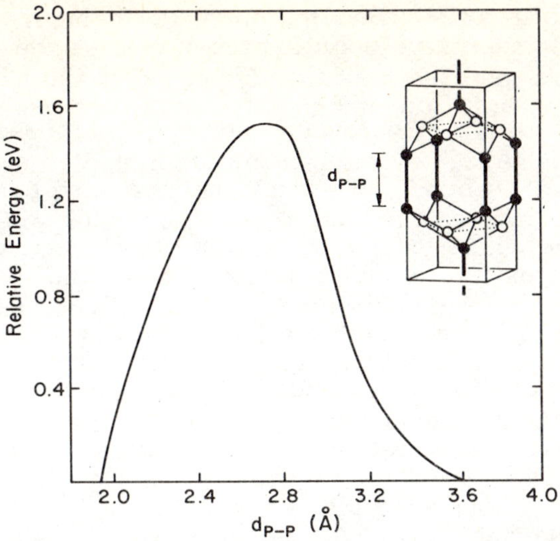

Figure 8. Relative energy per unit cell of the three-dimensional $Mn_2P_2^{2-}$ lattice as a function of phosphorus-phosphorus distance.

Note a maximum with two low energy regions on each side. The energy keeps dropping on the short P-P side, and there is a double minimum, whereas all the known compounds have a singlet one. These anomalies could be the result of deficiencies of the extended Hückel method, but actually they are not - they result from omitting the A^{2+} cations in these calculations. The counterions provide a Madelung energy to hold the slabs together on the long P-P side, and through their large size a countervailing force keeping the slabs apart on the short P-P side. Schematically these effects are shown in 16.

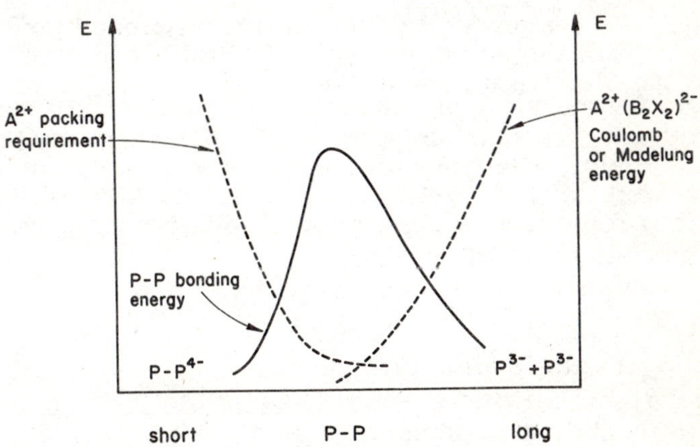

16

The position of the P–P bonding curve is the main variable factor here. The maximum in the curve could be thought of as the point of maximum σ^* filling before the electrons flow from σ^* to the metal. And this point moves to the right(longer P–P distance)as the metal Fermi level falls. Anyway the"bonding" "Coulomb" and "packing" requirements combine in the known AB_2X_2 structures to give a single minimum, either a made or a broken P–P bond. We think that for some compounds in this series, for some choice of atoms, there will in fact exist two minima, i.e. a possible phase transition is predicted. We would recommend a study of these materials under pressure along the c axis.

This is the end of this little story. There is much more, of course, for these AB_2X_2 compounds are quite remarkable. For instance there is an alternative structure, called $CaBe_2Ge_2$. This one is exactly the same in one layer as **1**, i.e. Be occupying square lattice sites and Ge above and below them. But in the other layer in the unit cell, the positions of Be and Ge are precisely switched relative to the first layer!

The main point of our analysis is not the specific result, the explanation of X···X pairing in one AB_2X_2 structure, but the methodology and philosophy. There is a chemical starting point (the geometrical structure, what it tells us about the bonding locally), then a band structure calculation, then a slow, paced working back to a chemical explanation. What could be more chemical than an oxidation-reduction reaction **15**, which is what is going on in the solid? Throughout this progression, one essential analytical tool of quantum chemistry, the population analysis, is used. There is nothing special here about the extended Hückel method, any and all better computational procedures may be used. But the essence of what we are doing is to tie this quantum chemical analytical tool to the solid state methodology of band structures and densities of states.

This can be done for any solid, of any dimensionality, for surface-adsorbate complexes just as for complex bulk and simple one-dimensional systems.

Acknowledgment. We are grateful to the National Science Foundation for its support of this work through grant CHE 8406119 and grant DMR 8217227 A02 to the Materials Science Center at Cornell University. Thanks are due to the members of our research group. C. Z. would especially like to thank Tim Hughbanks and Sunil Wijeyesekera for sharing their expertise with us. The computational procedures we use owe very much also to the work of Myung-Hwan Whangbo and Miklos Kertesz. We are grateful to Jane Jorgensen and Elisabeth Fields for their expert graphics; and to Eleanor Stagg for the typing.

References

(1) For some papers with leading references:
(a) Marchand, R.; Jeitschko, W.; J. Solid State Chem., **1978**, 24, 351; Jeitschko, W.; Jaberg, B. ibid **1980**, 35, 312; Hofmann, W. K.; Jeitschko, W. ibid **1984**, 51, 152.
(b) Pearson, W. B. ibid **1985**, 56, 278.

(2) A detailed discussion of these molecules, with further references, is given by R. Hoffmann and C. Zheng, J. Phys. Chem., 00, 0000 (1985); C. Zheng and R. Hoffmann, Inorg. Chem., to be published.

(3) Mewis, A. Z. Naturf., 1980, 35b, 141.

(4) The modern classical solid state physics text is a Cornell production: N. W. Ashcroft and N. D. Mermin, "Solid State Physics", Holt, Rinehart and Winston, New York (1976). Introductions eminently accessible to chemists may be found in T. A. Albright, J. K. Burdett, and M.-H. Whangbo, Orbital Interactions in Chemistry, Chapter 20, Wiley-Interscience, New York (1985).; J. K. Burdett, Progr. Sol. State Chem., 15, 173 (1984); M.-H. Whangbo, Accounts Chem. Res., 16, 95 (1983).

(5) Some other applications of the COOP curves may be found in:
(a) Wijeyesekera, S. D.; Hoffmann, R. Organometallics, 1984, 3, 949.
(b) Kertesz, M.; Hoffmann, R. J. Am. Chem. Soc., 1984, 106, 3453.
(c) Saillard, J.-Y.; Hoffmann, R. ibid 1984, 106, 2006.

(6) Minkin, V. I.; Minyaev, R. M. Uspekhi Khimii 1982, 51, 586; Russ. Chem. Rev., 1982, 51, 332.

(7) See Klemm, W. Proc. Chem. Soc. (London), 1959, 392: Schäfer, H.; Eisenmann, B.; Müller, W. Angew. Chem. Int. Ed., 1973, 12, 694 and references therein.

CLUSTER-SURFACE ANALOGY: NEW DEVELOPMENTS*

Evgeny Shustorovich
Research Laboratories
Eastman Kodak Company
Rochester, New York 14650
USA

ABSTRACT. The cluster-surface analogy for surface chemisorption states is discussed in light of recent developments in the theory of chemisorption. Emphasis is put on analytic modeling based on general principles such as perturbation theory and bond-order conservation. The chemisorption phenomena considered include the nature and periodic trends of the heat of atomic and molecular chemisorption, preferred adsorbate sites, and the activation barriers for surface migration and dissociation. The indirect character of cluster modeling is stressed, and some of its principal limitations are delineated.

1. INTRODUCTION

When the cluster-surface analogy for surface chemisorption states was articulated a decade ago, it served as a conceptual bridge between the mature realm of molecular coordination chemistry and the basically virgin lands of surface chemistry. Earl L. Muetterties was among the first to suggest and develop this analogy [1-4], which provided guidance in a choice of model systems and of criteria for comparison of the two regimes. It is well documented that there is no discontinuity along the series small clusters → large clusters → small metal particles → large metal particles → single-crystal surfaces [3,4]. This continuity is almost self-evident, however. What is not evident and constitutes the heuristic value of the cluster-surface analogy is the basic tenet that small transition-metal clusters geometrically resembling a tiny piece of a metal surface can model chemisorption phenomena. This belief came

*Dedicated to the memory of Earl L. Muetterties (1927-1984).

A. Veillard (ed.), Quantum Chemistry: The Challenge of Transition Metals and Coordination Chemistry, 445–464.

from the fact that there is reasonably close correlation between ligand-to-cluster and ligand-to-surface stereochemistry, bond distances, and bond energies [3,4]. On the other hand, the chemistry of these metal-ligand bonds (the subject the chemist cares most about) appears to be quite different in the two regimes [2-4]. A good example is heterogeneous catalytic reactions, most of which (e.g., the ammonia synthesis $N_2 + H_2 \rightarrow NH_3$ and the Fischer-Tropsch synthesis $CO + H_2 \rightarrow C_nH_m + C_nH_mO_p$) do not have analogs at the molecular (organometallic cluster) level.

Historically, most theoretical modeling of chemisorption was and is made on small metal clusters with the tacit assumption that what is good for clusters is good for surfaces. Accordingly, the model projections are considered to be the more reliable the more comprehensive and accurate is the computational technique used. But suppose that in the two regimes the rules of the game are different. Then, what is the good of polishing one's English if, after all, one should speak French? Or perhaps Chinese?

The recent years witnessed tremendous advances in studying chemisorption phenomena, which challenged current theoretical constructs. Several examples may illuminate the point. Atomic chemisorption patterns may differ greatly from the molecular patterns. For instance, for atomic radicals A such as H, O, and N, the heats of chemisorption Q_A monotonically decrease from left to right along a particular transition series and typically down a column [3-5], as illustrated in Table I. At the same time, for molecules AB such as CO or NO, the periodic changes in Q_{AB} are nonmonotonic and show even the reverse trend compared with adatoms (see Table I).

Cluster-type calculations can hardly address these periodic regularities at all, simply because at the molecular level the trends may be quite different. For example, the dissociation energy of diatomics M-H distinctly increases from Ni to Pt (68 vs. 80 kcal/mol [4], p. 92) but Q_H slightly decreases from Ni(111) to Pt(111) (63 vs. 60 kcal/mol in Table I). The current band-structure calculations were able to describe the monotonic atomic patterns but failed to reproduce the nonmonotonic molecular behavior [6]. Another problem is the preferred chemisorption site, which always corresponds to the highest (hollow) coordination for atoms [3-6] but typically to the lowest (on-top or twofold bridge) coordination for molecules such as CO or NO [3-7]. No satisfactory explanation was given to this conspicuous difference.

TABLE I. Heats of Chemisorption Q for Some Atomic and Molecular Absorbates. Examples are given for ordered transition-metal surfaces of the highest atomic density with the stated Miller indices. The first three metal entries compare changes within a column, and the latter three compare changes across the 5d series [3-5].

Metal surface	Q (kcal/mol) for				
	H	O	N	CO	NO
fcc Ni(111)	63	115[a]	135	27	25
fcc Pd(111)	62	87	130	34	31
fcc Pt(111)	60[b]	85	127	32	27
fcc Ir(111)	63	93	127	34	20
bcc W(110)	68	104-129	155	27	[c]

[a] J. B. Benziger (private communication). [b] R. J. Madix (private communication). [c] Dissociated.

Another theoretical challenge is the nature of activation barriers for surface migration and dissociation of adsorbates. Experimentally, these barriers show approximately linear dependence on the relevant heats of chemisorption. By itself, it is not very remarkable: everything that varies does it linearly to first approximation. The real puzzle is the values of the proportionality coefficients confined to very narrow ranges, 0.1-0.2 for atomic migration [4-6] and 1.4-1.7 for A_2 dissociation (cf. Table III). Moreover, the A-B dissociation barrier ΔE^*_{AB} shows no correlation with the molecular heat of chemisorption Q_{AB}, which, in turn, does not correlate with the A-B stretching frequency ν_{AB}. Thus, contrary to popular beliefs (and presumptions of many theoretical works on this subject), the propensity of surfaces to dissociate AB, say CO or NO, has nothing to do with how strongly AB is bound to the surface and how strongly AB is destabilized under chemisorption (see below).

The purpose of theory is to interrelate entities, not to simply describe them. Earl Muetterties firmly believed that there should be some unifying principle behind the variety of seemingly disparate chemisorption phenomena. During our collaborative work, Earl encouraged us to search for such a principle, and in this lecture I will briefly describe our attempts in this direction. I will show that the mentioned experimental puzzles as well as many other aspects of chemisorption can be explained coherently if we assume one ruling principle of chemisorption bonding, namely <u>bond-order conservation</u> (BOC). In doing so, I will emphasize the analytic modeling [6,8-10], whose results are rigorous and as such can be directly compared with experiment. Importantly, these results are expressed in terms of observables only, namely the atomic Q_A (Q_B) and molecular Q_{AB} heats of chemisorption, the most fundamental parameters that determine the course of surface reactions. Because knowledge of the variations of Q_A and Q_{AB}, especially their periodic behavior, is critical in this respect, we have also developed an analytic model [6,11], based on perturbation theory (PT), to treat these aspects as well.

Our modeling relies on some fundamental solid-state results, and I will summarize them first. Then I will highlight the major findings of the PT and BOC modeling and its conceptual implications. Finally, I will comment on the present status of the cluster-surface analogy.

2. SOME BASIC FACTS ABOUT TRANSITION-METAL SURFACES

Transition-metal M lattices are so densely packed that for both bulk and surface atoms, the M-M interactions are effectively quasi-spherical. The main supporting arguments are the following: (1) Both transition and simple s^1 (IA and IB group) metals have the same (hcp, fcc, bcc) crystal lattices. (2) For various cases of transition and simple metallic binding, including <u>chemisorption</u>, there exists an apparently universal relation between binding energy and interatomic spacing

[12,13]. (3) The d-orbital anisotropy is averaged out, and for many purposes the d band may be effectively represented by a degenerate s-type band [13,14]. (4) Details of the d density of states appear to be insignificant for the cohesion energy E_c (which is the integral over all the d-band states), so that the E_c regularities for transition metals and their alloys can be well reproduced within the constant d-density approximation [15].

The major feature of the metal band structure is the presence of the Fermi energy E_F separating the occupied part of the d band of width W^{occ} from the vacant part of width W^{vac} ($W = W^{occ} + W^{vac}$). The Fermi energy practically does not change under chemisorption. Therefore, the value of Q will include a contribution from the direct metal-adsorbate interaction, similar to that in clusters, and a contribution from the redistribution of electron density required to maintain the constant E_F, which is specific for chemisorption bonding. As a result, the structure of Q appears to differ for donor, acceptor, and radical adsorbates, as do their periodic variations.

3. PERTURBATION THEORY MODELING

3.1 Periodic Regularities of Q

Within the Hückel-type approximation and assuming the constant d-density [15], the first-order perturbation results are as follows [11]:

1) For a lone-pair donor adsorbate, Q^D is roughly proportional to the metal d-hole count $N_h = 10\, W^{vac}/W$ and inversely proportional to $E_F - \varepsilon_A$, namely,

$$Q^D \propto \frac{\beta^2 N_h}{E_F - \varepsilon_A} \tag{1}$$

where β is the Hückel resonance integral and ε_A is the lone-pair energy. Because $|E_F|$ increases only slightly from left to right along the transition series but N_h decreases significantly, one can expect Q to decrease monotonically in this direction.

2) For a vacant-orbital acceptor adsorbate, Q^A may depend on how close the orbital energy ε_A^* is to E_F and the value of the resonance integral β^*. In particular, for moderate acceptors where $\beta^*/(\varepsilon_A^* - E_F) \leq 1$, we have

$$Q^A \propto \frac{\beta^{*2} N_d}{\varepsilon_A^* - E_F} \qquad (2)$$

that is, Q^A is proportional to the d occupancy $N_d = 10\ W^{occ}/W$, which is just opposite to the dependence of Q^D on N_d in eq. (1) because $N_h = 10 - N_d$. For strong acceptors with $\beta^*/(\varepsilon_A^* - E_F) >> 1$, the dependence of Q^A on N_d and N_h appears to be nonmonotonic and may show parabolic character, namely,

$$Q^A \propto \beta^* N_d N_h \qquad (3)$$

3) For a k-valent radical adsorbate, in which singly occupied orbitals lie below the Fermi level ($\varepsilon_A < E_F$),

$$Q^R = (E_F - \varepsilon_A)^2/2U_A + kn\beta^2[\ln(W^{vac}W^{occ}/\beta^2) + 3/2]/W \qquad (4)$$

where U_A and n are scaled parameters. Quantitative estimates made with eq. (4) are highly accurate, errors being typically less than 5-10% [11b]. Because $|E_F|$ increases monotonically from left to right within groups VI-VIII of the periodic table, the E_F-dependent term in eq. (4) decreases in this direction. We can also foresee that Q^R will increase as k increases but at less than the first power, because only the second term in eq. (4) is k-dependent.

To summarize, periodic regularities of Q for atoms are expected to be simple, namely a monotonic decrease, but for molecules to be complex, depending on whether the admolecule is effectively a donor, a weak acceptor, or a strong acceptor. Thus all the observed periodic trends in Q (cf. Table I) can be understood in terms of eqs. (1)-(4).

3.2 The Crucial Role of the Antibonding Adsorbate Orbitals

In reality, a molecule AB has both occupied and vacant valence orbitals, the former being bonding, σ or π, but the latter antibonding, σ* or π*. To decide which contribution, donor or acceptor, will prevail, one should analyze both the numerator and the denominator in eqs. (1) and (2). For transition metals, typically $|E_F| = 4.5$ to 5.5 eV [11], so that $E_F - \varepsilon_A > \varepsilon_A^* - E_F$. Furthermore, for the middle and later transition metals, $N_d > N_h$. Finally, the antibonding σ* and π* orbitals overlap with the metal d orbitals more strongly than their bonding σ and π counterparts, $|\beta^*| > |\beta|$, because of the normalization LCAO MO conditions [6,16]. We conclude that both the numerator and the denominator in eqs. (1) and (2) favor acceptor bonding. Thus, not only π-bonded molecules AB such as N_2 or CO but also σ-bonded molecules such as H_2 (or the C-H bond in CH_4) will typically behave as acceptors on transition metal surfaces. In this case, the antibonding (out-of-phase LCAO MO) AB orbital, π* or σ*, will be the major source of the M-AB interaction, so that the additive M-A and M-B contributions to Q_{AB} will typically be of the opposite sign, the former being attractive but the latter repulsive. To minimize the M-B destabilizing contribution, the M-B distance should be as large as possible, so that on a flat metal surface AB is expected to be coordinated upright, namely, via the less electronegative end, say C in CO or N in NO [11c,16]. This stereochemical criterion of the acceptor AB/M bonding is in agreement with extensive experiment [3-5,7].

From eqs. (1) and (2) it follows that the Fermi level is an analog of the frontier d orbitals (HOMO-LUMO) in clusters. Since atomic d-orbital energies are much lower than E_F, the cluster modeling with conventional parameters will always underestimate the acceptor ability of AB and overestimate its donor ability.

4. BOND-ORDER-CONSERVATION MODELING

4.1 The Model Assumptions

Migration and dissociation of an adsorbate X = A or AB involves changes in its coordination mode M_n-X (where n is the coordination number) and in the M-X distances r accompanied by changes in the M_n-X total energy E. To describe these processes, one should use a general model potential relating E to r or some convenient function of r, for example, the two-center M-A bond order χ

$$\chi = \exp[-(r - r_o)/a] \tag{5}$$

where r_o and a are some constants. The possible chemisorption sites should correspond to the total energy minima, so that the model potential must include both attractive and repulsive forces. The simplest general function of the kind is the (two-center) Morse potential including only linear and quadratic terms in χ, namely,

$$E(\chi) = -Q_o(2\chi - \chi^2) \tag{6}$$

where Q_o is the M-A equilibrium bond energy. $E(\chi)$ [eq. (6)] has only one minimum at the equilibrium distance r_o when the bond order $\chi = 1$, by definition [eq. (5)]. Since $\chi = 1$ is the unique stationary point, we will assume the conservation of $\chi = 1$ (normalization to unity) to be essential for the Morse-type description of many-center M_n-A interactions, which we will treat as pairwise additive. Finally, to make our results analytically soluble, we will limit the M_n-A interactions to n nearest neighbors. For instance, for A/fcc(100), the maximum n = 4 can be reached in the hollow site but n = 2 or 1 in the bridge or on-top sites, respectively.

To summarize, our model assumptions are the following: (i) Each two-center M-A interaction is described by the Morse potential [eqs. (5) and (6)]. (ii) For a given M_n-A, n two-center M-A interactions are additive. (iii) Along a migration path up to dissociation, the total

M_n-X bond order is conserved and normalized to unity. (iv) For a given M_n-X, n is limited to nearest neighbors.

The last assumption reflects the known efficiency of the nearest-neighbor approximation in many problems of metallic binding [13]. One can add that in the parallel five-layer metal-film calculations where n exceeded 1000 [17], all our analytic (nearest-neighbor) conclusions have been corroborated. The critical model assumption is BOC in its pairwise additive form. For various three-center A...B...C gas-phase interactions, such a form ($x_{AB} + x_{BC} = 1$) was shown to be very reasonable [18]. We postulate the similar BOC for many-center M_n-A interactions because each two-center M-M and M-A interaction appears to be quasi-spherical (see above). Anyway, the rules of the game are strictly defined by the model assumptions (i)-(iv), and the results obtained are rigorous, sometimes even exact. Consider now the major findings.

4.2 Atomic Migration

When a single atom A is chemisorbed within a unit mesh M_n-A [8], the maximum bond energy Q_n (the minimum total energy $E_n = -Q_n$) is reached for equivalent two-center M-A interactions at the distance [cf. eq. (5)]

$$r_n = r_o + a \ln n \tag{7}$$

when

$$Q_n = Q_o(2 - 1/n) \tag{8}$$

Here Q_o is the maximum M-A bond energy in the on-top site. Because Q_n increases monotonically with n, on highly symmetric flat surfaces such as hcp(001), fcc(111), fcc(100), and bcc(100), adatoms will always occupy the highest-symmetry sites (C_{3v} or C_{4v}) in the hollow depressions. Indeed, on the surfaces in question, adatoms as varied as H, C, N, O, S, Se, Te, Cl, and I (as well as Na and Cd) have been invariably found in the hollow sites of maximum coordination [3-5]. From eq. (8) it follows that the on-top site (n = 1, $Q_n = Q_o$) is very unfavorable energetically, but the energy differences between various hollow sites (n = 3, 4, or 5 when Q_n = 1.67, 1.75, or 1.80 Q_o) are rather small

($\Delta Q/Q < 0.05$). Thus, the model predicts that for different surfaces of the same metal, the anisotropy $\Delta Q/Q$ should be of the order of a few percent, in good agreement with experiment [3-5].

Obviously, the lowest-energy pathway between two adjacent unit meshes M_n will be that along the twofold axis, namely, hollow (C_{nv}) → bridge (C_{2v}) → hollow (C_{nv}) with the barrier

$$\Delta E_n^* = kQ_n \quad \text{for } k = \frac{n - 2}{4n - 2} \tag{9}$$

The coefficient k monotonically increases with n when the surfaces become more open. For the typical coordinations n = 3-5, k = 0.10-0.17, in excellent agreement with the experimental range 0.09-0.18 [4,5,19].

4.3 Molecular Migration

Now let us turn to the more complex and interesting case of molecular AB chemisorption [8,10]. Although the analytic expressions are rather cumbersome (and will not be reproduced here), they permit some unambiguous conclusions. First of all, the migration M_n-AB energy profile can have several energy minima, so that the preferred coordination mode becomes parameter dependent. Importantly, the bond-energy difference ΔQ between the n-fold M_n-A-B and on-top M-A-B sites (an upright AB with the A end down) is the sum of two opposite terms, $\Delta Q^{(1)} > 0$ and $\Delta Q^{(2)} < 0$, namely,

$$\Delta Q = \Delta Q^{(1)} + \Delta Q^{(2)} \tag{10}$$

Thus, if $\Delta Q > 0$, AB will prefer the n-fold site, but if $\Delta Q < 0$, the on-top site will be preferred. The absolute values of both terms in eq. (10) increase as n increases, so that ΔQ may be of any sign and $|\Delta Q|$ should be small. Since $|\Delta Q^{(2)}|$ increases with n more rapidly, the highest coordinations $n > 3$ appear to be unfavorable. Thus, contrary to adatoms, admolecules AB will typically prefer low-coordination sites such as on-top (n = 1) or bridge (n = 2), with an easy interchange between possible chemisorption sites.

Another important point to make [10] is that, for a given M, as n increases, the equilibrium AB bond order χ_{AB} will always decrease monotonically and so will the stretching frequency ν_{AB}. The conclusion is that the value of ν_{AB} is indicative for the coordination site M_n-AB but the value of Q_{AB} is not, and there cannot be definitive interrelations between Q_{AB} and ν_{AB}.

The best defined and most extensively studied AB/M systems are those of chemisorbed CO. On flat surfaces, CO has nearly always been found to stand perpendicular to the surface [3-5,7,20]. Table II lists registries of CO on various flat surfaces of hcp Re, Ru, and Os and fcc Ni, Pd, Pt, Ir, and Rh. We see that the on-top sites are typical but the hollow (n = 3) sites are exceptionally rare [and have never been found for n > 3, say for fcc(100) or bcc(100)]. The energy differences among the on-top, bridge, and hollow (n = 3) sites, however, are usually so small that at higher coverages and temperatures some (or all) these

TABLE II. CO Registries on Some Flat Surfaces

Surface	Coverage Monolayer	Preferred Site[a,b]	Initial Q, kcal/mol[b]
hcp Re(001)	<0.67	T	
Ru(001)	<0.67	T	34
Os(001)	<0.63	T	
fcc Ni(111)	<0.5	B + T	27
	>0.5	T	
(100)		T	30
Pd(111)	<0.33	H	34
	0.33-0.50	B	
(100)	<0.50	B	37
Pt(111)	<0.17	T	32
	>0.17	T + B + H	($\Delta Q < 1$)
(100)	>0.05	T	38
Ir(111)	<0.67	T	30
fcc Rh(111)	<0.33	T	31
	0.33-0.75	T + B	
Rh(100)		T	29

[a]T (on-top), B (bridge), H (hollow). [b]Refs. 3-5, 7, 21, 23.

sites may coexist [5,7,20,21]. For example, for CO/Pt(111), ΔQ is <1 kcal/mol in the order of Q on-top > bridge > hollow [21a]. Since here Q_{CO} = 32 kcal/mol [21b], the anisotropy $\Delta Q \simeq 3\%$, which makes all the CO sites on Pt(111) practically isoenergetic. From Table II it is clear that the values of Q_{CO} are not indicative of the coordination M_n-CO sites. At the same time, the values of ν_{CO} show distinct correlations with n, commonly ν_{CO} = 1810-1836, 1870-1950, and 1990-2110 cm^{-1} for the hollow (n = 3), bridge, and on-top sites, respectively [3,4,7]. All these observations are in agreement with the model predictions.

We said that the M_n-AB total energy should change nonmonotonically with n and there can be several minima very close in energy. If so, the migration barrier ΔE^* should be larger than just the energy difference ΔQ between two relevant chemisorption sites (minima). This model conclusion is consistent with the recent findings for CO/Pt(111) where the barrier ΔE^* = 7 kcal/mol [21c] conspicuously exceeds ΔQ = 1 kcal/mol [21a].

4.4 The Atomic vs. Molecular Heats of Chemisorption

The BOC analysis also allows one to interrelate the molecular Q_{AB} and atomic Q_A (Q_B) heats of chemisorption via the gas-phase dissociation energy D_{AB}. When AB stands upright with the A end down, the contribution to Q_{AB} from the B end can be neglected, and then [10]

$$Q_{AB} \leq \frac{Q_{0A}^2}{D_{AB} + Q_{0A}} \tag{11}$$

where Q_{0A} is the maximum two-center M-A bond energy, which relates to the atomic heat of chemisorption Q_A as [cf. eq. (5)]

$$Q_A = Q_{0A}(2 - 1/n) \tag{12}$$

Typically, D_{AB}/Q_A = 1-2, so that for the usual coordinations n = 3-5 we have $Q_{AB} \simeq (0.1\text{-}0.2)Q_A$. In other words, Q_{AB} will be smaller than Q_A by a factor of 5-10, in general agreement with experiment [4,5]. Because the neglected B contribution to Q_{AB} [eq. (11)] is repulsive for an acceptor AB (see above), the periodic changes in Q_{AB} may be nonmonotonic, which is consistent with experiment (cf. Table I) and the earlier PT arguments [11]. From eqs. (11) and (12) it immediately follows that AB will be coordinated to M through the atom whose heat of chemisorption is larger, namely, M-A-B if $Q_A > Q_B$. For adatoms within the first periodic row, the values of Q_A increase as the atomic valency (the number of unpaired electrons) increases [5,11], in the order $O < N < C$. Thus, CN and CO should always be coordinated via C and NO via N, in full agreement with experiment [3-5,7] and with the earlier PT arguments [11c]. For CO and NO chemisorption, eq. (11) gives the values of Q_{AB} with the typical error 10-15% [10]. The same estimates show that the coordinations via oxygen (M-O-C or M-O-N) would be less favorable by 10-15 kcal/mol. Equation (11) provides a quick way to estimate Q_{AB} if Q_A is known, and vice versa. This is of special importance when the values of Q are difficult to obtain. For example, the carbon heat of chemisorption Q_C has been measured accurately only for Ni(111) and Ni(100) [22], but the molecular CO heats of chemisorption are available for many surfaces [3-5].

4.5 Molecular Dissociation Under Chemisorption

Now we turn to molecular AB → A + B dissociation. At the dissociation point, by definition, the molecular AB energy hypersurface intersects the atomic A + B hypersurface. At this point, $x_{AB} = 0$, so that the dissociation barrier ΔE^*_{AB} can be defined in terms of the atomic variables only, namely [9],

$$\Delta E^*_{AB} = D_{AB} - (Q_A + Q_B) + Q_AQ_B/(Q_A + Q_B) \qquad (13)$$

In the homonuclear case A = B, eq. (13) reads as

$$\Delta E^*_{A_2} = D_{A_2} - kQ_A \quad \text{for } k = 3/2 \qquad (14)$$

that is, $\Delta E^*_{A_2}$ is linearly dependent on Q_A with slope k = 3/2. Table III summarizes experimental values of k for dissociation of H_2, N_2, and O_2 on various surfaces of Fe, Ni, Cu, W, and Pt. Those values lie within the range 1.4-1.7, close to the theoretical value of 1.5. Unfortunately,

TABLE III. Some Empirical Estimates of k[a]

A_2	Surface	Q_A	D_{A_2}	$\Delta E^*_{A_2}$	k
H_2	Fe(111)	62	103	≈0	1.66
	Ni(111)	63		2	1.60
	Cu(100)	58		5	1.69
N_2	W(110)	155	228	10	1.44
	Fe(110)	140		8	1.57
	Fe(100)	140		2.5	1.60
	Fe(111)	140		-0.8	1.63
O_2	Pt(111)	85	119	-1	1.43

[a]From: $\Delta E^*_{A_2} = D_{A_2} - kQ_A$. All energies in kcal/mol [9].

almost no experimental data on the heteronuclear ΔE^*_{AB} are available to directly verify eq. (13). However, the relevant estimates for dissociation of NO (shown in Table IV) nicely correlate with the observations that at room temperature NO dissociation can easily be detected on Ni,

TABLE IV. Estimates of ΔE^*_{NO} for NO Dissociation[a]

Surface	Q_N	Q_O	ΔE^*_{NO}	Exper. Behavior	Q_{NO}
Ni(111)	135	115	-25	Diss.	25
Rh(111)	128	102	-10	Diss.	
Ir(111)	127	93	-3	Diss.	20
Pd(111)	130	87	≈0	Diss. ⇄ Non-Diss	31
Pt(111)	127	85	2	Non-Diss	27

[a]Eq. (13), D_{NO} = 163. All energies in kcal/mol [9].

Rh, and Ir surfaces but adsorption of NO is predominantly molecular on Pd and especially on Pt surfaces [23].

From eqs. (13) and (14), it follows that for a given D_{A_2} (D_{AB}), the larger Q_A (Q_B), the smaller $\Delta E^*_{A_2}$ (ΔE^*_{AB}). Clearly, the periodic trends for ΔE^* will be just the reverse of those for Q_A (Q_B), namely ΔE^* will typically increase from left to right along the transition series and down the column, the trends being most pronounced for the multiply bonded molecules such as N_2, CO, and NO. Indeed, for these molecules, the experiment shows distinct border lines within the periodic table separating active metals (typically, upper and left) from the inactive metals (typically, close to the right bottom corner, especially Pt, which is the low extreme of Q_A [23,24]).

Within the BOC model, ΔE^*_{AB} [eq. (13)] does not explictly depend on Q_{AB}. Although this conclusion looks paradoxical at first glance, it is consistent with experiment. We stressed above that molecular heats of chemisorption (unlike the atomic ones!) vary slightly from left to right along the transition series. Moreover, they vary nonmonotonically and can even increase (cf. Table I), which shows no correlation whatsoever with the discussed periodic trends in ΔE^*_{AB}. In particular, both NO and CO have smaller Q_{AB} on Ni(111) than on Pt(111) (cf. Table I), though the former surface is much more active for AB dissociation than the latter [23,24] (cf. also Table IV). One should add that the promoter and poisoning effects on AB dissociation can also be explained by the relevant BOC changes in the atomic A and B energy profiles [10]. In particular, the presence of a modifier D in the vicinity of AB should decrease χ_{AB} and thus ν_{AB}. This intrinsic AB destabilization will be especially pronounced for the attractive D-AB interactions when the D-AB distance is short, for example, for alkali modifiers such as D = Na or K [10].

This BOC conclusion provides an alternative to the common notion that a decrease in ν_{AB}, say ν_{CO}, under coordination is due to the $M_n \rightarrow$ CO electron transfer q_n into the antibonding $2\pi^*$ orbitals of CO (the larger n, the larger q_n) assisted by electropositive (donor) atoms such as K [3-5]. The difficulty with this explanation is that the vibrational frequency of CO^- (with one electron in the $2\pi^*$ orbital) is reduced by only ~300 cm^{-1} [25a], but in clusters [3,4] and especially on surfaces this decrease, depending on the environment, may be much larger, up to ~600 cm^{-1} [25b] or even ~1000 cm^{-1} [20]. Clearly, the $M_n \rightarrow CO$ electron transfer cannot be the only or even major factor responsible for such a dramatic destabilization of CO. Consistent with this is the growing evidence that the $M_n \rightarrow A$ charge transfer is insignificant even for chemisorption of strongly electronegative atoms such as H, Cl, or S [26]. Here the surface polarization effects, being opposite to the electrostatic ones, appear to be comparable and sometimes even more important [6,27].

5. CONCEPTUAL IMPLICATIONS FOR THE CLUSTER-SURFACE ANALOGY

Now we return to the key questions: (1) Why are the metal-ligand bond energies and distances similar for small clusters and surfaces but their chemistries so different? (2) What can be done for better cluster modeling of chemisorption?

To answer the first question, let us remember that for the surface M_n-A bonding, r_n and Q_n [eqs. (7) and (8)] are rather insensitive to the number of interacting metal atoms when $n > 2$. In clusters, the qualitative picture seems to be similar. By contrast, the dissociation barrier ΔE^*_{AB} apppears to be very sensitive to the size and geometry of the surface active site. At the dissociation point $\chi_{AB} = 0$, the A and B atoms should be far apart in space, presumably within different unit meshes M_n. Thus, it may be inadequate in principle to simulate AB dissociation within small metal clusters, especially if no constraints are imposed on where the molecular AB and atomic A + B potential profiles can intersect, not to mention the accuracy of the profiles themselves. Remember also that ΔE^*_{AB} shows no definitive correlation with Q_{AB} and/or ν_{AB}, so that these characteristics cannot be used as surrogates to judge the propensity of surfaces to dissociate AB.

The second question is critical for the heuristic value of the cluster-surface analogy. Here the good news is that the nearest-neighbor approximation is efficient in describing a variety of chemisorption phenomena [8-10,17]. The bad news, however, is that these nearest neighbors, being the tip of the metal iceberg, may have not much in common with their free-cluster prototypes. Remember that in small transition-metal clusters, the d-levels are significantly split in energy and their occupancies are different, so that the metal-metal and metal-ligand interactions may be far from quasi-spherical. From this viewpoint, the conventional wisdom of calculating the model clusters--the more sophisticated the method, the better--may not be the best way to go. Modeling is an art that reflects the reality in some indirect and transformed way, not naturalistic but rather surrealistic. For example, the cluster analogs of eqs. (1) and (2) would look like a mermaid, the upper part (the β terms in the numerators) coming from real clusters but the lower part (the E_F terms in the denominators) from real surfaces. Indeed, in the semiempirical cluster modeling it is usually beneficial to adjust the atomic d-orbital energies to fit experiment for the metal bulk (surface) [28].

Another example of the "cluster surrealism" is the use of the spherically averaged d^{n-1} pseudopotentials, reducing the valence shell of each metal atom to the s^1 configuration, as was done, for instance, to model chemisorption on various Ni clusters [29]. In my opinion, the success of this s^1 modeling (e.g., in projecting the hollow preferred sites for adatoms) is due to the quasi-spherical character of the M-M interaction (as we discussed above), but the latter is not literally the s-s bonding. Not surprisingly, this s^1 pseudopotential modeling has also produced serious errors, both quantitative and qualitative [30]. Again, the surrealistic irony is that the "most accurate" all-electron calculations of the cluster Ni_5-CO to mimic CO/Ni(100) gave Q_{CO} = 13 kcal/mol, which fell short of the experimental value of 30 kcal/mol, but the "less accurate" pseudopotential calculations produced almost perfect fit [29c,30].

Ultimately, the value of the cluster-surface analogy is determined by how much surface chemistry can be understood and projected from the knowledge of molecular coordination chemistry. I hope that the chemisorption modeling described may further illuminate the grounds and boundaries of the analogy. I cannot find a better way to conclude than to refer to Earl Muetterties' attitude: "Coordination chemistry principles, especially those derived from molecular cluster chemistry, can be invaluable conceptually in attempts to delineate the molecular details of metal surface chemistry--provided that the application of such concepts is done with considerable caution and reserve" ([4], p. 122).

References

1. E. L. Muetterties, Bull. Soc. Chim. Belg. **84**, 959 (1975); Science **196**, 839 (1977); Angew. Chem. Int. Ed. Engl. **17**, 545 (1978); Recherche **11**, 1364 (1980); Isr. J. Chem. **20**, 84 (1980); ACS Symp. Ser. **155**, 273 (1981).

2. E. L. Muetterties and J. Stein, Chem. Rev. **79**, 479 (1979).

3. E. L. Muetterties, T. N. Rhodin, E. Band, C. F. Brucker, and W. R. Pretzer, Chem. Rev. **79**, 91 (1979).

4. E. L. Muetterties and R. M. Wexler, Surv. Progr. Chem. **10**, 61 (1983).

5. G. Ertl in: 'The Nature of the Chemical Bond,' T. N. Rhodin and G. Ertl, Eds. (North-Holland, Amsterdam, 1979), Chapter 5.

6. See a discussion in: E. Shustorovich and R. C. Baetzold, Science **227**, 876 (1985).

7. (a) F. M. Hoffmann, Surf. Sci. Rep. **3**, 107 (1983); (b) J. P. Biberian and M. A. Van Hove, Surf. Sci. **118**, 443 (1982); **138**, 361 (1984).

8. E. Shustorovich, J. Am. Chem. Soc. **106**, 6479 (1983).

9. E. Shustorovich, Surf. Sci. **150**, L115 (1985).

10. E. Shustorovich, Surf. Sci., in press.

11. (a) E. Shustorovich, Solid State Commun. **44**, 567 (1982); (b) E. Shustorovich, J. Phys. Chem. **88**, 1927, 3490 (1984); (c) E. Shustorovich, R. C. Baetzold, and E. L. Muetterties, ibid. **87**, 1100 (1983).

12. (a) J. H. Rose, J. R. Smith, F. Guinea, and J. Ferrante, Phys. Rev. B **29**, 2963 (1984); (b) D. Spanjaard and M. C. Desjonqueres, ibid. **30**, 4822 (1984).

13. G. C. Abell, Phys. Rev. B **31**, 6184 (1985).

14. H. L. Skriver, Phys. Rev. B **31**, 1909 (1985).

15. (a) J. Friedel in: 'The Physics of Metals,' J. M. Ziman, Ed. (Cambridge University Press, London, 1969); (b) D. G. Pettifor, Phys. Rev. Lett. **42**, 846 (1979).

16. E. Shustorovich, J. Phys. Chem. **87**, 14 (1983).

17. R. C. Baetzold, Surf. Sci. **150**, 193 (1985); J. Chem. Phys. **82**, 5724 (1985).

18. See a discussion in: J. R. Murdoch, J. Am. Chem. Soc. **105**, 2667 (1983).

19. R. Gomer, Vacuum **33**, 537 (1983).

20. One of the few exceptions: for CO/Cr(110), both perpendicular and parallel geometries have been found: N. D. Shinn and T. E. Madey, Phys. Rev. Lett. **53**, 2481 (1984).

21. (a) B. E. Hayden and A. M. Bradshaw, Surf. Sci. **125**, 787 (1983); (b) B. Poelsema, R. L. Palmer, and G. Gomsa, ibid. **136**, 1 (1984); (c) B. Poelsema, L. K. Verheij, and G. Comsa, Phys. Rev. Lett. **49**, 1731 (1982).

22. Q_C = 171 kcal/mol: (a) L. C. Isett and J. M. Blakely, Surf. Sci. **43**, 493 (1974); (b) J. M. Blakely, personal communication.

23. B. E. Nieuwenhuys, Surf. Sci. **126**, 307 (1983).

24. G. Broden, T. N. Rhodin, C. Bruker, R. Benbow, and Z. Hurych, Surf. Sci. **59**, 593 (1976).

25. (a) H. Ehrhardt, L. Langhans, F. Linder, and K. S. Taylor, Phys. Rev. **173**, 222 (1969); (b) W. Eberhardt, F. M. Hoffmann, R. de Paola, D. Heskett, I. Strathy, E. W. Plummer, and H. R. Moser, Phys. Rev. Lett. **54**, 1856 (1985).

26. P. J. Feibelman and D. R. Hamann, Surf. Sci. **149**, 48 (1985).

27. (a) E. Shustorovich, J. Phys. Chem. **86**, 3114 (1982); (b) E. Shustorovich and R. C. Baetzold, Appl. Surf. Sci. **11/12**, 693 (1982).

28. See, for instance: (a) H. Tatewaki, E. Mioshi, and T. Nakamura, J. Chem. Phys. **76**, 5073 (1982); (b) B. Bigot and C. Minot, J. Am. Chem. Soc. **106**, 6601 (1984); (c) J.-Y. Saillard and R. Hoffmann, ibid. **106**, 2006 (1984).

29. (a) C. F. Melius, T. H. Upton, and W. A. Goddard, Solid State Commun. **28**, 501 (1978); (b) T. H. Upton and W. A. Goddard, Phys. Rev.

Lett. **42**, 472 (1979); (c) J. N. Allison and W. A. Goddard, Surf. Sci. **110**, L615 (1981).

30. P. S. Bagus, C. W. Bauschlicher, C. J. Nelin, B. C. Laskowski, and M. Seel, J. Chem. Phys. **81**, 3594 (1984).

CLUSTER MODEL CALCULATIONS OF THE INTERACTION OF H AND CO WITH SMALL Pd CLUSTERS

Gianfranco Pacchioni[§] and Jaroslav Koutecký
Institut für Physikalische Chemie
Freie Universität Berlin
1000 Berlin 33
Federal Republic of Germany

ABSTRACT. The interaction of hydrogen and CO with small Pd_n clusters (n = 1-4) was investigated by means of pseudopotential Hartree-Fock calculations followed by configuration interaction. The three-hollow position is found to be the most favourable site for chemisorption of H and CO on the Pd (111) surface. The MO spectra obtained from the present calculations are in excellent agreement with the photoemission spectra of H and CO on Pd. The strength and the nature of the Pd-H and Pd-CO interactions are discussed.

1. INTRODUCTION

The study of CO and H chemisorption on Pd surfaces is of fundamental importance for the deep understanding of the reaction mechanisms in Fischer-Tropsch catalysis. In this work we have studied this problem by means of the cluster-model approach performing non-empirical pseudopotential (PP) configuration interaction (CI) calculations on Pd_n-CO (n = 1-4) and Pd_m-H (m = 1-3) systems.

2. COMPUTATIONAL DETAILS

The theoretical method employed is the pseudopotential version (1) of the Multireference Doubly-Excited Configuration Interaction (MRD CI) procedure (2). The basis sets adopted are of double-zeta (DZ) quality plus polarization functions (DZP) as described in Ref. (3,4). The 5p polarization functions on Pd are found to be unimportant for the description of the Pd-H and Pd-CO bondings. Therefore, while Pd_n-CO clusters (n = 1,2) were computed with both DZ and DZP basis sets, the study of the larger Pd_n-CO systems (n = 3,4) were performed without to include p polarization functions in the basis set.

In general, the use of DZ instead of minimal basis sets is necessary in order to avoid the occurence of basis set superposition errors (BSSE). In particular, the use of single-zeta (SZ) basis sets yields

[§]Permanent address: Dipartimento di Chimica Inorganica e Metallorganica, Università di Milano, I-20133 Milano (Italy)

A. Veillard (ed.), Quantum Chemistry: The Challenge of Transition Metals and Coordination Chemistry, 465–475.

strongly bonded nacked Pd clusters as a consequence of BSSE. On the other hand, DZ basis sets correctly give weakly bonded Pd clusters in CI, while the SCF potential energy curves are repulsive (5).

One big computational problem in our study is represented by correlation. The maximum number of electrons we can correlate in our approach is 30. Therefore, for systems containing a larger number of valence electrons (e.g. Pd_4-CO) it is necessary to freeze some MOs which, for symmetry reasons, cannot interact with the CO MOs. Nevertheless, the correlation of 30 electrons is a formidable task. The MRD CI procedure makes use of an extrapolation technique which allows to treat explicitly in the secular equation only a limited number of configurations which are selected on the basis of their estimated contribution to the total energy of the system. The final CI energy is then extrapolated to the whole CI space (2). The reliability of this procedure and the error introduced in the final CI energy have been carefully analyzed for the Pd-CO and Pd_m-H systems and the results can be considered as very satisfactory (4).

The calculations for Pd_n-CO clusters were performed with 6 reference configurations. The contribution of the SCF configuration to the final CI expansion is always more than 90%. The other reference configurations chosen to generate the CI space are $\pi \rightarrow \pi^*$ excitations internal to the CO molecule. The calculations for Pd_m-H clusters are of single-reference CI (SR CI) quality.

3. CHEMISORPTION OF CO ON Pd CLUSTERS

The chemisorption of CO on Pd surfaces has been the subject of various experimental and theoretical studies. Nevertheless, there is still controversy about the mechanism of the bonding between transition metals and CO (6). According to some recent theoretical studies, the commonly accepted σ-donation π-back donation mechanism originally proposed for the rationalization of the bonding in metal carbonyl complexes, is no longer valid since the σ-interaction has prevailing repulsive character (7,8). This contradicts a recent experimental work on Pd carbonyl clusters were it was concluded that the π-interaction is unimportant and that the stabilization of the system is mainly due to the σ-donation mechanism (9).

Also the role of the metal-metal bonds in carbonyl clusters where the metal atoms are in zero-oxydation state is matter of discussion (10). The fact that every molecular Pd cluster characterized up to now always contains bridging CO ligands (11) can be interpreted as a sign of the weakness of the Pd-Pd bond and of the importance of the CO molecule for the stabilization of the molecular cluster.

3.1. Interaction energies and geometries

The clusters chosen to simulate the CO chemisorption on Pd surfaces are, beside Pd-CO, the Pd_2-CO (C_{2v}), Pd_3(equilateral triangle)-CO (C_{3v}), and Pd_4(rhombus)-CO (C_{2v}) clusters which are sections of the (111) surface of Pd and the Pd_3(isosceles triangle)-CO (C_{2v}) cluster section of the Pd (100) surface (Fig. 1). The Pd-Pd bond length was

Table 1 - Stabilities and geometries of Pd_n-CO clusters

	Pd-CO $C_{\infty v}$	Pd_2-CO C_{2v}	Pd_3-CO C_{3v}	Pd_4-CO C_{2v}	Pd_3-CO C_{2v}	exp.
D_e [a] kcal/mol	5.3	10.2	17.0	8.1	9.2	34
R_e (Pd-C) au	3.98	4.09	4.20	4.09	4.31	3.75 [b]
R_e (C-O) au	2.19	2.20	2.19	2.19 [c]	2.19 [c]	2.17 [b]

a) From CI extrapolated energies (see text). b) From Ref. (12).
c) Distance not optimized.

fixed to the nearest neighbour distance of fcc Pd crystal (5.2 a.u.). In Pd_3-CO (C_{2v}) the distance between the two Pd atoms directly interacting with CO is therefore 7.3 a.u. (Fig. 1).

The dissociation energy (D_e) of the Pd-CO molecule is 5.3 kcal/mol (Table 1) in agreement with more accurate calculations (8). In Pd_2-CO we found an interaction energy, 10.2 kcal/mol, which is about twice that of Pd-CO. The use of a DZP basis set yields only slightly larger D_e's (5.6 and 13.6 kcal/mole for Pd-CO and Pd_2-CO, respectively).

The negligible role of 5p orbitals in the bonding with CO is demonstrated also by the values of the Mulliken population analysis: the 5p orbitals are occupied by only 0.06-0.08 electrons (Table 2). The interaction energy increases in the case of Pd_3-CO (C_{3v}) cluster were the CO molecule is in a three-hollow coordination site. The D_e = 17.0 kcal/mol suggests that the interaction energy is more or less proportional to the number of Pd atoms directly bonded to CO. In Pd_3-CO (C_{2v}) the bonding is weaker than in Pd_3-CO (C_{3v}) since the CO molecule is interacting with only two Pd atoms. The third atom is practically "inert" and does not significantly contributes to the bonding. Finally, also the interaction between the Pd_4 rhombic cluster and CO is comparable with that of Pd_2-CO since the two Pd atoms at the corners of the

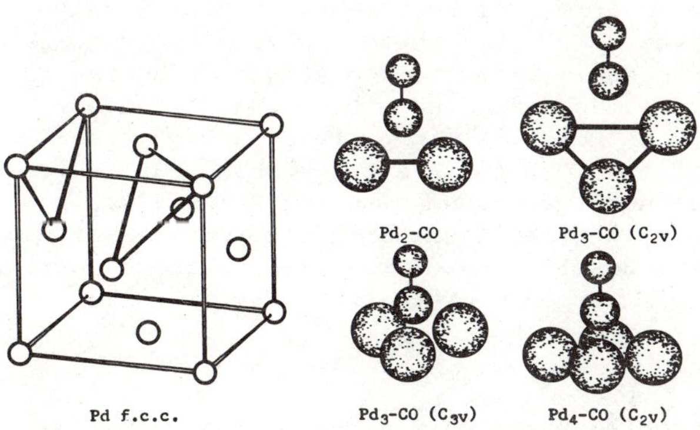

Fig. 1 - Geometries of the Pd_n-CO clusters. The two Pd_3 clusters considered are sections of the (100) and (111) Pd surfaces as shown on the left of the figure.

Table 2 - Mulliken population analysis for Pd_n-CO clusters

	Pd-CO $C_{\infty v}$	Pd_2-CO C_{2v}	Pd_3-CO C_{3v}	Pd_4-CO[a)] C_{2v}	Pd_3-CO[a)] C_{2v}	free CO $C_{\infty v}$
Pd 5s	0.173	0.149	0.153	0.139	0.094	-
5p	0.084	0.064	-	-	-	-
4d	9.768	9.781	9.799	9.780	9.837	-
net charge	-0.024	+0.006	+0.048	+0.081	+0.068	-
CO_σ	5.833	5.768	5.803	5.844	5.920	6.000
CO_π	4.142	4.244	4.342	4.281	4.217	4.000

a) The population of the two Pd atoms directly interacting with CO has been reported (see Fig. 1).

rhombus are too distant from the CO molecule to significantly overlap with the CO MOs. Therefore, the bonding in this bridge site has similar character as in Pd_2-CO.

The sequence of the computed D_e's, hollow site > bridge site> top site, is consistent with the experimental finding that at low coverages CO prefers to occupy the three-hollow positions of the (111) Pd surface; increasing the degree of coverage the CO molecules occupy the bridge and then the top positions (12).

The optimization of the distance of CO from the Pd cluster was carried out in CI. We found that the Pd-CO molecule has the shortest Pd-C distance (3.98 a.u.). In the two bridge sites considered, namely Pd_2-CO and Pd_4-CO, the optimal metal-carbon distance is the same, R_e = 4.09 a.u. . In Pd_3-CO (C_{3v}) the Pd-C bond length is 0.1 a.u. longer than in the bridge sites and even longer is R_e for the Pd3-CO (C_{2v}) cluster (Table 1). All these values are close to those found in Pd carbonyl clusters (3.8 - 4.2 a.u.) but much longer than the Pd-C bond distance 3.65 a.u. observed for CO chemisorbed on Pd (100) (13).

The simultaneous optimization of the Pd-C and of the C-O bond distances was attempted for Pd-CO, Pd_2-CO and Pd_3-CO (C_{3v}) systems. We found that, according to the experimental knowledge, the CO distance is slightly elongated in coordinated CO with respect to free CO (compare the values of Table 1 with the equilibrium bond distance 2.18 a.u. computed for the free CO molecule). However, the vibrational frequency calculated for coordinated CO in Pd-CO is higher than that of free CO, indicating a strengthening of the CO bond. A similar behaviour, which contradicts the common experience, was found also for Pt-CO (14). Recently it was shown that a full vibrational analysis is necessary in order to obtain a quantitative description of the small changes occuring in the stretching frequency of CO upon coordination (15).

3.2. Electronic structure

In Pd_n-CO we can distinguish three main group of interactions: $Pd(d_\pi)$-$CO(2\pi^*)$, $Pd(d_\sigma)$-$CO(5\sigma)$, and Pd-Pd interaction (Fig. 2).

The Pd-Pd bonding occurs mainly through d-orbitals. However, due to the closed shell nature of the Pd atom (^1S $4d^{10}$ ground state), the metal-metal bonding is very weak and is found only at CI level. The Pd_n-CO complexes studied are stable with respect to dissociation into Pd atoms and CO mainly because of the CO molecule which acts as a "glue" for the whole system. The Pd atoms in the clusters have an electronic configuration which is very close to that of the atomic ground state. In fact, not only the 5p but also the 5s orbital is weakly populated (Table 2).

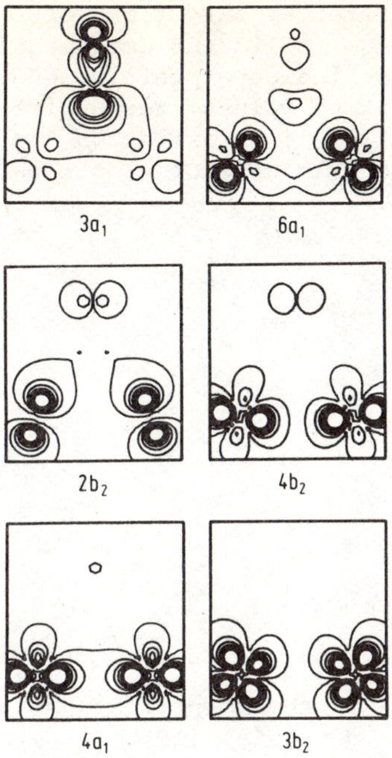

Fig. 2 - Electron density contour maps of some MOs of Pd_2-CO: σ-interaction ($3a_1$, $6a_1$); π-interaction ($2b_2$,$4b_2$); Pd-Pd interaction ($4a_1$, $3b_2$).

The nature of the σ-π interaction in Pd_n-CO clusters is very similar as in Pd-CO (8). The metal atoms are neutral or slightly positive as a consequence of the compensation of the donation back-donation mechanism. The changes in the population of the σ and π MOs of CO indicate that, while the σ-population does not change significantly with the number of Pd atoms directly bonded to CO (Pd-CO = 5.83, Pd_2-CO = 5.77, Pd_3-CO (C_{3v}) = 5.80 electrons, respectively), the total π population increases with the increasing coordination (4.14, 4.24 and 4.34 electrons, respectively)(Table 2). Therefore, we found a direct correlation between the population of the π orbitals of CO and the interaction energy.

The interpretation of the decrease of σ-population of CO, which formally corresponds to a σ-donation from the CO molecule to the metal, is controversial. Some authors have emphasized the limitation of the population analysis for this purpose. Recently it was concluded that there is no σ-donation (7). More reliable than the population analysis are the electron density difference contour maps. These maps, obtained as the difference between the electron density of the molecule and those of the separated Pd_n and CO fragments, clearly show the existence of two different regions (Fig. 3). The electron distribution, which is shown in the Fig. 3 for the Pd_2-CO system, is completely similar as in Pd-CO (8). In the region corresponding to the σ-bonding we found a decrease of the electron density while an accumulation of electronic charge can be observed in the π-bonding region. Moreover, the metal orbitals polarize away from the σ-region in order to minimize the repulsion with the 5σ electron cloud (Fig. 3).

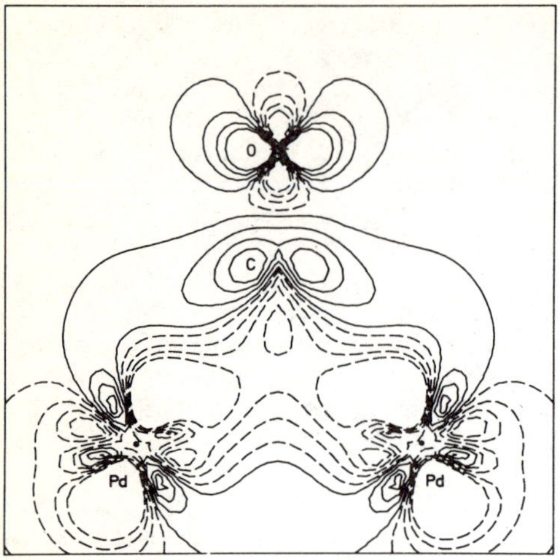

Fig. 3 - Electron density difference contour map of Pd_2-CO. Solid and dot-dashed lines indicate positive and negative values, respectively. The lines are drawn in intervals of 0.01 electrons/a.u.^{3}.

3.3. Ionization potentials

One of the commonly accepted features of CO chemisorption on metal surfaces is that the increase in the ionization potential (IP) of the 5σ MO of CO is due to the CO 5σ donation to the metal. The stabilization of the 5σ MO is an effect observed also for CO on Pd surfaces (13). In particular, the 5σ and 1π MOs, which are separated by 2.9 eV in free CO, are practically degenerated when CO is chemisorbed on Pd metal. The IP of the low-lying 4σ MO on the contrary is not influenced by the chemisorption process (13). Therefore, the values IP(1π)-IP(5σ) and IP(4σ)-IP(1π) are useful quantities for the comparison of the MO spectra of Pd_n-CO clusters with the photoemission spectra of CO on Pd.

Table 3 - Ionization potentials (eV) for Pd_n-CO clusters

	free CO		Pd-CO	Pd_2-CO	Pd_3-CO	Pd_4-CO	Pd_3-CO	exp.[a]
	exp.	calc.	$C_{\infty v}$	C_{2v}	C_{3v}	C_{2v}	C_{2v}	
IP(5σ)	14.0	15.1	16.6	16.8	16.9	16.8	15.9	∿13.2
IP(1π)	16.9	17.4	17.1	16.8	16.6	16.7	16.7	∿13.2
IP(4σ)	19.7	21.7	21.5	21.2	21.0	21.1	21.1	∿16.2
IP(1π)-IP(5σ)	2.9	2.3	0.5	0.0	-0.3	0.1	0.8	0.0
IP(4σ)-IP(1π)	2.8	4.3	4.4	4.4	4.4	4.4	4.4	3.0

a) From Ref. (13). The values are computed with respect to the vacuum by using a work function for Pd ∿5.0 eV.

The Table 3 reports the IPs of the 5σ, 1π, and 4σ MOs of free and coordinated CO. The IP(5σ) increases from 15.1 eV in gas-phase up to 16.9 eV in Pd_3-CO (C_{3v}). In Pd_3-CO (C_{2v}) cluster, where the overlap between the Pd atoms and CO is small because of the large Pd-Pd and Pd-C distances, the IP(5σ) is 15.9 eV. On the other hand, the IP(1π) decreases from 17.4 eV (free CO) to 16.6 eV (Pd_3-CO (C_{3v})). The extent of the 5σ stabilization seems to be proportional to the metal-CO (5σ) overlap. It follows that the IP(1π)-IP(5σ) value, which is 2.3 eV in free CO, is practically zero in Pd_n-CO systems. The two levels are coincident and, in one case, their order is reversed (Table 3). On the contrary, the IP(4σ)-IP(1π) quantity remains constant on going from free to coordinated CO.

Despite the limitations of this discussion based on IPs obtained without allowing any electronic relaxation (Koopmans's theorem), the trend in computed IPs is in full agreement with the photoemission spectra of CO on Pd (13).

A possible interpretation of the stabilization of the 5σ MO in chemisorbed CO is that the interaction between the very diffuse 5σ lone pair and the Pd σ-orbitals induces a polarization of the metal orbitals away from CO (Fig. 3). Because of the nuclear attraction from the Pd atoms the 5σ MO can partially expand and delocalize over the metal centers reducing in this way the Coulomb repulsion. This mechanism stabilizes the 5σ MO and produces a bonding effect which partially compensates the σ-repulsion. This explains the data of population analysis which indicate the existence of a charge transfer from CO to Pd. However, this cannot be interpreted as a sign of the formation of a σ-bonding in classical sense having the σ-interaction repulsive character (7,8).

The true and determinant bonding mechanism is the donation of electrons from the filled $d_{\parallel}$ metal orbitals into the empty $2\pi^*$ MO of CO. This mechanism is reinforced when CO is coordinated to more metal centers as shown by the present results as well as by the experimentally observed red-shift in the vibrational frequencies of μ_2-CO and μ_3-CO ligands with respect to terminal CO in transition metal carbonyl clusters.

4. CHEMISORPTION OF H ON Pd CLUSTERS

The mechanism of dissociative chemisorption of H_2 on Pd surfaces has been investigated with different spectroscopic techniques (16-18) and some of the salient features of the experimental measurements have been rationalized on the basis of the results of band structure calculations (18-20). Less intensive theoretical work on this problem has been carried out with the cluster-model approach (4,21,22).

In order to model the possible chemisorption sites on the (111) surface of Pd three different clusters have been used: a Pd_2 cluster with the H atom in a bridge position, a linear Pd_3 cluster with H directly bonded to the central Pd atom (top site), and an equilateral triangle Pd_3 cluster with H perpendicular to the center of the triangular face (three-hollow site).
Also in this case the Pd-Pd bond distances have been fixed to the value 5.2 a.u. which corresponds to the nearest neighbor separation in fcc Pd bulk.

4.1. The Pd-H molecule

The molecular ground state of the Pd-H diatomic molecule is $^2\Sigma^+$. For obvious symmetry reasons only the $4d_\sigma$, $5p_\sigma$, and 5s AOs of Pd can interact with the 1s orbital of the H atom. The interaction of $4d_\sigma$ with 1s AO generates a bonding 1σ and an antibonding 2σ MO (Fig. 4) with different character. The lowest doubly occupied 1σ MO has large 1s H and 4d Pd participations with a small 5s Pd contribution. On the other hand, the 2σ MO has mainly d_σ character with small 5s Pd and 1s H contributions. This is in agreement with the ESR study of the matrix-isolated Pd-H molecule which shows that the unpaired electron is predominantly $4d_\sigma$ with about 30% of s character (23).

A significant internal rearrangement in the electronic structure of the metal atom is shown by the energy lowering of the "passive" d_π and d_δ orbitals of Pd-H with respect to free Pd atom. These energy levels lie between those of the d-shells of the ^1S ($4d^{10}$) and of the ^3D ($4d^9 5s^1$) Pd atomic states (Fig. 4). This is mainly due to the partial occupation of the 5s AO and therefore to the lower population of the $4d_\sigma$ orbital which results in a reduced Coulomb repulsion and in a consequent stabilization of the "inert" d_π and d_δ orbitals.

To summarize, the bonding is due to the d_σ-1s interaction, but it is accompanied by a substantial rehybridization of the metal atom which assumes an atomic configuration intermediate between $4d^{10}$ and $4d^9 5s^1$.

The Mulliken population analysis indicates the existence of a small flow of charge from the metal to hydrogen. However, the population analysis is not reliable in the case of very diffuse orbitals like the 5s orbital of Pd. More significant is the value of the dipole moment, 3.2 Debye. Since the Pd-H distance is 3.02 a.u., this corresponds to a formal negative charge -0.4 on hydrogen which is much more pronounced than that resulting from the population analysis (-0.1).

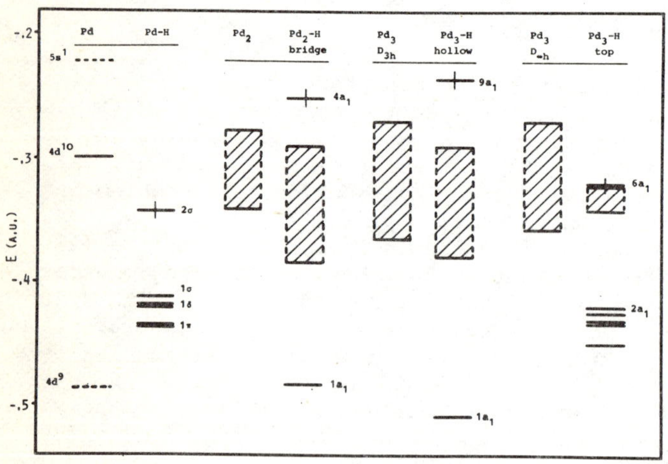

Fig. 4 - Orbital energies of the free Pd clusters and of the Pd_m-H systems. The cluster MOs with prevailing d-character have been represented by a d-band (4).

4.2. Pd_2-H and Pd_3-H clusters

A comparison of the interaction energies of the three adsorption sites considered indicates the hollow site as the preferred one. The strength of the chemisorptive bond for this position is 10 kcal/mol larger than that of the bridge site which, in turns, is slightly more favourable than the top site (Table 4).

In the optimal geometry of the Pd_3-H hollow cluster the H atom lies 1.26 a.u. above the surface with a Pd-H bond length of 3.25 a.u. . This result agrees with the angle-resolved photoelectron study of H on Pd (111) (18) in which it was shown that the H atom occupies three-fold surface sites with a Pd-H bond legnth of 3.19 a.u. . The further approach of the H atom to the cluster plane is unfavourable. An energy barrier of 10.6 kcal/mol is found when H passes the center of the Pd_3 face.

The electronic structure of the Pd_3-H top cluster resembles that of Pd-H. The bonding involves mailny the d_σ and, to a smaller extent, the 5s orbital of the central atom of Pd_3. The two terminal atoms of the linear Pd_3 cluster are not involved in the bonding with hydrogen. On the other hand, the electronic structure of Pd_2 and Pd_3 clusters is strongly influenced by chemisorption of H in bridge or hollow positions. The interaction of 1s H orbital with an appropriate combination of metal orbitals produces a very stable $1a_1$ bonding MO energetically close to the H orbital and with the largest weight of 1s H orbital. The contribution of the 5s orbital of Pd in this MO is negligible and the bonding is originated through metal d and H 1s orbitals. The singly occupied MO (HOMO) of the cluster has prevailing metal character (Fig. 4). The existence of a charge transfer from the cluster to hydrogen is found also in the case of Pd_2-H and Pd_3-H systems as shown by the data of population analysis, by the dipole moments and by the electron density deformation maps (Fig. 5).

The appearance of a well separated low-lying level in the MO spectra of Pd_2-H bridge and Pd_3-H hollow clusters present some analogies with the photoemission spectra of H on Pd (111). In the present cluster calculations the $1a_1$ MO lies about 6-7 eV below the analogous of the Fermi level (the HOMO of the cluster). Experimentally, the most noticeable effect of H chemisorption on the photoemission spectra of Pd (111) surface is the presence of a state density peak clearly separated from the d-band at 6.5 eV below the Fermi level (24,25). Calculated density

Table 4 - Electronic properties of Pd_m-H clusters (CI results)

	Pd-H	Pd_2-H	Pd_3-H hollow	Pd_3-H top	exp. H on Pd
Ground state	$^2\Sigma^+$	$^2A_1$	$^2A_1$	$^2A_1$	-
R_e au	3.02	3.11	3.25	3.04	3.19
D_e kcal/mol	31.2	36.1	45.9	33.8	62
μ_e Debye	3.2	1.2	0.7	3.1	0.02

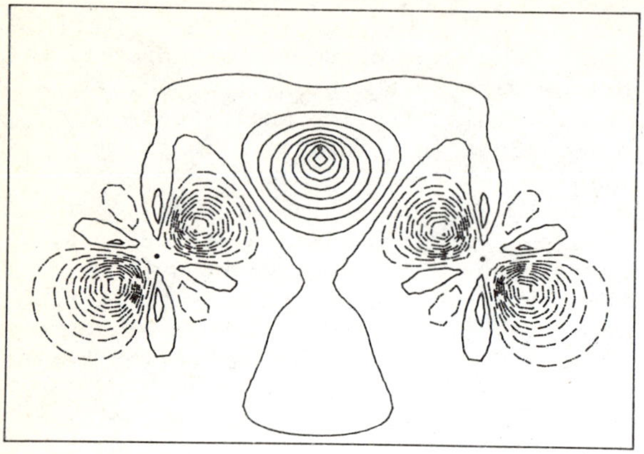

Fig. 5 - Electron density deformation map of Pd_2-H bridge cluster. Solid and dot-dashed lines indicate positive and negative values, respectively. The lines are drawn in intervals of 0.01 electrons/a.u.^{3}.

of states for H on Pd (111) agree with the experimental curve when hydrogen occupies three-fold hollow sites (18,19). If the monolayer of H is directly on top of every surface Pd atom the peak at -6.5 eV is not observed and a different peak at -4.5 eV is found (18, 19). This is consistent with the MO spectra of the Pd_3-H top cluster (Fig. 4).

The clusters used to model the Pd (111) surface necessarily represent very crude approximations of the real physical situation. Nevertheless, the parallelity between the qualitative results of cluster calculations and the experimental data is very interesting. The surprisingly high number of items where experiment and theory show similar qualitative features allows us to speculate that the bonding between hydrogen and Pd surface is of relatively local nature and the surrounding atoms play only a secondary effect in the determination of the chemisorption properties. A similar conclusion was drawn by other authors on the basis of both theoretical and experimental results (18-22).

5. REFERENCES

(1) Ph. Durand and J.C. Barthelat, Theoret. Chim. Acta **38**, 283 (1975).
(2) R.J. Buenker and S.D. Peyerimhoff, Theoret. Chim. Acta **35**, 33 (1974).
(3) G. Pacchioni, J. Koutecký and P. Fantucci, Chem. Phys. Letters **92**, 486 (1982).
(4) G. Pacchioni and J. Koutecký, Surface Sci. **154**, 126 (1985).
(5) E. Miyoshi, Y. Sakai and S. Mori, Chem. Phys. Letters **113**, 457 (1985).
(6) R.G. Pearson, Inorg. Chem. **23**, 4675 (1984).
(7) P.S. Bagus, C.J. Nelin and C.W. Bauschlicher Jr., Phys. Rev. B **28**, 5423 (1983); J. Vacuum Sci. Technol. A **2**, 905 (1984).
(8) J. Koutecký, G. Pacchioni and P. Fantucci, Chem. Phys., in press (1985).
(9) D. Belli Dell'Amico, F. Calderazzo, C.A. Veracini and N. Zandonà, Inorg. Chem. **23**, 3030 (1984).
(10) P. Fantucci, G. Pacchioni and V. Valenti, Inorg. Chem. **23**, 247 (1984).

(11) R.D. Feltham, G. Elbaze, R. Ortega, C. Eck and J. Dubrawski, Inorg. Chem. 24, 1503 (1985) and references therein.
(12) A.M. Bradshaw and F.M. Hoffmann, Surface Sci. 72, 513 (1978).
(13) R.J. Behm, K. Christmann, G. Ertl and M.A. Van Hove, J. Chem. Phys. 73, 2984 (1980).
(14) H. Basch and D. Cohen, J. Am. Chem. Soc. 105, 3856 (1983).
(15) H. Basch, Chem. Phys. Letters 116, 58 (1985).
(16) H. Conrad, G. Ertl and E.E. Latta, Surface Sci. 41, 435 (1974).
(17) R.J. Behm, K. Christmann and G. Ertl, Surface Sci. 99, 320 (1980).
(18) W. Eberhardt, S.G. Louie and E.W. Plummer, Phys. Rev. B 28, 465 (1983).
(19) S.G. Louie, Phys. Rev. Letters 42, 476 (1979).
(20) C.T. Chan and S.G. Louie, Phys. Rev. B 27, 3325 (1983).
(21) R.P. Messmer, D.R. Salaub, K.H. Johnson and C.Y. Yang, Chem. Phys. Letters 51, 84 (1977).
(22) H.G. Fritsche, Z. Naturforsch. 38a, 1118 (1983).
(23) L.B. Knight and W. Weltner Jr., J. Mol. Spectr. 40, 317 (1971).
(24) J.E. Demuth, Surface Sci. 65, 369 (1977).
(25) H. Conrad, G. Ertl, J. Kuppers and E.E. Latta, Surface Sci. 58, 578 (1976).

ACKNOWLEDGMENTS. This work has been supported by the Deutsche Forschungsgemeinshaft Sonderforschungsbereich 6 "Structure and Dynamics of Interfaces".

THEORETICAL STUDY ON THE CATALYTIC ACTIVITIES OF PALLADIUM FOR THE HYDROGENATION REACTION OF ACETYLENE

Hiroshi Nakatsuji and Masahiko Hada

Division of Molecular Engineering
Graduate School of Engineering
Kyoto University
Kyoto 606
Japan

ABSTRACT. Small palladium clusters show catalytic activity for the hydrogenation reaction of acetylene to form ethylene. The catalytic effect is important in both the dissociative adsorption step of hydrogen molecule and in the surface reaction step between acetylene and hydrogen. Energetically, the effect is larger for the first step. In the second step, we considered the Eley-Rideal mechanism and the Langmuir-Hinshelwood mechanism. In any case, the hydrogen adsorbed on palladium is very reactive despite of the PdH bond. Among the three modes of the reaction studied here, the two step Langmuir-Hinshelwood mechanism involving vinyl radical as a surface intermediate seems to be the most probable mechanism. We explained the selectivities in the hydrogenation of acetylene included as impurities in ethylene gas.

1. INTRODUCTION

Electronic processes on catalytic surfaces are of fundamental importance in both chemistry and chemical industry. Palladium shows especially a variety of catalytic activities.[1-3] We report here a theoretical study on the catalytic activities of palladium for the hydrogenation reaction of acetylene. This reaction consists of two steps. One is the chemisorption of hydrogen molecule on a palladium surface and next is the subsequent attack of hydrogen to olefins on a metal surface. Experimentally, it is known that the first step, namely the dissociative adsorption of hydrogen molecule, is the necessary step for the occurrence of the second hydrogenation reaction step.[4]

Previously, we studied theoretically the chemisorption of a hydrogen molecule on a small palladium cluster.[5-7] We have shown that even the Pd_2 fragment shows a catalytic activity for the dissociative adsorption of the hydrogen molecule.[6,7] The H_2 molecule with a binding

A. Veillard (ed.), Quantum Chemistry: The Challenge of Transition Metals and Coordination Chemistry, 477–487.

energy of 104 kcal/mol is dissociated, *with almost no barrier*, into two atomic hydrogens on the Pd_2 'surface', like on an extended surface. We found a very smooth equilibrium $H_2 \rightarrow H\cdot + H\cdot$ about 1.5Å apart from the Pd_2 'surface'. We clarified the electronic mechanism of the catalytic activity and showed that the $4d_\delta$ and secondly the 5s AO's of the Pd_2 fragment, which constitute the so-called 'dangling' bond of the metal surface, play an essential role. This mechanism may be sketched as

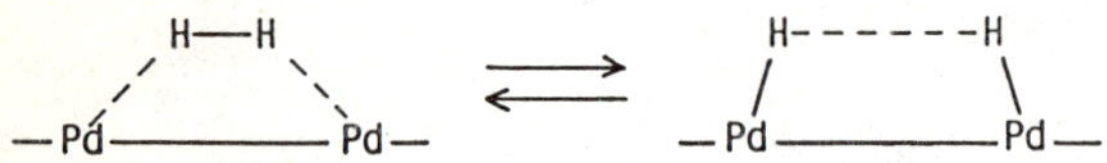

(1)

Note that in this bond alternation mechanism, the Pd-Pd bond is not weakened. This is important, suggesting a stability of the catalytic surface. Otherwise, the Pd atom would be exfoliated as a PdH molecule from the surface. This mechanism of the catalytic activity of palladium is different from the one proposed for a nickel surface,[8,9] and also for a calcium surface.[9]

After we have thus been able to produce the dissociatively adsorbed hydrogens on the Pd_2 fragment on a purely theoretical ground, so to speak, in a file of a computer, instead of a test tube, the first thing we wanted to do next was to investigate the activity of this system for the hydrogenation reaction. We have chosen acetylene as a reactant because the catalytic reaction

$$C_2H_2 + H_2 \rightarrow C_2H_4 \qquad (2)$$

is one of the most popularly used reactions in chemical industry to convert acetylene included as impurities in ethylene gas.[1,10] Palladium is a good catalyst of this reaction.[10] As far as acetylene exists in the mixture, it is hydrogenated selectively to ethylene, and ethane is not formed. Another reason we have chosen this reaction is that it is typically a symmetry-forbidden reaction.[11-13] Without an existence of the catalyst the barrier of this reaction is too high to occur smoothly. Then, an actual occurrence of this reaction on a palladium surface is essentially due to the catalytic activity of palladium. We want to know the electronic origin of this catalytic activity.

We consider here two modes of the reaction. One is that acetylene in a gas phase or in a van der Waals layer of catalyst reacts with the hydrogen molecule dissociatively adsorbed on palladium, namely,

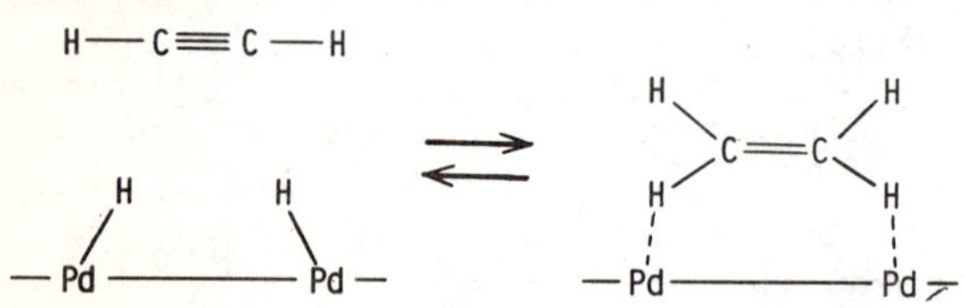

(3)

We considered this mode to solve the following question. Is the hydrogens on Pd_2 we obtained active enough, despite of the newly formed Pd-H bonds in Eq.(1)? We will call this pathway as Eley-Rideal (ER) mode. Experimentally this mode is not realistic since acetylene is more easily adsorbed on palladium than hydrogen.[1] However, Bond and Wells[10] considered this mechanism for the hydrogenation of ethylene when excessive amount of H_2 molecule exists.

Another mode of the reaction we considered is that the hydrogen dissociatively adsorbed on palladium attacks acetylene also adsorbed on a palladium surface. We call this mode as Langmuir-Hinshelwood (LH) mode. Experimentally, this mode is natural.[1,10,14] Theoretically, this mode is more difficult to study, because we have to clarify the mode of the adsorption of acetylene and further the mode of the interaction between acetylene and hydrogen both on a palladium surface. We will explain our way of investigation later in the corresponding section.

2. CALCULATIONAL METHOD

The calculations were carried out by the CAS(complete active space)-MC-SCF method. We considered electron correlations within the active MO's of the reaction. The SCF process was performed within an adequate space of the MO's. The gaussian basis set we used are (3s3p3d)/[3s2p2d] set for Pd and the Kr core was replaced by the relativistic effective core potential.[15] For hydrogen we used (4s)/[2s] set[16] and for carbon 4-31G set.[17] In the calculations of the ER mode, we added further the derivatives of the basis set for the carbons and hydrogens, so that the Hellmann-Feynman theorem is essentially satisfied for the forces acting on these nuclei.[18]

3. ELEY-RIDEAL MODE

We have shown in Figure 1 the assumed pathway of the reaction. The acetylene molecule attacks from a side-on orientation the two hydrogen atoms dissociatively adsorbed on the Pd_2 fragment. For the Pd_2-H_2 system, we used the optimum geometry calculated previously by the SAC method.[6,7] For acetylene, the geometry is assumed to change gradually as a function of R, shown in Figure 1, and at R = 0 we adopted the geometry of ethylene. The dependence of the geometrical parameters on R was assumed to be the same as the change of the H-H distance in the molecular adsorption step of the hydrogen molecule on Pd_2.[7]

The potential energy curve of the system along this

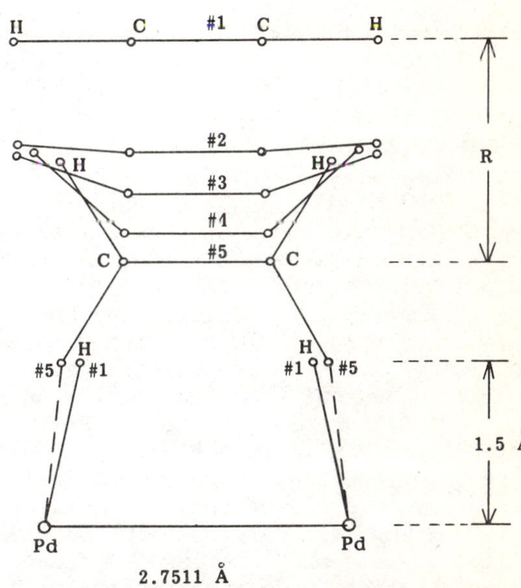

Figure 1. Assumed reaction pathway for the Eley-Rideal mode of the reaction

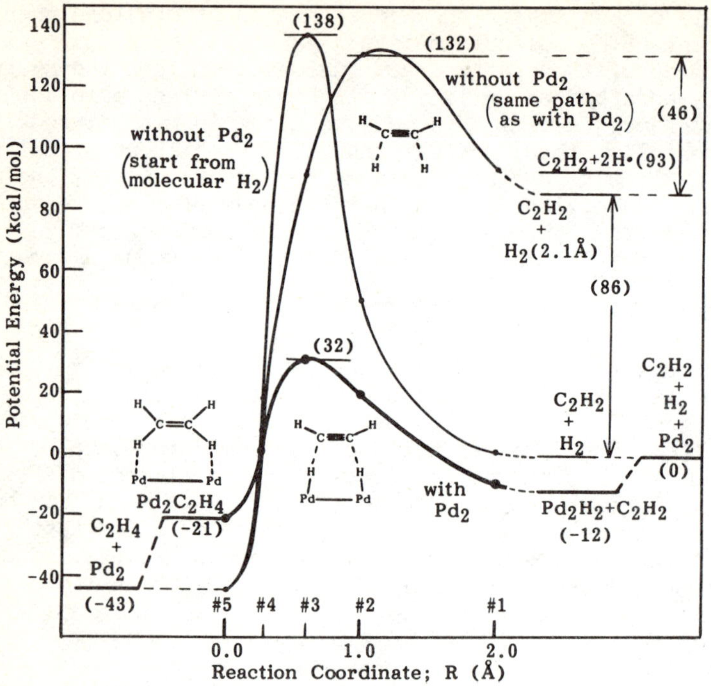

Figure 2. Potential energy curves for the hydrogenation reaction of acetylene with Pd_2 in the Eley-Rideal mode and without Pd_2. For the reactions without Pd_2, see the text. The numbers show relative energies in kcal/mol.

reaction coordinate is shown in Figure 2. It also shows the potential curves for the same reaction without Pd_2. The curve starting from the level, $C_2H_2 + H_2$(2.1Å), is the potential curve of the system entirely the same as that shown in Figure 1 except for the non-existence of the Pd_2 fragment. Another sharp curve is also without Pd_2 starting from C_2H_2 and molecular hydrogen. The dependence of the H-H distance on R is the same as that used in defining the reaction coordinate shown in Figure 1.

The barrier of the reaction with the existence of Pd_2 is about 32 kcal/mol with respect to the free system. When ethylene is formed on Pd_2, it leaves out from the complex automatically, since there, C_2H_4 and Pd_2 are coplanar so that the system is more unstable than the free system by 22 kcal/mol. The Pd_2 fragment thus generated again adsorbs H_2 as reported previously and enters again into the reaction cycle. This is the catalytic cycle of the hydrogenation reaction in the ER mode involving Pd_2 as a catalyst. The barrier of the reaction in this cycle, 32 kcal/mol, is much smaller than that of the same reaction without Pd_2. When we start from C_2H_2 and molecular hydrogen, the barrier is as large as 138 kcal/mol. When we start from C_2H_2 and H_2 with the H-H distance of 2.1Å , the H-H distance of the dissociatively adsorbed Pd_2H_2 system, the barrier is about 46 kcal/mol, but to elongate H_2 up to 2.1Å , about 86 kcal/mol is necessary so that the sum is 132 kcal/mol. Thus, the existence of Pd_2 is essential to reduce the barrier of the reaction. This is an important role of the catalyst.

It is interesting to compare the barriers of the reaction starting from C_2H_2 + H_2(2.1Å) with and without Pd_2. It is 46 kcal/mol without Pd_2 as shown above but it is also 44 kcal/mol even with Pd_2. This result implies two facts. First, the most important step, energetically, in this catalytic process is the dissociative adsorption of the H_2 molecule on the Pd_2 fragment. Since this process occurs without barrier,[6,7] we see that the catalytic activity of Pd_2 reduces the barrier of this step by more than 86 kcal/mol. Second, we notice that the hydrogen dissociatively adsorbed on Pd_2 is essentially as reactive as the free atomic hydrogen, despite of the existence of the Pd-H bond in the Pd_2-H_2 system as shown in Eq.(1). This is surprising indeed and shows the catalytic activity of Pd_2 in the hydrogenation step. If the hydrogens were tightly bound by the Pd_2, much larger energy barrier would have been resulted in the hydrogenation reaction step!

Now, what is the origin of this catalytic role of Pd_2 in the hydrogenation reaction step? In Figure 3, we have shown the orbital correlation diagram of this system in the earier stage of the reaction. It is based on the analysis of the natural orbitals of the CAS-MC-SCF method. The left-hand side shows the π and π* MO's of acetylene. The right-hand side shows the active MO's of the Pd_2-H_2 system. They were depicted in the previous paper[7] as the key MO's of the Pd_2-H_2 system. Here again, the existence of the b_2 MO in the occupied space of the Pd_2 H_2 system is very important. This MO can interact with the π* MO of acetylene to form new C-H bonds. However, this interaction alone does not explain the repelling role of H_2 from the Pd_2 fragment. It is due to the second b_2 MO in the unoccupied space of the Pd_2-H_2 system. This MO is very important to describe the electron correlation of the system, so that the occupation number is 0.14 even in the free Pd_2-H_2 system. As the interaction increases, the second a_1 MO of the interacting system becomes more and more unstable. This is the origin of

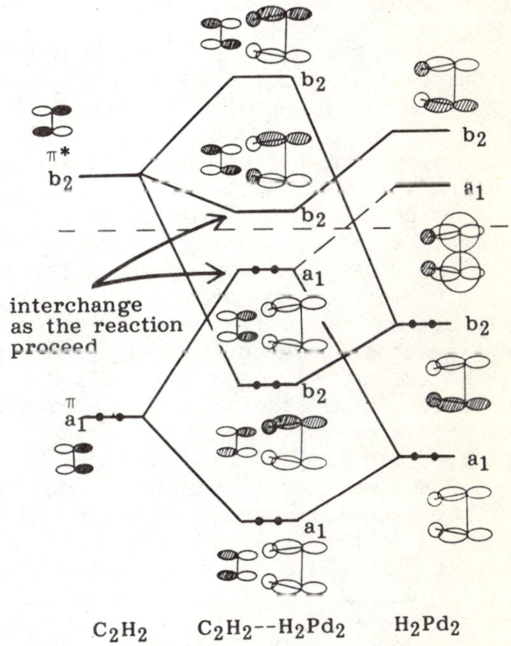

Figure 3. Orbital correlation diagram for the earier stage of the reaction, C_2H_2 + H_2-Pd_2 → C_2H_4--Pd_2.

the existence of the barrier. The spacing between the occupied a_1 and unoccupied b_2 MO's becomes closer and closer and the electron correlation involving these two orbitals increases and works to reduce the activation barrier. At the transition state, the order of the occupation numbers of these MO's is inverted. We thus understand that in the hydrogenation reaction step, the b_2 orbital interactions and the electron correlation are important in reducing the barrier. We note that in these b_2 MO's the dominant palladium orbitals are the d_δ AO's, which constitute the 'dangling' bonds of a palladium surface.

4. LANGMUIR-HINSHELWOOD MECHANISM

In the Langmuir-Hinshelwood mechanism, the hydrogen dissociatively adsorbed on palladium reacts with acetylene also adsorbed on the surface. Theoretically, this mode of the reaction includes much more freedoms than the ER mode. To construct a model of the surface reaction complex within a minimum number of necessary freedoms, the previous experimental and theoretical informations are valuable.

The stable adsorption site of atomic hydrogen on palladium is reported to be a hollow site which is on the center of the Pd_3 triangle of the surface.[19,20] The Pd-H bond is 1.69Å.[19c] When H is trapped at the bridge site of Pd_2, the binding energy is reduced by about 10 kcal/mol.[20b]

The geometry of the surface complex of acetylene on palladium is rather complicated. It seems to depend on temperature and the nature of a surface.[21] Fischer and Kelemen[21c] reported that on the (100) surface the electron spectrum of acetylene differs only moderately from the gas phase spectrum, but on the (111) surface, acetylene forms an olefinic complex which strongly interacts with the neighbouring Pd atoms through π and σ bonds. Ozin, Goddard, et al.[22] reported an experimental and theoretical study on ethylene-nickel cluster complexes. They showed that ethylene forms a π-complex probably with only one of the Ni atoms.

The mode of the attack of hydrogen to acetylene on the surface is interesting but little seems to be known. Smith[23] reported from the experimental study on the hydrogenation of olefins that hydrogen adds almost exclusively from the surface to the surface side of chemisorbed molecules. Experimentally, the elementary steps of the hydrogenation reaction are considered to involve vinyl radical formation as an intermediate step of the reaction.[10]

Based on these informations, we constructed two models of the surface reaction complex of the LH mechanism as illustrated in Figure 4. We refer to the left and right ones as complexes A and B, respectively. Complex B is the model leading to a vinyl radical formation and is a simplified one of complex A. We assumed a side-on π-complex interaction for acetylene. The distance between acetylene and Pd was settled to 2.4 Å which was obtained by optimizing $Pd-C_2H_2$ system. The initial position of the hydrogen on palladium was optimized for Pd_3H_2 system. Besides of the central bridging position,[20b] there was another slightly lower minimum near the outer Pd atoms. The H-Pd(outer)-Pd(central) angle was 50° with the Pd(outer)-H distance fixed at 1.535Å.[24] The bending angle θ of acetylene was 6° when optimized for the $Pd-C_2H_2$ system, but

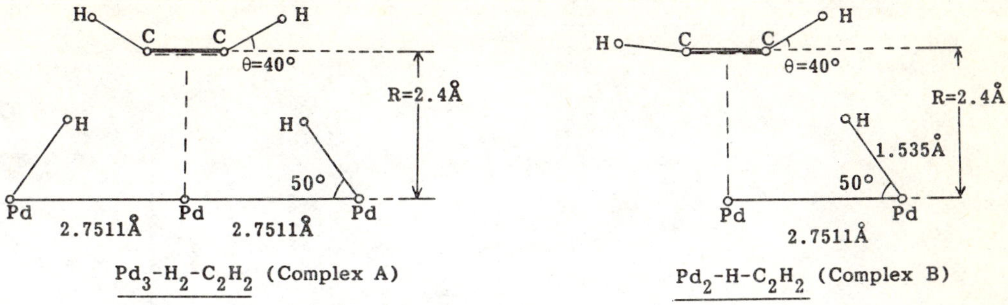

Figure 4. Model of the surface reaction complex in the Langmuir-Hinshelwood mechanism.

40° when optimized for complex B. We used the same angle $\theta = 40°$ for complex A, too. The energy of the reaction complex depends largely on this angle θ. The difference $E(6°) - E(40°)$ was 34 kcal/mol and 18 kcal/mol for complexes A and B, respectively.

In Figures 5 and 6, we have shown the energetics of the hydrogenation reaction of acetylene in the LH mechanism. Starting from the geometries shown in Figure 4 (#1 on the coordinate), we let the systems to react and to form ethylene and vinyl radical on the Pd_3 and Pd_2 'surfaces', respectively (#3 on the coordinate). Figure 5 is for the reaction involving complex A and Figure 6 for complex B. The reaction involving complex B consists of two steps, one corresponding to the change from acetylene to vinyl radical and the other to the change from vinyl radical to ethylene. These figures also show the energetics for the systems without including palladium. The curves were obtained for the same reaction paths just by deleting Pd_3 and Pd_2 from complexes A and B, respectively.

From Figure 5 we see that complex A is higher in energy than the separated system by 28 kcal/mol. Namely, when two hydrogens attack the adsorbed acetylene simultaneously from left and right, the energy barrier would be in this order. Afterwards, the reaction proceeds barrierlessly and leads to ethylene coplanar with Pd_3. In this geometry, ethylene is repulsive from Pd_3 and released automatically out of the reaction complex. The remaining Pd_3 is again involved in the reaction cycle by adsorbing acetylene, and so on. This is the cycle of the reaction involving the reaction complex A. In comparison with the same reaction without Pd_3, the catalytic activity of Pd_3 seems to exist essentially in the dissociative adsorption step of the H_2 molecule. Without Pd_3, we need 93 kcal/mol (experimentally 104 kcal/mol) to dissociate the H_2 molecule into two atomic hydrogens.

The two step LH mechanism involving the reaction complex B seems to proceed more smoothly than the one step simultaneous LH mechanism involving the reaction complex A. In Figure 6, the energy of complex B is only 7 kcal/mol higher than the separated system. Afterwords, the reaction proceeds smoothly to form vinyl radical adsorbed on Pd_2. The

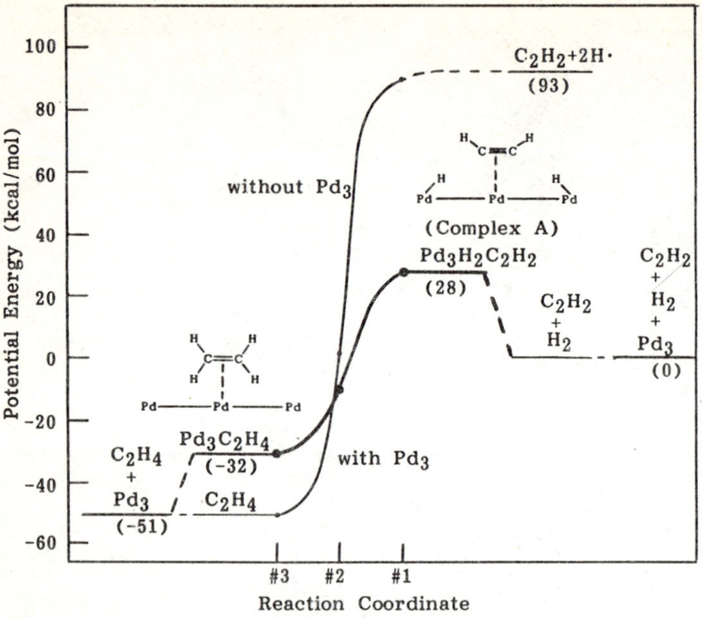

Figure 5. Energetics for the hydrogenation reaction of acetylene in the Langmuir-Hinshelwood mechanism involving complex A shown in Figure 4 with and without Pd_3. The numbers show relative energies in kcal/mol.

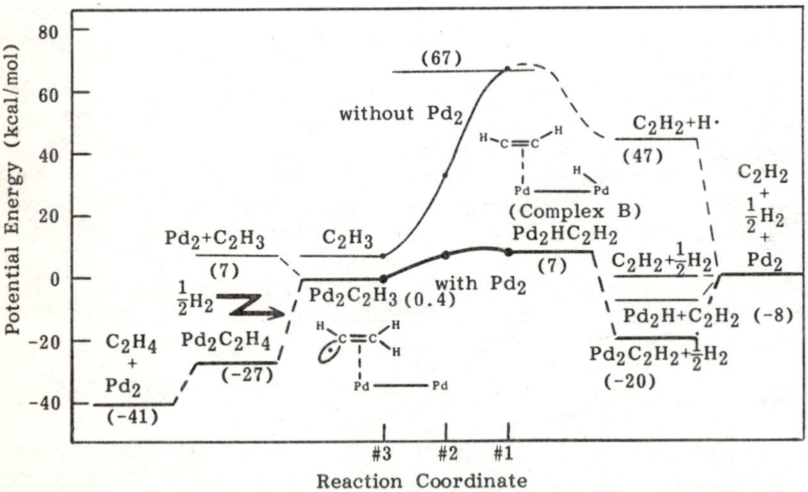

Figure 6. Energetics for the hydrogenation reaction of acetylene in the Langmuir-Hinshelwood mechanism involving complex B shown in Figure 4 with and without Pd_2. The reaction proceeds in two steps and involves vinyl radical as the reaction intermediate. The numbers show relative energies in kcal/mol.

present model is incomplete for this intermediate complex because the σ radical lobe on carbon cannot have a sufficient interaction with palladium (we need at least one more Pd atom on the left). Nevertheless, the present complex of the vinyl radical is more stable than the separated system, $Pd_2 + C_2H_3$, by 7 kcal/mol. In a more complete model this stabilization energy would be larger. Therefore, the vinyl radical still remains on the surface and receives a second attack of adsorbed hydrogen to form ethylene. In this final product form, ethylene is coplanar with Pd_2 so that it is repelled from the surface and goes out of the reaction cycle. This is the completion of the catalytic cycle. The naked site of palladium thus released adsorbs acetylene and enters again into the cycle. In comparison with the energetics obtained without including Pd_2, the catalytic role of Pd_2 is observed in two steps. One is in the dissociative adsorption step of H_2 on the palladium and the other in the surface reaction step to form complex B. Thus, the two step LH mechanism involving an intermediate formation of vinyl radical is most favorable within the three modes of the hydrogenation reaction studied here.

5. CONCLUDING REMARKS

Lastly we explain briefly the selectivities in the hydrogenation reaction of acetylene included as impurities in ethylene gas.[10] They are as follows. (1) Hydrogenation of ethylene does not occur until all acetylene impurities are converted to ethylene. (2) Ethane is generated only scarcely from acetylene. These selectivities, which are of special importance in chemical industry, occur despite of the fact that palladium is even a better catalyst for the hydrogenation reaction of ethylene. A possible explanation based on the present calculations is as follows. For the first selectivity, we think that the differences between acetylene and ethylene in the heat of adsorption and in the sticking probability to the catalyst are important. The stabilization energy of acetylene was calculated to be 7 kcal/mol for the Pd atom and 20 kcal/mol for the Pd_2 fragment. The corresponding values for ethylene were 12 kcal/mol and 19 kcal/mol, respectively. Therefore, for an extended surface, the heat of adsorption should be larger for acetylene than for ethylene as observed experimentally.[1] Further, the sticking probability of acetylene should be larger than ethylene because acetylene has the active π orbitals in all angles around the C=C axis but ethylene has the π orbitals only in the plane perpendicular to the molecular plane. The second selectivity is explained from all the surface reaction mechanisms studied here. The product ethylene generated by the hydrogenation reaction of acetylene is coplanar with the active palladium atoms. In this configuration ethylene is repelled from the surface and goes out of the reaction cycle so that the hydrogenation reaction up to ethane does not occur.

In this study, we have shown that the small palladium clusters show catalytic activity for the hydrogenation reaction of acetylene to form ethylene. The reaction proceeds in a smooth cycle involving the two main steps. One is the dissociative adsorption of the H_2 molecule on the Pd surface and the other is the surface reaction step between acetylene and hydrogen. Palladium shows catalytic activity for both steps. Energetically, the catalytic role of palladium seems to be more important in the first step than in the second step. The occurrence of the first step is necessary for the occurrence of the second step.[4] In the second step, the hydrogen dissociatively adsorbed on palladium is very reactive despite of the Pd-H bond with the surface metal. This activity is due to the catalytic role of palladium. Further, the second step is more important than the first step to manifest a selectivity, which is another important role of catalyst. An example is explained above. Within the three mechanisms of the hydrogenation reaction studied here, one in the ER mode and two in the LH mode, the two step LH mechanism involving vinyl radical as a surface intermediate is the most probable mechanism. This result seems to agree with the mechanism frequently used by experimental chemists in the hydrogenation reactions of olefins.[1,10,14] It would be interesting to observe vinyl radical on a Pd surface in the presence of acetylene and hydrogen.[21e]

ACKNOWLEDGEMENTS

We are grateful to Professors T. Yonezawa and S. Yoshida for some valuable discussions. The calculation was carried out with the HITAC M200H computer at the Institute for Molecular Science and with the FACOM M382 and VP 100 computers at the Data Processing Center of Kyoto University. We acknowledge the computer center of IMS for the grant of computer time. Part of this study was supported by the Grant-in-Aid for Scientific Research from the Ministry of Education, Science, and Culture.

REFERENCES

1. G. C. Bond, *Catalysis by Metals*, Academic Press, New York, 1962.
2. F. A. Lewis, *The Palladium-Hydrogen System*, Academic Press, New York, 1967.
3. P. M. Maitlis, *The Organic Chemistry of Palladium*, Academic Press, New York, 1971, Vols.1 and 2.
4. B. J. Wood and H. Wise, J. Catalysis, 5, 135 (1966).
5. H. Nakatsuji and M. Hada, Croatica Chemica Acta, 57, 1371 (1984).
6. H. Nakatsuji and M. Hada, in *Proceeding of the Nobel Laureate Symposium on Applied Quantum Chemistry (PAC' CHEM '84 Honolulu)*, Ed. by V. H. Smith, Jr, Reidel, Dordrecht, 1985.
7. (a) H. Nakatsuji and M. Hada, J. Am. Chem. Soc. to be published.
(b) H. Nakatsuji, M. Hada, and T. Yonezawa, to be published.
8. (a) C. F. Melius, Chem. Phys. Lett. 39, 287 (1976).
(b) C. F. Melius, J. W. Moskowitz, A. P. Mortola, M. B. Baillie, and M. A. Ratner, Surf. Sci. 59, 279 (1976).
9. C. Satoko and M. Tsukada, Surf. Sci. 134, 1, 1983.
10. G. C. Bond and P. B. Wells, J. Catalysis, 4, 211 (1965); 5, 65 (1965).
11. R. B. Woodward and R. Hoffmann, *The Conservation of Orbital Symmetry*, Academic Press, New York, 1970.

12. G. Henrici-Olive and S. Olive, *Coordination and Catalysis*, Verlag Chemie, Weinheim, 1977.
13. (a) R. G. Pearson, Acc. Chem. Res. 4, 152 (1970).
(b) F. D. Mango, Topics in Current Chem. 45, 39 (1974).
14. J. Horiuti and M. Polanyi, Trans. Faraday Soc. 30, 1164 (1934).
15. P. J. Hay, J. Am. Chem. Soc. 103, 1390 (1981).
16. S. Huzinaga, J. Chem. Phys. 42, 1293 (1965).
T. H. Dunning, Jr., J. Chem. Phys. 53, 2823 (1970).
17. R. Ditchfield, W. J. Hehre, and J. A. Pople, J. Chem. Phys. 54, 724 (1971).
18. H. Nakatsuji, K. Kanda, and T. Yonezawa, Chem. Phys. Lett. 75, 340 (1980); H. Nakatsuji, T. Hayakawa, and M. Hada, Chem. Phys. Lett. 80, 94 (1981); H. Nakatsuji, K. Kanda, M. Hada, and T. Yonezawa, J. Chem. Phys. 77, 3109 (1982).
19. (a) H. Conrad, G. Ertl, and E. E. Latta, Surf. Sci. 41, 435 (1974).
(b) R. J. Behm, K. Christmann, and G. Ertl, Surf. Sci. 99, 320 (1980).
(c) W. Eberhardt, S. G. Louie, and E. W. Plummer, Phys. Rev. B28, 465 (1983).
20. (a) R. P. Messmer, D. R. Salahub, K. H. Johnson, C. Y. Yang, Chem. Phys. Lett. 51, 84 (1977)
(b) G. Pacchioni and J. Kouteckey, *Theoretical Investigation of the Interaction between the Hydrogen Atom and Pd Clusters*, private communication.
21. (a) J. E. Demuth and D. E. Eastman, Phys. Rev. Lett. 32, 1123 (1974).
(b) L. L. Kesmodel, R. C. Baetzold, and G. A. Somorjai, Surf. Sci. 66, 299 (1977).
(c) T. E. Fischer and S. R. Kelemen, Surf. Sci. 74, 47 (1978).
(d) H. Ibach and S. Lehwald, J. Vac. Sci. Technol. 15, 407 (1978).
(e) J. E. Demuth, Surf. Sci. 80, 367 (1979).
(f) J. E. Demuth and H. Ibach, Surf. Sci. 85, 365 (1979).
22. G. A. Ozin, W. J. Power, T. H. Upton, and W. A. Goddard, J. Am. Chem. Soc. 100, 4750 (1978).
23. G. S. Smith, J. Catalysis, 5, 152 (1966).
24. K. P. Huber and G. Herzberg, *Molecular Spectra and Molecular Structure. IV. Constants of Diatomic Molecules*, Van Nostrand Reinhold, New York, 1979.

THEORETICAL STUDY OF THE SUCCESSIVE HYDROGENATION OF SMALL PLATINUM CLUSTERS

B. BIGOT
Laboratoire de Chimie Théorique, E.N.S. de Saint-Cloud
92211 SAINT-CLOUD Cedex FRANCE
and
Institut de Recherche sur la Catalyse
69626 VILLEURBANNE Cedex France

and

C. MINOT
Laboratoire de Chimie Théorique
Université de Paris-Sud Orsay
91405 ORSAY Cedex FRANCE

ABSTRACT: The successive hydrogenation steps, up to saturation of small platinum clusters Pt_n with n between 2 and 13 are studied by the Extended Huckel Method (EHT) with and without spin-orbit coupling. For each hydrogenation step, a large number of possible structures has been calculated in order to determine the stereochemistry of the best stepwise hydrogenated compounds. The resulting structures can be viewed as the progressive build-up of a three-dimensional network of square planar and/or octahedral monometallic complexes. The energy of the first successive steps of dihydrogen adsorption decreases steeply with the hydrogen rate for the smallest clusters. At the contrary, for some of the larger clusters, the first hydrogenation steps are more and more stabilizing before a decrease occurs. The energy of first adsorption depends on the cluster size. The largest value is found with the Pt_3 and Pt_4 species. These results confirm the interpretation proposed by Fraissard et al. for explaining the observations recently made on platinum clusters supported on zeolites.

I - INTRODUCTION.

The transition metal clusters at low coordination rate are nowadays the subject of a large number of both experimental[1] and theoretical[2] studies. Their large metallic surface/volume ratio makes them potentially very active catalytic species. Added to this property, their conditions of synthesis make them attractive model

A. Veillard (ed.), Quantum Chemistry: The Challenge of Transition Metals and Coordination Chemistry, 489–499.

compounds to get insight into the origin of the catalytic properties of metallic surfaces. Recently, Fraissard et al.[3] detected unusually small platinum clusters by using high resolution NMR techniques. From the study of the first hydrogenation steps, they estimate the average number of metallic atoms per cluster to be less than 8. They observe that, on a given cluster, the polyhydrogenation is stepwise. A cluster does not adsorb a (n+1)th dihydrogen molecule before all clusters have already adsorbed n previous molecules. In a preceding paper[4], a systematic study of the stability of the purely metallic clusters Pt_n has been reported. The present report concerns the progressive hydrogenation of some of the most stable structures. It intends to provide detailled informations on the structures and energetics of the hydrogenated species. It assumes no relaxation of the metallic structures. The analysis is based on Extended Huckel calculations performed with the weighted approximation for estimating the off-diagonal hamiltonian matrix elements[5]. The spin-orbit contribution to this matrix and the used parameters are described elsewhere[4]. Throughout the study, the platinum-platinum and platinum-hydrogen bond lengths are maintened at 2.77 A and 1.6 A respectively. The hydrogen-hydrogen distance in molecular dihydrogen is kept fixed to 0.74 A.

II - THE OXIDATIVE AND MOLECULAR HYDROGENATION MECHANISMS.

Each of the successive hydrogenation step is associated with the occurence of two new platinum-hydrogen bond-lenghts of 1.6 Å. Depending on the distance between the two adsorbed hydrogen atoms(>1.2 Å or =0.74 Å), the dihydrogen molecule dissociate or does not during the adsorption process. This fundamental difference defines the two mechanisms of oxidative or molecular adsorption. Let us consider them in molecular orbital terms. In all clusters, the energies of the the occupied orbitals lie in the interval (-14,-11 eV) which spreads around the position of the d levels of atomic platinum (-12.6 eV). The lowest unoccupied molecular orbital (LUMO) is always at less than 1 eV above the upper limit of this range. The σ occupied orbital of H_2 lies at -17.6 eV, the σ^* unoccupied one at +4.25 eV and the 1s H atomic orbital at -13.6 eV. In these conditions, it is convenient to consider either the H_2 or the H ligand as two-electron donor ligands. For the H ligand, this electron occupation results from an electron transfer from the cluster to the ligand. The consequence is that, formally, the dissociation of H_2 to 2H during the adsorption process is associated to a +II oxidation step for the platinum cluster.

There is a common point and a clear-cut difference in the interaction between H_2 or 2H and the clusters. According to the fragment analysis, the interaction between an occupied orbital of a fragment and an unoccupied orbital of another fragment is stabilizing and related to the expression $S^2/\delta E$. S is the overlap and δE the energy difference between the interacting orbitals.

The common point is that the dominant stabilizing

interaction occurs in both case between the occupied orbitals of the ligands and the unoccupied orbitals of the cluster. There is now an electron transfer from the ligand to the cluster. Its magnitude is proportional to the strength of the interaction. The difference is that, in the oxidative mechanism, this interaction is mainly controlled by the energy gap between these orbitals while in the molecular addition it is controlled by the overlap. An H_2 ligand has to interact with the cluster MO which offers the best overlap even if it is not the LUMO while an H- ligand intends to interact with the lowest unocuppied orbital even if the overlap is moderately good.

In order to determine whether it is an incoming H- or H_2 ligand which can be bound and in which position it will be fixed, another factor intervenes. Indeed, the occupied orbitals of the ligands interact with all the valence orbitals of the clusters which overlap significantly with them, regardless of their occupation. The interaction with an occupied cluster orbital is destabilizing. It results that the electron count on the cluster is a critical point to determine if the incoming ligand-cluster is attractive or repulsive on the whole. A non-oxidized platinum atom has 5 occupied d orbitals. Necessarily, the incoming two-electron ligand overlap with one of them. This generates a repulsive interaction. A platinum atom oxidized at the +II state has a vacant d orbital. When the incoming ligand is introduced in the main directions of this orbital, the previously repulsive interaction is replaced by an attractive one.

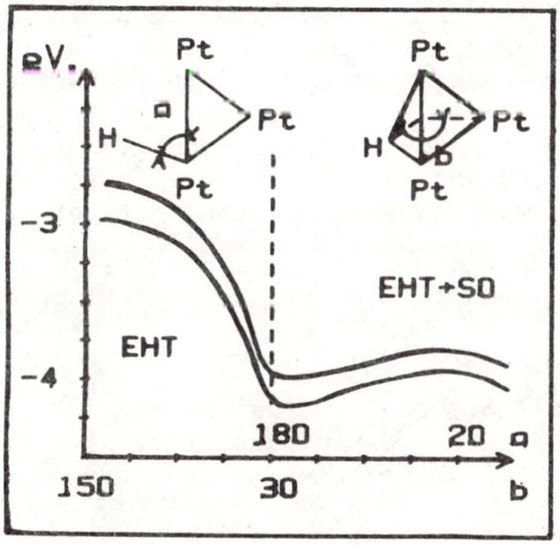

Fig. 1 : Energy curves of Pt_3H-

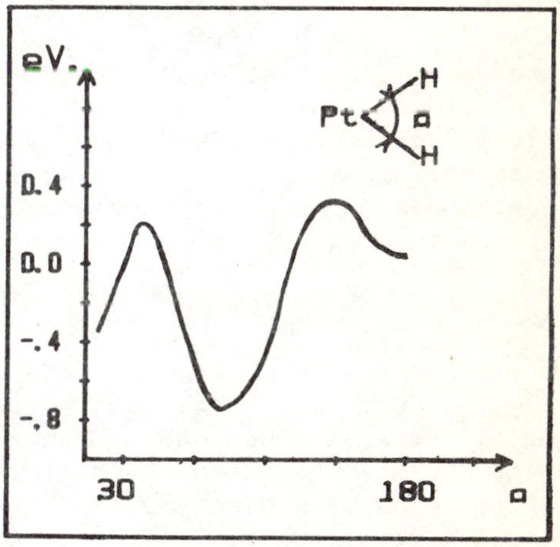

Fig. 2 : Energy curves of PtH_2

From this analysis, is is concluded that a clearly favorable interaction with H- or H_2 occurs only if the cluster has some vacant d orbital. The best position for the incoming ligand is in the region where this orbital has its main contributions.

Up to this point, we have considered the factors which determine the position of a ligand on a metallic center without taking into account the influence of the cluster structure as a whole. In fact, the positions depends on the near-by existing metal-metal and metal-ligand bonds. If we consider that the Pt-H bonds are 10 times stronger that the Pt-Pt bonds, it is reasonable to estimated that the position of the incoming ligand depends more on the orientation of the other ligands than on the structure of the metallic network. Thus, the structure of the hydrogenated clusters can be analyzed as a set of monometallic platinum complexes with common ligands which impose the relative orientation of the coordination structure of the adjacent complexes.

We distinguish three kinds of position for the hydride ligands relatively to the cluster structure. The terminal, bridged and face-centered ligands are bound to one, two or three platinum atoms respectively. The smallest cluster for which the three kinds of ligands may be encountered is the Pt_3 cluster. Figure 1 depicts the energy of Pt_3H^- as a function of the position of the ligand. In the left part, the Pt_3H^- structure is planar; H^- moves from a terminal to a bridged position. The increase in stability (-1.1 eV) is due to the interaction of the hydride orbital with the unoccupied s and p orbitals of the second metallic center. The preferred location does not depend on the oxidation state since the energy level of Pt_3H^- is insensitive to the hydride position. In the right part of Figure 1, H^- moves from a planar bridged postion to a face-centered position. The energy variations are small, but make the bridged position slightly better than the centered one. The hydrogenation at a tetrahedral site has been tested on the Pt_4 cluster. Pt_4H^- is found to be more stable when the hydride is in the bridged position than in the face centered (+0.2 eV), the terminal (+0.21 eV) or the tetrahedral (+0.35 eV) positions. These results agree with the fact that, for any cluster, no first adsorption step involves face-centered or tetrahedral positions. However, the energy difference between these positions and the bridged one are small enough to make them better in some polyhydrogenated clusters when they allow good matches of the structures of adjacent complexes.

III - RESULTS.

Before we report on the results of the hydrogenations of some clusters, let us consider the hydrogenation of an isolated atom to set the main features of the local coordination steps of a given platinum atom in a cluster.

Pt_1

The first hydrogenation step is oxidative (0.73) eV). A molecular adsorption would involve 0.42 eV only. The charge on the metal center is +0.5 a.u in $Pt(H)_2$ and -0.2 a.U in $Pt(H_2)$. Thus, the term of oxidative addition is justified even if the formal charge of +2.0 a.u. is reduced by the formation of two mainly covalent Pt-H bonds. Figure 2 shows the energy of $Pt(H)_2$ as a function of the HPtH angle or H-H distance. The curve has 3 minima (27°, 90°, and 180°).

The first one corresponds to a molecular adsorption; the others to oxidative processes. The energy barrier at 44° or 1.2 A defines the border between the two mechanisms. The 90° and 180°minima correspond to the best positions of the two hydride ligands to interact with the same d vacant orbital. Any distortion from these angles is strongly unfavorable.

The $Pt(H)_2$ complex has an unoccupied d orbital of which two main directions are still available to allow the fixation of two H_2 ligands. This generates a square planar structure $Pt(H_2)_2(H)_2$ where platinum is in a formal +II oxidation state. The stabilization energies associated with this two molecular adsorptions are 0.47 and 0.42 eV. For the next step, it is not possible to add a supplementary ligand without overlap with a occupied orbital. An oxidation step is required. It leads to an octahdral complex $Pt(H_2)_2(H)_4$ with a stabilization energy of 0.2 eV. Platinum is now associated with a format +IV oxidation state. The best structure with two cis H_2 ligands is shown on Figure 3. The trans isomer is 0.13 eV less stable.

Three main conclusions can be drawn from the study of the single atom hydrogenation. 1) The relative orientation of the ligands is of prime importance. It is determined by the shape of the vacant d orbitals. 2) The stable hydrogenated structures are part of the planar square d8 complex or octahedral d6 complex. 3) the energies of the successive hydrogenation steps are in decreasing order. The reason is bound do the upward move of the vacant orbitals of the cluster associated with the progressive fixation of two electron-donor ligands.

Pt_2-Pt_3-Pt_4-Pt_5

The saturated hydrogenated compounds of the 4 Pt_2-Pt_3-Pt_4-Pt_5 clusters have in common to have structure in full agreement with what was reported for Pt_1. They can all be viewed as made of the association of d6 local octahedral complexes. All metallic edges are bridged. The formula are respectively $Pt_2(H)_6(H_2)_2$, $Pt_3(H)_{12}(H_2)_3$, $Pt_4(H)_{16}(H_2)_2$,$Pt_5(H)_{20}(H_2)$. The structure are depicted on Figure 3. The energies of the successive adsorptions are depicted on Figure 4. We notice that every first hydrogenation step is significantly more stabilizing than the following ones. The most stabilizing first step occurs for Pt_4 at the EHT level and for Pt_3 at the EHT-SO level. This change is discussed later on. After the first step, a regular decrease is observed, except for a few steps. The analysis of these steps at the molecular orbital level shows that these irregularities are associated with the impossibility to avoid, at the former step, secondary repulsive interactions between the occupied orbitals of the incoming ligands and some of the clusters. This occurs when several occupied MO's of the clusters have important contributions in the same region of space. It is impossible to empty all of them by a single oxidation step. Some further oxidation step annihilates the repulsion and brings a large stabilization. Such a cooperative effect is found for the third hydrogenation step of Pt_3.

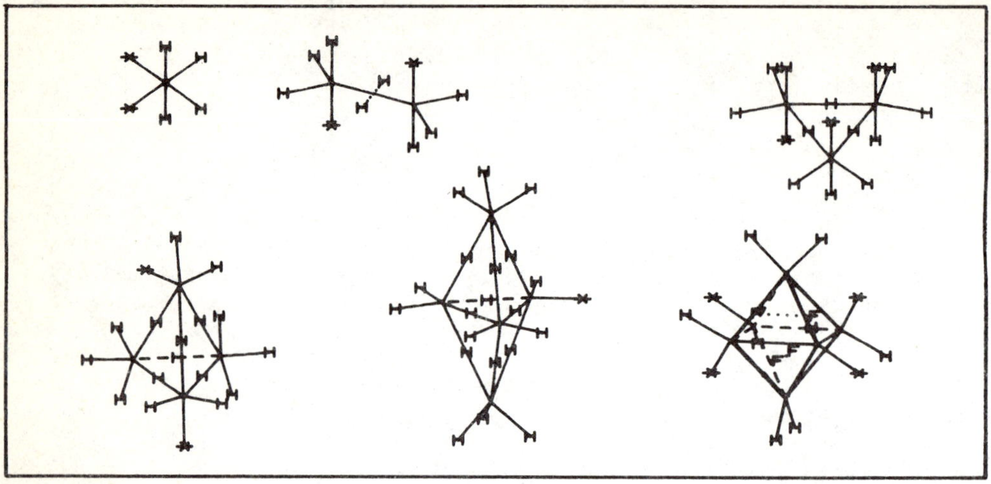

Fig. 3 : Structures of the fully hydrogenated $Pt_{1 \longrightarrow 6}$ clusters

The build-up of the fully hydrogenated clusters agrees well with what was found for Pt_1. The first steps are bridging oxidative reactions in adequate number and positions in order to empty the d orbitals of a metallic center on which molecular hydrogen ligands could be fixed to comptlete the square planar d8 or octahedral d6 coordination arrangement. The mean adsorption energy per adsorbed dihydrogen molecule increases with the cluster size. It is 0.53, 0.58, 0.74 and 0.76 eV respectively for the 4 above mentionned clusters.

Pt_6

The Pt_6 regular octahedron is the first cluster to present some special features in its progressive hydrogenation. The first hydrogenation step is oxidative and places two bridging ligands on adjacent Pt-Pt bonds in a planar arrangement. The bridging of all the metal-metal bonds does not allow the formation of undistorded local square planar or octahedral coordination arrangements for all the metallic centers. On the contrary, four face-centered hydrides located on non-adjacent triangular external faces are well suited to build local square planar arrangements since the H-Pt-H angles (70.5°) can be considered as an acceptable distorsion from the optimal 90° value. This skeleton generates a $Pt_6(H)_{12}(H_2)_4$ cluster made of 6 slightly distorded d8 square planar complexes (Figure 3). The introduction of two new terminal hydride ligands in directions perpendicular to the plane of the complexes would lead to a structure $Pt_6(H)_{24}(H_2)_4$ made of 6 d6 octahedral platinum complexes. Although the local electronic and geometric requirements thus would be fullfilled, this structure is unstable for steric reasons (too short H-H and H-Pt distances between non-bonded atoms). The move of the hydrides out of the triangular faces along the local ternary symmetry axis decreases the angular distorsion in the $Pt_6(H)_{12}(H_2)_4$ structure. The largest stabilization is 0.7 eV when the hydrides are moved 0.24 A above the faces. The H-Pt-H angle is then 90°. This favorable move is at variance with what is observed for an hydrogen atom when only a face is involved (see Figure 1). It illustrates the importance of the relative orientation of the ligands around a platinum center.

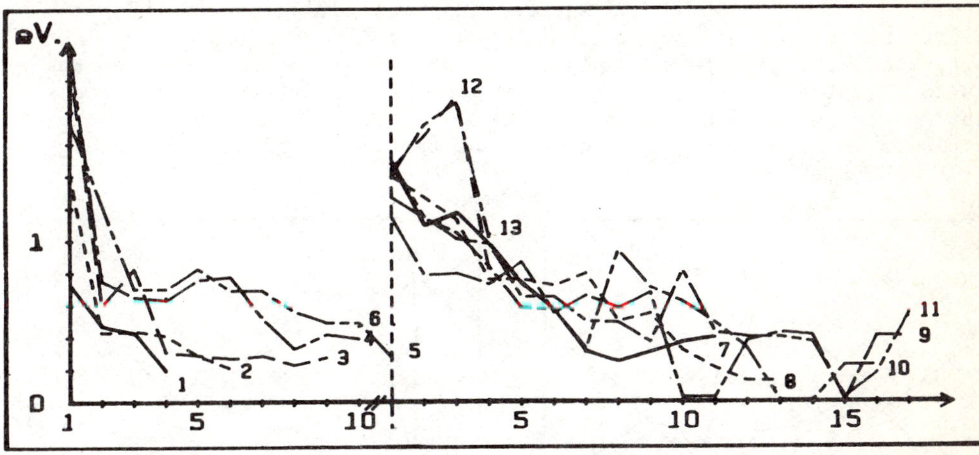

Fig. 4 : Energies of the successive hydrogenation steps of Pt_n clusters

The positions of the first fixed hydrogen atoms do not anticipate the final structure. Indeed, the first four hydrogenation steps place eight bridging hydrides around two of the three Pt_4 square planar fragments which exist in the Pt_6 octahedron. Next, there is a progressive switch from bridged positions to centered ones. This switch induces oscillations in the evolution of the stabilization energies of the successive steps. However, the first step remains the most stabilizing.

We will not comment in details here on the successive steps of hydrogenation of the cluster Pt_7. Pt_8 and Pt_9. indeed, they are similar to what has been already reported for the smaller clusters. Their general features will be analyzed in the discussion of chapter IV. Let us now consider the cluster Pt_{10} which brings some new behaviors which exist in the cluster Pt_{11}, Pt_{12} and Pt_{13}.

Pt_{10}

The most stable Pt_{10} cluster can be described as the sum of two adjacent Pt_6 octahedra sharing a common edge in such a way it can also be considered as made of a Pt_6 equatorial rectangle plus two Pt_2 pairs located above and below this plane. The structure of the saturated compounds $Pt_{10}(H)_{28}(H_2)_2$ (Figure 5) is derived from the saturated structure of the Pt_6 octahedral cluster. It has eight face centered hydrides. Two of these hydrides are on the internal triangular faces of the cluster. These faces are adjacent and coplanar. They share the edge common to the two octahedra in which Pt_{10} can be splitted. It is the first example of hydride ligand located within the volume of the cluster. The structure is completed with 18 terminal and 2 bridging hydride ligands in addition to the H_2 ligands. The bridging ligands are located on the Pt-Pt bonds of the Pt_2 pairs located above and below the equatorial plane. In such couple, one atom is d8 with a square planar coordination and the other d6 with an octahedral coordination. The four atoms located at the corner of the equatorial plane are d8. The two last atoms of the equatorial plane are d6. All have the coordination arrangements adapted to their oxidation state. The good match of the local complexes is illustrated by the value of the mean adsorption energy per H_2 (0.58 eV).

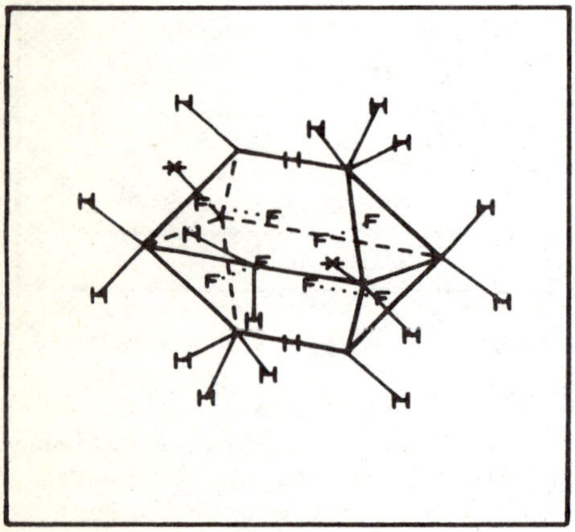

Fig. 5 : Structures of the fully hydrogenated Pt_{10} cluster

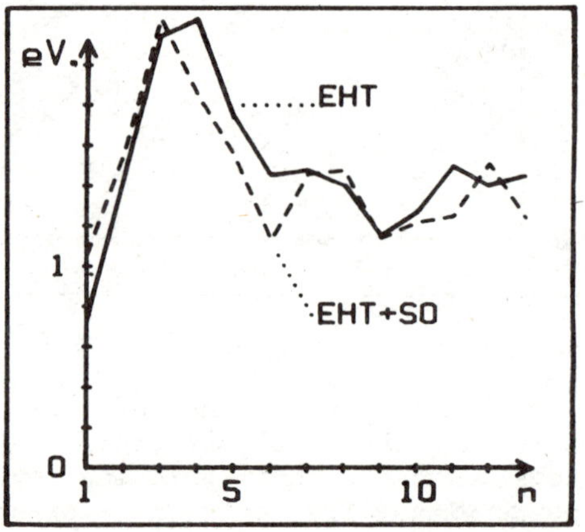

Fig. 6 : Stabilization energies of the first hydrogenation steps.

The seven first steps place bridging ligands with reaction energies decreasing regularly from 1.27 eV to 0.33 eV. The next step induces a move of the hydrogen already adsorbed from the bridging position to the face centered ones. This move is associated with a large increase of the reaction energy (0.94 eV) as it was observed for Pt_6. From this stage, the successive enthalpies decrease again regulary (see Figure 4).

The C_{3v} Pt_{10} cluster made of the stacking of a Pt_7 centered hexagon and a Pt_3 equilateral triangle has also been considered. It is equivalent to an incomplete cubooctahedron with three adjacent atoms removed. It is only less stable than the best structure by 0.3 eV. The three first hydrogenation steps place bridges on all the edges of the hexagon. For the first time, they occur with values which increase with the hydrogenation rate(1.50, 1.75, 2.07 eV). This striking result can be interpreted as associated with the cooperative effect already mentionned for Pt_3. Here, the occupied hydrid ligands which bridge the hexagon interact with three d MO's of the cluster (the combinaisons of d orbitals in phase along the edges). In the first step, the two introduced hydrides interact dominantly with the vacant MO of this set but cannot avoid some repulsive interactions with the two still occupied other ones. The second and third adsorption not only involve smaller repulsion for the associated incoming ligand, but also reduce the repulsion associated with the already fixed hydrides. The resulting $Pt_{10}(H)_6$ isomer is 1.56 eV more stable than its isomer derived of the other Pt_{10} structure. The next step which involves other d MO's shows no cooperative effect. It has an energy of 1.1 eV only. The eleven further hydrogenation leads to a structure $Pt_{10}(H)_{30}$ with 18 bridging and 12 terminal ligands. It is 3.86 eV more stable than the equivalent isomer of the other platinum skeleton. It has an exceptional mean adsorption energy per H_2 of 0.87 eV. This is the highest value found for a saturated cluster. The above results show that the most stable purely metallic cluster do not necessarily lead to the structure of the best hydrogenated compounds. The hydrogenation can induce profound structure changes in the cluster structure.

IV - DISCUSSION.

The energies of the first hydrogenation step versus the cluster size are depicted on Figure 6. There is a peak for Pt_4 at the EHT level and for Pt_3 when SO coupling is included. The extremum for Pt_4 can be interpreted as it follows. As the cluster size increases up to Pt_4, the energy of the lowest "6s,6p" MO decreases progressively, but it remains unoccupied. The interaction with the hydride orbital increases due to the the reduction of the energy gap. beyond Pt_4, an increasing part of the "6s,6p" MO's is occupied. This occupation contributes to the metal-metal bonding and stabilizes the purely metallic cluster. However, these orbitals are no longer available for a favorable interaction with the ligands. Worse, they induces a destabilizing interaction. All the first hydrogenation steps are oxidative and place bridging hydride ligands.

An interesting point is to know if they occurs preferentially on sites of low or high metallic coordination. The study of the results obtained for the whole Pt_2-Pt_{13} serie shows that in every case the low coordination sites are favored to be bridged first. It is worth noting that the low values observed on the curve of Figure 6 are associated with an high coordination.

In most cases, the successive hydrogenation steps occur with energies of decreasing magnitude (see Figure 3). The exceptions occur when there is a structural change or a cooperative effect. The latter is more and more frequent with the increase of cluster size. The experimental observations on the hydrogenation behavior of the clusters studied by Fraissard et al[1] are in agreement with sizes less than 10 atoms.

Let us now consider the spin orbit influence. The curves

with and without SO for Pt_3H- (Figure 1) are parallel. They illustrates that the SO coupling does not modified the position of the preferred site of adsorption significantly. It has been previously shown[4] that the stabilization induced by the SO coupling increases with the depopulation of the d orbitals of the metal atoms of the clusters, but decreases with the rate of coordination. A stabilization for the oxidative addition results. The contrary is observed for the molecular adsorption. This is illustrated on the successive hydrogenation of the small clusters (Figure 4). The stabilizing change induced by the SO coupling for the first oxidative step decreases rapidly with the increasing size of the clusters. Beyond Pt_4, it becomes even destabilizing as illustrated on Figure 6. As a consequence, the energies of the first hydrogenation step of the smallest clusters are increased by the SO coupling while those of the largest clusters are reduced. This result reinforces the Fraissard's[3] conclusions on the cluster size.

V - CONCLUSION.

In zeolite, Fraissard et al.[3] observed that dihydrogen is adsorbed in a stepwise way by small platinum clusters. From the present study, it is concluded that such a process is specific of the small Pt_2-Pt_8 clusters since the energy of the first adsorption is significantly larger (>+0.5 eV) that the energy of the following step. For larger clusters, the reverse is observed. These results suggest that the cluster size used in the above mentionned experiments is really around 2-8. Since the first hydrogenation step of the Pt_3-Pt_4 clusters has energy larger (>+0.4 eV) than that of the other small clusters, a first hydrogenation takes place on all the Pt_3 and Pt_4 clusters before any cluster of a different size is hydrogenated. The structures of the fully hydrogenated clusters can be considered as a network of monometallic complexes with square planar or octahedral coordination. The bonds between the network units are made of the hydride ligands which are common to two, or more, metallic centers. The local square planar and octahedral coordination structures are associated with formal +II (Pt d8) and +IV (Pt d6) oxidation states respectively. They result from one or two oxidative additions (H_2 -> 2H) per platinum atom and an appropriate number of molecular adsorptions.

References

1) B. F. G. Johnson, "Transition Metal Clusters" Wiley Interscience, Chichester, Engl., (1980). E. Muetterties, T. N. Rhodin, E. Band, C. F. Brücker, W.R. Pretzer, Chem. Rev., 79, 91 (1979). P. Gallezot, A. Bienenstock, M. Boudart, Nouv. J. Chim. 2, 263 (1978).
2) See references 3 therein the present reference 4.
3) J. P. Fraissard, T. Ito, L. C. de Ménorval, 8th Int. Congress on Catalysis, Berlin (1984). J.P. Fraissard, T. Ito, L. C. de Menorval,

M. A. Springuel-Huet. Metals Microstructures in Zeolites, P.A. Jacobs Ed, Elsevier, Amsterdam (1982). Zeolites, P. A. Jacobs Ed, Elsevier, Amsterdam (1982).
4) B. Bigot, C. Minot, J. Amer. Chem. Soc., 106, 6601 (1984).
5) R. Hoffmann, J. Chem. Phys., 39, 1397 (1963). R. Hoffmann, P. Hofmann, J. Amer. Chem. Soc., 98, 598 5 (1976). J.H. Ammeter, H-B. Burgi, R. Hoffmann, J. Amer. Chem. Soc., 100, 3686 (1978).

PARTICIPANTS

T. Albright
Department of Chemistry
University of Houston
Houston, Texas 77004
U.S.A.

S. Alvarez
Facultad de Quimica
Universidad de Barcelona
Diagonal, 647
E-08028 Barcelona
Spain

J. Anglada
Mallorca 210-202[a]
E-Barcelona - 8
Spain

E. Baerends
Scheikundig Laboratorium
der Vrije Universiteit
De Boelelaan 1083
NL-1081 HV Amsterdam
The Netherlands

N. Balcioglu
Hacettere University
Chemistry Dept.
Beytere - Ankada
Turkey

V. Barone
Université de Naples
Departement de Chimie
Via Mezzo Cannone 4
I-80134 Napoli
Italy

J.C. Barthelat
Laboratoire de Physique Quantique
Université Paul Sabatier
118 route de Narbonne
F-31062 Toulouse cedex
France

C. Bauschlicher
STC-230-3
NASA Ames Research Center
Moffett Field, CA 94035
U.S.A.

A. Bencini
I.S.S.E.C.C., C.N.R.
Via F.D Guerrazzi 27
I-50132 Firenze
Italy

A. Bernier
Départment Physico-Chimie
C.E.N. Saclay
F-91191 Gif sur Yvette Cedex
France

B. Bigot
Laboratoire de Chimie Théorique,
ENS St Cloud
Grille d'Honneur, le Parc
F-92211 Saint Cloud Cedex
France

M. Blomberg
University of Stockholm
Theoretical Physics
Vanadisvagen 9
S-11346 Stockholm
Sweden

P. Bruna
University of New Brunswick
Dept. of Chemistry
Bag Service Nb. 45222
Fredericton
Canada E3B6E2

M.J. Calhorda
Centro Quimica Estrutural,
Complexo I, I.S.T.,
Av. Rovisco Pais
P-1096 Lisboa Codex
Portugal

E. Canadell
Laboratoire de Chimie Théorique
Bât. 490 - Université de Paris-Sud
F-91405 Orsay Cedex
France

P. Chaquin
Chimie Organique Théorique
Université Paris VI, Bât F
4, Place Jussieu
F-75230 Paris Cedex 05
France

H. Chermette
Institut Physique Nucléaire Lyon
43, boulevard du 11 novembre 1918
F-69622 Villeurbanne Cedex
France

E. Coronado
ENSCS-DSM
Rue Blaise Pascal
F-67008 Strasbourg
France

C. Daniel
Laboratoire de Chimie Quantique
Institut Le Bel
Université Louis Pasteur
4, rue Blaise Pascal
F-67000 Strasbourg
France

J.P. Daudey
Université Paul Sabatier
118, route de Narbonne
F-31062 Toulouse Cedex
France

A. Dedieu
Laboratoire de Chimie Quantique
Institut Le Bel
Université Louis Pasteur
4, rue Blaise Pascal
F-67000 Strasbourg
France

J. Delhalle
Facultés ND de la Paix
Département de Chimie
61, Rue de Bruxelles
B-5000 Namur
Belgium

A. Demolliens
Laboratoire de Chimie Théorique
Batiment 490 - Centre d'Orsay
F-91405 Orsay Cedex
France

M. Drillon
ENSCS-DSM
Rue Blaise Pascal
F-67008 Strasbourg Cedex
France

N. Dupré
DRF/LCH/CEA CENG
85 X
F-38041 Grenoble
France

P. Durand
Laboratoire de Physique Quantique
Université Paul Sabatier
F-31062 Toulouse Cedex
France

S. Durmaz
Ondokuzmayis Universitesi
Fen-Edebiyat Fakûltesi
Kimya Bölümü
Samsun-Turkey

O. Eisenstein
Laboratoire de Chimie Théorique
Batiment 490 - Centre de Paris-sud
F-91405 Orsay Cedex
France

D. Evans
School of Chemistry
Cantock's Close
Bristol
B58 ITS, England

K. Faegri
Department of Chemistry
University of Oslo
P.O. Box 1033, Blindern,
N-0315 Oslo 3
Norway

J. Gaillard
C.E.A. DRF/SCPM
85 X
F-38041 Grenoble Cedex
France

B. Gimarc
Department of Chemistry
University of South Carolina
Columbia, South Carolina
U.S.A.

W. Goddard III
California Institute of Technology
Noyes Laboratory (127-72)
Pasadena, CA 91125
U.S.A.

D. Gonbeau
Laboratoire de Physico Chimie
Moléculaire
UA 474 I.U.R.S.
Avenue de l'Université
F-64000 Pau
France

R. Gonzalez Luque
Dpto Quimica Fisica
Facultad de Ciencias Quimicas,
Dr. Moliner-50
Burjasot (Valencia)
Spain

A. Goursot
Laboratoire du Prof. Gineste
Ecole de Chimie
8, rue de l'Ecole Normale
F-34075 Montpellier Cedex
France

J. Green
Inorganic Chemistry Lab.
South Parks Road
Oxford OXI 3QR
England

C. Guimon
Laboratoire de Physico-Chimie
Moléculaire
UA 474 - I.U.R.S.
Avenue de l'Université
F-64000 Pau
France

J.F. Halet
Laboratoire de Cristallochimie
LA 254,
Université de Rennes I
F-35042 Rennes Cedex
France

M. Hall
Dept. of Chemistry
Texas A+M University
College Station
Texas 77843
U.S.A.

J. Hay
Los Alamos Natl. Lab.
Theoretical Chemistry Group (J-12)
Mailstop J569
Los Alamos NM 87545
U.S.A.

I. Hillier
Chemistry Dept.
University of Manchester
Manchester M139PL
U.K.

M. Hliwa
Laboratoire de Physique Quantique
Université Paul Sabatier
118, route de Narbonne
F-31077 Toulouse
France

R. Hoffmann
Dept. of Chemistry
Cornell University
Baker Laboratory
Ithaca NY 14853
U.S.A.

P. Hofmann
Anorganisch-chemisches Institut
der TU München
Lichtenbergstraße 4
D-8046 Garching
R.F.A.

Y. Jean
Laboratoire de Chimie Théorique
Bât 490 - Université Paris XI
Centre d'Orsay
F-91405 Orsay Cedex
France

G. Jeung
Physics Department
Dan-Kook University
Cheon-An, Choong-Nam-Do
330-00 South Korea

R. Johnston
Inorganic Chemistry Laboratory
South Parks Road
Oxford OXI 3QR
England, UK

H. Jörg
Lehrstuhl für Theoretische Chemie
Technische Univ. München
Lichtenbergstrasse 4
8046 Garching
West Germany

O. Kahn
Laboratoire de Spectrochimie
des Eléments de Transition
Batiment 420 - Université Paris Sud
F-91405 Orsay Cedex
France

C.W. Kern
Chemistry Division
National Science Foundation
Washington D.C. 20550
U.S.A.

N. Koga
Institute for Molecular Science
Myodaiji
Okasaki 444
Japan

A. Le Beuze
Laboratoire de Chimie Théorique
Université de Rennes I
Campus de Beaulieu
F-35042 Rennes Cedex
France

R. Lissillour
Laboratoire de Chimie Théorique
Université de Rennes I
Campus de Beaulieu
F-35042 Rennnes Cedex
France

J. Lluch Lopez
Dpto. Quimica Fisica
Universidad Autonoma de Barcelona
Bellaterra (Barcelona)
Spain

C. Marian
Universität Bonn
Wageterstr. 12
D-5300 Bonn
Germany

J. Mascetti
Laboratoire de
Spectroscopie Infrarouge
351, cours de la Libération
F-33405 Talence Cedex
France

M. Merchan
Dpto Quimica Fisica
Facultad de Ciencias Quimicas,
Dtor. Moliner 50,
Burjassot (Valencia)
Spain

G. Mignani
Rhône-Poulenc Recherches
85, avenue des Frères Perret
F-69190 St Fons
France

P. Millié
Departement de Physico-Chimie
CEN/Saclay
F-91191 Gif sur Yvette Cedex
France

D. Mingos
Inorganic Chemistry Lab.
University of Oxford
South Parks Road
Oxford OXI 3QR
U.K.

C. Minot
Laboratoire de Chimie Théorique
Bât. 490
F-91405 Orsay Cedex
France

D. Moncrieff
Applications Group - USG
University of Manchester Regional
Computer Centre (UMRCC)
Oxford Road
Manchester M13QPL
U.K.

H. Nakatsuji
Division of Molecular Engineering
Graduate School of Engineering
Kyoto University
Kyoto 606,
Japan

T. Nguyon
Laboratoire de Chimie Théorique
Batiment 490 - Centre de Paris-Sud
F-91405 Orsay Cedex
France

M. Nguyen
Department of Chemistry
University College
Belfield
Dublin 4,
Ireland

G. Nicolas
Laboratoire de Physique Quantique
Université Paul Sabatier
118, route de Narbonne
F-31062 Toulouse Cedex
France

C. Nicolaides
Hellenic Research Foundation
Theoretical Chemistry Institute
48 Vassileos Constantinou Avenue
Athens 50/1
Greece

G. Pacchioni
Dip. Inorg. Chemistry
Università di Milano
via Venezian 21
I-20133 Milano
Italy

M. Pelissier
Université Paul Sabatier
Laboratoire de Physique Quantique
118, route de Narbonne
F-31077 Toulouse Cedex
France

G. Pfister-Guillouzo
Laboratoire de Physico-Chimie
Moléculaire
U.A. 474. Avenue de l'Université
F-64000 Pau
France

G. Pourroy
ENSCS-DSM
Rue Blaise Pascal
F-67000 Strasbourg
France

H. Rabaâ
Laboratoire de Cristallochimie
U.E.R. "S.P.M"
Campus de Beaulieu
F-35042 Rennes Cedex
France

F. Renneke
Schering AG Arzneimittelchemie
Mullerstrasse 170-178
D-1000 Berlin 65
West Germany

G. Rius
C.E.A./ C.E.N. Grenoble
DRF-G/Service de Physique
85 X
F-38041 Grenoble Cedex
France

J.L. Rivail
Université de Nancy I
BP 239
F-54506 Vandoeuvre Nancy Cedex
France

M.-M. Rohmer
Laboratoire de Chimie Quantique
Institut Le Bel,
Université Louis Pasteur
4, rue Blaise Pascal
F-67000 Strasbourg
France

N. Rösch
Lehrstuhl für Theoretische Chemie
Technische Univeristat Munchen
Lichtenbergstrasse 4,
D-8046 Garching
R.F.A.

A. Rossi
University of Connecticut
Department of Chemistry,
U-60 Storrs,
Connecticut 06268
U.S.A.

J.Y. Saillard
Laboratoire de Cristallochimie
L.A. 254, Université de Rennes I
F-35042 Rennes Cedex
France

S. Sakaki
Kumamoto University, Kurokami,
Kumamoto 860
Japan

P. Sautet
Laboratoire de Chimie Théorique
Batiment 490
F-91405 Orsay Cedex
France

A. Sawaryn
Institute of Physics
Ratzeburger Allee 160
D-2400 Lübeck
Germany

U. Schubert
Institut für Anorganische Chemie
der Universität
Am Hubland
D-8700 Würzburg
Germany

A. Sevin
Université de Paris VI
Batiment F, 6ème étage
4, place Jussieu
F-75230 Paris Cedex 05
France

E. Shustorovich
Research Laboratories
Eastman Kodak Company
Rochester, NY 14650
U.S.A.

P. Siegbahn
Institute of Theoretical Physics
University of Stockholm
Vanadisvagen 9
S-11346 Stockholm
Sweden

D. Simon
Laboratoire de Chimie Théorique
E.N.S. de Saint-Cloud
Grille d'Honneur du Parc
F-92211 Saint Cloud Cedex
France

F. Stefani Semprini
C.N.R.
Ist. Teoria e struttura
Elettr. e Compor. Spettroch.
Via Salaria Km. 29.500 - C.P. 10
I-00016 Monterotondo - Stazione
Italy

H. Stoll
Institut für Theoret. Chemie
Universität Stuttgart
Pfaffenwaldring 55
D-7000 Stuttgart 80
Germany

C. Teichteil
Laboratoire de Physique Quantique,
118, route de Narbonne
F-31062 Toulouse Cedex
France

I. Tkatchenko
Institut Rech. Catalyse
2, avenue A. Einstein
F-69626 Villeurbanne
France

M. Tranquille
Université de Bordeaux 1
Laboratoire de spectroscopie IR
351 Cours de la Libération
F-33405 Talence Cedex
France

G. Trinquier
Laboratoire de Physique Quantique
Université Paul Sabatier
F-31062 Toulouse
France

D. Urch
Chemistry Dept., Queen Mary College
Mile End Road
London EI 4NS
England

L. Vanquickenborne
University of Leuven
Department of Chemistry
Celestijnenlaan 200 F
3030 Heverlee-Leuven
Belgium

A. Veillard
Institut Le Bel
Université Louis Pasteur
4, rue Blaise Pascal
F-67000 Strasbourg
France

M. Verdaguer
Laboratoire de Spectrochimie des
Eléments de Transition
Bât. 420
F-91405 Orsay Cedex
France

F. Volatron
Laboratoire de Chimie Théorique
Bât. 490
F-91405 Orsay Cedex
France

S. Walch
NASA Ames Research Center
STC 230-3, Moffett Field, Ca 94035
U.S.A.

U. Wedig
Institut für Theoretische Chemie
Universität Stuttgart
Pfaffenwaldring 55
D-7000 Stuttgart 80
Germany

T. Ziegler
University of Calgary
Department of Chemistry
Calgary, Alberta
Canada T2N 1 N4

INDEX